TRINITY

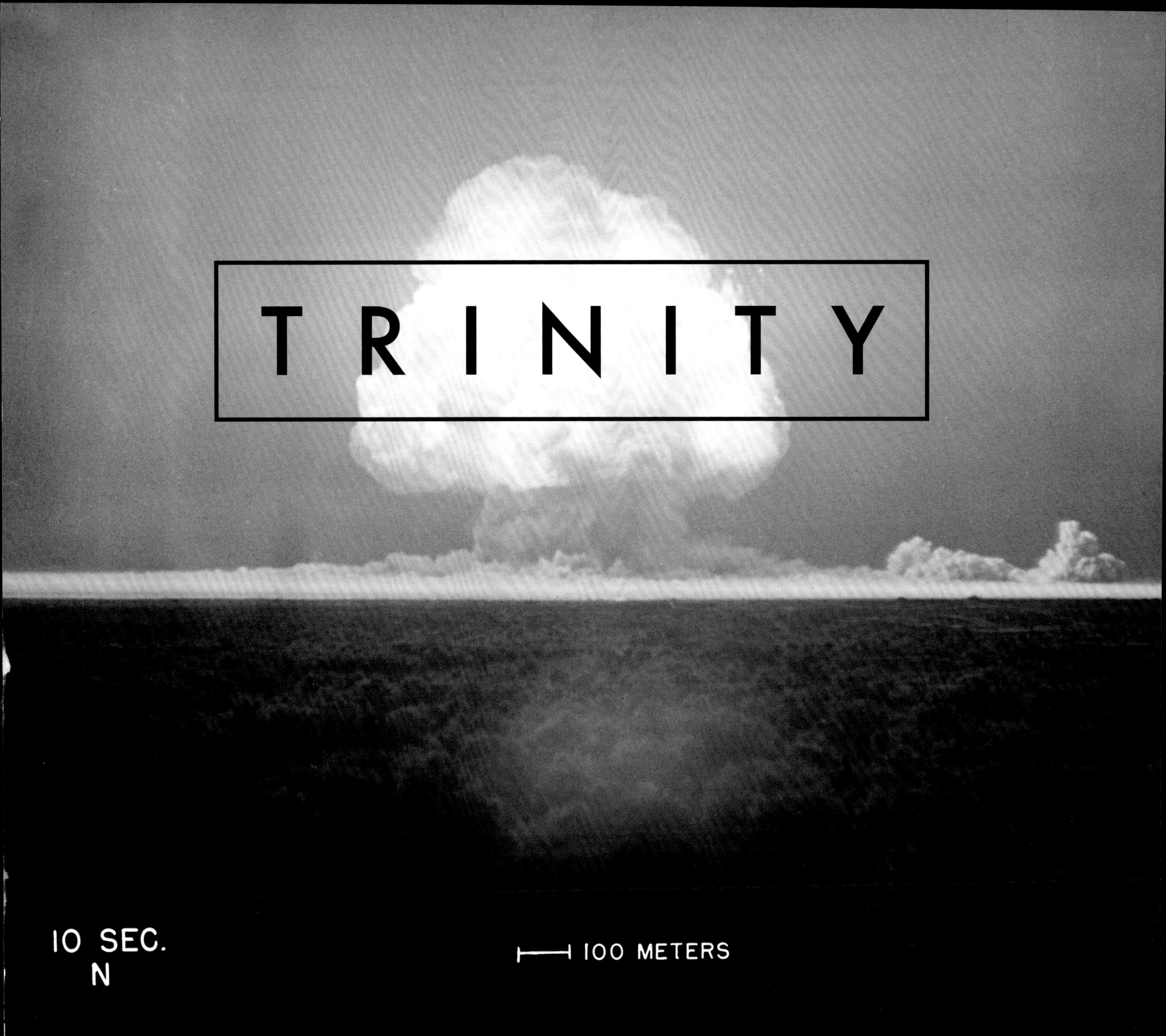

10 SEC.
N

100 METERS

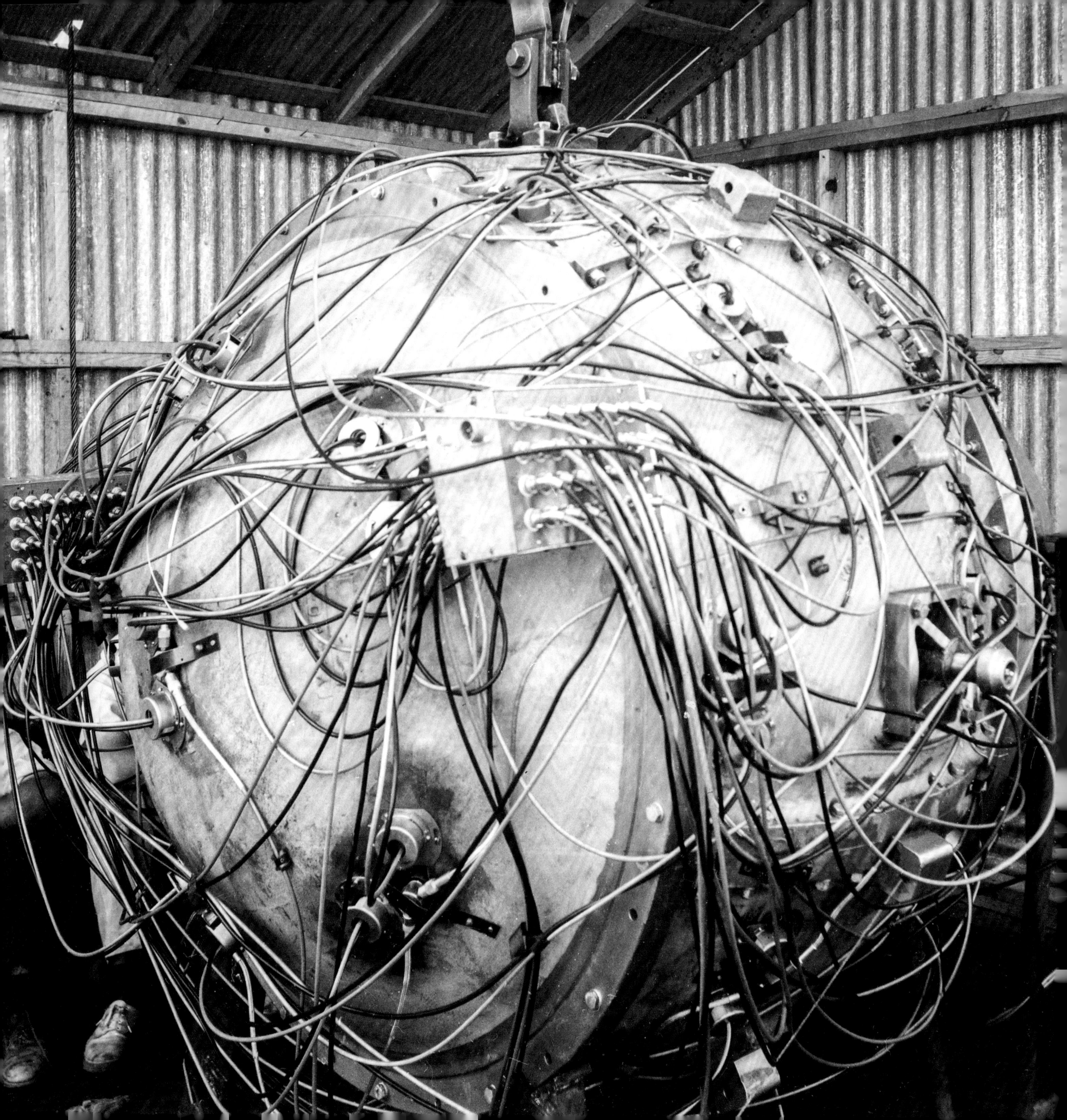

TRINITY

AN ILLUSTRATED HISTORY OF THE WORLD'S FIRST ATOMIC TEST

Emily Seyl

with contributions by **Alan B. Carr**

and illustrations by **Paul Ziomek**

FOREWORDS by

James W. Kunetka and Jim Eckles

AFTERWORD by

Charles Oppenheimer

PHOTO EDITING by

Rizwan Ali

ARCHIVAL SUPPORT from

Angie Piccolo

The University of Chicago Press | Chicago and London

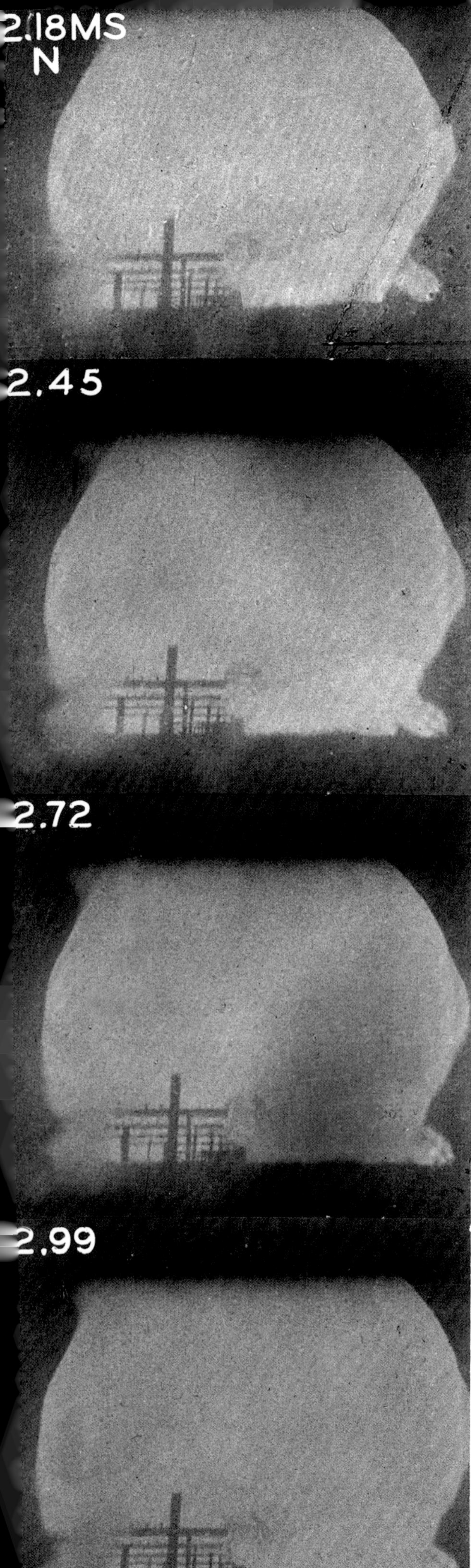

The University of Chicago Press, Chicago 60637
The University of Chicago Press, Ltd., London

Published 2026
Printed in China

35 34 33 32 31 30 29 28 27 26   1 2 3 4 5

ISBN-13: 978-0-226-84840-2 (cloth)
ISBN-13: 978-0-226-85078-8 (ebook)
DOI: https://doi.org/10.7208/chicago/9780226850788.001.0001

Cataloging information is on file with the Library of Congress.

Book design and illustrations by Paul Ziomek.

Cover image: Armored camera used to record small-scale implosion tests at Los Alamos;
photo by David Woodfin.
Page xx: Illustration adapted from a 1946 charcoal drawing by Melvin A. Miller. (University of Chicago Archives, courtesy of AIP Emilio Segrè Visual Archives.)
Page 264: Illustration adapted from August 6, 1945, front page of the *Santa Fe New Mexican*. (© 1945 The New Mexican, Inc. Reprinted with permission. All rights reserved.)
Page 296: Photo courtesy of Library of Congress, Manuscript Division, Washington, DC.

Except where otherwise noted, photographs and documents are courtesy of Los Alamos National Laboratory.

♾ This paper meets the requirements of ANSI/NISO Z39.48-1992 (Permanence of Paper).

Authorized Representative for EU General Product Safety Regulation (GPSR) queries: **Easy Access System Europe**—Mustamäe tee 50, 10621 Tallinn, Estonia, gpsr.requests@easproject.com
Any other queries: https://press.uchicago.edu/press/contact.html

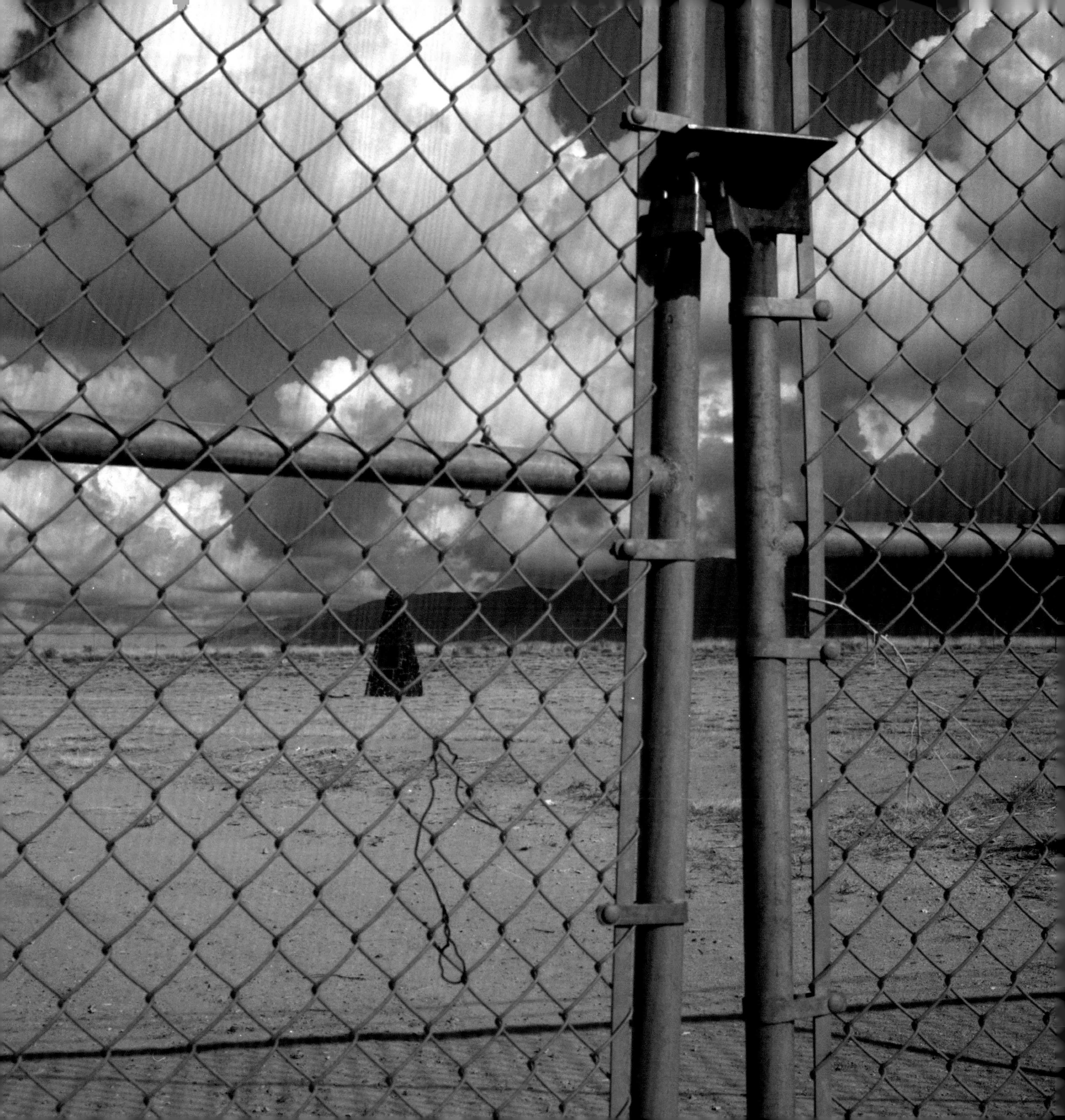

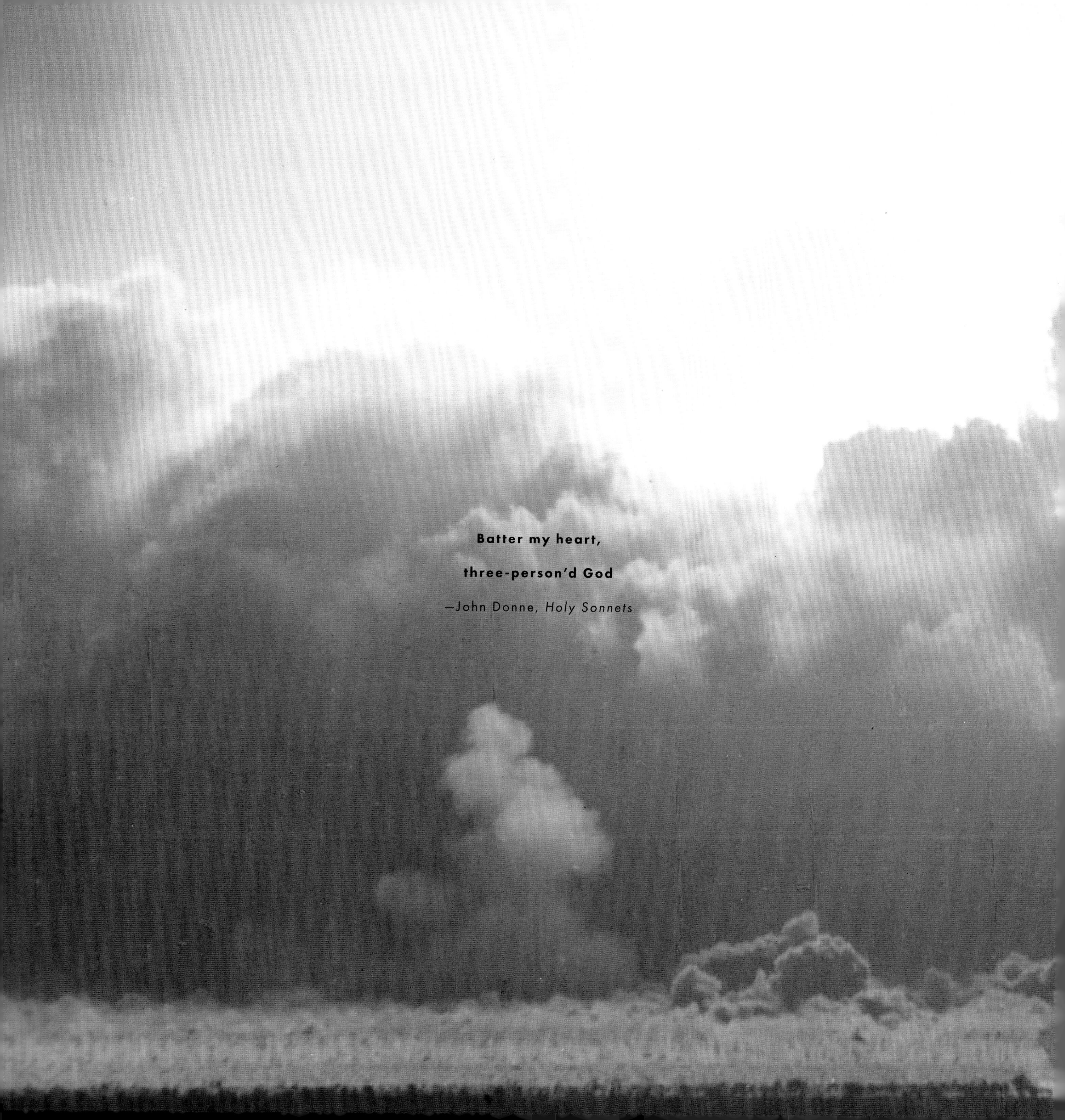

Batter my heart,

three-person'd God

—John Donne, *Holy Sonnets*

Contents

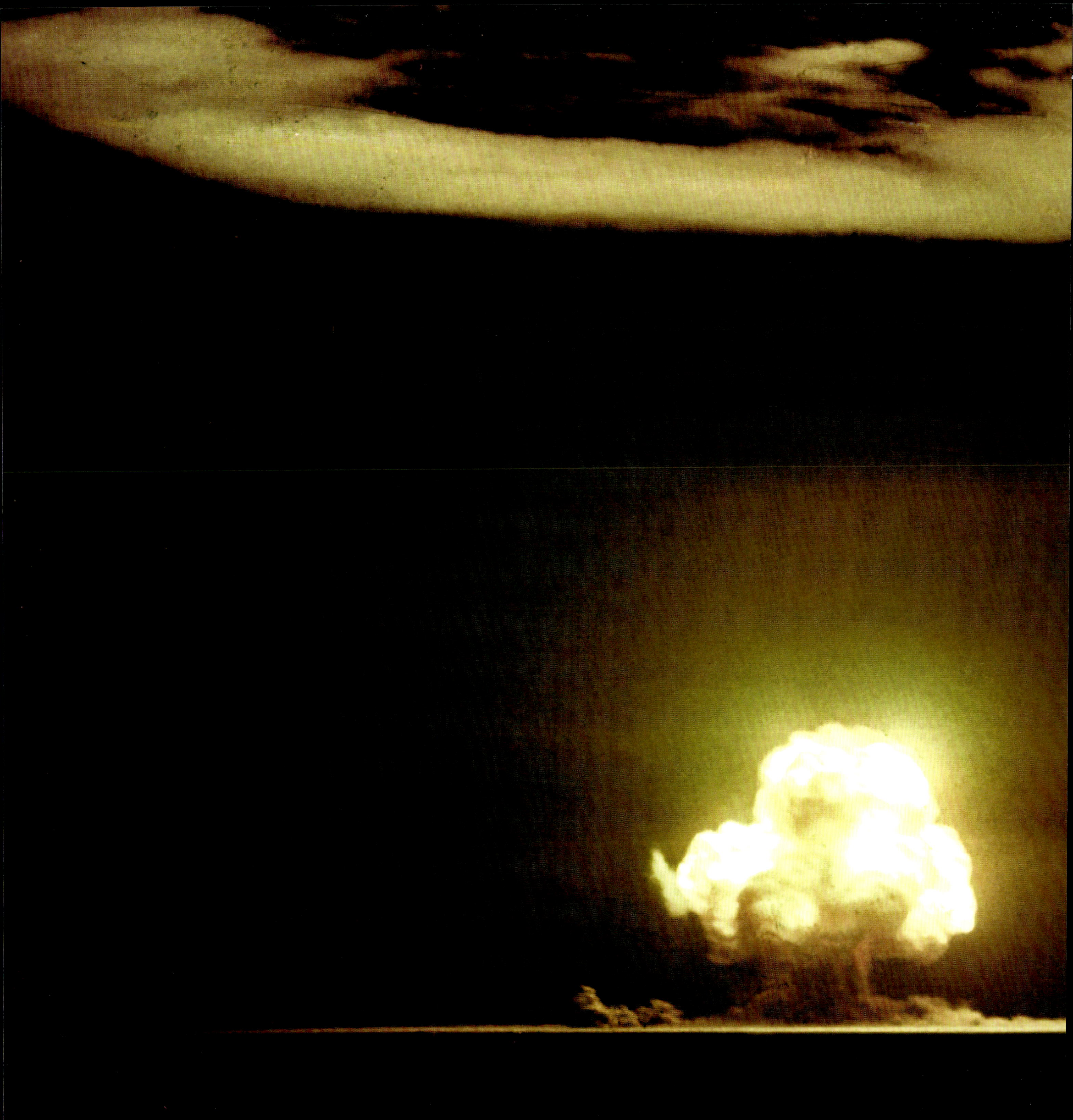

Foreword

From night to day was instantaneous. Darkness disappeared in an explosion of noiseless, brilliant light that lit up distant mountains. Observers rose from their sheltered positions to see a rising ball of fire that resembled an infant sun. Less than a minute later the silence was shattered by a loud, thunderous crack accompanied by a blast of wind mixed with dirt, broken vegetation, and whatever else the shock wave brought with it. The fireball then began to fade and dissolve into what would become an iconic image in the years to follow: a growing, drifting mushroom cloud. It was Monday morning, July 16, 1945, in an isolated stretch of the southern New Mexico desert codenamed "Trinity."

That dazzling, terrifying display was the product of the world's first atomic bomb. Among the witnesses were the two men most responsible for the morning's success. The first was J. Robert Oppenheimer, a theoretical physicist and university professor with no previous management or leadership experience before the war. Nonetheless, his wartime direction of the secret laboratory at Los Alamos delivered two versions of an atomic bomb. The other was Gen. Leslie R. Groves, the man in charge of the Manhattan Project. He oversaw the largest scientific and industrial undertaking in history. By mid-1945, his Manhattan Project had spent nearly two billion wartime dollars.

Both men had come to Trinity to witness a demonstration of the more inventive but unproven of the two types of atomic bombs developed at Los Alamos. If multiple bombs were needed to end the Asia-Pacific War, only this design could be produced in quantity. Nicknamed the Gadget, it was technically complex and utilized imploding explosive waves to compress nuclear material. Oppenheimer and his team had to prove that the design would work before the military could plan on its use.

What happened that day in the desert was the first of its kind, a carefully orchestrated and executed scientific experiment on a grand, unprecedented scale and at the same time a very human story. The technical details of the test are well documented, but the look and feel of Trinity as a makeshift laboratory and bustling desert community inevitably began to fade as men packed up and returned to Los Alamos. Fortunately for contemporary generations, much of the activity both before and after the test was faithfully captured by a team of photographers. With a few exceptions, the photos are unposed and there is an air of casual intimacy as men go about their work. A thin, hatless Oppenheimer nervously watches the final assembly of the bomb. Norris Bradbury later stands by the Gadget within its protective cupola on top of a 100-foot-high steel tower. A young Donald Hornig sits vigilantly next to the device, packed with thousands of pounds of high explosives. There is a brief glimpse of Harry Daghlian; now largely forgotten in history, he and Trinity veteran Louis Slotin would both be killed in nuclear criticality accidents within eleven months.

It has been eight decades since that early July morning. Any physical legacy of Oppenheimer and Groves and their colleagues, their footprints and tire tracks, has largely weathered away. Fortunately, photographic records testify to the extensive preparations, the network of roads and buildings and instrument bunkers, and the many who labored during that hot New Mexico summer.

The images assembled in this book are invaluable because they tell the story of Trinity that words alone cannot convey, preserved in photos and motion picture films now carefully restored by Los Alamos National Laboratory and its National Security Research Center. Trinity the place, and what remains of it, will always resonate with the human capacity to imagine and invent. *Trinity* the book, and its images, preserves for future generations that remarkable day when history changed in an instant.

—James W. Kunetka,
author of *The General and the Genius: Groves and Oppenheimer—The Unlikely Partnership That Built the Atomic Bomb*

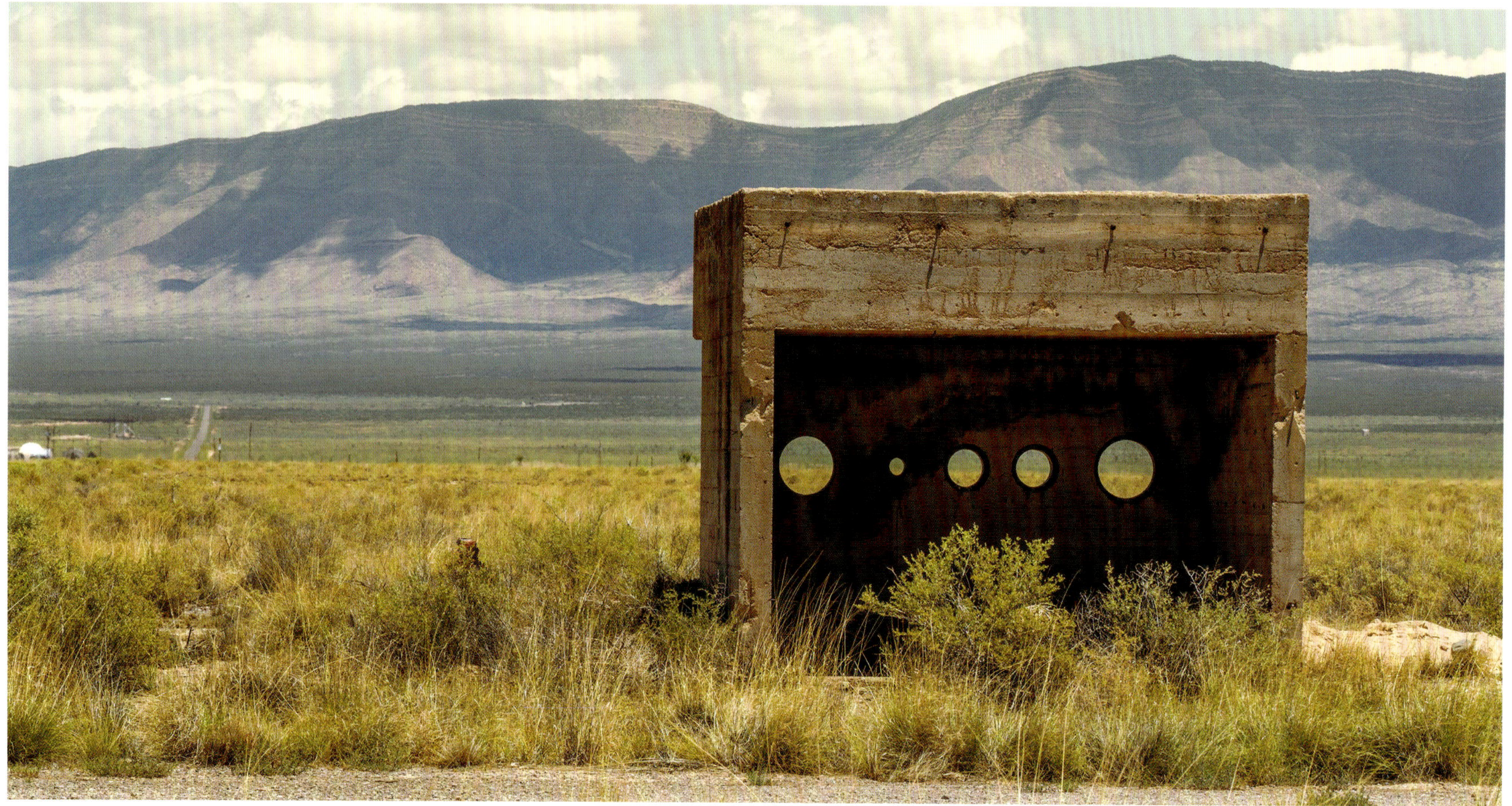

A photography bunker 10,000 yards (about five-and-a-half miles) west of the detonation site—just visible at the end of the paved road—is one of the only structures remaining from 1945. Much of the usable footage of the test shot was taken from here and an identical bunker at north 10,000 by cameras mounted on the roofs and in the portholes.

Foreword

It seems I've been waiting for this book most of my life. When I started work in the Public Affairs Office at White Sands Missile Range in 1977, we ran the annual Trinity Site open houses. Not having much information of our own, we harvested facts from Los Alamos publications. It was worse for the photos. We tried to display photos of test preparations and blast events, but they were fuzzy, dull, and dark, obvious copies of copies of copies. As a photographer I always wondered if the originals were that bad. Now I know. There is no comparison, and this book reveals how they should appear.

Another terrific aspect of this book is the variety. Until Los Alamos made hundreds of scanned images available, we were stuck with the same ten to twenty photos that showed up in every newspaper and magazine article, every book, and most documentaries. That made for a very narrow view of this world-changing event. Surely there was more going on. Now, the blinds have been opened and we can readily see most of the landscape outside.

Along those same lines, as the Trinity event grows dimmer in the past, people focus mainly on the very tip of the iceberg. Most questions directed my way now are about Oppenheimer and Groves. Nearly forgotten are the likes of Lt. Howard Bush, who was camp commander and head of security. His after-action report is one of my favorites. He was at south 10,000, the control bunker, along with J. Robert Oppenheimer, director of Los Alamos, and Kenneth Bainbridge, Trinity test director. Bush elected to experience the test outside the bunker. He reported that he was facing away from ground zero and down in a crouch. When the landscape lit up with that blinding white light at detonation, Bush said he was painfully aware of the glare and touched his eyelids to make sure he really had them closed. After the light faded, he stood up and faced the fireball starting to rise at ground zero more than five miles away. After a few seconds, the shock wave arrived and knocked him on his backside. He was uninjured.

In my years working Trinity Site open houses, logistics were often a top priority. There were some big questions, like were there going to be enough port-a-potties if an extra thousand people showed up, and was there going to be enough space in the parking lot to handle an extra hundred vehicles? One year the toilets were not delivered at all, and we scrambled to find eight or nine scattered around the missile range that could be commandeered for the day. One cannot imagine the chaos that might have ensued if we'd had none for our 1,500 visitors.

For me, the site became a portal for meeting some really interesting people. Early on I got the assignment to escort Walter Cronkite of CBS News to ground zero so he could do a standup at the obelisk. It was a windy, cool day and the crew had difficulty keeping Cronkite's thin, wispy hair in place. They used most of a can of hair spray to glue it down. You miss that in the final aired feature.

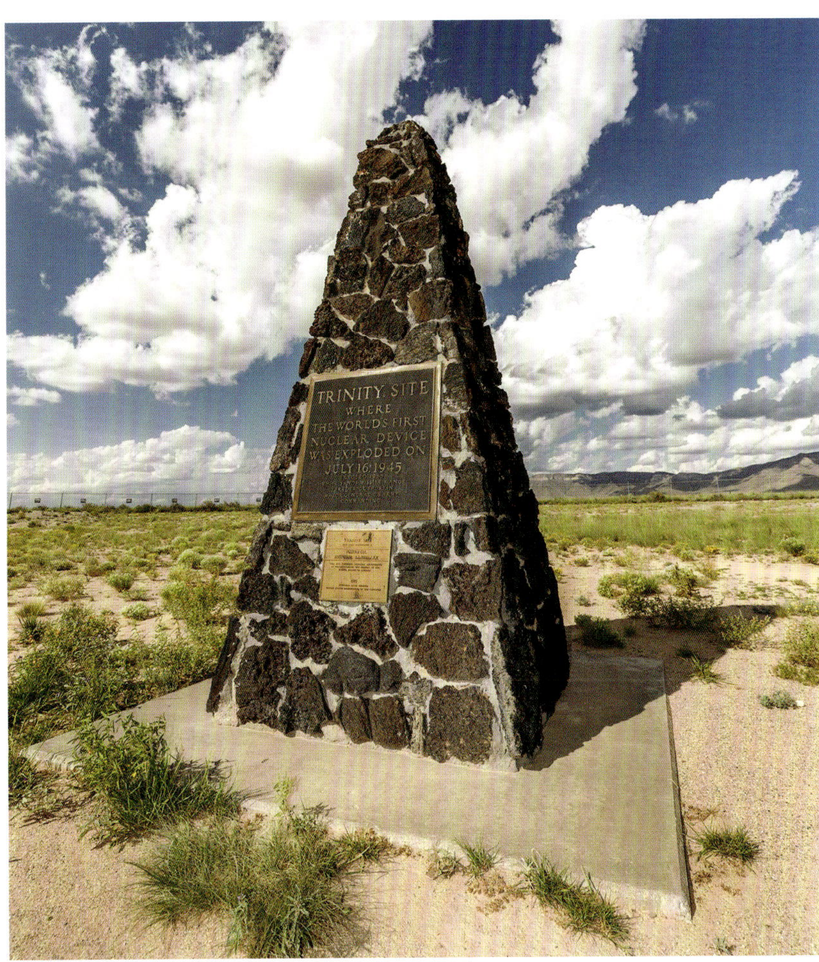

Since 1965, a lava rock obelisk has marked the location of the shot tower, shown on the following page, that once elevated the Trinity test device 100 feet above the ground.

I got to meet many of the scientists and personnel who worked at the site and who witnessed the explosion. I spoke with leaders like Robert Wilson and Kenneth Bainbridge and technicians like Berlyn Brixner, who took much of the footage in chapter 6 of this book, and Jack Aeby, who captured the only good color image of the explosion. Then there were the soldiers like Marvin Davis and Felix DePaula who were stationed at the site for months. Davis described feeling the heat from the test at ten miles away and compared it to opening the oven back home to check on a pie. Army clerk Dave Rudolph provided one of the best descriptions of working for the Manhattan Project in New Mexico. He said, "Life on the hill (Los Alamos) was isolated; life at Trinity was penal."

One might think the story of the Trinity test has already been told and we just repeat ourselves endlessly. Actually, we are constantly discovering lost information and correcting misinformation that has developed. For instance, for decades we (and everyone else, for that matter) talked about the formation of trinitite (green glass near the crater) in terms of the crème brûlée effect—the fireball melted the sand on the ground to form pools of liquid glass. Now, because of the curiosity of Robb Hermes and Bill Strickfaden, we know trinitite was created in the fireball and then fell like fiery-hot raindrops.

The sequence of photos at two to five seconds are some of my favorites because they show the two barrage balloons carrying instruments and their steel anchor cables being vaporized by the incredible heat of the fireball. Couple that with their back story of an attempt to capture data at the beginning of the fission reaction, during the first ten one-billionths of a second, and you have a fascinating component of the test that is now forgotten.

There is very little to see at Trinity, and this book will make a great memento for visitors to the site. The photos provide a glimpse of what a team of dedicated human beings, working for a common goal, accomplished in just a few months. The world changed around them in the blink of an eye. Now it is up to new generations of smart and sensible people to deal with the consequences.

—Jim Eckles

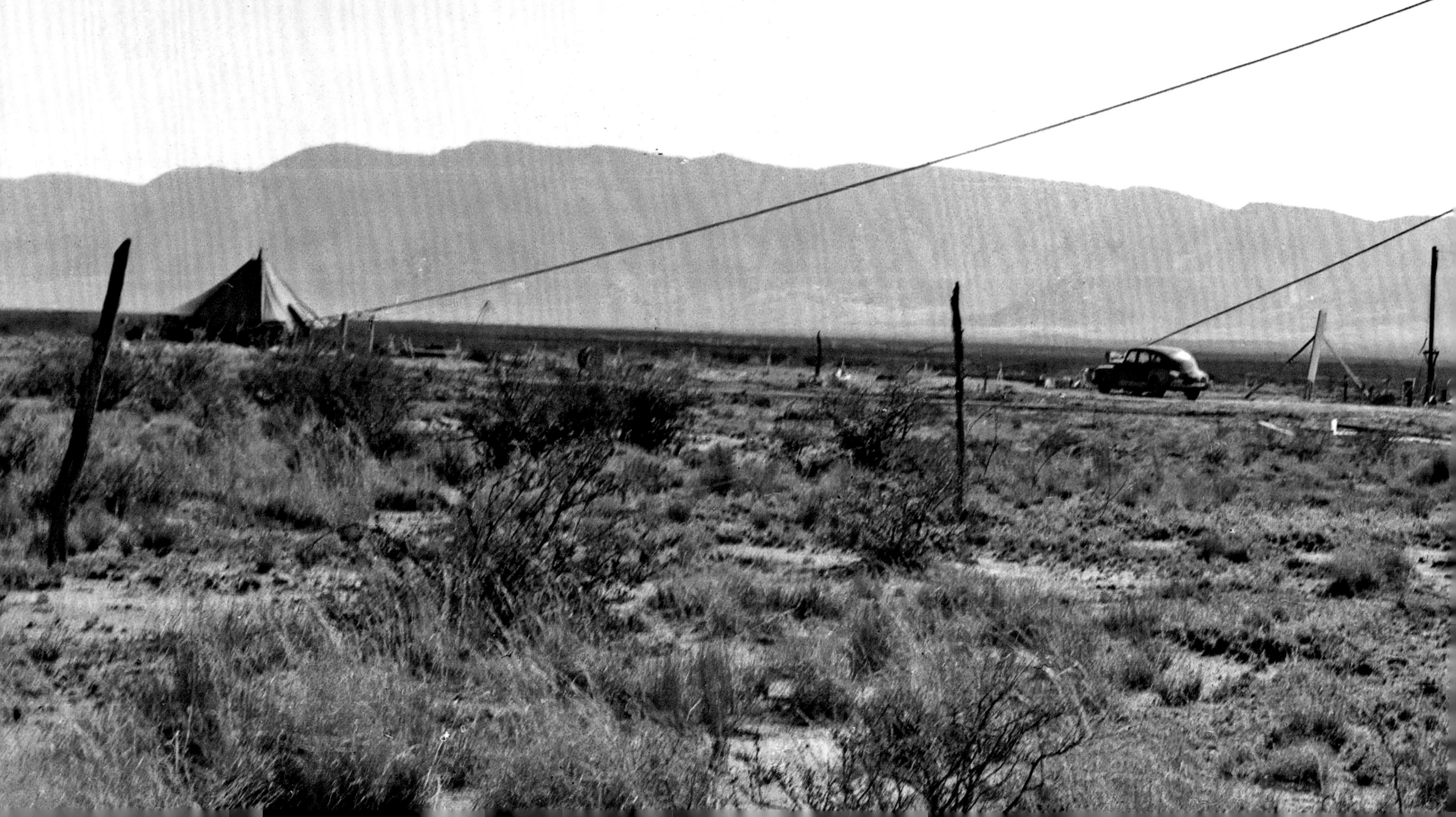

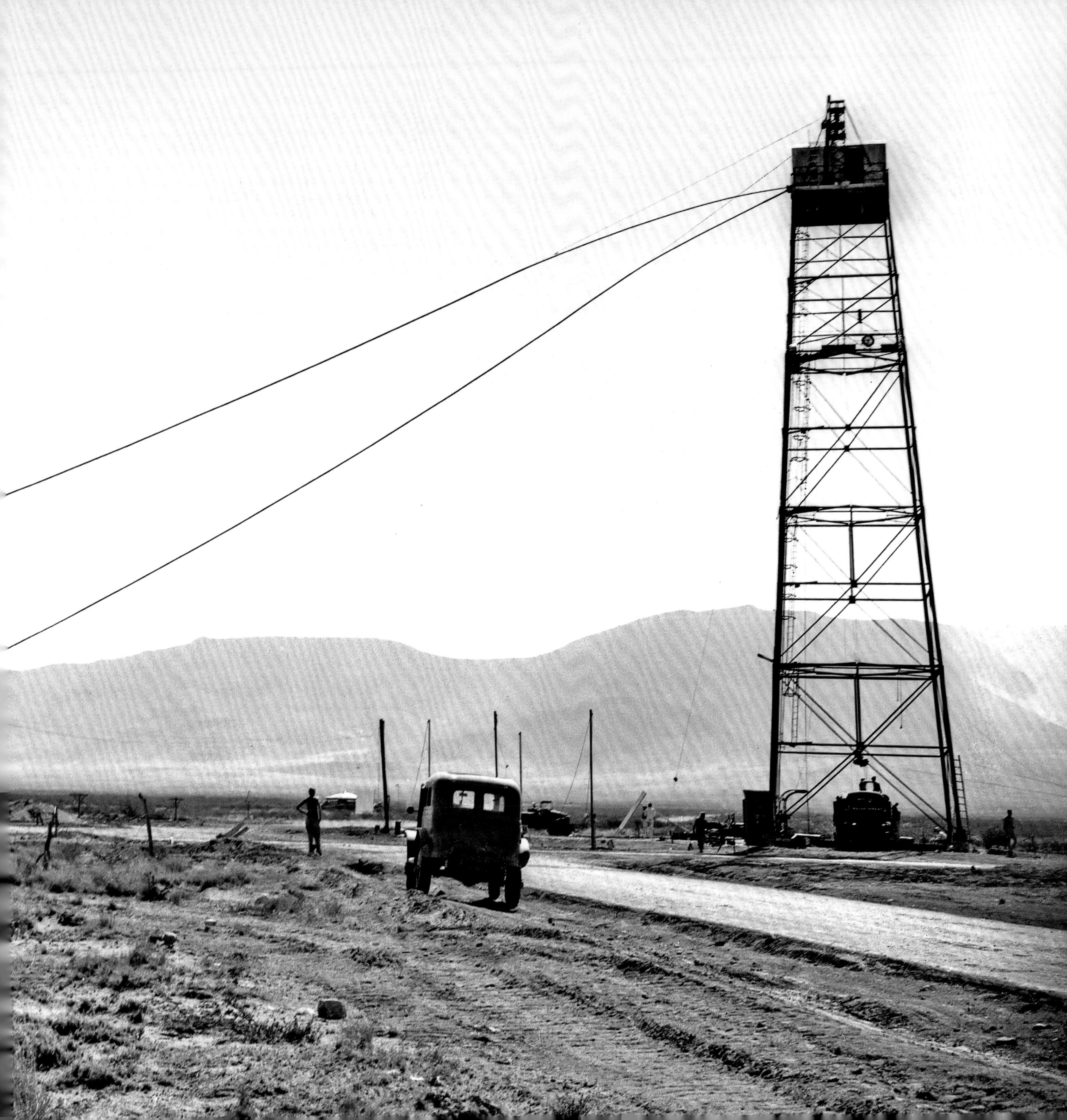

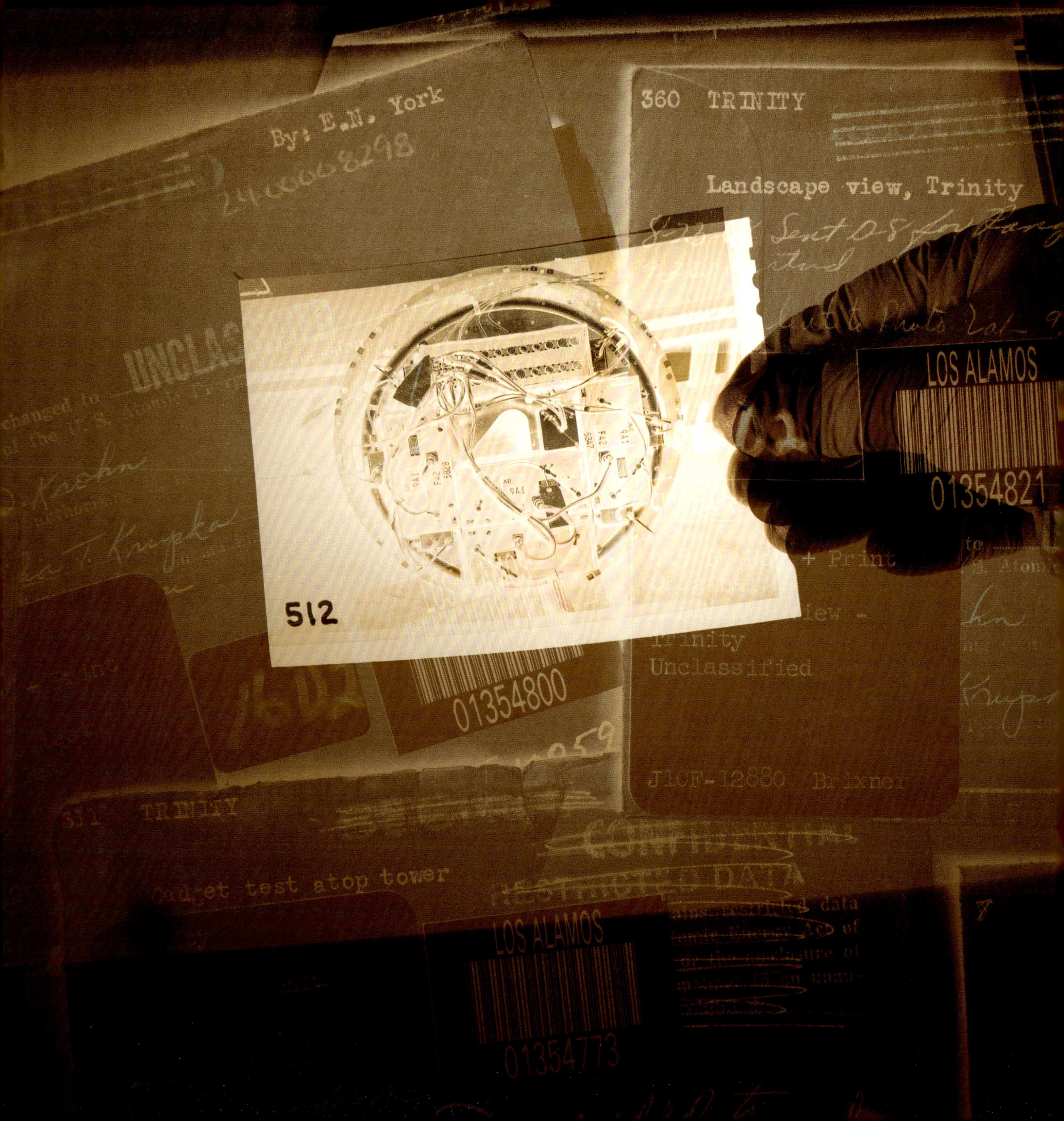

By: E.N. York

24000 8298

360 TRINITY

Landscape view, Trinity

UNCLAS

changed to

of the U.S. Atomic Energ

LOS ALAMOS

01354821

512

01354800

Trinity

Unclassified

J10F-12880 Brixner

311 TRINITY

Gadget test atop tower

CONFIDENTIAL

RESTRICTED DATA

LOS ALAMOS

01354773

Preface

Weeks before the world would learn of the atomic bomb, upon its wartime use against Japan, a team of scientists and military personnel detonated the first nuclear device on a remote bombing range in south-central New Mexico, in an operation codenamed Trinity. What was painstakingly orchestrated to be a secret moment lit up the predawn sky for hundreds of miles as the explosion shattered projections of its magnitude. Both a military proof-test and an elaborate, well-documented science experiment, the July 16, 1945, trial shot brought under the control of humankind a new fire—the energy of the atom. The chapters that follow trace the history of this first nuclear explosion through carefully restored images housed for eighty years in the archives at Los Alamos National Laboratory, which under its original, wartime codename of Project Y invented the atomic bomb and conducted the test known as Trinity.

The foundation of this book is a series of more than 800 photographs and still frames captured at three locations in 1945: the Trinity site, during the setup, execution, and aftermath of the test; on Tinian Island, where two atomic bombs—including a weaponized version of the Trinity device—were prepared for deployment; and in postwar Japan, in the two cities devastated by those bombs on August 6 and 9. These visual artifacts are complemented by select photographs from personal and public archives along with other historical images and original documents held at Los Alamos.

Photography was a pivotal part of research at Project Y, especially as a diagnostic tool for early, scaled tests of nuclear weapon components. Many of the innovators behind the high-speed and high-impact cameras used in these groundbreaking experiments (including the armored camera on the cover of this book) were members of the laboratory's Optics Group. Led by physicist Julian Mack, the group, under a new designation—Spectrographic and Photographic Measurements (TR-5)—went on to tackle the nebulous task of filming the full-scale detonation at Trinity.

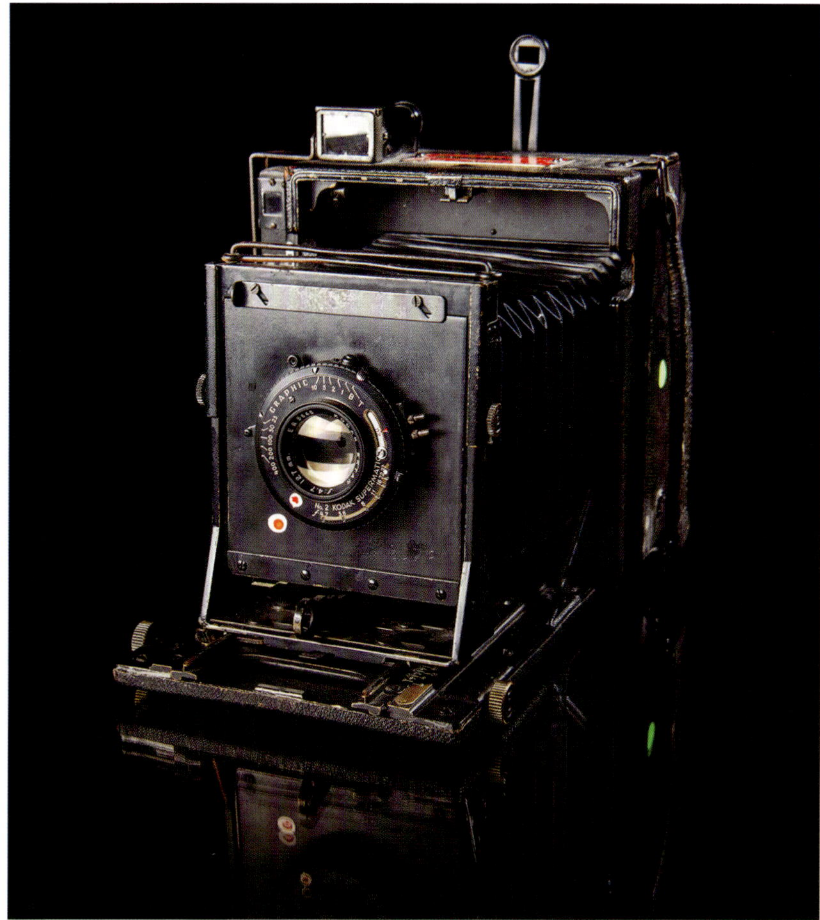

One of the cameras used to photograph Trinity activities was the large-format Graflex Speed Graphic, which used 4 × 5 film sheets. A photographer could set the focus by extending and retracting its accordion-like bellows. This Manhattan Project unit (identified by the red "fried egg" marking) is in the collections of Los Alamos National Laboratory's Bradbury Science Museum.

In addition to arranging dozens of cameras to record the main event, Mack also solicited requests from leaders of the six other groups involved with test preparation to record what they deemed "significant things and events." This service, which produced the eclectic group of photographs taken before and after the test, was handled largely by Ernest Wallis, a commercial photographer gaining his first experience shooting technical subjects. From mid-March 1945 onward, when scientists began arriving in the desert basin to set up experiments, Wallis dutifully captured their work along with notable structures and happenings around the test site—taking artistic liberties along the way as he shot landscapes and candid moments and experimented with angles.

Today, very little remains of the bustling base camp, the shelters housing humans and instruments, and the constellation of diagnostic tools that once dotted the landscape. The foresight to memorialize these details has proven crucial to reconstructing the history of the test, as before long the harsh environment would sweep the basin clean of most Trinity artifacts. Wallis's images and those captured by the many automated motion picture and still cameras pointed at ground zero on July 16 well preserve what became a turning point in science and history.

After the war, the original photographic negatives were numbered and labeled by hand with "TR," for Trinity, and stored in paper envelopes—with a copy, or print, mounted on each envelope for identification. Over the past eight decades, these images have been used for various purposes, from scientific analysis immediately after the test shot to informing the National Park Service's efforts to turn the site into a national monument.

The idea to compile the series in a book originated almost twenty years ago with former Los Alamos National Laboratory archivist Dan Comstock and *Trinity* contributor Alan Carr, now senior historian at the Laboratory, during a 2006 effort to digitize the negatives in their entirety for the first time and share them with the public. Comstock, who passed away in 2022, took on the task of scanning the negative envelopes to create a database and a searchable index of the photographs in tandem with the digitization project. He, along with Henry Johnson (then leader of the Records Management and Media Services

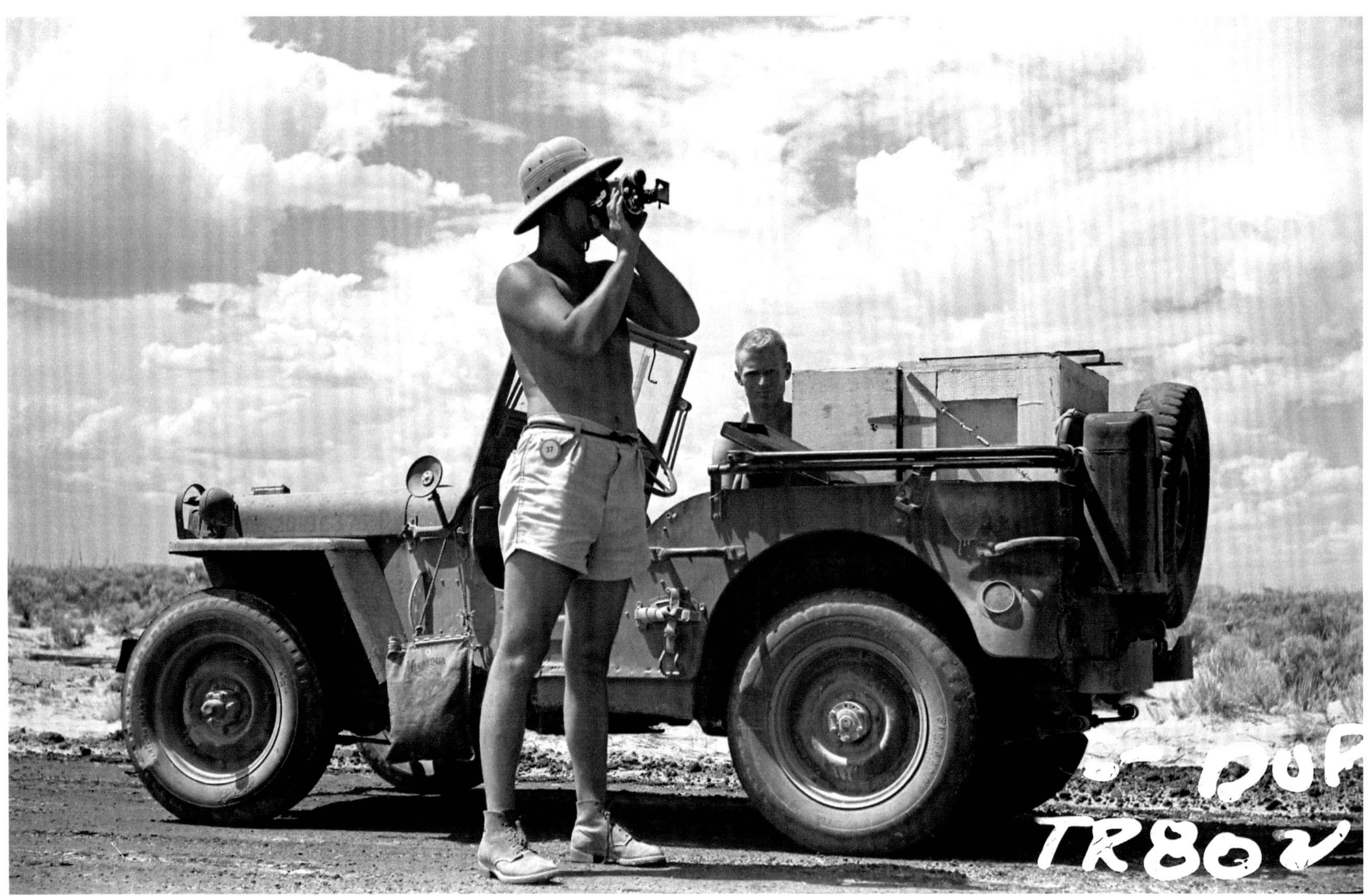

Ernest Wallis using a Contax camera equipped with a telephoto lens to capture preparations at the Trinity site, 1945. Wallis was also known to capture rattlesnakes while on duty, stowing them in a box (pictured) on the back of his jeep.

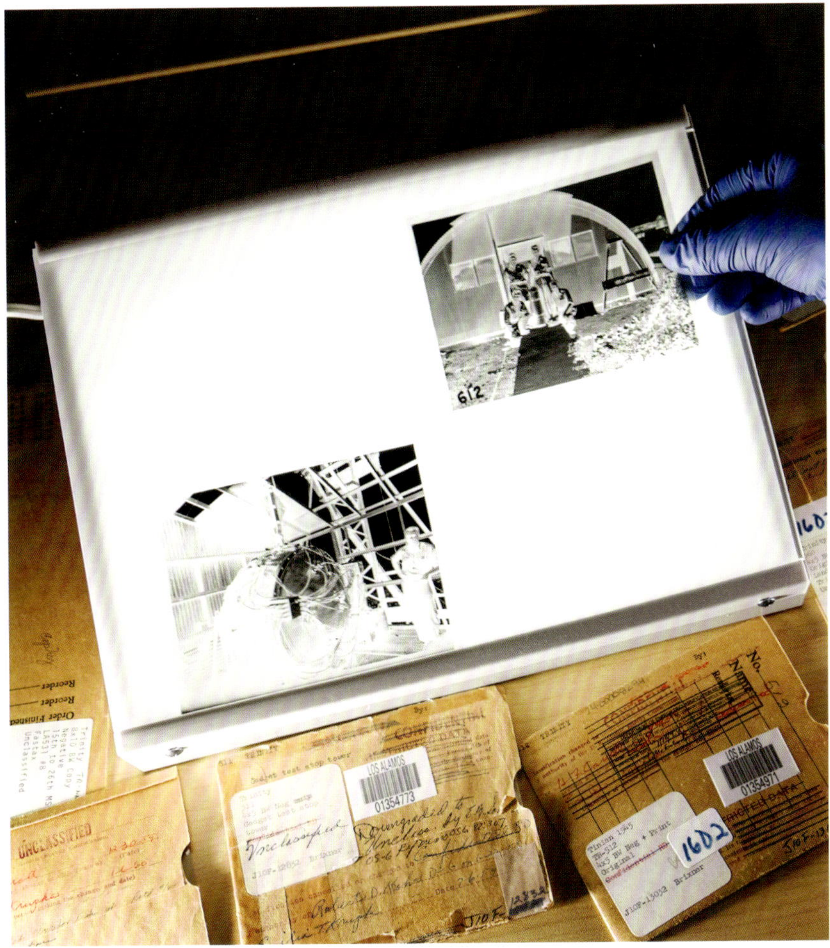

A selection of TR series negatives and envelopes, archived at the National Security Research Center. Negatives are color reversed, meaning the lightest areas of the image look darkest and vice versa.

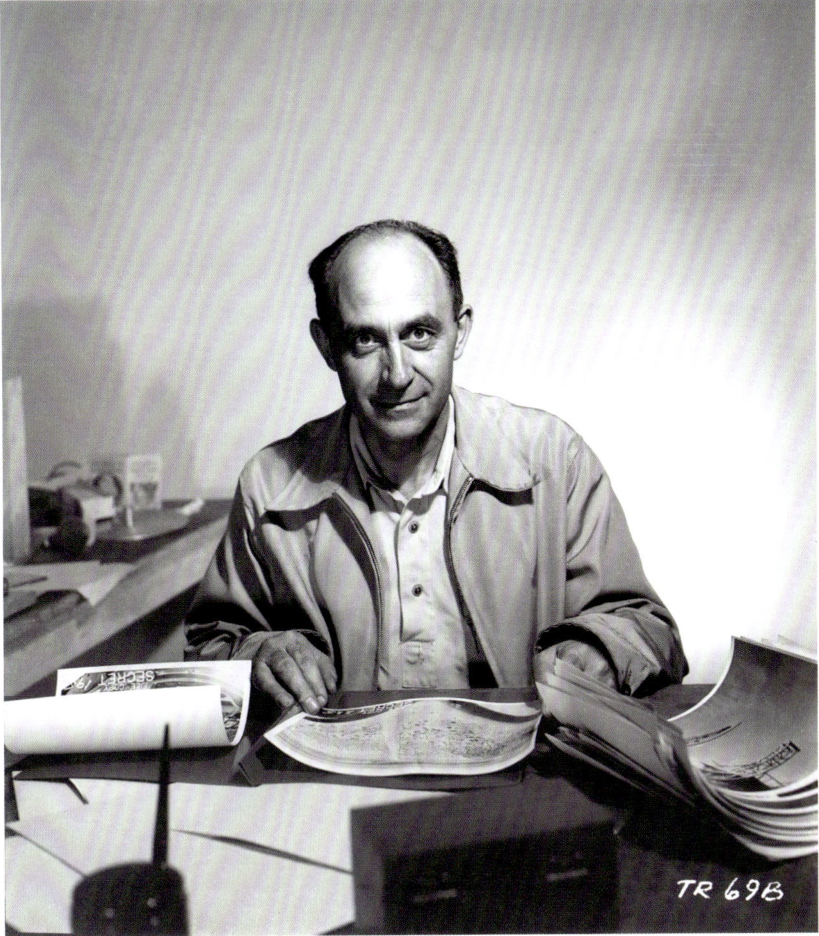

Manhattan Project physicist Enrico Fermi with a set of 8 × 10 TR series prints, 1945. Making a print or scan of a negative reverses the colors again, producing an image that appears normal.

and Operations Group) and the Los Alamos Historical Society, which partnered with Los Alamos to scan the images, are due much gratitude for recognizing the value of the TR series and having the foresight—like the Spectrographic and Photographic Measurements Group in 1945—to ensure the preservation of this important history.

The resources to build on their efforts and create this book came into being with the founding of the National Security Research Center (NSRC) in 2019—today's iteration of a now-vast technical library begun by the Laboratory's original director J. Robert Oppenheimer. Although its millions of holdings contain mostly classified material, the NSRC also curates unclassified records and artifacts of historical value, including memos; reports; drawings; and photographs, like the TR series, that contribute to the story of the atomic bombs' creation and the test at Trinity.

In the years since 1945, the TR series negatives have been accessed by historians, scientists, former Trinity workers, and others. As copies were made and original films reviewed, the repeated handling led inevitably to scratches, smudges, and tears, while deterioration due to aging gradually obscured the negatives beneath graininess and haze. Rizwan Ali, former director of the NSRC, undertook the delicate task of restoring the images reproduced in this book, working from a combination of the 2006 scans and new, higher-resolution scans of the negatives. The goal of restoration was to show readers how the photographs likely looked in 1945, as opposed to how they happened to look at the time of this writing.

In some instances, it was hard to distinguish the effects of time from flaws original to the negatives, which might be attributable to equipment, techniques, or lighting conditions. Many times, film was exposed imperfectly in the field when an image was captured, or oddities were introduced during the development process: when film was soaking in chemicals, for example, there might be more activity near the edges of an image, resulting in a shadow around the perimeter.

Depending on the circumstances and an image's intended use, darkroom technicians may have adjusted variables like lighting and contrast and masked significant artifacts like tears and dust when making print copies from the original negative. With the intention to replicate these practices in modern editing programs, our team took great care and a minimalist approach to align the final images presented in the book with how a print made in 1945 may have looked, in our estimation. The exception to this is footage of the nuclear blast (in chapter 6), which was not retouched for this project since the effects of the explosion and the highly experimental techniques used to capture it from various angles and distances contributed to the uneven quality of the negatives.

During the eighteen years since it was digitized and cleared for public release, the TR series has been circulated both as a set and, mostly, in bits and pieces. At Trinity Site itself, some of the images feature prominently in exhibits at the historical highlights, which visitors can see at the annual open houses: an obelisk erected in 1965 that marks the exact spot of the explosion, surrounded by a fence hung with footage of the fireball (shown at right); the colossal remnant of a steel container called Jumbo, destroyed and abandoned after the test; a bunker 800 yards from the blast that was built to house high-speed cameras; and a ranch house, fully restored in 1984, where the material that would fuel the test device was prepared for detonation.

In the photographic history that follows, over two hundred of the TR series images have been carefully restored, organized, and placed in their full context as a complete representation of the Trinity test. The series constitutes many important pieces of a larger and fascinating story; in this book, with many helping hands, that story finds its way. Our hope is that in exploring anew how the test was executed, why it mattered then, and why it still matters eighty years later, *Trinity* embodies the unprecedented environment in which Manhattan Project scientists lived and invented; the wartime inextricability of violence and progress, of hope and fear; and the complicated reverberations of the July 16, 1945, crescendo that continue to shape the world today.

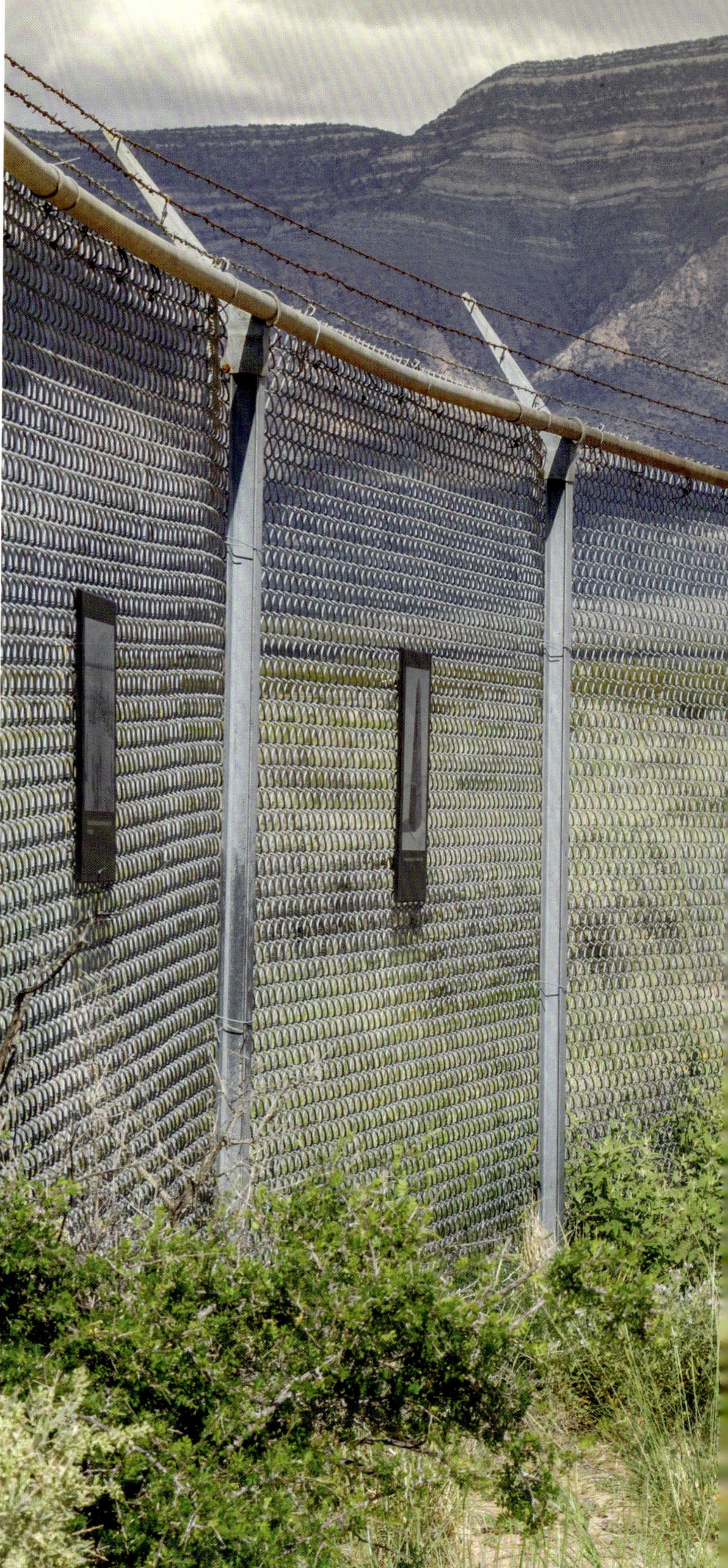

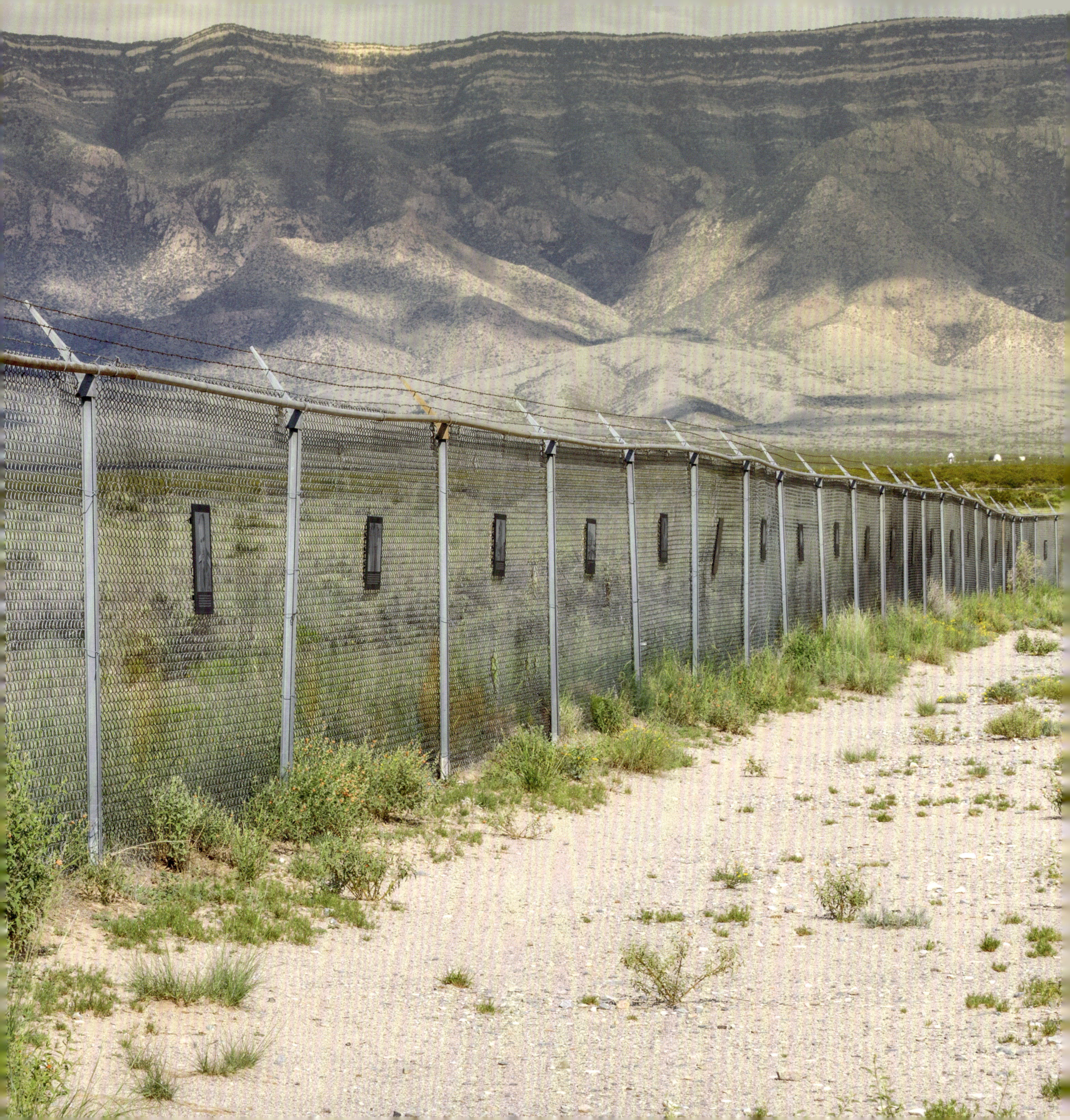

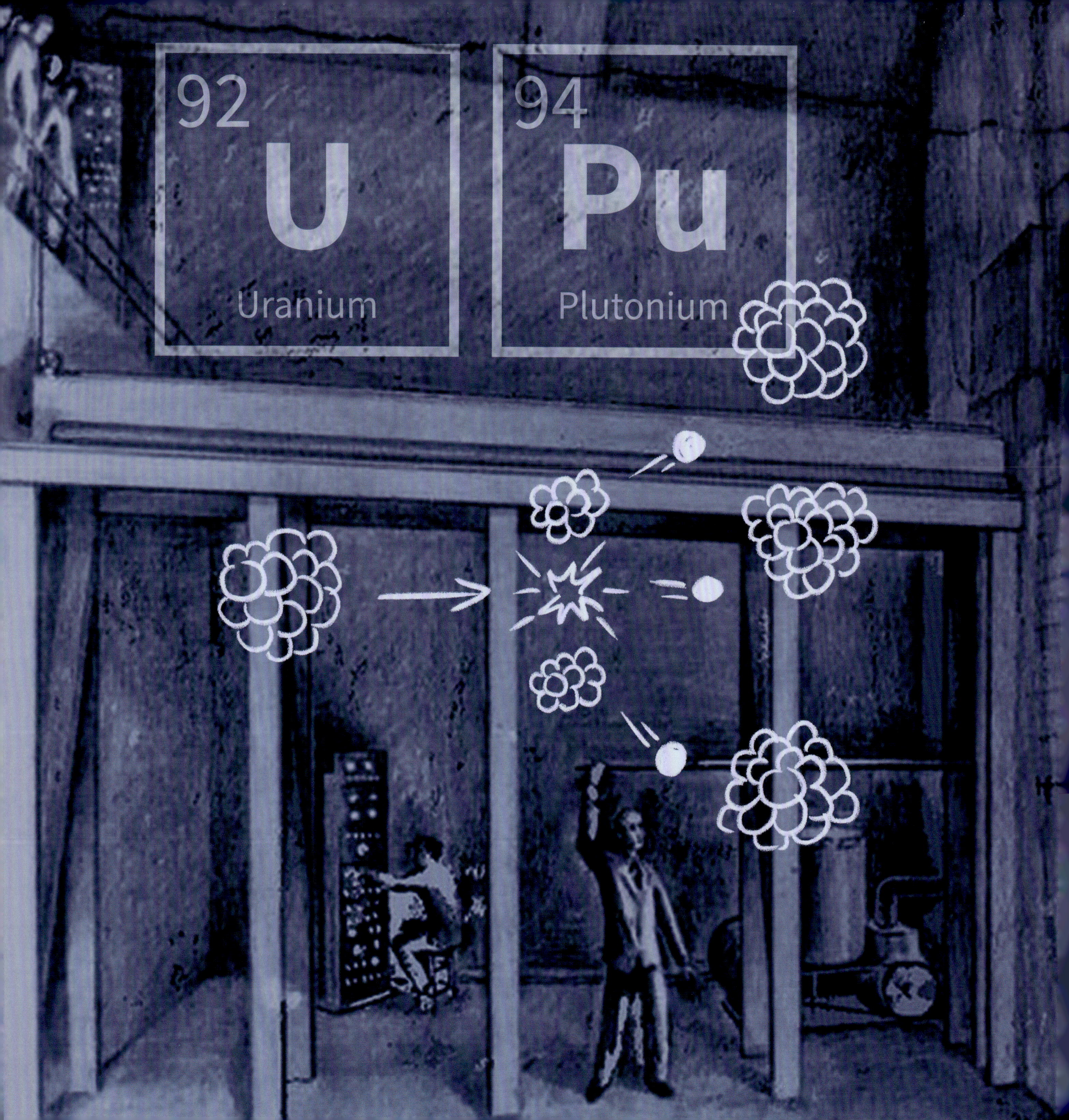
92
U
Uranium
94
Pu
Plutonium

Introduction Splitting the Atom

Less than one year before the start of World War II, as the threat of conflict hovered over Europe, physicist Lise Meitner—who had fled Nazi Germany for Sweden—and her visiting nephew, physicist Otto Frisch, shared one of the most consequential "eureka" moments in history as they walked through the December snow. For several years, at Meitner's former laboratory in Germany, she and chemist Otto Hahn had been firing neutrons at uranium—then the heaviest known element—to study what happened when a neutron was absorbed by a uranium atom.

The elemental identity of an atom is determined by the number of protons in its nucleus. Every atom of the element uranium, for example, contains ninety-two protons. The other type of particle in a nucleus is the neutron. Unlike with protons, the number of neutrons in the nucleus does not affect the atom's identity. But, when a nucleus absorbs an additional neutron, it will typically become unstable; the usual outcome, observed for many of the elements through experiments like Meitner and Hahn's, is that the nucleus will subtly rearrange its constituent particles, often transforming into a nearby element on the periodic table with slightly more or fewer protons.

Hahn and Meitner had found that uranium did not follow this trend, however; when the heavy atom absorbed a neutron, unidentifiable substances continuously appeared. Hahn—working with their assistant Fritz Strassmann after Meitner's departure from Germany—was finally able to confirm that one of the substances produced was the much lighter element barium, an implausible answer that only raised another question: How could absorption of a single neutron cause uranium, an element with ninety-two protons, to transform into barium, with only fifty-six protons? At the time, Hahn later wrote, the finding "contradict[ed] all previous experience in physics."

After receiving a puzzled letter from Hahn, Meitner and Frisch solved the mystery from afar. They hypothesized—correctly—that when a uranium nucleus absorbs a neutron, the event causes the atom to become so unstable that it cannot correct the imbalance through the subtle kind of change described above. Instead, the nucleus splits apart, and its protons form two new atomic nuclei roughly half the size of uranium. It seemed at first like there was no mechanism to explain this behavior, but a few simple calculations of mass and energy later, the pair sat together on a tree trunk looking at a discovery that would define the next century. It was Frisch who gave the nuclear reaction a name: *fission*.

As word spread throughout Europe and the United States, physicists were gripped by the new world of fission research. Many were quick to realize that this major scientific breakthrough could have dramatic consequences. A force of tremendous strength—the strong force—binds protons and neutrons to each other within a nucleus. When the grip of the strong force is broken, as when a nucleus splits, an enormous burst of energy is released. As the protons from the original nucleus recombine into smaller nuclei, the energy burst slings neutrons away at high speeds. These neutrons can enter and split the nuclei of other nearby uranium atoms, sending more neutrons hurtling away to cause additional, energy-releasing fissions: a *chain reaction*. The implications were both exciting and unnerving: If a nuclear chain reaction could be initiated and sustained in a controlled manner, the energy freed by fission might be harnessed to generate electricity. If the same energy was released in an uncontrolled manner—a runaway chain reaction—fission might produce an explosion.

The mechanism underlying a designed nuclear explosion was alarmingly simple, especially to scientists in prewar Europe. Considering the extreme amount of energy released by a single instance of fission (relative to the size of an atom), a chain reaction–based explosive should be significantly more efficient than any existing weapon: The heaviest and largest conventional bomb that would be used during the impending war, Britain's Grand Slam, produced a yield equivalent to that of about 6.5 tons of TNT. In comparison, a bomb fueled by only a small amount of uranium should produce the same yield as *thousands* of tons of TNT.

As scientists rushed to learn more, however, national governments were slower to appreciate the potential applications—both peaceful and military—of atomic energy. By the time Germany invaded and declared war on Poland in September 1939, fears had mounted within the scientific community that, as other governments hesitated to pursue fission research, Nazi Germany might, secretly, be gaining a head start. Otto Frisch, by this time a refugee in the United Kingdom, partnered with German-born Jewish physicist Rudolf Peierls to warn the British government of fission's military potential. Their March 1940 memo imagined an air-deliverable "super-bomb" that was possibly

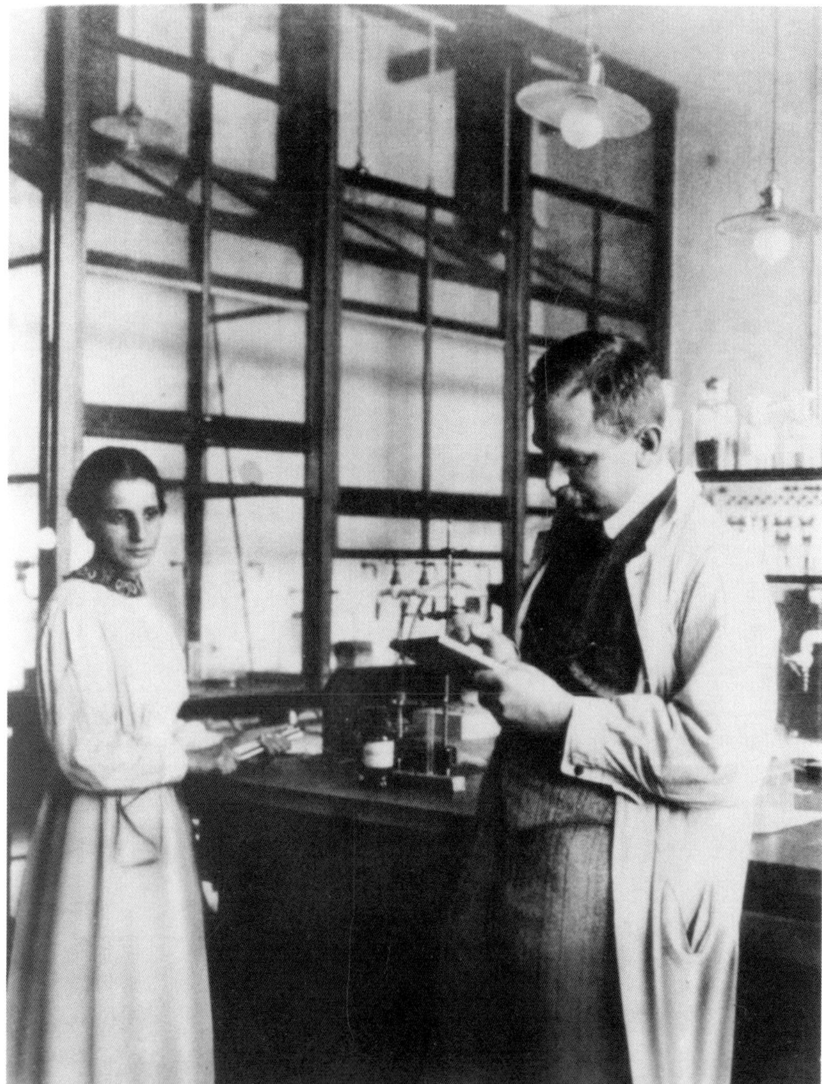

Lise Meitner and Otto Hahn at the Kaiser-Wilhelm Institute for Chemistry in Berlin, Germany, 1913. (Smithsonian Institution Archives, Accession 90-105, Science Service Records, Image No. SIA2008-3209.)

capable of destroying the heart of a major city, and they included calculations to support an astounding projection: "The energy liberated in the explosion of such a super-bomb . . . will, for an instant, produce a temperature comparable to that in the interior of the sun." The memo continued ominously, "We have no information that the same idea has also occurred to other scientists, but since all the theoretical data bearing on this problem are published, it is quite conceivable that Germany is, in fact, developing this weapon."

Though it's now well known that Germany never came close to developing an atomic bomb—in part because many of its top scientists had fled the country—the anxiety that Adolf Hitler might be the first to harness nuclear technology was well founded. The basic ingredients for producing a nuclear explosive were present—a heritage of innovative engineering, an advanced industrial capability, and access to uranium. Moreover, the country was home to premier institutes of physics and chemistry that attracted top scientists to teach and research. It was expected that scientists remaining in Nazi Germany, such as Nobel Prize–winning nuclear physicist Werner Heisenberg, would be asked to urgently explore the possibility of weaponizing fission.

This fear and the urging of American scientists and their European counterparts—many of whom, including Albert Einstein, were then working at American universities—is ultimately what prompted the US government to greenlight the early activities that would evolve into the atomic bomb crash program known as the Manhattan Project. From late 1939 until the Manhattan Project's 1942 inception, while much of continental Europe was falling under control of dictators, the early American enterprise unfolded as a long period of feasibility research. Guided initially by President Franklin D. Roosevelt's Advisory Committee on Uranium—which was formed in response to a warning letter sent to the president by Einstein and Hungarian physicist Leo Szilard—scientists studied the process of fission and prospect of a chain reaction.

Uranium was at that time a familiar material with well-understood chemical and physical properties, having been discovered over a century earlier. But much research and experimentation were needed to demystify the element's nuclear characteristics and begin to reliably predict its behavior. Physicists had already determined, shortly after fission's discovery, that not all uranium atoms have equal ability to split. As with every element, there are different types of uranium atoms, each called an isotope. Isotopes have the same number of protons in their nuclei but different numbers of neutrons. In a chunk of uranium, 99 percent of the atoms are actually the non-fissile isotope uranium-238 (fissile, here, meaning able to sustain a chain reaction); only about 0.7 percent are the fissile form, uranium-235, which has three fewer neutrons. Although both isotopes can absorb a neutron, the more stable uranium-238 will not undergo the dramatic nuclear process of fission with the same regularity.

Szilard and Nobel laureate Enrico Fermi, a refugee from fascist Italy then at Columbia University, took on the work of determining whether a fission chain reaction could be produced and sustained in a large sample of natural uranium, even though only a small portion (0.7 percent) of the uranium could fission. Fermi had been the first scientist to bombard atoms with neutrons, shortly after the existence of the neutral particle was confirmed in 1932; it was his pioneering experimentation, which included the chance discovery that slowing down a neutron could increase its likelihood of being absorbed into a nucleus, that had spurred Hahn and Meitner's later research into uranium nuclear reactions.

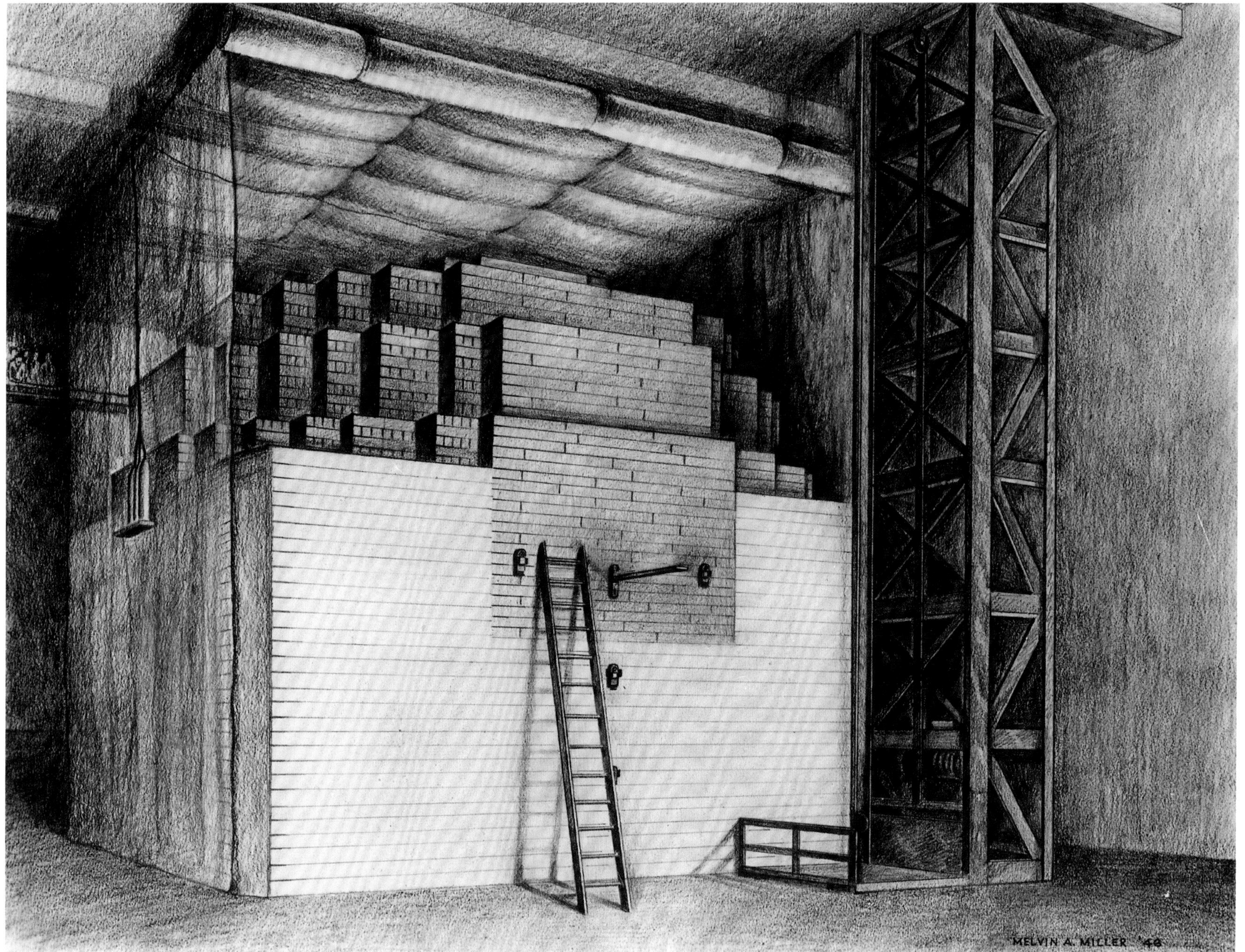

Chicago Pile-1, Melvin A. Miller, 1946. (Digital Photo Archive, Historian's Office, US Department of Energy, courtesy of AIP Emilio Segrè Visual Archives.)

Fermi and Szilard believed that if enough natural uranium was amassed in a huge pile (what we now call a nuclear reactor), the total number of fissile uranium-235 atoms would at some point become sufficient to sustain a fission chain reaction. They planned to surround the uranium in the pile with a moderator, which would act like a bumper, slowing down the neutrons ejected during fission and increasing their chances of entering nearby uranium nuclei to keep a chain reaction going. In one of its earliest acts of financial support for nuclear research, the US government distributed $6,000 in funding for acquisition of graphite to use as a moderator in the nuclear reactor. The structure, two years later (in 1942), became known as the Chicago Pile—quite literally a pile of uranium and graphite—when efforts to centralize fission research led to the experiment moving from Columbia to a squash court underneath the University of Chicago's Stagg Field.

Although it would serve as an important step in nuclear science, if successful, Fermi's proposed "pile" method was designed only to achieve the first fission chain reaction and then facilitate its study. Creating a militarily useful device was another task entirely. For one thing, any chain reaction driven by slowed neutrons, as in Fermi's graphite-moderated design, would not proceed quickly enough to generate explosive force before a weapon melted down or blew itself apart, producing harmful radiation but little destructive power. In addition, the pile structure eventually built in Chicago—fueled by six tons of uranium metal and fifty tons of uranium oxide—was necessarily enormous, whereas a deliverable weapon must fit within the size and weight limitations of an aircraft. In the course of developing a bomb, scientists would have to create a significantly smaller chain reaction environment with a more rapid proliferation of fissions. To do so, irrespective of a weapon's design, required separating the rare, fissile uranium-235 from the predominant isotope uranium-238. Of every 140 atoms of natural uranium, only 1 is an atom of uranium-235; the other 139 atoms would weigh down a weapon without contributing meaningfully to the chain reaction.

Isolating the tiny percentage of fissile atoms found in natural uranium was a monumental, expensive, and uncertain task that would require construction of massive factories outfitted with novel equipment. The difficulty lay in the nature of isotopes. Whereas two different elements, with differing numbers of protons (and electrons), can be separated from each other via chemical reactions, separating isotopes of the same element—which have identical chemical properties—requires physically distancing atom from atom. Uranium-238, though practically indistinguishable from uranium-235, is very slightly heavier—by the mass of its three additional neutrons. The way to separate the two isotopes was by exploiting this tiny mass difference. One of several methods explored involved using a machine called a mass spectrometer to send a stream of uranium atoms in circles through a magnetic field, with the idea being that the lighter isotope would be deflected ever so slightly more by the magnetic field; two streams of particles, differentiated by atomic mass, would gradually form and could then be collected separately. A modified, more powerful version of the mass spectrometer, dubbed the calutron by its inventor, University of California physicist Ernest Lawrence, would later make the electromagnetic method more viable—but in 1940, using then-available technology and equipment, it was projected to require some 27,000 years to separate just one gram of uranium-235, while the amount of fissile uranium needed for a weapon would be in the tens of thousands of grams.

In February of 1941, as the task of uranium separation loomed, the unexpected discovery of another fissile element—plutonium-239—introduced an advantageous wrinkle into the quest for an atomic bomb. Plutonium, with ninety-four protons, is the end result of a series of nuclear reactions that occur when non-fissile uranium-238 absorbs a neutron. It was one of the products Hahn, Meitner, and others had been expecting when they bombarded uranium with neutrons throughout the 1930s. While uranium-238 could not itself fuel a weapon, the abundant isotope could be used to synthesize plutonium—potentially in some version of the uranium pile that Fermi envisioned. There were now parallel pathways to obtaining fuel for a fission bomb.

A group of British scientists conducting their own, secret fission research across the Atlantic arrived by the summer of 1941 at the same juncture then facing the United States: theoretical studies were reaching the end of the runway. Though the basic architecture of an atomic bomb seemed straightforward, to continue pursuit of a weapon, someone would have to invest in the massive industrial facilities and teams of operators required to produce usable nuclear fuel. Seeking collaboration and to instill a greater sense of urgency in their powerful ally, the British shared their fifteen months of feasibility studies with the United States in a document called the MAUD Report (named for the group of scientists, called the MAUD Committee, who produced it). The report indicated with certainty that a weapon would work so long as a "critical amount" of fissile material—the minimum amount required for a self-sustaining chain reaction to begin—could be assembled. "All that is necessary to detonate the bomb," the report stated, "is to bring together two pieces of the active material each less than the critical size but which when in contact form a mass exceeding it." A weapon, it projected, could be ready in time to produce decisive results during the war—for whomever might be trying to build one.

President Roosevelt responded by authorizing Vannevar Bush, who was then leading the Office of Scientific Research and Development and had overseen uranium research since June of 1940, to begin exploratory conversations with the military to determine the cost of taking next steps. Shortly thereafter, on December 7, 1941, Japan's surprise attack on Pearl Harbor brought the distant war to United States soil. On the home front and abroad, the full might of American military and industry rose up against Japan and its allies, Italy and Germany. As the country mobilized, the second phase of the nuclear enterprise was formally launched under great secrecy, moving at last from theory and possibility into physical reality.

Bush determined that designing, constructing, and operating plutonium and uranium production facilities was beyond the scope of what policymakers and academics could accomplish. He delegated the next steps in the United States' still fledgling nuclear research program to the US Army Corps of Engineers. The initiative established to conduct this work was created in August 1942 to appear as nothing more than a standard geographic district of the Corps, which had carried out military and civil construction projects since the American Revolution. Despite what the name suggests, there were no boundaries to the new Manhattan District (headquartered in New York City), and its activities, which came to be known as the Manhattan Project, would take place at secret locations spread around the country.

The project came under command of Col. Leslie R. Groves, a West Point graduate, in September of 1942, after what Bush complained was a sluggish start. Groves had been in the Corps since 1918 and became deputy to the chief of construction in 1940, overseeing creation of depots, air bases, munitions plants, and hospitals. The Pentagon, the largest office building ever constructed at the time, was completed in sixteen months under his direction. Groves's record of success made him a logical choice to manage the growing number of facilities, as did his ongoing involvement with the Manhattan Project as a consultant for the previous overseer, Col. James C. Marshall. After receiving a hasty and purposeful promotion, the new brigadier general immediately tackled the objectives of building and operating the production plants, successfully lobbying the War Production Board to grant the project the highest level of emergency materials procurement priority.

It was still not known how much of either uranium or plutonium would be necessary to make a bomb. When Groves sought an estimate from Fermi and others in Chicago to scale production facilities to the appropriate size, he was told that their calculations were precise by a factor of ten, which the general later equated to a caterer being asked to serve between 10 and 1,000 guests. Forging ahead anyway, Groves finalized a contract with the chemical company DuPont for a small plutonium pilot reactor at Oak Ridge, Tennessee—a site he acquired within days of taking over the Manhattan Project that would also host several uranium separation facilities—and a sprawling complex at Hanford, Washington, that would include four nuclear reactors and three chemical separation plants.

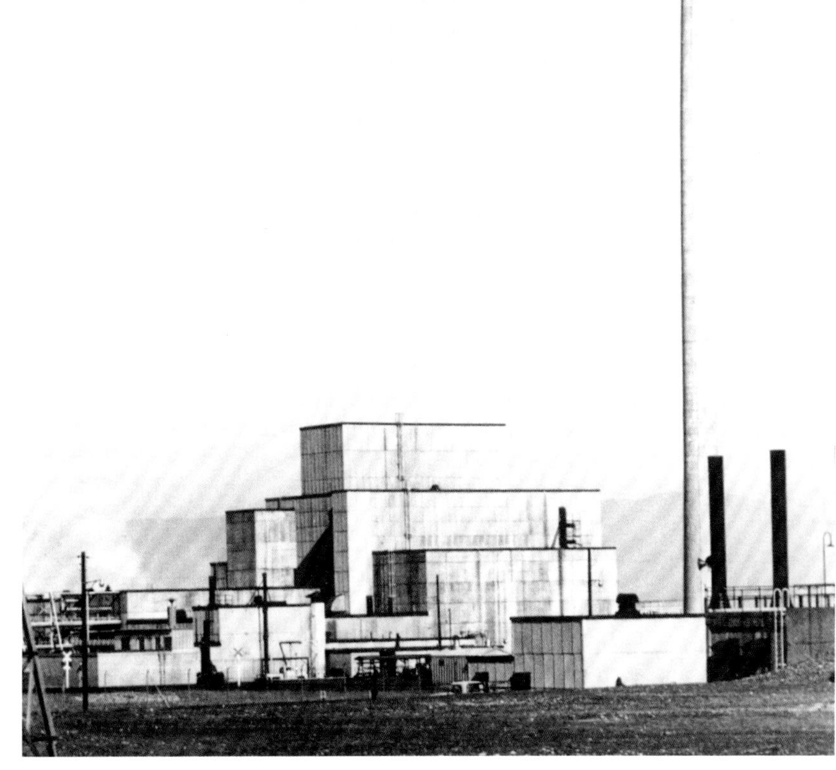

Construction of B Reactor—modeled after the Chicago Pile—in Hanford, Washington, began in October 1943. The plutonium synthesized here was used in the Trinity test device and in the atomic bomb detonated over Nagasaki.

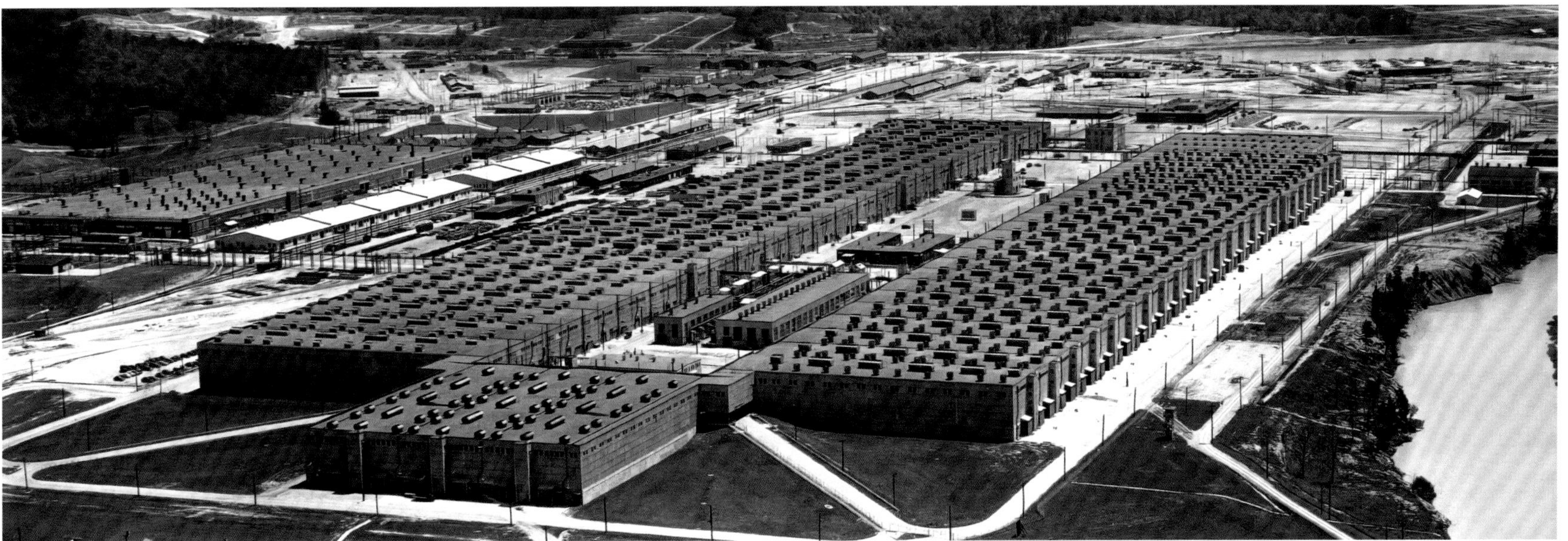

Three uranium isotope separation facilities were constructed at Oak Ridge, Tennessee, beginning in February of 1943: the Y-12 Electromagnetic Isotope Separation Plant (which housed cascading series of calutrons known as "racetracks"), the S-50 Liquid Thermal Diffusion Plant, and the U-shaped K-25 Gaseous Diffusion Plant (*above*; photo courtesy of US Department of Energy).

A fission chain reaction had yet to be demonstrated, but Groves believed that "nothing would be more fatal to success than to try to arrive at a perfect plan before taking any important step."

On December 2, 1942, the general's act-now philosophy was justified when, at the University of Chicago, the most important milestone before the Trinity test was reached. Fermi achieved a critical mass of uranium in the Chicago Pile; running for about thirty minutes, the successful experiment provided the first demonstration of a self-sustaining fission chain reaction. Proof of concept was finally in hand: as feared, as hoped, and as Groves had already decided to assume, a nuclear weapon was inevitable.

Because the work of translating theories and calculations into a usable weapon was an endeavor viewed with high optimism by most scientists, compared to the monumental task of processing nuclear fuel, the practical issues of designing a weapon had thus far received little attention. Groves brought a necessary sense of urgency to this aspect of the project, which he would soon entrust to J. Robert Oppenheimer, a California-based physicist who had been guiding the widespread theoretical research that would underlie a bomb.

Though a brilliant theoretician, Oppenheimer lacked administrative experience and experimental acumen, both of which Groves thought important qualities for the leader of the weapons development program that would build upon Oppenheimer's research. As well, there was consensus among Manhattan Project scientists that the role should go to a Nobel laureate, which Oppenheimer was not and never became. However, the theoretician had perhaps the deepest knowledge of anyone about the task ahead, and Groves ultimately could not find a better available person for the job. He appointed Oppenheimer to direct the laboratory, codenamed Project Y, that would be built to consolidate design and fabrication of a device itself.

Oppenheimer and Groves together selected a location, and the new director began recruiting for the mission, calling upon many of the brightest scientists of the current generation to join him. For Oppenheimer, this was an opportunity to blend his two great loves, physics and high desert country. For potential recruits, the ask was not so appealing, given that the laboratory was to be built on a remote plateau in the New Mexico wilderness and its workforce would be walled off from the outside world. Many scientists were young and had families. Some, like Fermi in Chicago, had already relocated once in support of the Manhattan Project.

As a further complication, the laboratory was planned as a military outpost, requiring all workers to join the army. Even Oppenheimer, who had already ordered the uniforms of a lieutenant colonel, was not exempt. Fortunately, his employer, the University of California, agreed to administer the first phase of the

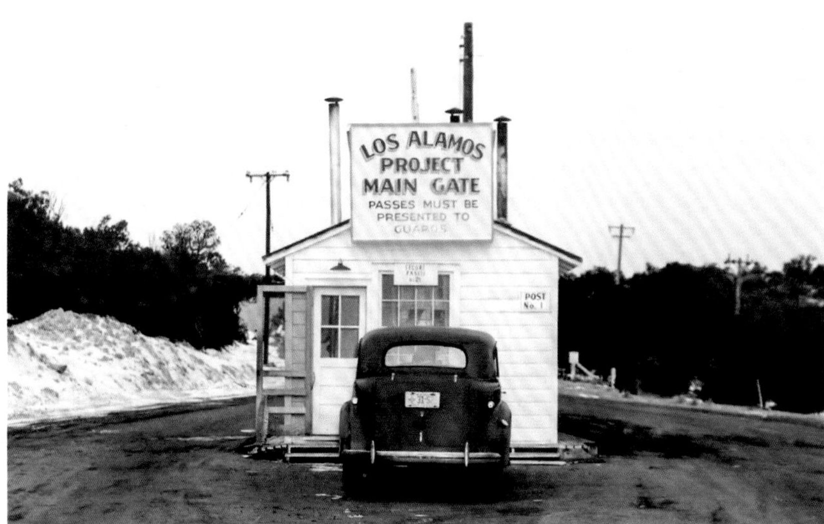

laboratory's work; instead of going to boot camp, Oppenheimer's recruits would join the institution's faculty. Under these improved circumstances and invoking the gravity of the mission at hand, the director succeeded in gathering a critical mass of theoreticians, experimentalists, and engineers to the Manhattan Project's newest site, a secret town and research facility constructed around the former Los Alamos Ranch School in northern New Mexico.

The scientists and engineers in the laboratory's workforce, which would grow to more than 2,500 technical and support staff over the next two years, were initially organized into four divisions: Under direction of Joseph Kennedy, a co-discoverer of plutonium, the Chemistry Division was responsible for fabricating reliable weapons components from rare and sometimes hazardous materials. The Theoretical Division, led by Hans Bethe, proposed different weapons designs and, while the laboratory awaited plutonium and enriched uranium for experiments, performed calculations predicting how those crucial materials might be used. Robert Bacher, one of Bethe's research partners at Cornell University, led the Physics Division, which was tasked with inventing methods and tools to study new substances and processes through unprecedented experiments. Finally, the Ordnance (or Engineering) Division, under Navy Captain William S. "Deak" Parsons, would take the components devised and fabricated in other divisions and integrate them into deliverable weapons.

Across the laboratory, the different groups tasked with co-creating an atomic bomb formed a winding assembly line, without a clear beginning or end but with a single common goal: to harness the power of fission for the United States military. As their efforts were beginning on the home front, the United States, Great Britain, and an unlikely ally—the Soviet Union, once in a nonaggression pact with Hitler—together turned the tide of war throughout 1943: German

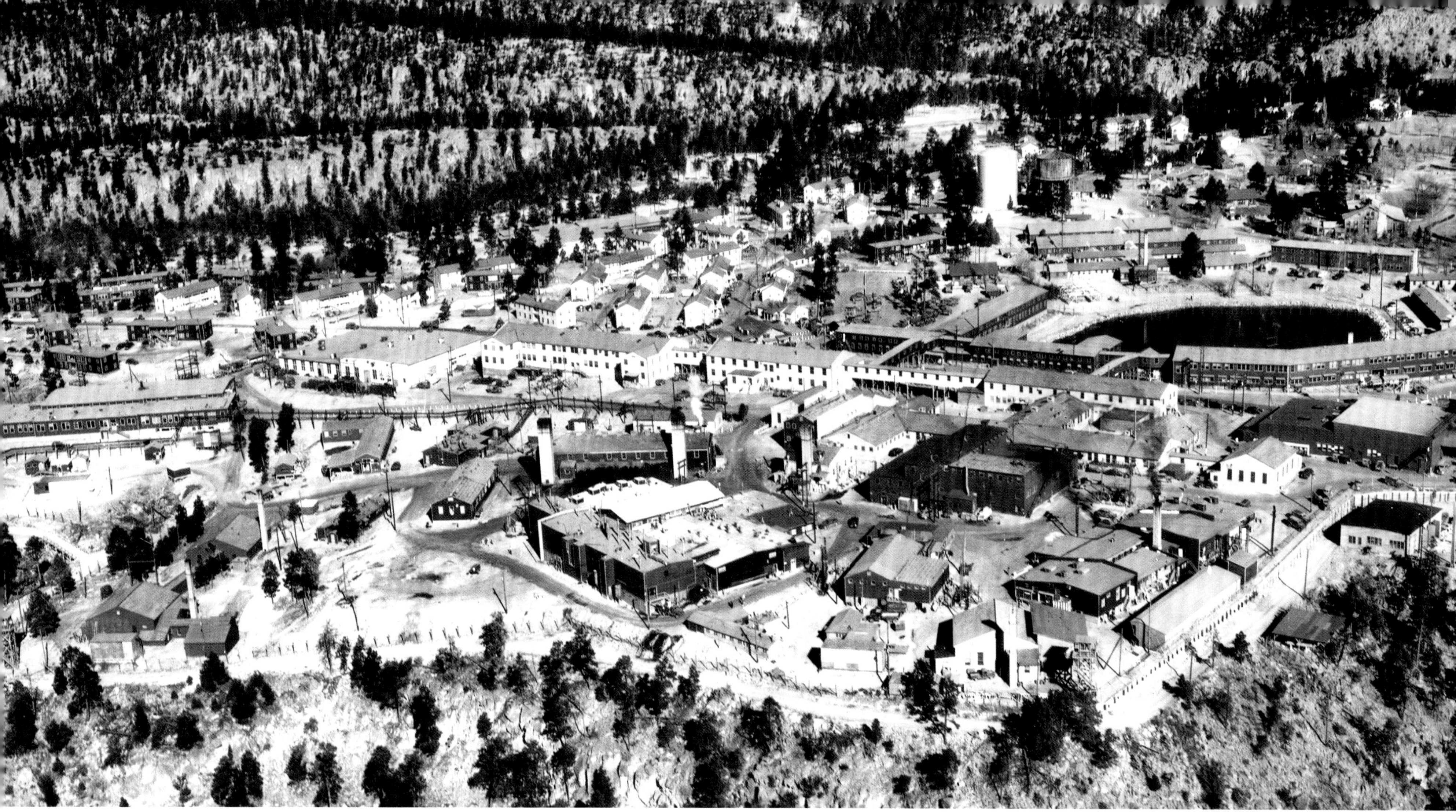

Project Y, called "the Hill" by scientists who lived and worked there during the Manhattan Project. The main gate of the secret installation is shown in the photo at left.

forces retreated from northern Africa and were halted on their march across the Soviet Union. Meanwhile, Italy's fascist regime had fallen by midyear, and the new government made peace with the Allied powers. Despite these victories, much of Europe remained occupied by Hitler's forces, and the fierce battles won by the United States against Japan in the Pacific added to a death toll that by war's end would climb to at least forty million civilians and fifteen million military personnel. The possibility weighed heavily that Allied progress, made at great cost, could be swiftly undone by a nuclear weapon in the enemy's hands.

As the pressure to succeed held steady, the turn of the year saw Los Alamos in pursuit of two very different designs for an atomic bomb, each with its own unique mechanism of assembling a critical mass—or the amount of fissionable material required, under a given set of conditions, for a chain reaction to begin and sustain itself. The gun-type design, premised upon firing one subcritical portion of fissile material at another, was universally regarded with optimism and commanded most of the laboratory's focus; simple from an engineering perspective, it relied on existing technology from the realm of conventional weapons to perform the novel function of starting a fission chain reaction. The other design, taking the form of a spherical device, relied on a technique called implosion, which, in addition to being far more complex, was a completely unprecedented idea drawn from the fledgling field of precision high explosives.

It was this implosion design that by mid-1945 would produce the first nuclear explosion, at the hands of a weathered group of soldiers and scientists preparing dutifully for months at a secret site in the New Mexico desert, 200 miles south of Los Alamos. The story of their mission, known as Project Trinity—its inception, execution, and enduring consequences—is the story herein: a return to the rudimentary beginning of the nuclear age, when the race to attain peace through power culminated with the rise of an atomic cloud—and the messages it carried—not once, but three times, into a shared human sky.

TRINITY

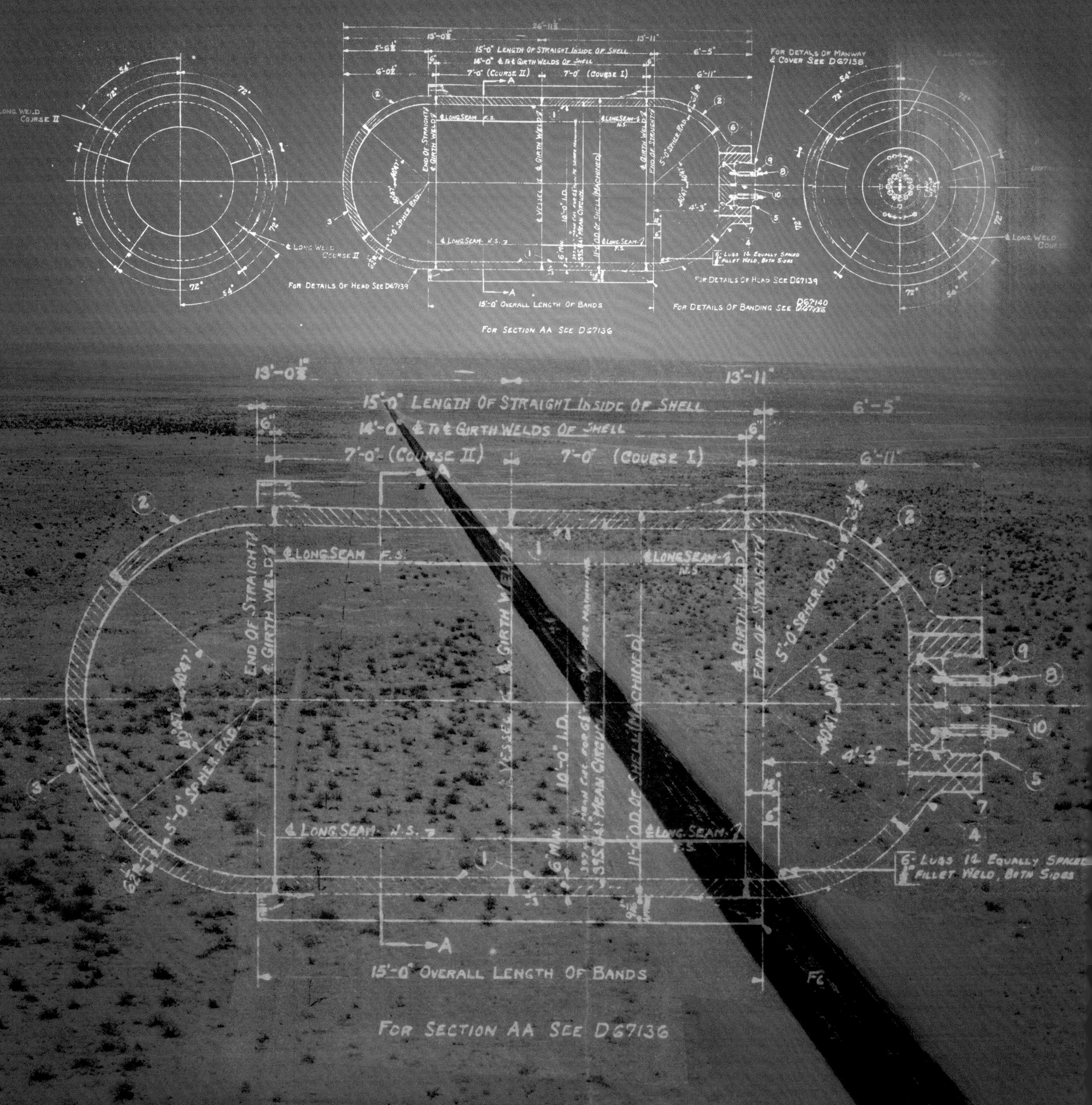

26'-11⅞"
13'-0⅞"
13'-11"
5'-6⅞"
15'-0" Length Of Straight Inside Of Shell
6'-5"
14'-0" ℄ To ℄ Girth Welds Of Shell
6'-0⅞"
7'-0" (Course II)
7'-0" (Course I)
6'-11"
For Details Of Manway & Cover See DG7138
Long Weld Course II
℄ Long Weld Course II
℄ Long Seam F.S.
℄ Long Seam N.S.
End Of Straight & Girth Weld
℄ Vessel & Girth Weld
10'-0" I.D.
11'-0 O.D. Of Shell (Machined)
395¼" Mean Circum.
5'-0" Spher. Rad.
40°47'
4'-3"
6" Min.
6 - Lugs 1¼ Equally Spaced
Fillet Weld, Both Sides
℄ Long Weld Course
For Details Of Head See DG7139
15'-0" Overall Length Of Bands
For Details Of Banding See DG7140
For Section AA See DG7136

01 Project Trinity

Through the first half of 1944, the prototype device that would be detonated in the Trinity test was still in the early stages of development at Los Alamos. According to its basic design, a small, central sphere of fissile material would be compressed by a much larger, outer sphere of conventional explosives to form a supercritical mass at the device's core, generating a nuclear explosion. Whereas techniques like mathematical modeling and rigorous components testing afforded high confidence in the more straightforward gun-type design, scientists believed that for the implosion design, only a full-scale trial shot of a completed weapon could reveal whether all possible flaws had been discovered and resolved; they could not otherwise rule out the possibility of a fizzle—a stunted, low-yield detonation.

Under ordinary circumstances, successfully detonating both the gun-type and the implosion-type bomb would be a clear prerequisite to declaring them ready for military use. However, conducting test shots would require expending uranium or plutonium, which were not yet available from the production facilities. Dedicating the first batches of usable nuclear fuel to a test device might therefore delay the schedule for completed, deliverable weapons. Despite this drawback, it was a strategic imperative that if a nuclear bomb was going to fail, it do so in secrecy. Thus, as the implosion method gained traction in the early part of 1944, Gen. Leslie R. Groves, the leader of the Manhattan Project, agreed tentatively to the full-scale trial shot proposed by the laboratory.

Prompted by fuel scarcity and low confidence in the implosion design, however, Groves placed a condition on the operation: scientists could move forward with planning, but the test must be conducted in such a way that if the device failed to produce a nuclear explosion, they could retrieve and reuse its unfissioned nuclear material. Accepting this requirement, laboratory director J. Robert Oppenheimer appointed physicist Kenneth Bainbridge to lead a new group that would begin working on the various facets of a test shot. He codenamed the operation Project Trinity. "Why I chose the name is not clear," Oppenheimer later wrote, recalling only that a line from John Donne's *Holy Sonnets* was in his mind at the time: "Batter my heart, three person'd God."

In addition to identifying and securing a test location and building out the necessary infrastructure, Bainbridge and his group were to ensure a complete photographic record was made of the attempted detonation, along with comprehensive measurements of its effects. If the implosion device worked, the information gained would be invaluable to the army and the field of nuclear science; if it didn't, scientists could hopefully determine from the records what went wrong and try again.

While many decisions loomed for the newly instated test director, the burden of meeting Groves's condition fell to Roy W. Carlson, a pioneer in concrete technology. His challenge was to devise a way of confining the test device's large conventional explosion. If the test device was a dud—meaning the conventional explosives component detonated but failed to trigger an explosion of the fissile component—the fissile material could then, theoretically, be salvaged. Putting forth an extraordinary effort in parallel with other test planning, Carlson's group ultimately commissioned a supersized detonation chamber, dubbed "Jumbo," into which twelve million wartime dollars and half a million pounds of steel were invested.

Meanwhile, development of an implosion device and plans for testing it proceeded mostly in a two-steps-forward, one-step-back manner. Then, an unexpected turn of events catapulted both to new importance within the nuclear program: the gun-type method, much preferred for its simplicity and significantly farther along in development, was determined to be compatible only with uranium—not plutonium. At this juncture, a billion dollars had been invested in synthesizing plutonium, which was suddenly unusable without a viable implosion-type weapon. Additionally, because the Manhattan Project's uranium enrichment facilities were riddled with technical difficulties and the gun-type design suffered from inherently low fuel efficiency, there would not be enough uranium available to quickly replicate uranium-fueled combat units. If the United States was to firmly establish a nuclear arsenal, and make use of its investment in plutonium, it seemed it would have to rely upon the straggling implosion technique.

The new circumstances not only redefined the significance of the forthcoming implosion test but also spurred Oppenheimer to reorganize personnel and activities at the laboratory. Study of the Gadget, as the first full-scale implosion device has come to be called (all weapons under development were referred to as "gadgets"), was previously led by a group housed within the Ordnance Division. That division became responsible for completion of a uranium-fueled gun-type weapon, codenamed Little Boy, while two new divisions were created to fast-track the design and fabrication of a plutonium-fueled implosion Gadget. It was August 1944, just short of a year and a half into the laboratory's existence.

Groves reported to US Army Chief of Staff Gen. George C. Marshall that a deliverable implosion weapon could be available in the second quarter of 1945, provided the test shot of the Gadget was satisfactory. For the first Little Boy unit, with uranium fuel, he estimated delivery in August 1945. As scientists rushed throughout the remainder of the year and into the next to develop the two very different weapons, the laboratory prepared simultaneously for the Manhattan Project's scientific moment of truth: a secret, full-scale field test that would reveal either a complete failure, a fizzle, or the world's first nuclear explosion.

* * * * * * *

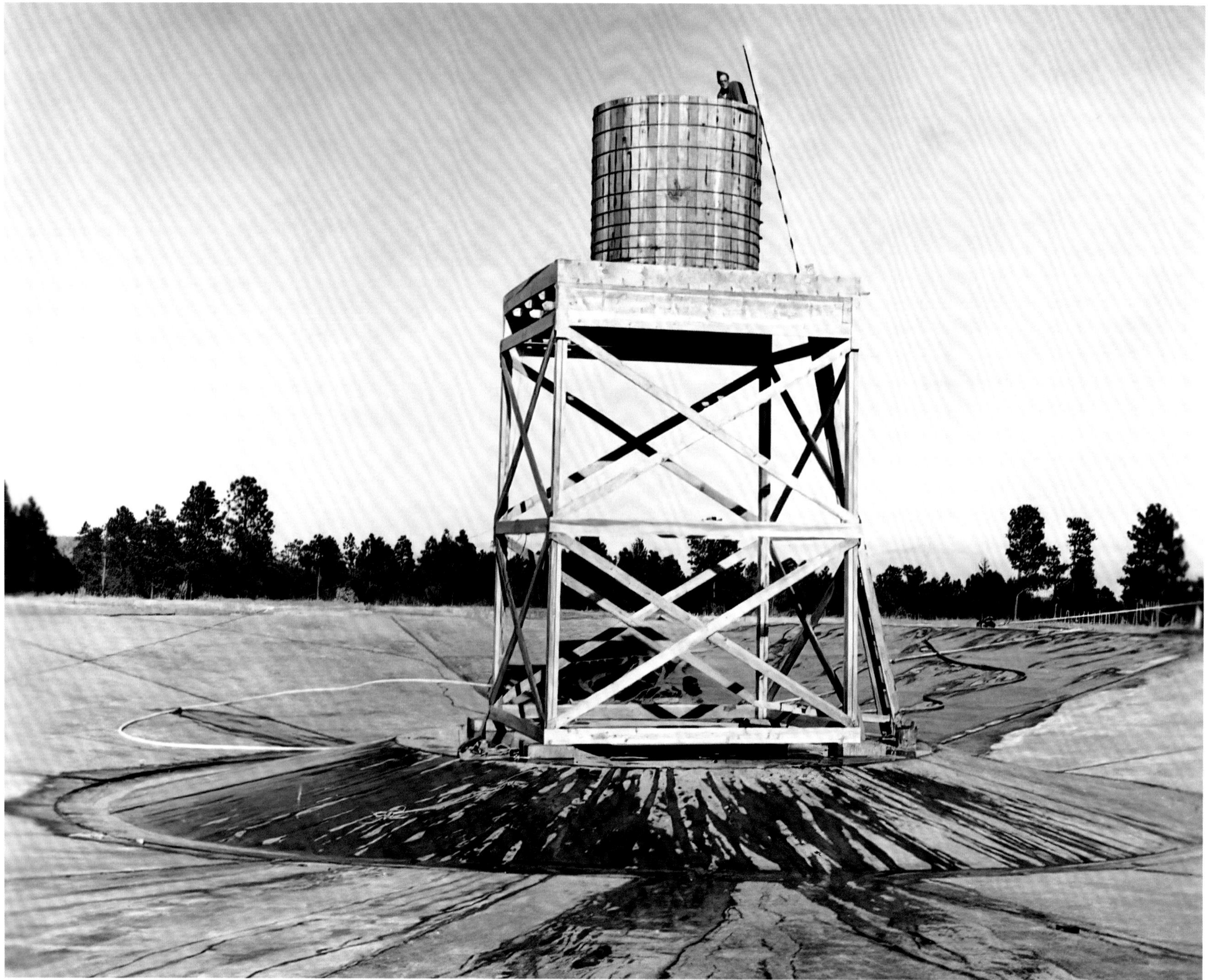

The first steps toward the Trinity test were taken early in 1944, as R. W. Carlson's group began exploring methods to stop the large blast produced by the Gadget's outer layer of chemical explosives from potentially blowing precious, unfissioned plutonium across the landscape. This conventional explosion would be several orders of magnitude weaker than the nuclear explosion it was meant to initiate, but fully containing its force still amounted to a strange and difficult assignment. The quest for solutions birthed not only the steel vessel known as Jumbo but a number of other ideas for how to confine a large blast and recover fissile material—such as burying the test device in a reinforced concrete bunker or covering it with a massive cone of sand. The most promising alternative to Jumbo would have involved detonating the Gadget in an elevated tank filled with thousands of gallons of water; a scale model built at Los Alamos is shown above. The muffled explosion would spread debris in a concrete bowl draining toward the center, from which a recovery team could separate out the fissile atoms for reuse.

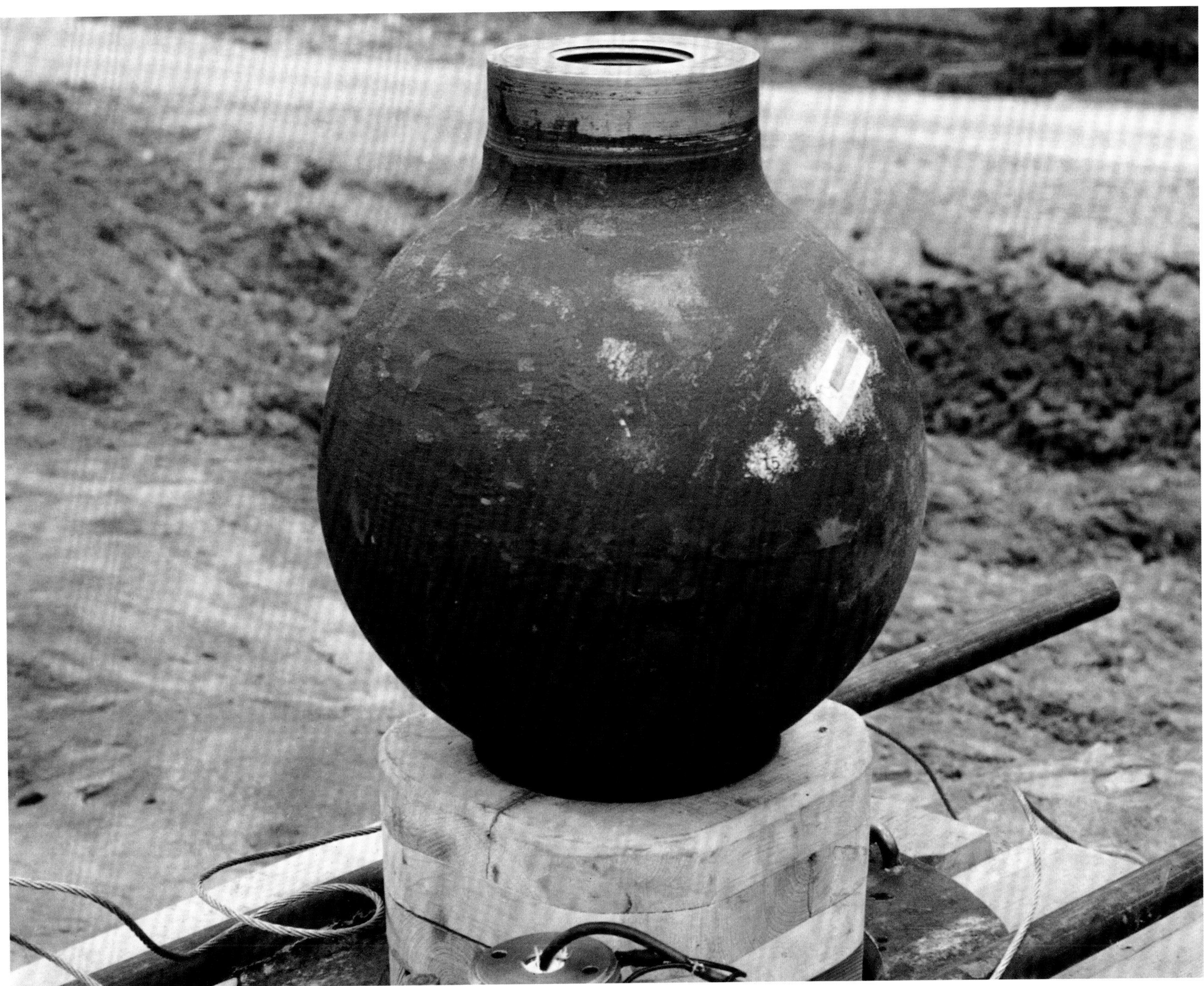

If constructed at the Trinity site, the tower and bowl pictured would have been ten times larger. However, of the various methods entertained, Jumbo alone was destined for full-scale development. Recovering scattered material from inside a relatively small and fully confined space was logistically easiest; but, engineering such a vessel also posed the most significant challenges. Namely, there were very few data in 1944 pointing to what exactly would happen inside the container or what materials were best suited to withstand the force of a multi-ton explosion. Starting at square one, the road to Jumbo began with firing tests inside one-tenth scale prototypes called "Jumbinos" (*above*).

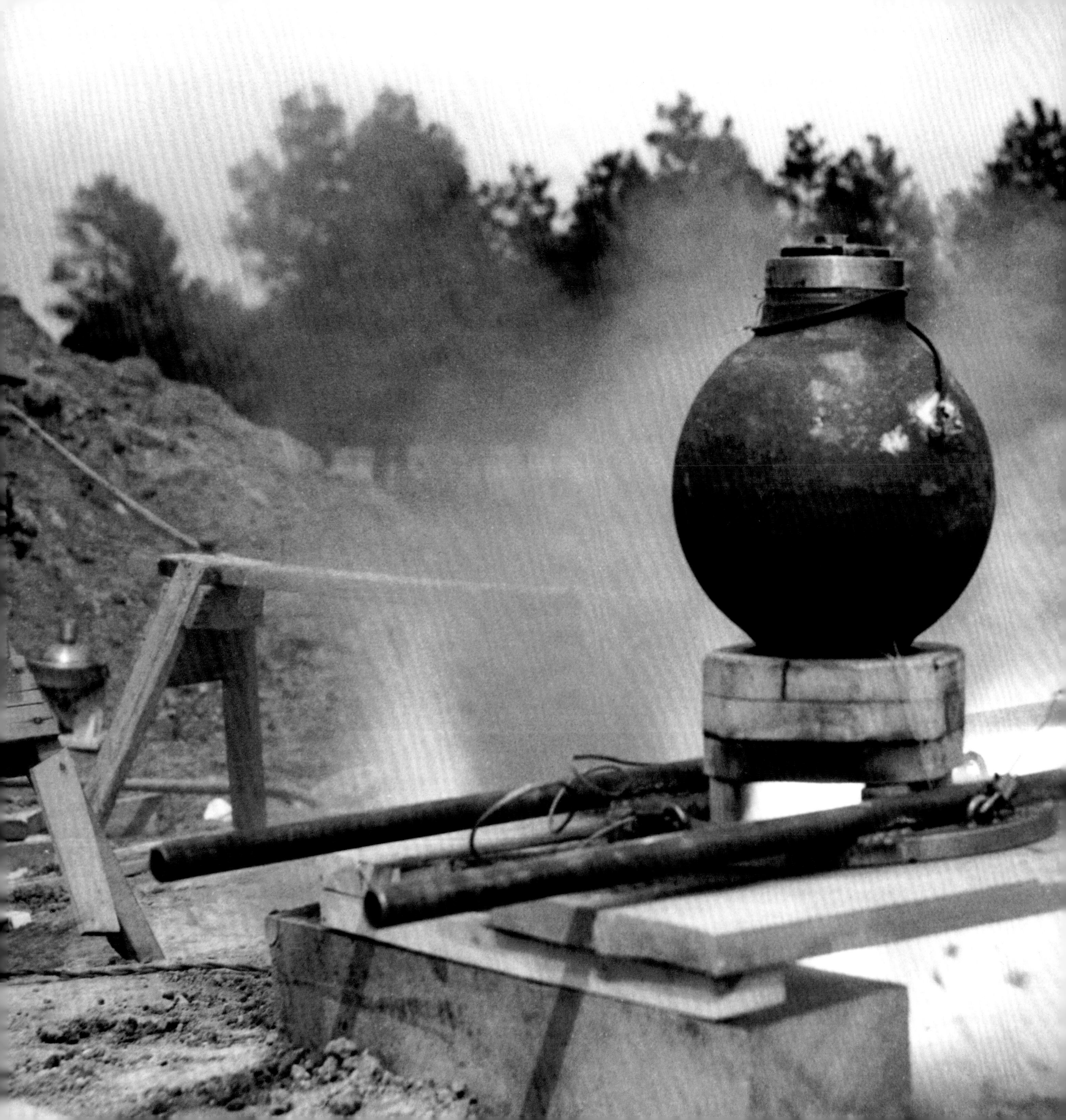

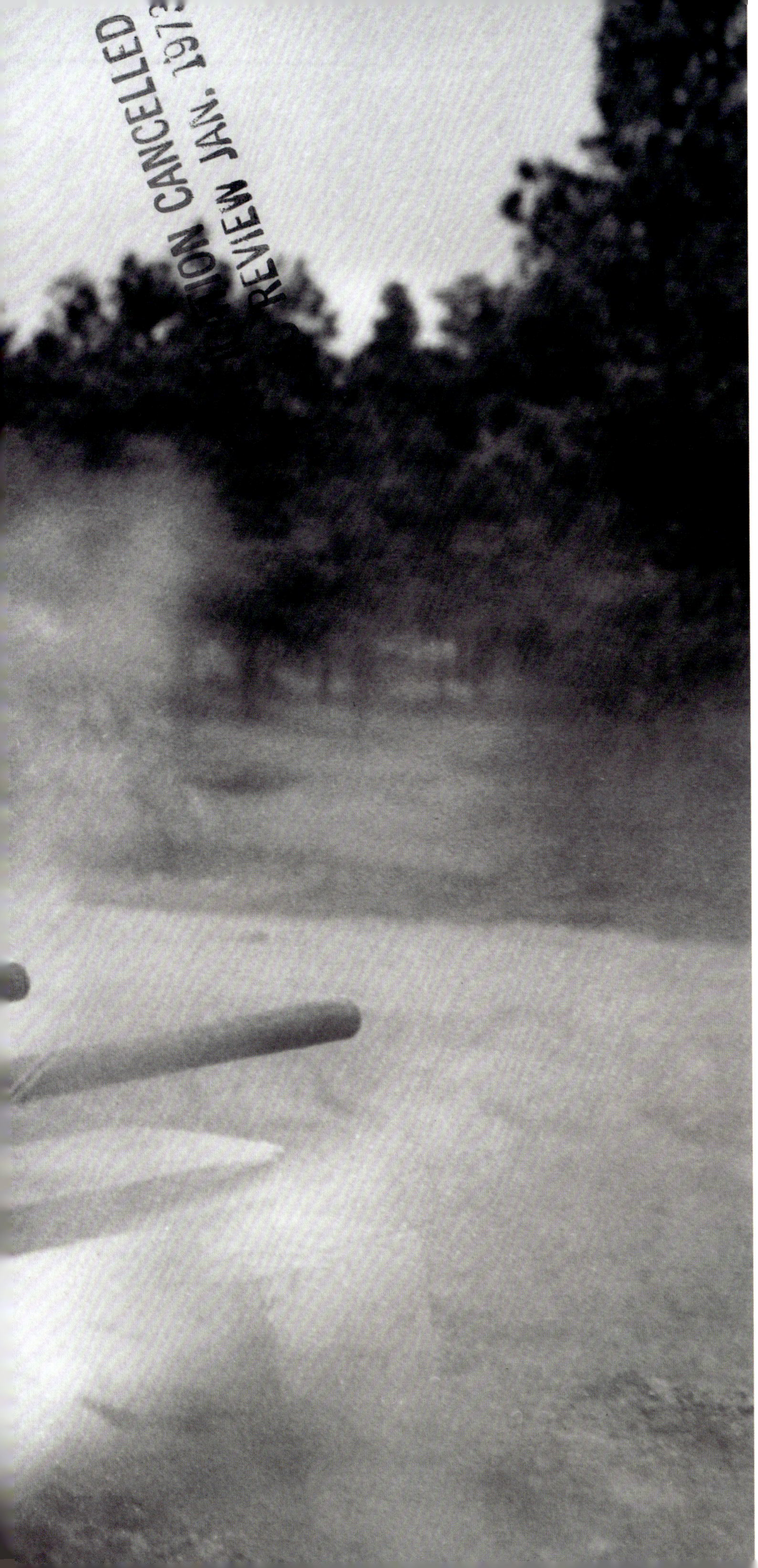

The round, steel vessels, ranging from five to twenty-four inches in interior diameter, hosted hundreds of experimental test shots. A fragment from one of the largest Jumbinos, recovered in 2020, is pictured below. A smaller containment vessel is shown failing a firing test at left.

After destruction of many sacrificial Jumbinos, and on the other side of a low point in March when the containment program appeared doomed to fail, a workable design finally emerged for a full-scale container to house the Gadget. The vessel was to be over thirteen feet long, with walls two feet thick. Upon reaching out to manufacturers, however, Oppenheimer was met with heavy concern about the design specifications for the mysterious, cast steel object, especially a daunting thermite weld that would encircle the giant sphere like an equator. As one chief engineer cautioned, the job would be "of a pioneering nature."

Unable to arrange for its fabrication, the laboratory abandoned the round vessel, thereafter dubbed "Jumbo #1," and turned to another one-tenth-scale prototype that had been waiting in the wings: the "elongated" Jumbino.

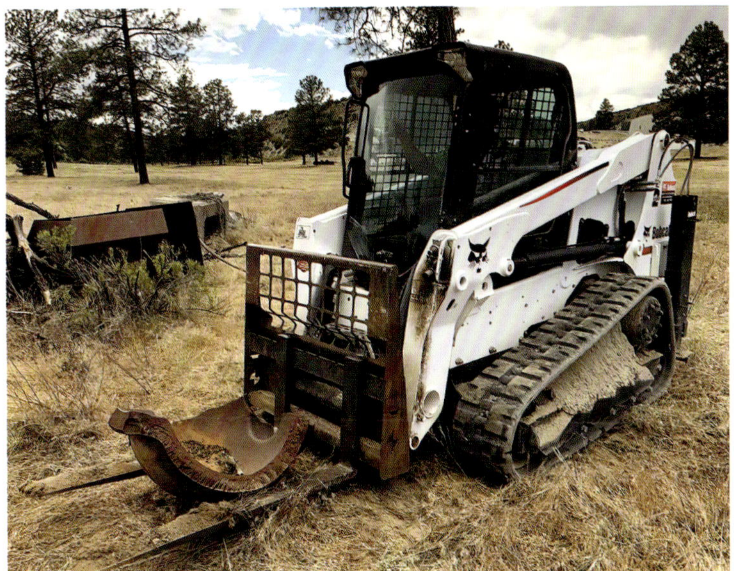

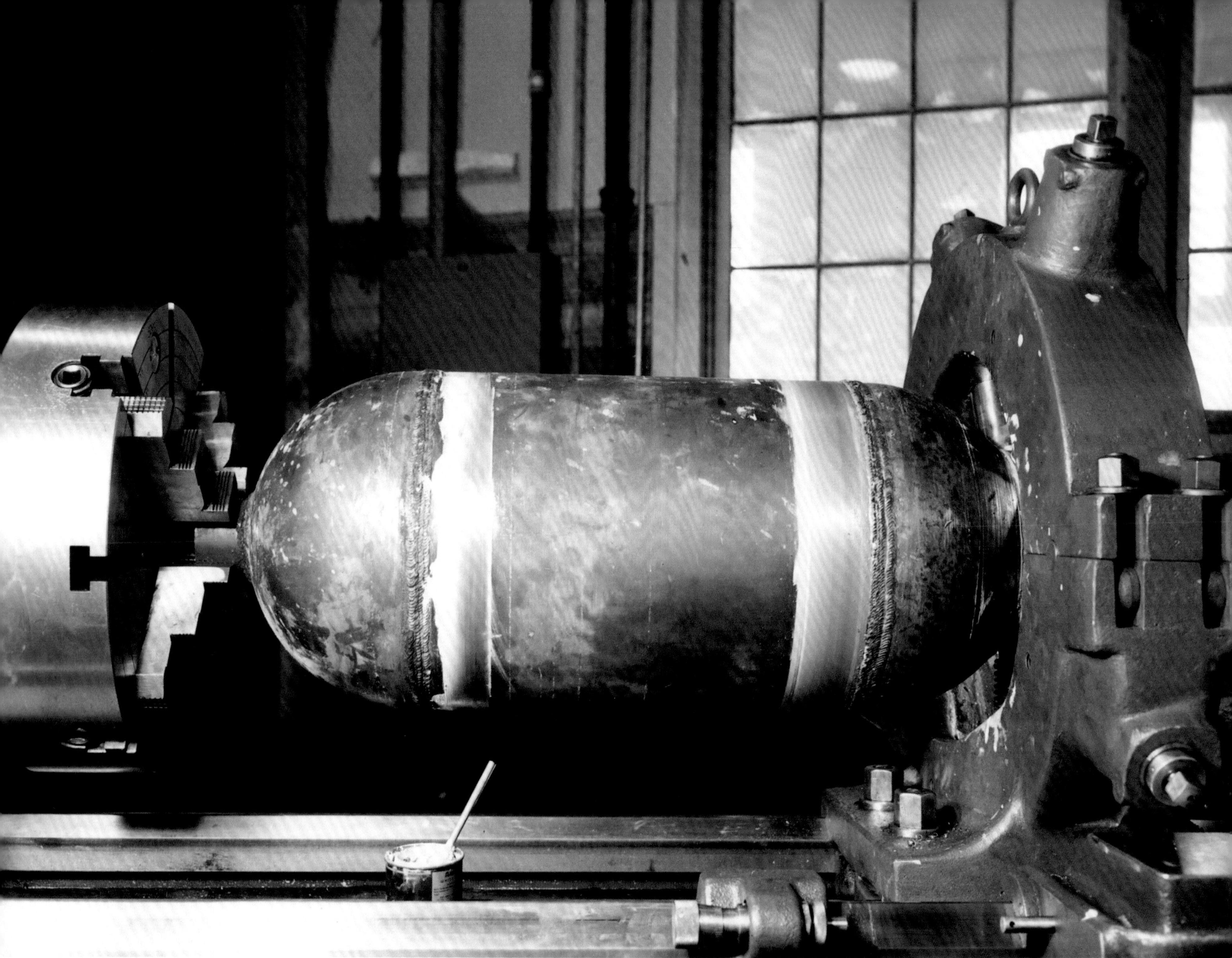

The new shape drew from the spherical original but was reconfigured as two hemispheres, separated by a long cylinder and reinforced with steel banding. Engineers hoped the changes would give more room to the explosion by increasing the volume of the containment vessel and its elasticity, or capacity to bend but not break. The bottle-like design also presented fewer fabrication and transportation challenges than its predecessor. Codenamed the "Straight Length Banded Accumulator," this second iteration of the steel container became known simply as "Jumbo."

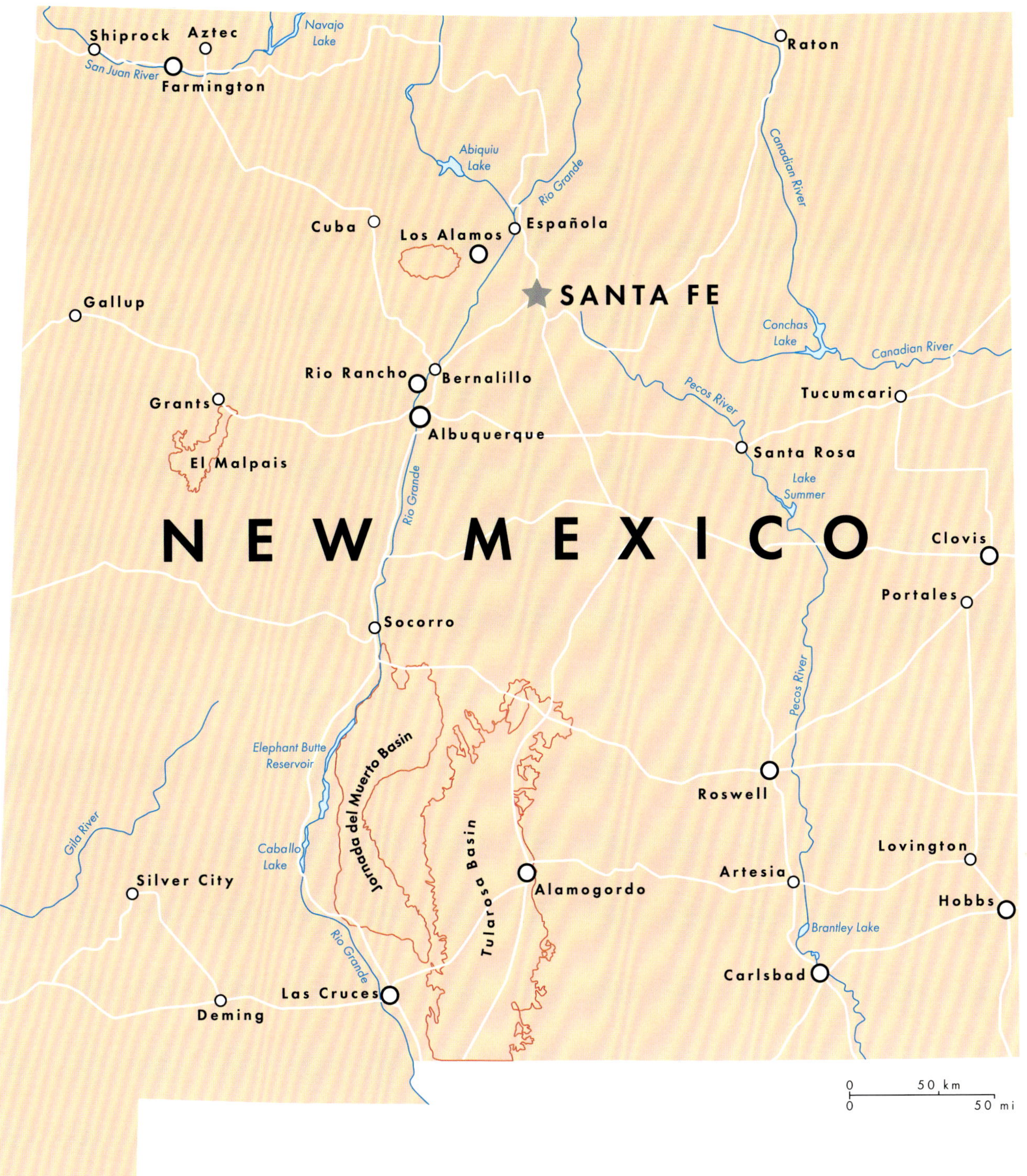

Surveying trips to find a site for Jumbo began in early May. The laboratory grounds, where smaller-scale explosives tests were common, were quickly ruled out given the unknown yield and effects of the weapon (should it be successful) and the possibility of attracting unwanted attention to the secret installation. The short list initially included eight locations all around the western United States, from California to Texas. However, Groves emphasized that the site must still be accessible enough for personnel and equipment from Los Alamos to travel back and forth. The options located in New Mexico thus became front-runners. These included a lava field in El Malpais National Conservation Area, a plateau due west of Los Alamos, and the Tularosa and Jornada del Muerto Basins in the northern Chihuahuan Desert.

Jumbo's engineers along with Bainbridge, Oppenheimer, and others toured New Mexico in search of a suitable site to place the giant container. In the image at right, taken at Valle Grande in May 1944, Oppenheimer is on the far left, next to Los Alamos security officer Major Peer de Silva and Major Wilber A. Stevens. Below, the laboratory director is emptying a thermos. (Photos by Kenneth Bainbridge, courtesy of AIP Emilio Segrè Visual Archives.)

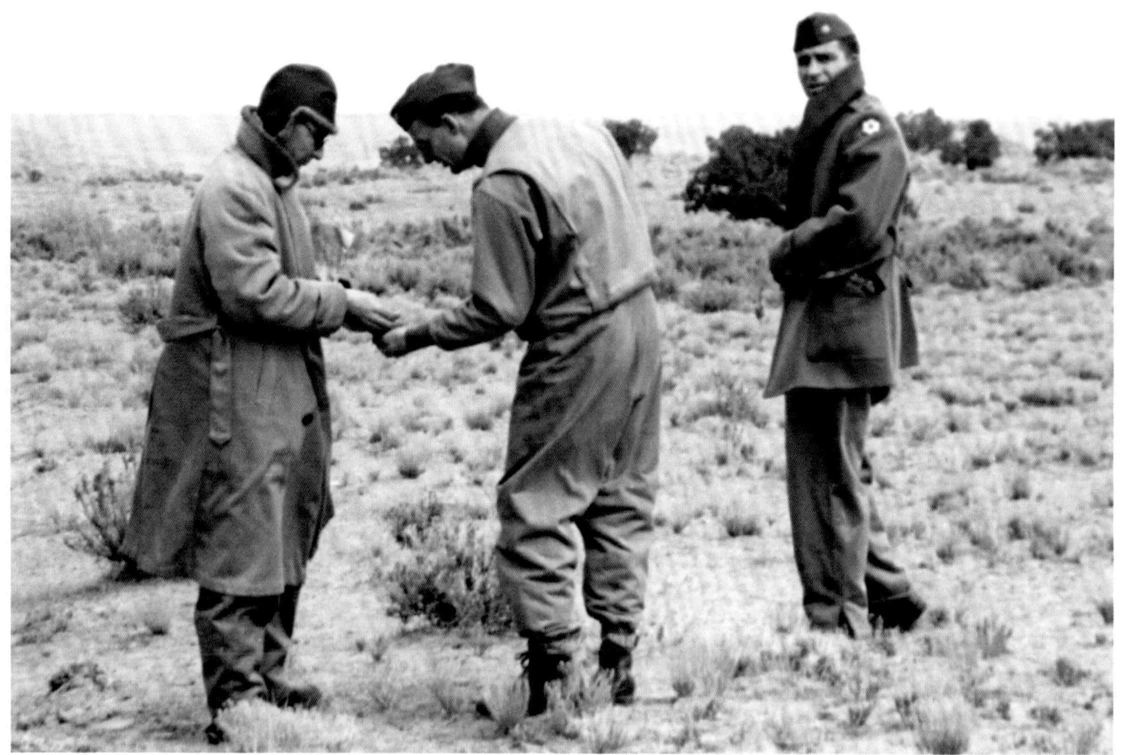

For the siting parties, a primary consideration was that the surrounding terrain and rail network had to allow for transportation of the "weighty albatross," as Bainbridge would later describe the 214-ton Jumbo. El Malpais was ruled out upon inspection because of the exceptionally rough terrain of the hardened lava flow. Ruling in a location proved to be much more difficult.

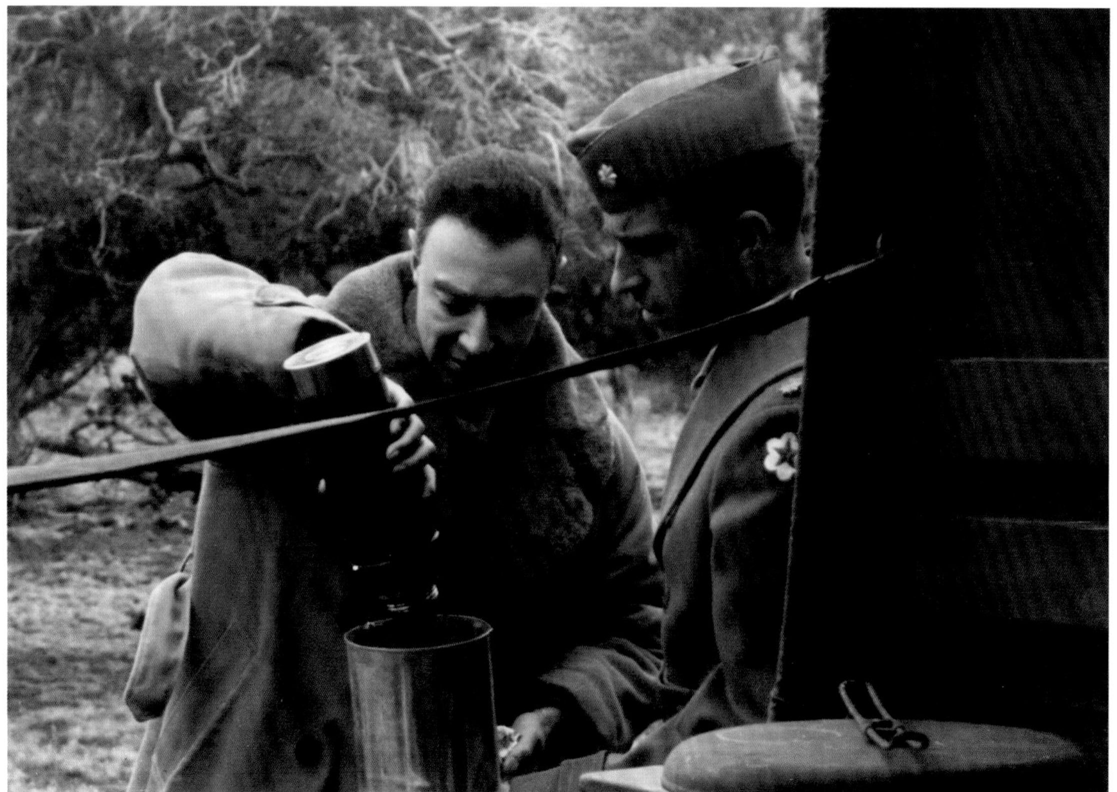

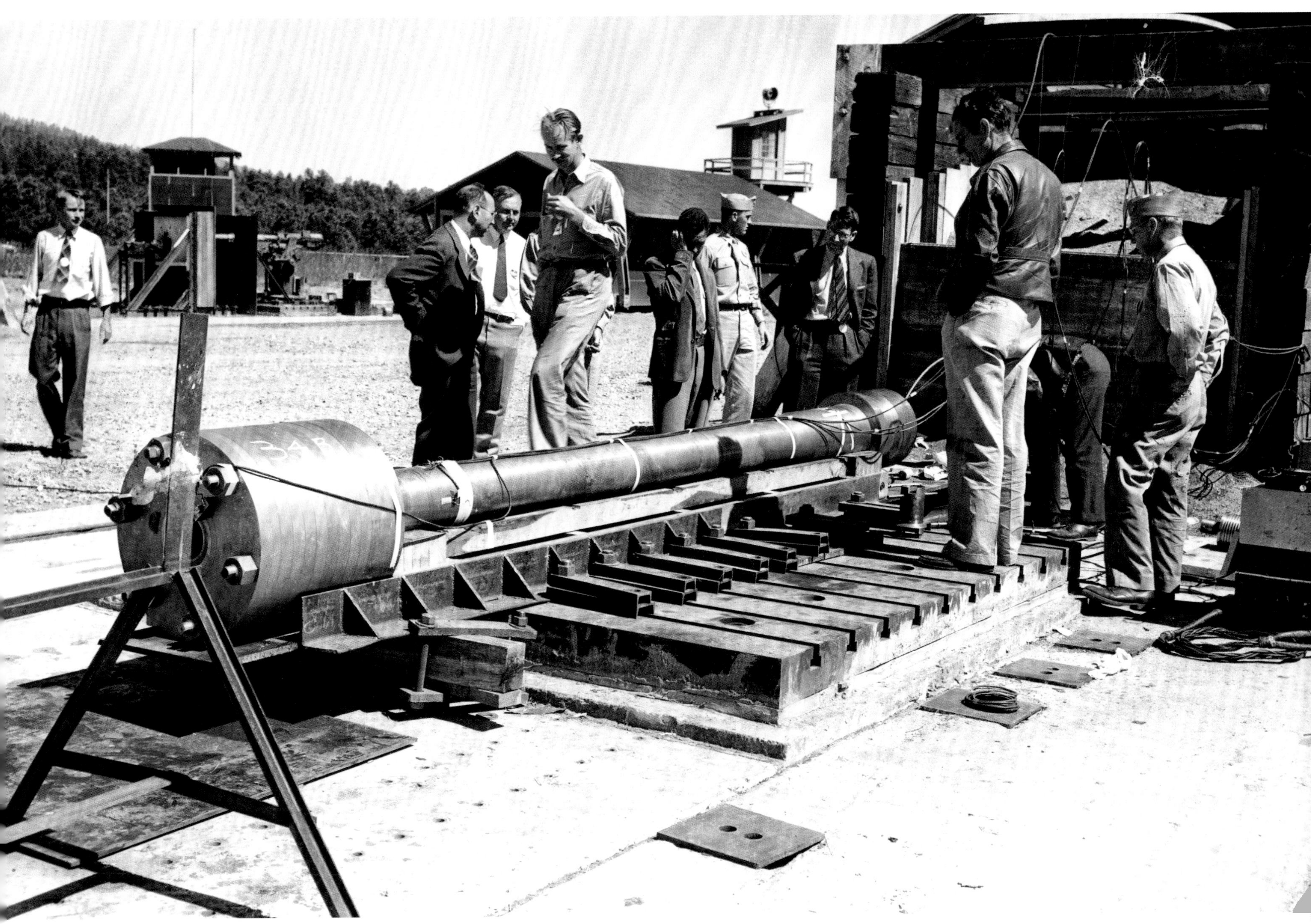

The loosely planned Project Trinity was soon to see action in leaps and bounds, however, as a major milestone at the laboratory—the arrival of the first batch of reactor-made plutonium—carried with it bad news about the gun-type plutonium weapon, the long-barreled Thin Man (*above*). Scientists discovered another isotope of plutonium, plutonium-240, mixed in with the desirable plutonium-239. Because plutonium-240 has a much higher likelihood of spontaneous fission than plutonium-239, its presence meant the plutonium fuel would have a higher neutron background than scientists had accounted for. This increased the risk that a nuclear chain reaction would start prematurely as the weapon fired, ending with a fizzle. There was no practical way to filter out the unwelcome isotope or to modify the weapon and avert the problem. With pressure still mounting to produce a combat-ready weapon, all work on the device that had been the main focus of the laboratory—Thin Man—was discontinued in late July, leaving the complex, unproven implosion Gadget as the new priority.

With the surefire option suddenly off the table, the implosion device became a crucial component of the Manhattan Project's mission—and yet, arrangements continued for a possible failure of the impending test. It was time to turn the idea of Jumbo into a full-scale reality. Blueprints and a schedule for fabricating the steel detonation chamber were finally solidified with a willing manufacturer—the Ohio-based Babcock & Wilcox (B&W), which was well-versed in creating and testing pressure vessels.

B&W was at the time producing thirty boilers per week for United States ships, and the company would later manufacture the reactor vessel for the nation's first commercial nuclear power plant, which bears a striking resemblance to Jumbo.

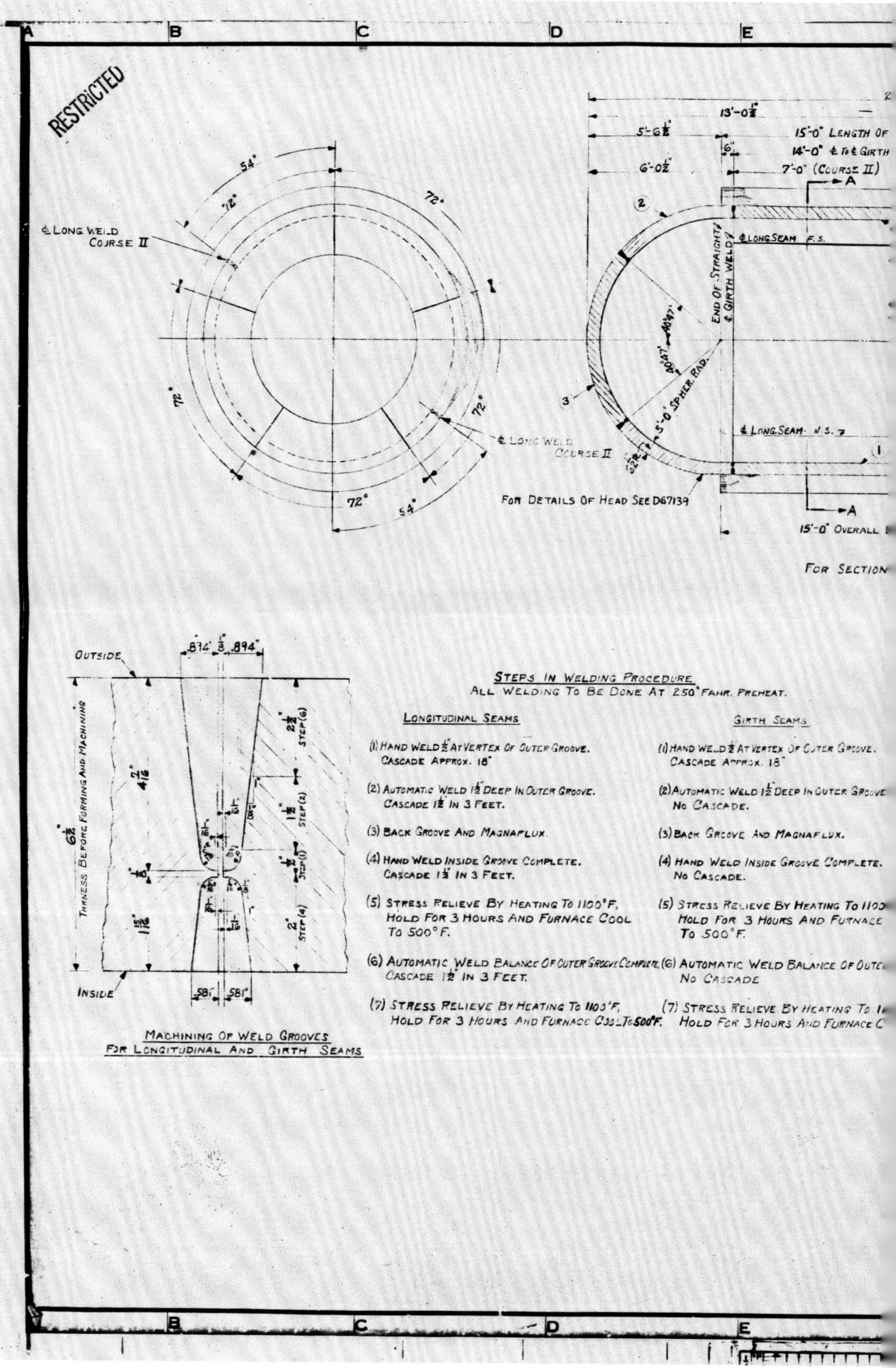

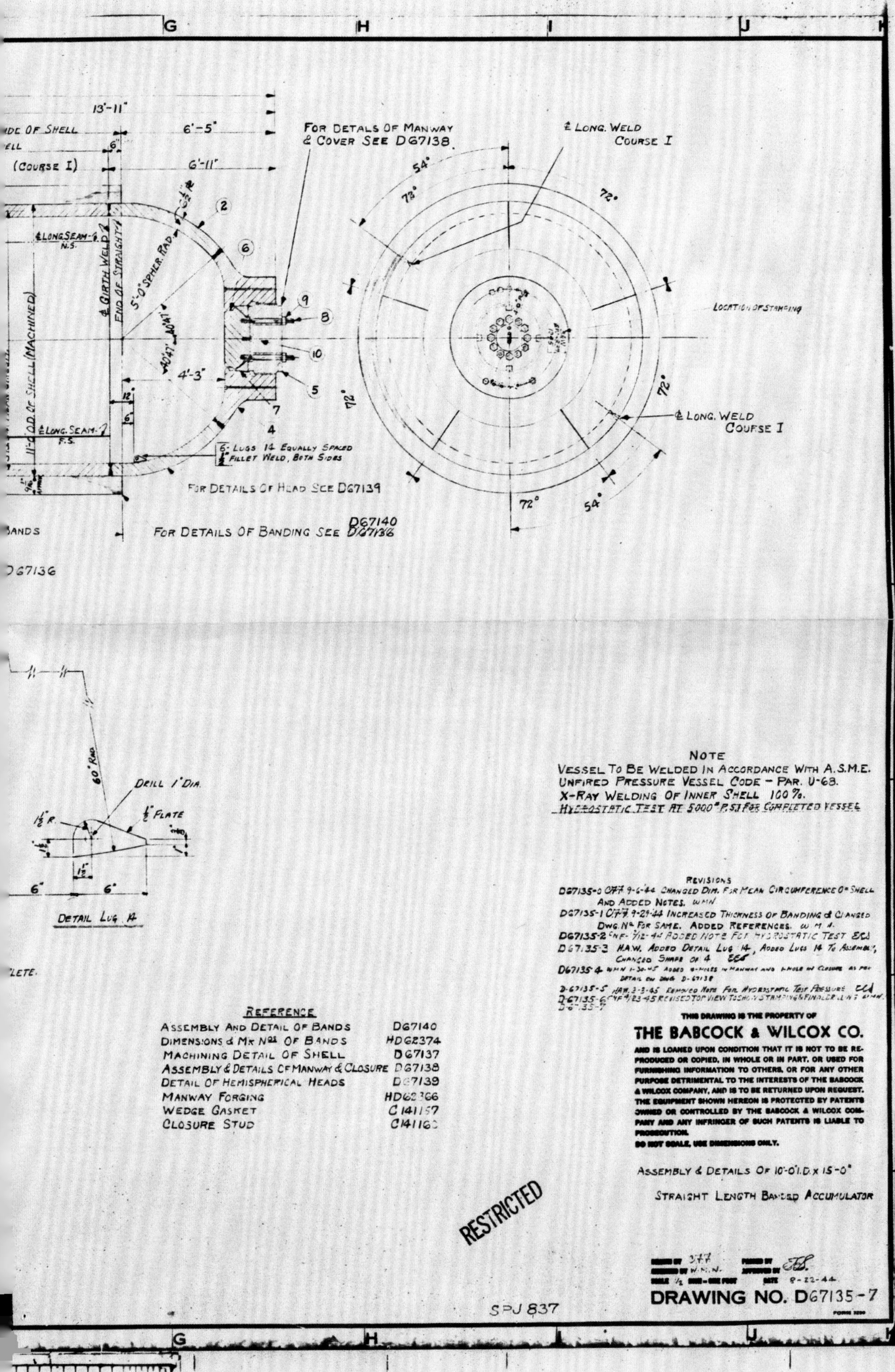

Per the final blueprints, the massive, mysterious "accumulator"—as Jumbo was known to B&W—would stand twenty-five feet tall, fashioned from thick steel plates and reinforced over its twelve-foot-round midsection with thirty-six layers of steel banding. This part of the container would absorb the brunt of the explosion, with the test device suspended in its center. The build was scheduled to take five months.

As the laboratory pursued implosion at full throttle and Groves was sorting out problems with steel procurement for Jumbo, Bainbridge selected and secured a site for the test shot: a military training ground called the Alamogordo Bombing and Gunnery Range, in the Chihuahuan Desert of south-central New Mexico. The commanding officer of the attached Alamogordo Army Air Field agreed to relinquish a tract in the northwest corner of the bombing range (*below*), spanning the northern Jornada del Muerto Basin and the edge of the neighboring Tularosa Basin.

About a nine-hour drive south from the laboratory, the flat, empty basin was believed to be adequately distant from populated areas and would serve as a perfect stage for taking films and measurements of the explosion. Though the area had always been desolate, a rail siding lay only thirty miles away. The abandoned stretch of track, named Pope Siding, was suitable for bringing Jumbo and other heavy shipments to the vicinity of the test site.

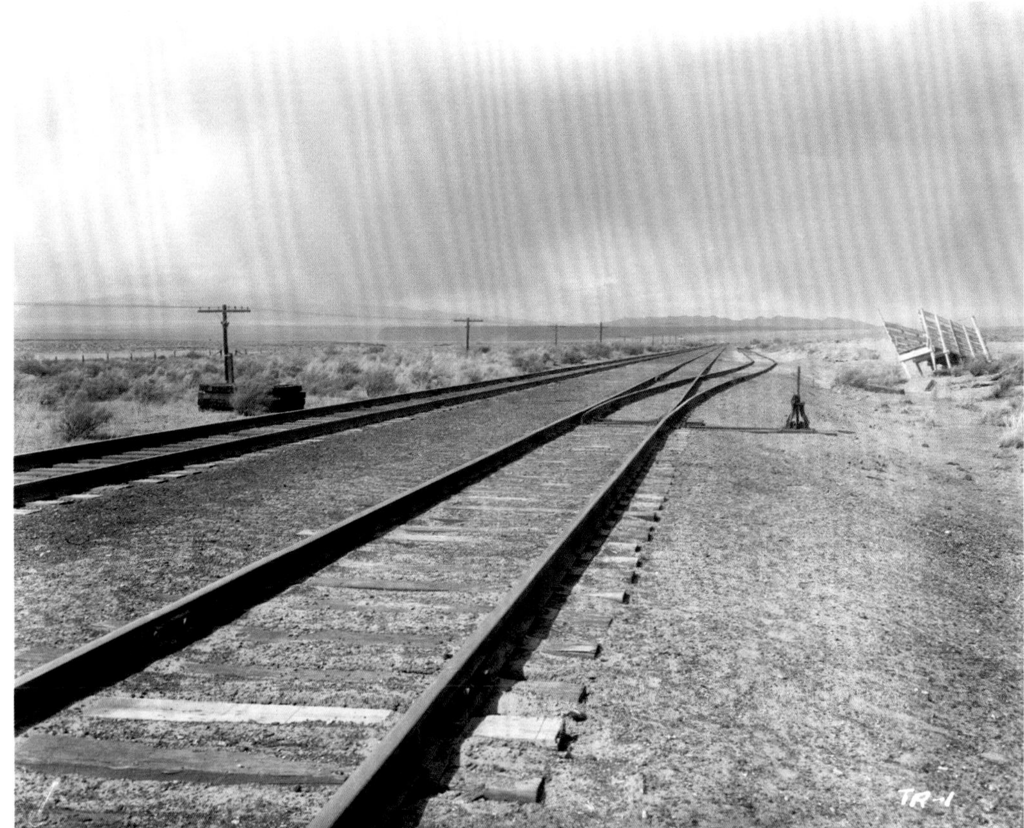

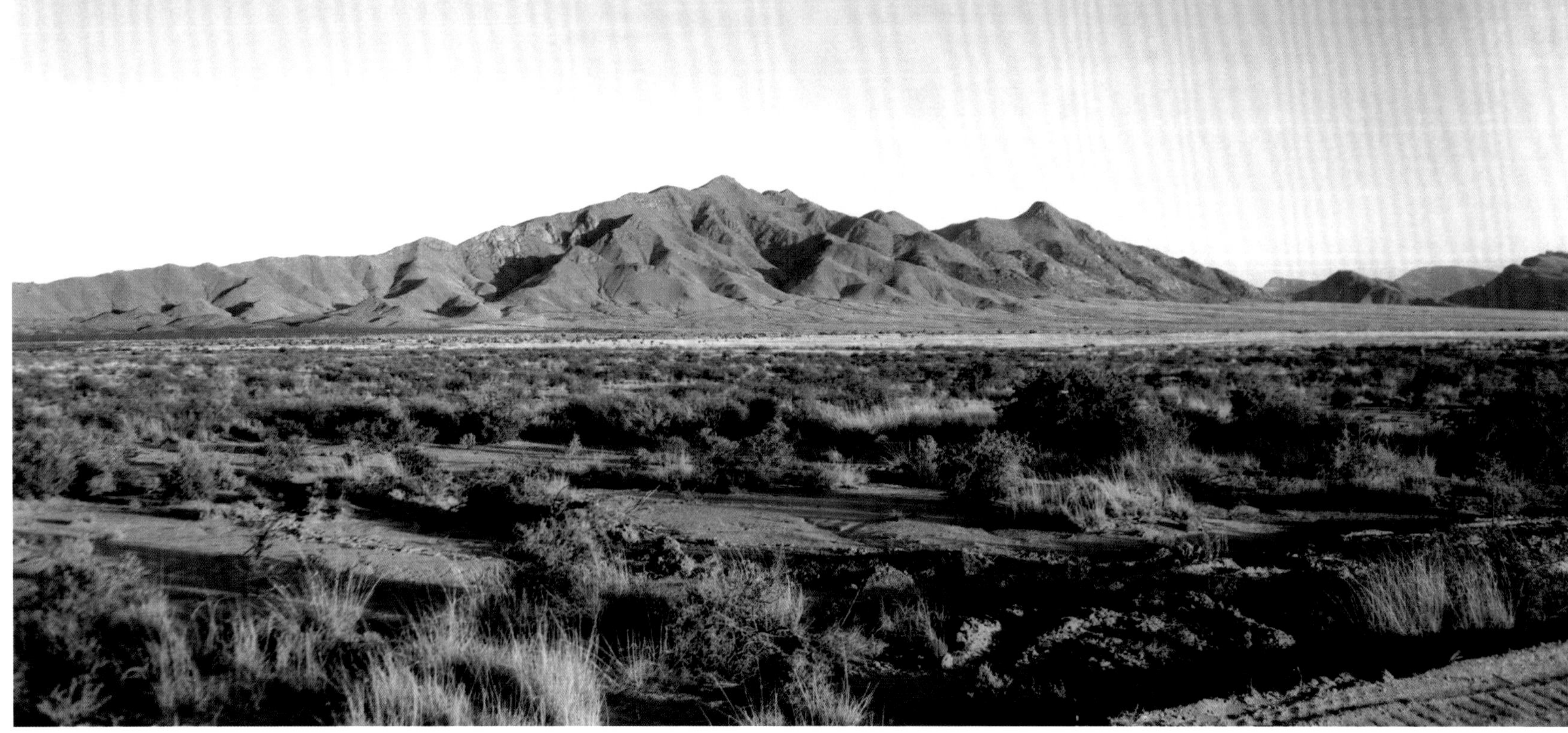

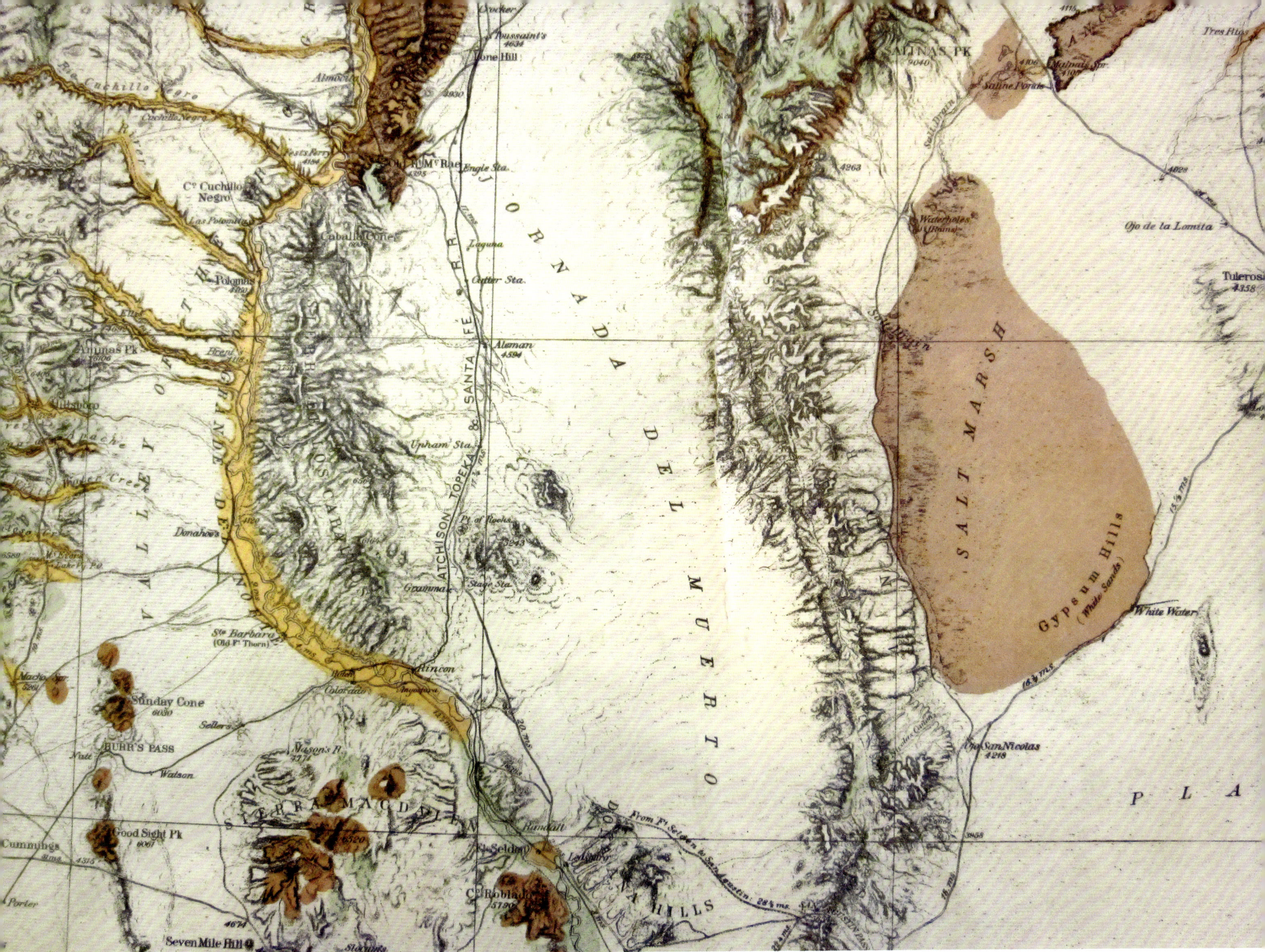

As depicted in this 1881 US Geological Survey—one of the maps that may have been available in 1944—the Jornada del Muerto ("Journey of the Dead Man") Basin lies between the Caballo Mountains and Fra Cristóbal Range in the west and the Oscura and San Andres Mountains to the east. The Tularosa Basin encompasses the salt marsh and gypsum hills. The term *jornada* was used in colonial New Spain to refer to a single day's journey between two *parajes*, or camps. The path through the desert basin was a treacherous, ninety-mile stretch of trail, where travelers were cut off for several days from the waters of the Rio Grande. (Annual report upon the geographical and topographical surveys of the territory of the United States west of the 100th meridian, in the states and territories of California, Colorado, Kansas, Nebraska, Nevada, Oregon, Texas, Arizona, Idaho, Montana, New Mexico, Utah, Washington, and Wyoming, in charge of George M. Wheeler, 978 W563 1881, Fray Angélico Chavez History Library / New Mexico History Museum, Santa Fe, New Mexico.)

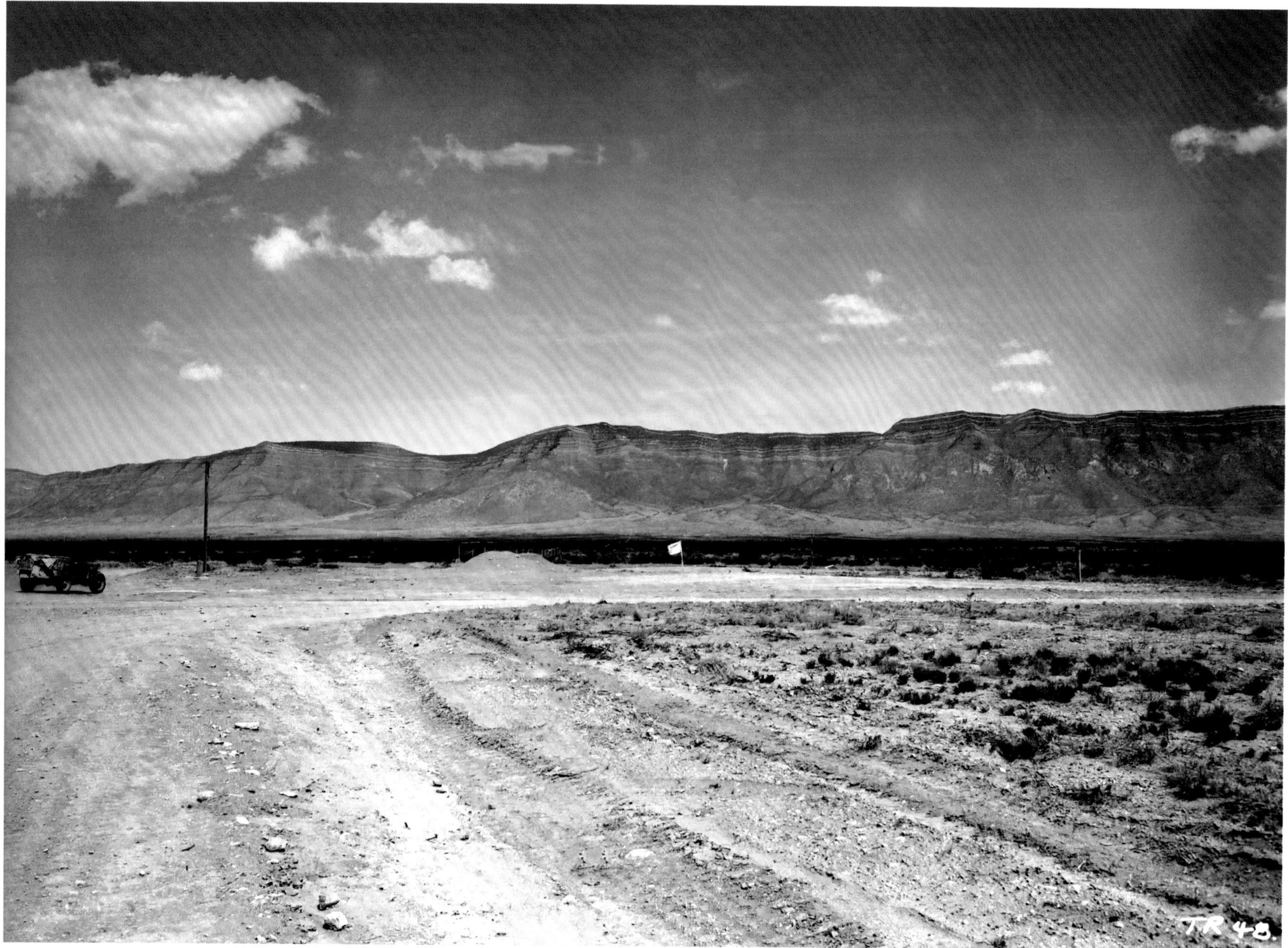

In October of 1944, what few maps existed of the desolate region were hard to procure or woefully inadequate, which complicated final surveying of the 432-square-mile parcel. Fortunately, one of the officers at Alamogordo was an expert in aerial photography. He instructed his men to photograph a strip at the center of the site, from which he pieced together a six-inch-to-the-mile scaled mosaic. According to Bainbridge, the resulting map was "marvelously sharp and detailed, every trail, road, bombing target, arroyo, and even fence lines plainly identifiable." The siting party selected the best terrain for ground zero (marked above) and a series of surrounding shelters. With that, the test site was official.

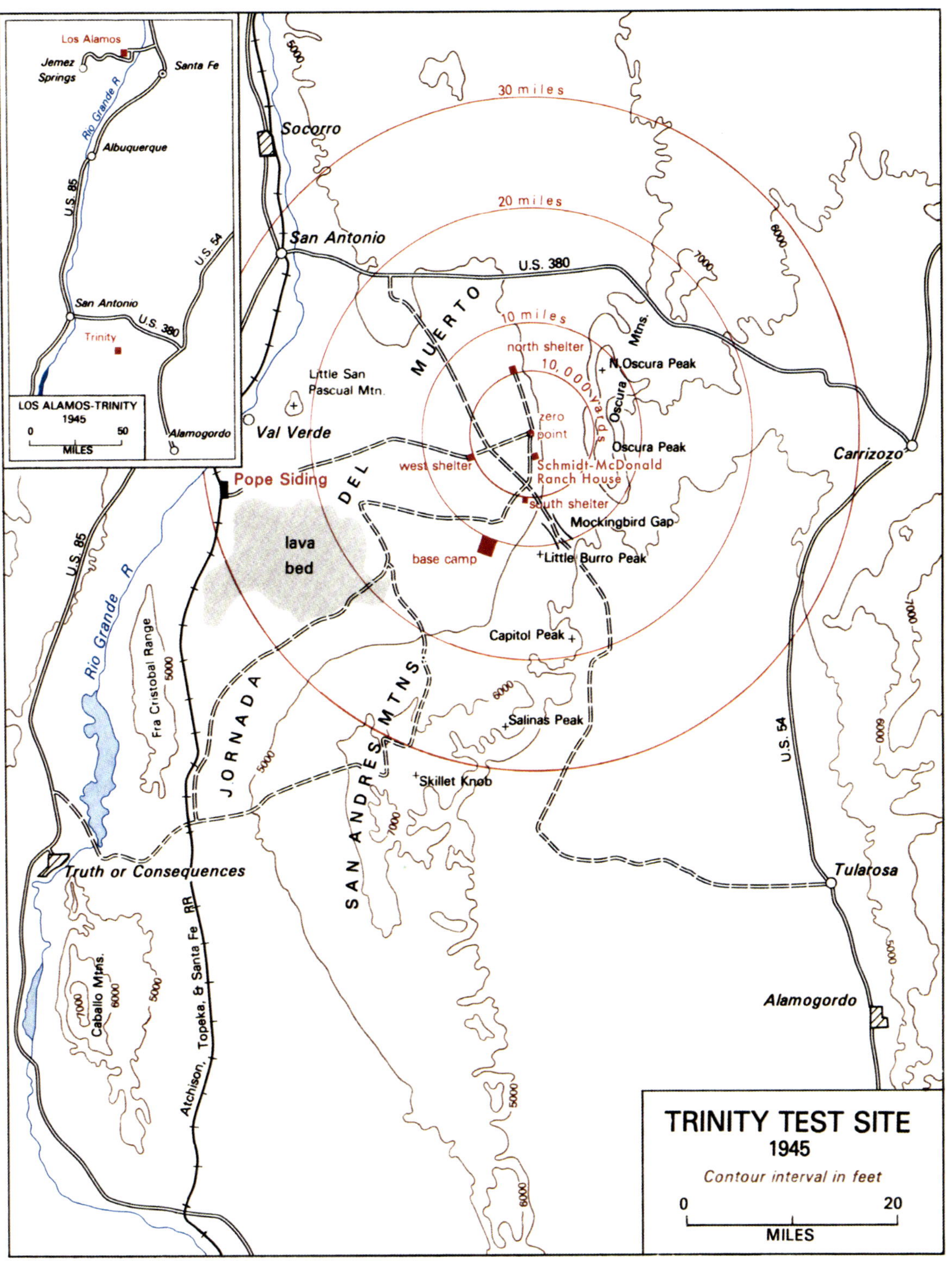

With begrudging approval from Groves, who was worried that building a test site would distract from building the Gadget, a second secret outpost began to rise from the New Mexico wilderness, cobbled together by the Army Corps of Engineers and civilian contractors in the same hasty and spartan style as Los Alamos 200 miles to the north. Members of the Special Engineer Detachment, a US Army program that funneled soldiers with technical backgrounds to Manhattan Project locations, would arrive at the test site in January to continue preparations.

The map at left shows locations of interest around the eighteen-mile by twenty-four-mile Trinity plot, including notable landmarks such as Pope Siding, the Oscura Mountains (also visible in the photograph of ground zero), and the nearest town, San Antonio, New Mexico. (Map adapted from Vincent C. Jones, *Manhattan: The Army and the Atomic Bomb*, 1985.)

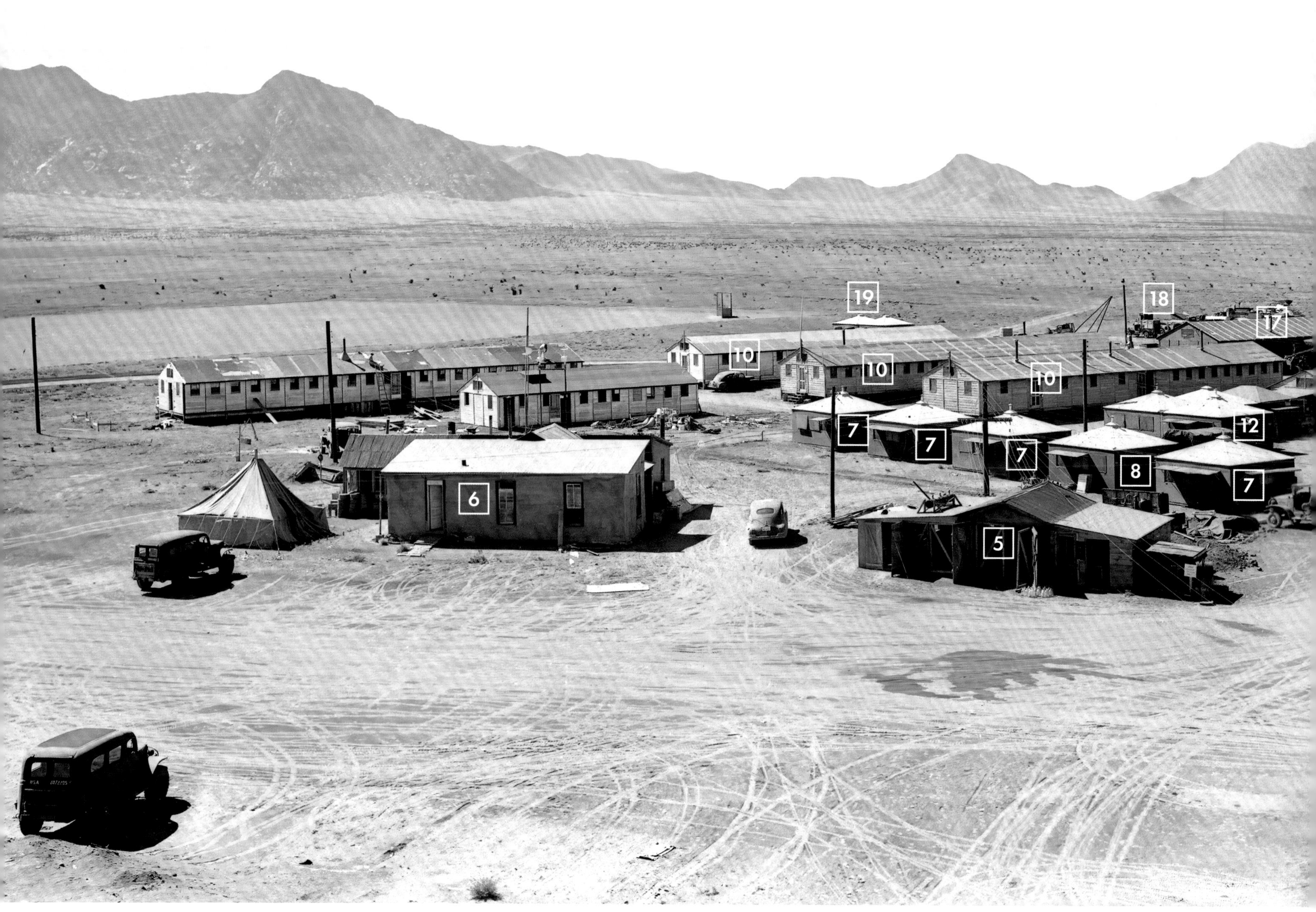

As the first order of business, a field headquarters was established about ten miles southwest of ground zero at a cluster of ranch buildings, inhabited before the army's 1942 arrival by rancher Dave McDonald and his uncle, Ross McDonald. The original structures (numbered 1 through 6) were quickly expanded into a base camp that would house the personnel setting up infrastructure and diagnostic equipment for the test shot. (Adapted from a composite by Jo Ruprecht.)

A third ranch complex, belonging to George McDonald (brother of Dave) and originally built in 1913 by Franz Schmidt, was located about two miles southeast of ground zero.

1. Pump House
2. Hay Barn
3. Ross McDonald Ranch House
4. Storage Shed
5. Generator Shed

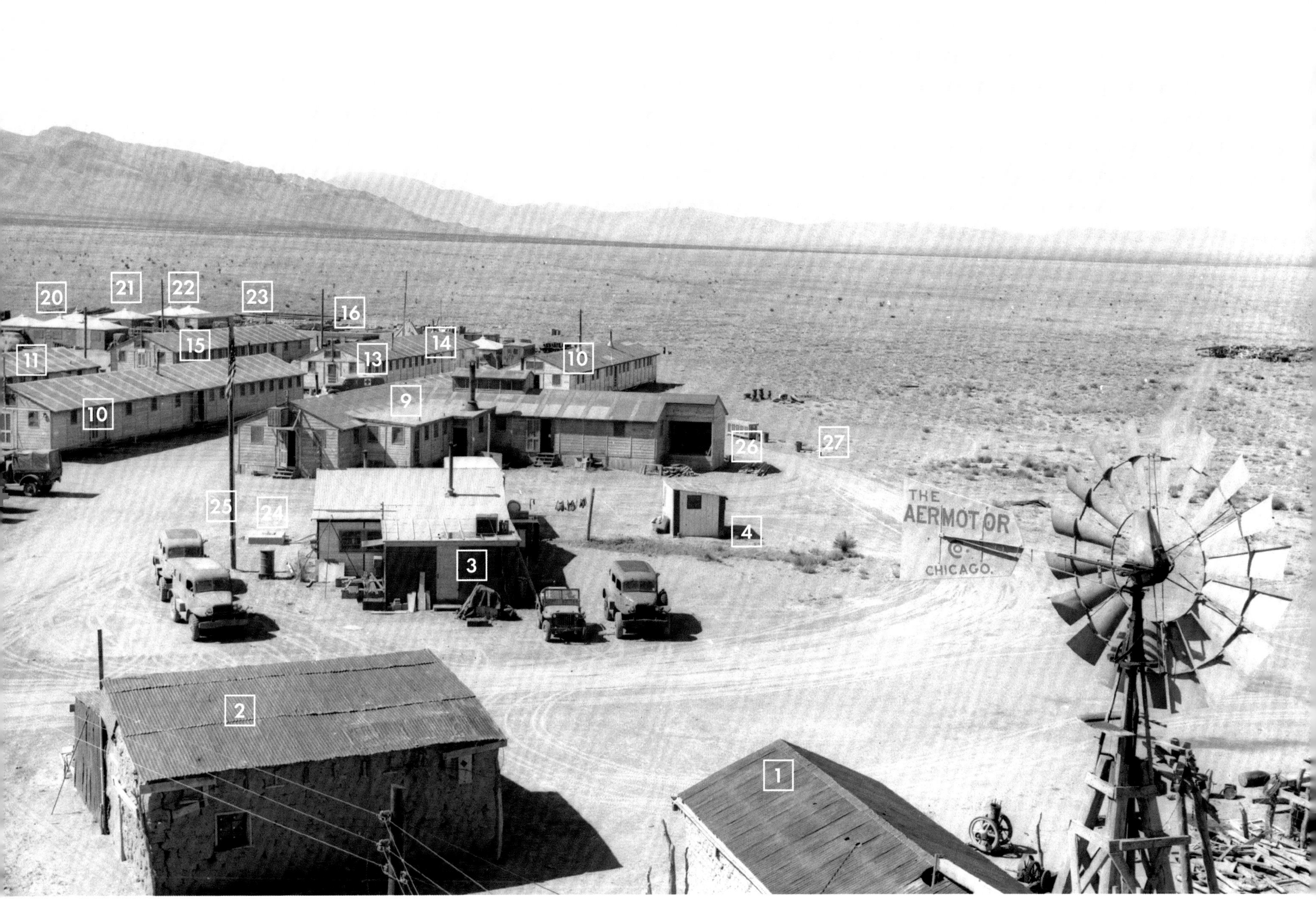

6. Dave McDonald Ranch House

7. Officers' Quarters

8. Supply Room

9. Mess Hall

10. Barracks

11. Latrine

12. PX and Recreation Room

13. Infirmary

14. Laboratory

15. Technical Warehouse

16. Office

17. Garage

18. Two Gas Storage Tanks

19. Fire Station

20. Engineering Office and Warehouse

21. Plumbing Shop

22. Electrical Shop

23. Carpentry Shop

24. Cistern

25. Flagpole

26. Grease Trap

27. Septic Tank

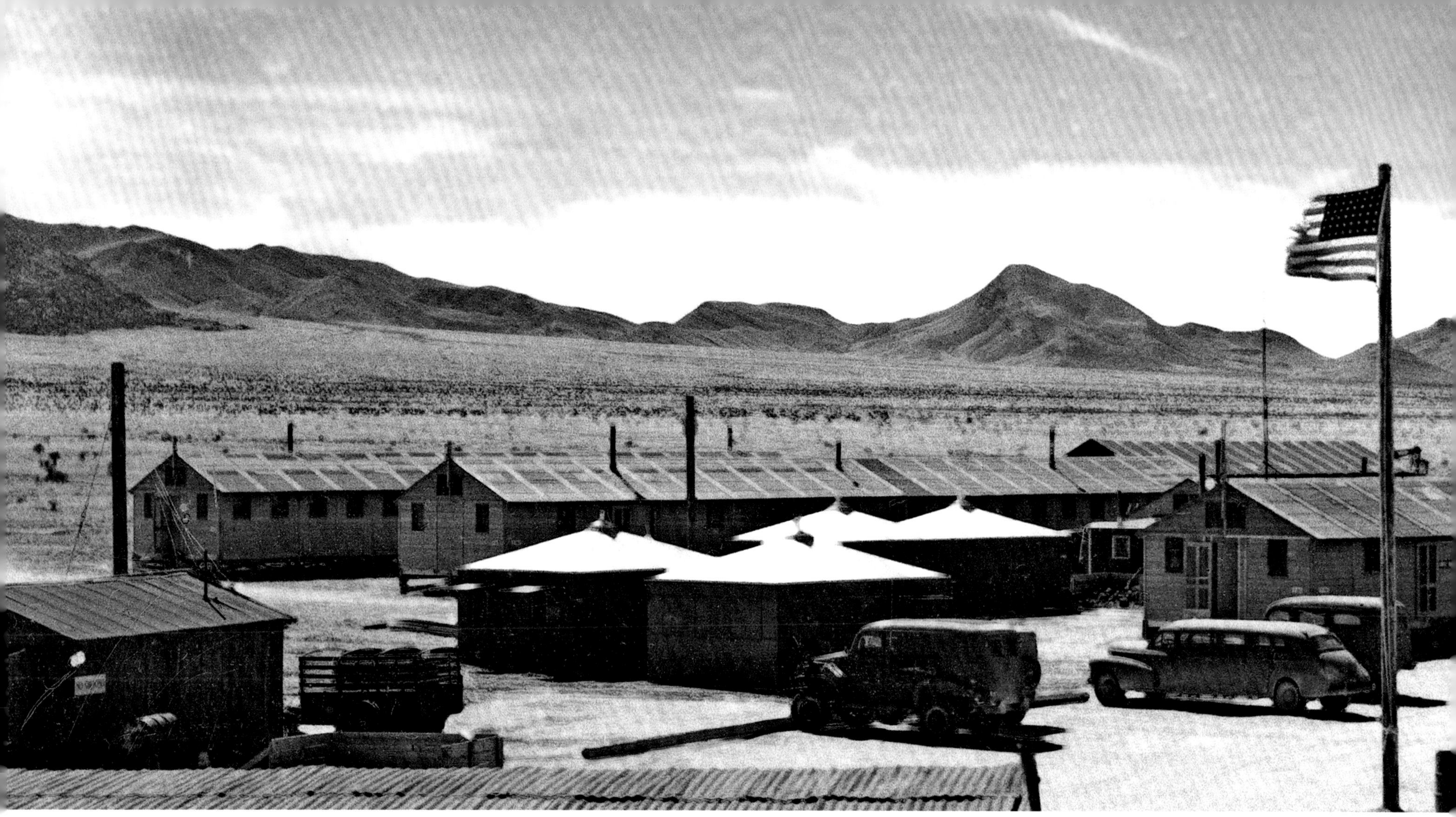

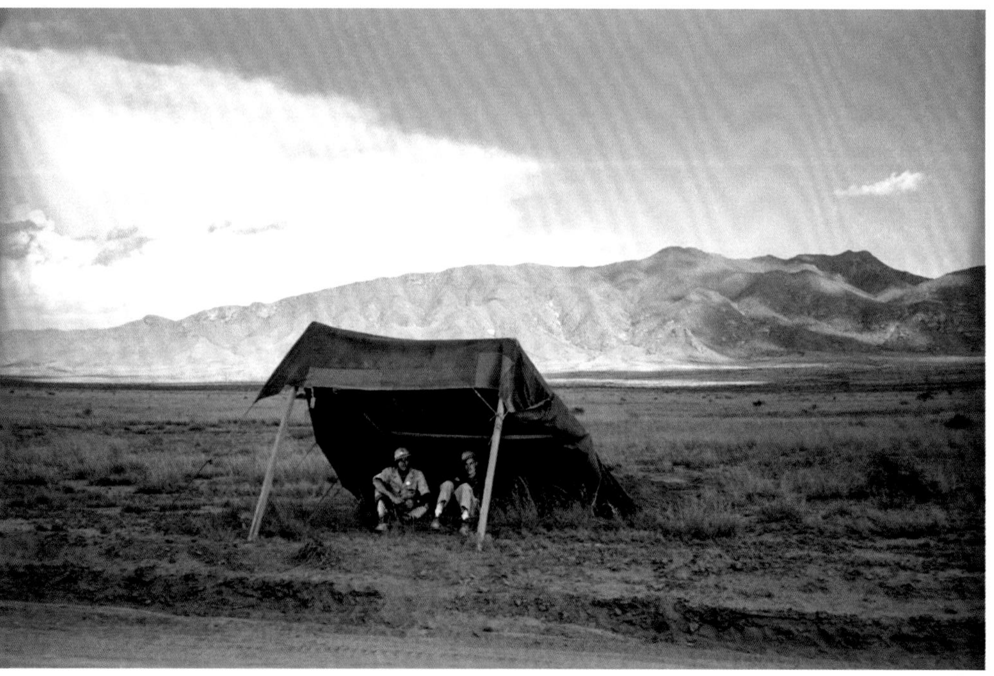

A small force of Military Police under the command of Lt. Howard C. Bush were the first to arrive on-site. The group began occupying base camp by the end of the year, while essential facilities and roads were still under construction. They maintained perimeter security on horses—before transitioning to jeeps and trucks—and kept watch from guard towers and ground stations.

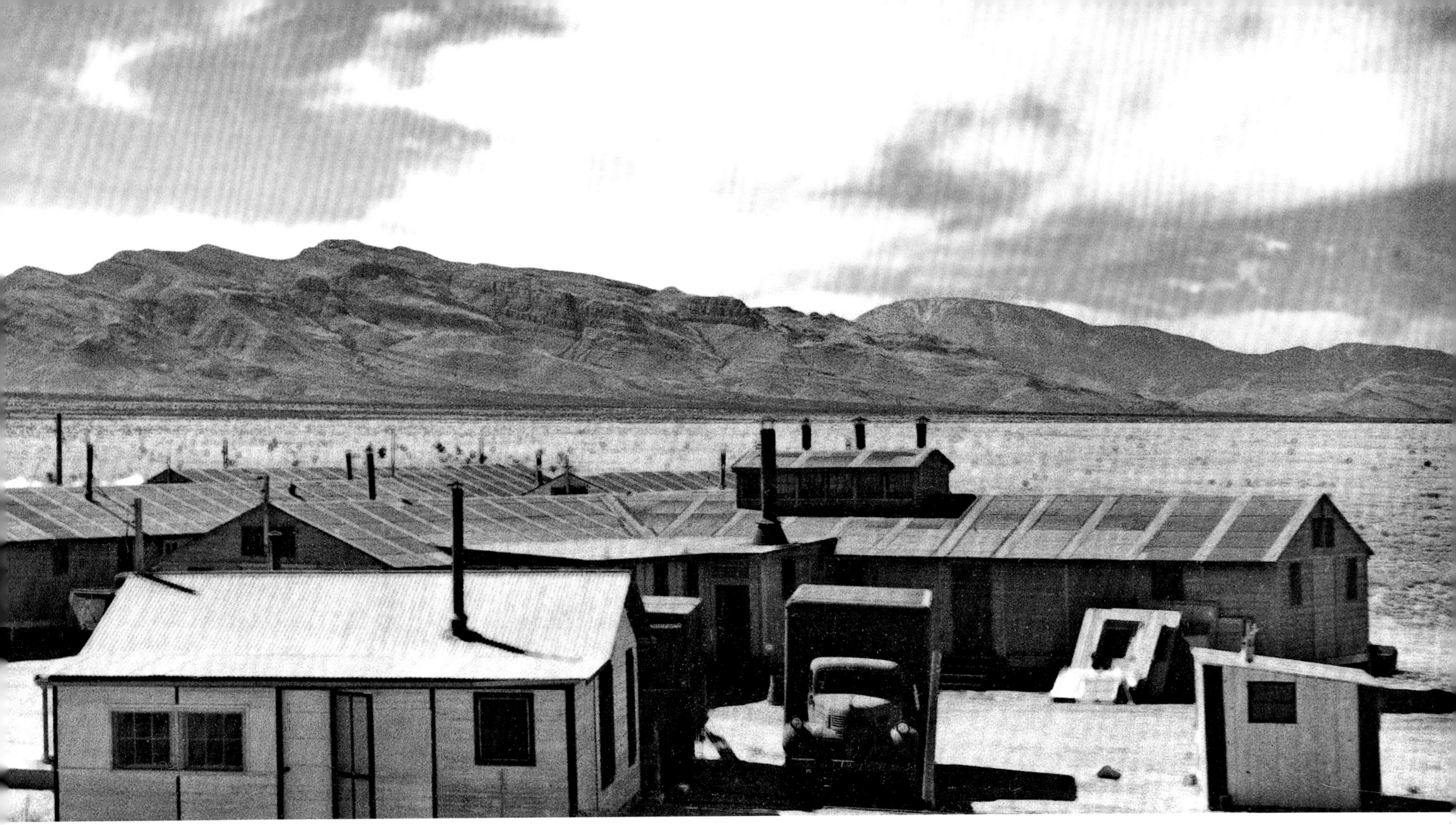

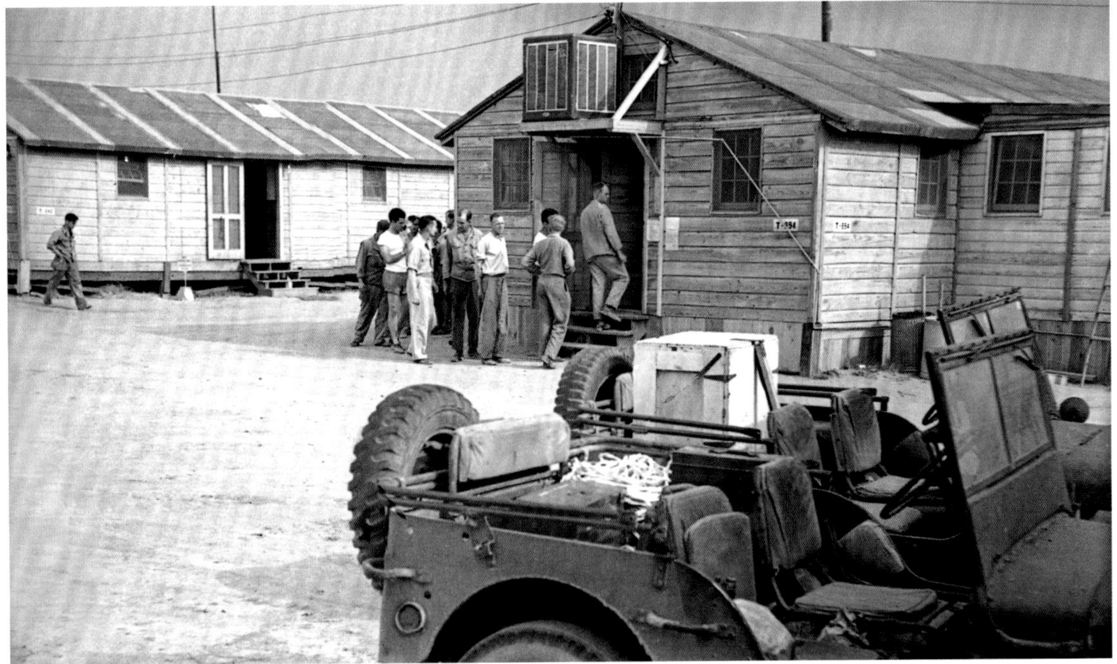

Many base camp buildings, such as the barracks and the mess hall (*left*), were brought in from campsites of the defunct Civilian Conservation Corps. Other structures, such as a latrine, were built up from new, poured-concrete foundations. All facilities had one thing in common: they were completed and outfitted with the utmost efficiency and cost savings. "Luxuries" were disallowed by Groves.

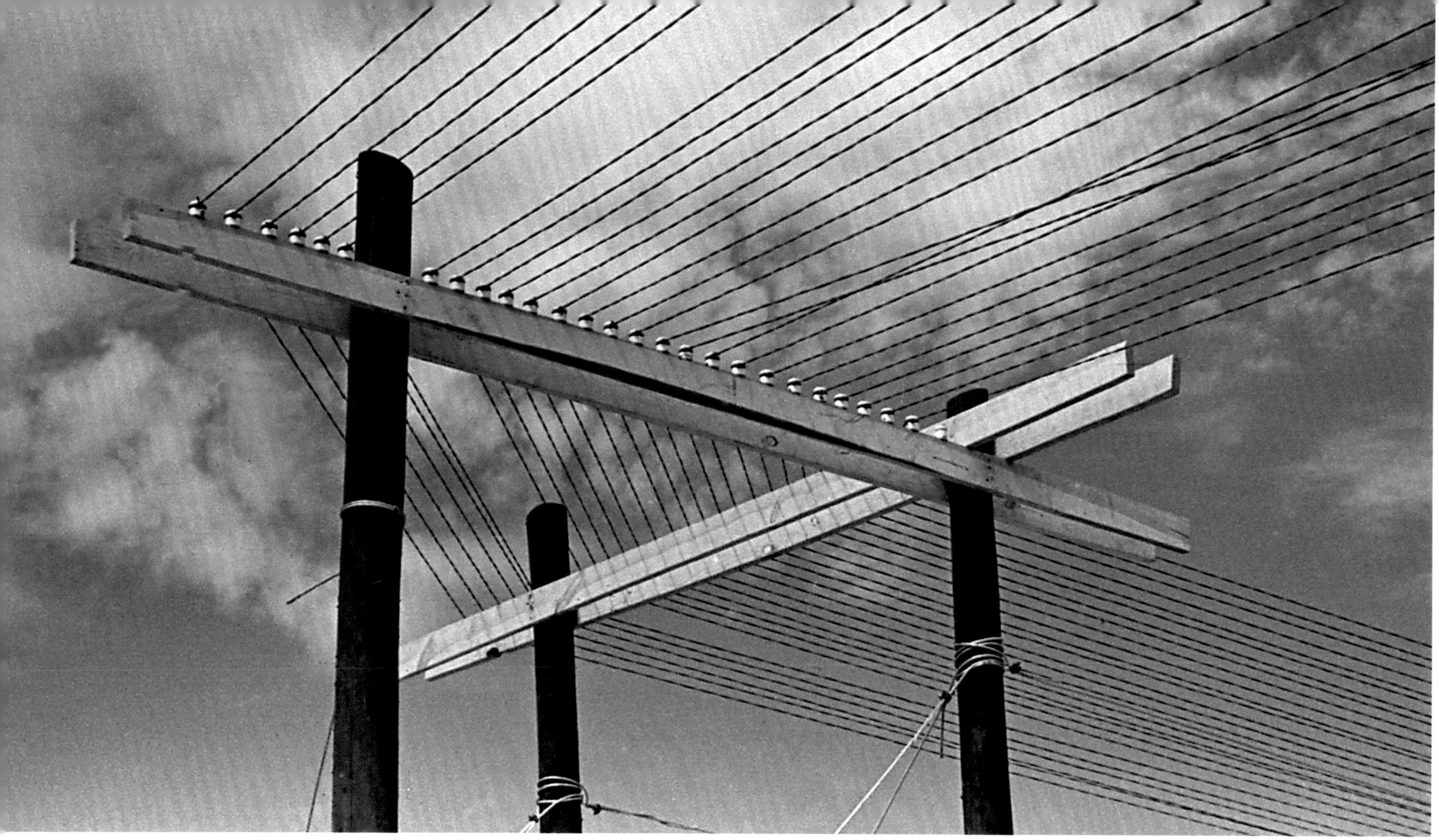

While base camp had materialized at a record pace, much work remained to be done elsewhere at the Trinity site. The test plans called for small and large bunkers, various towers (including one for erecting Jumbo), and always-evolving infrastructure to support an extensive program of diagnostics.

As islands of activity began to dot the empty basin, workers graded dirt and gravel roadways and laid a sprawling network of power lines and data cables to enable communication and travel between instrument stations, personnel shelters, ground zero, and base camp. The jeep at right, with a spool attachment, was used to run long spans of cable.

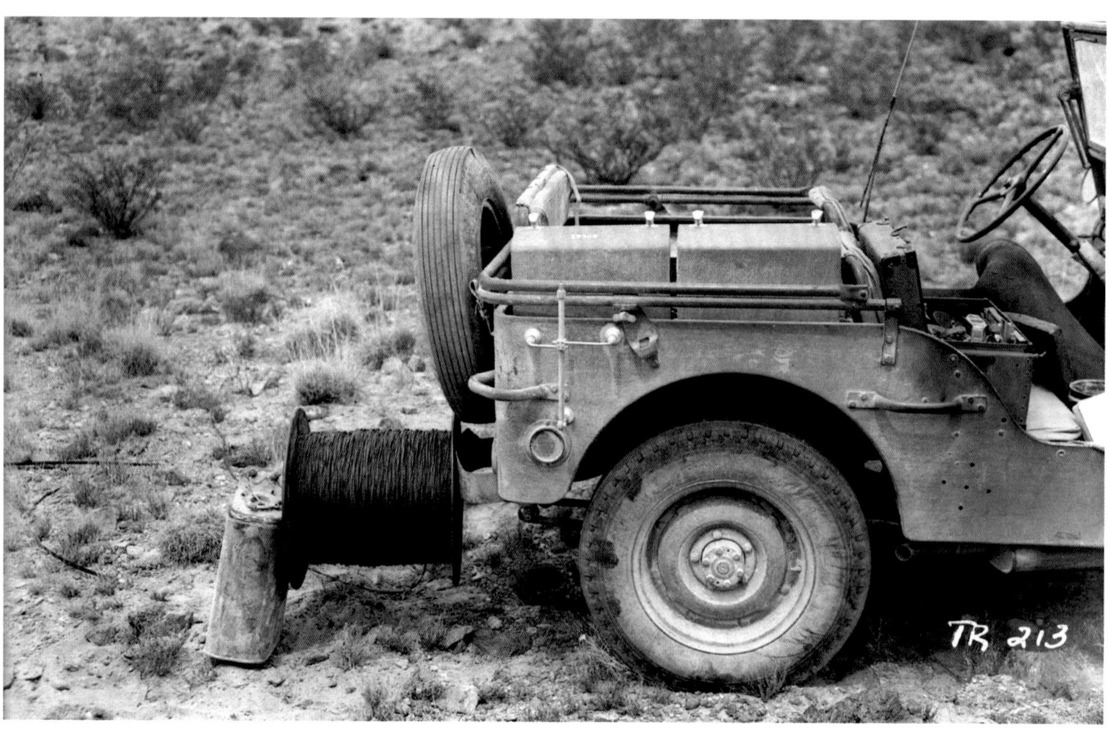

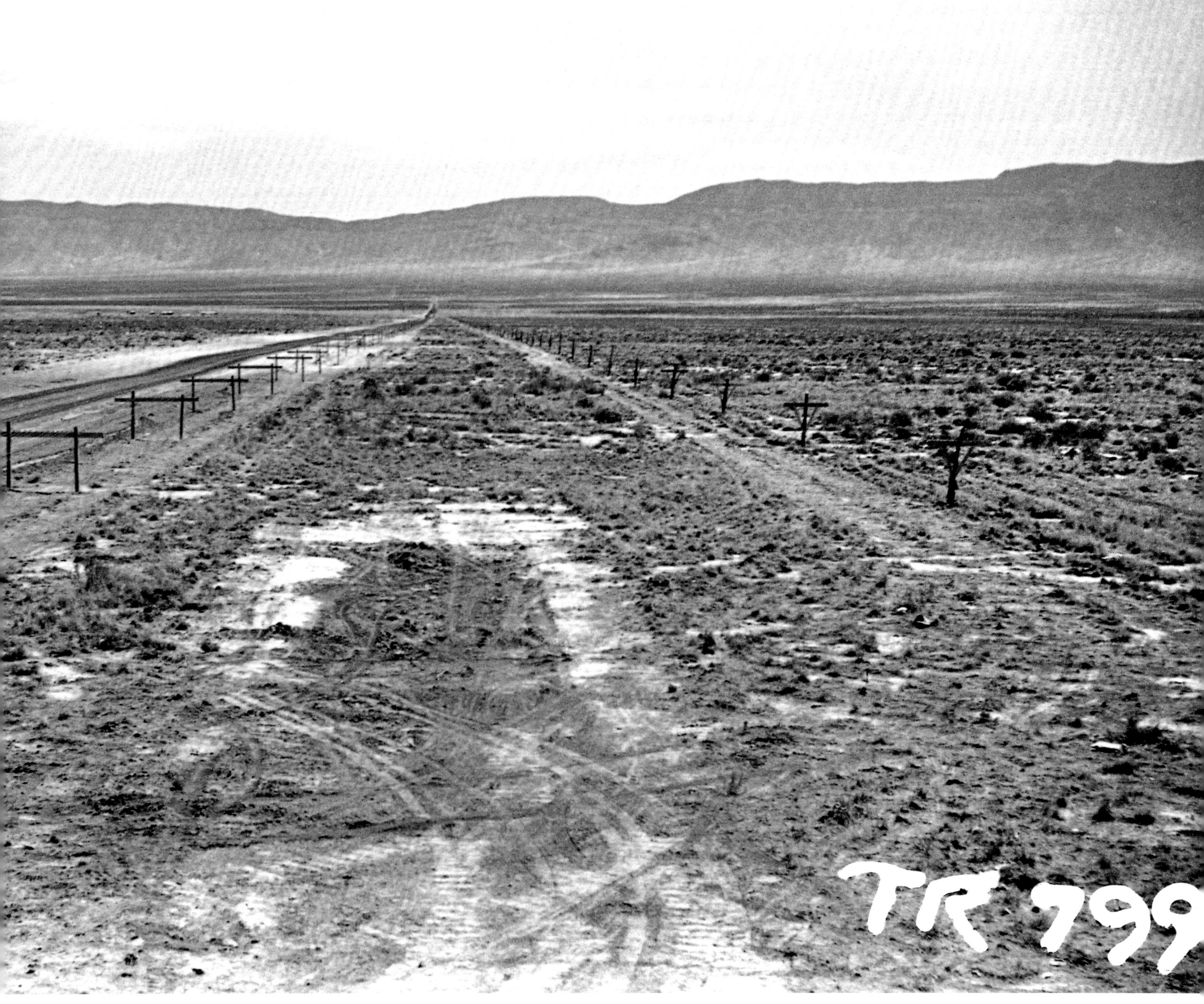

The twenty-eight-mile Pope Road, shown above running east (toward ground zero), was built to bring in Jumbo and other large deliveries from the rail siding.

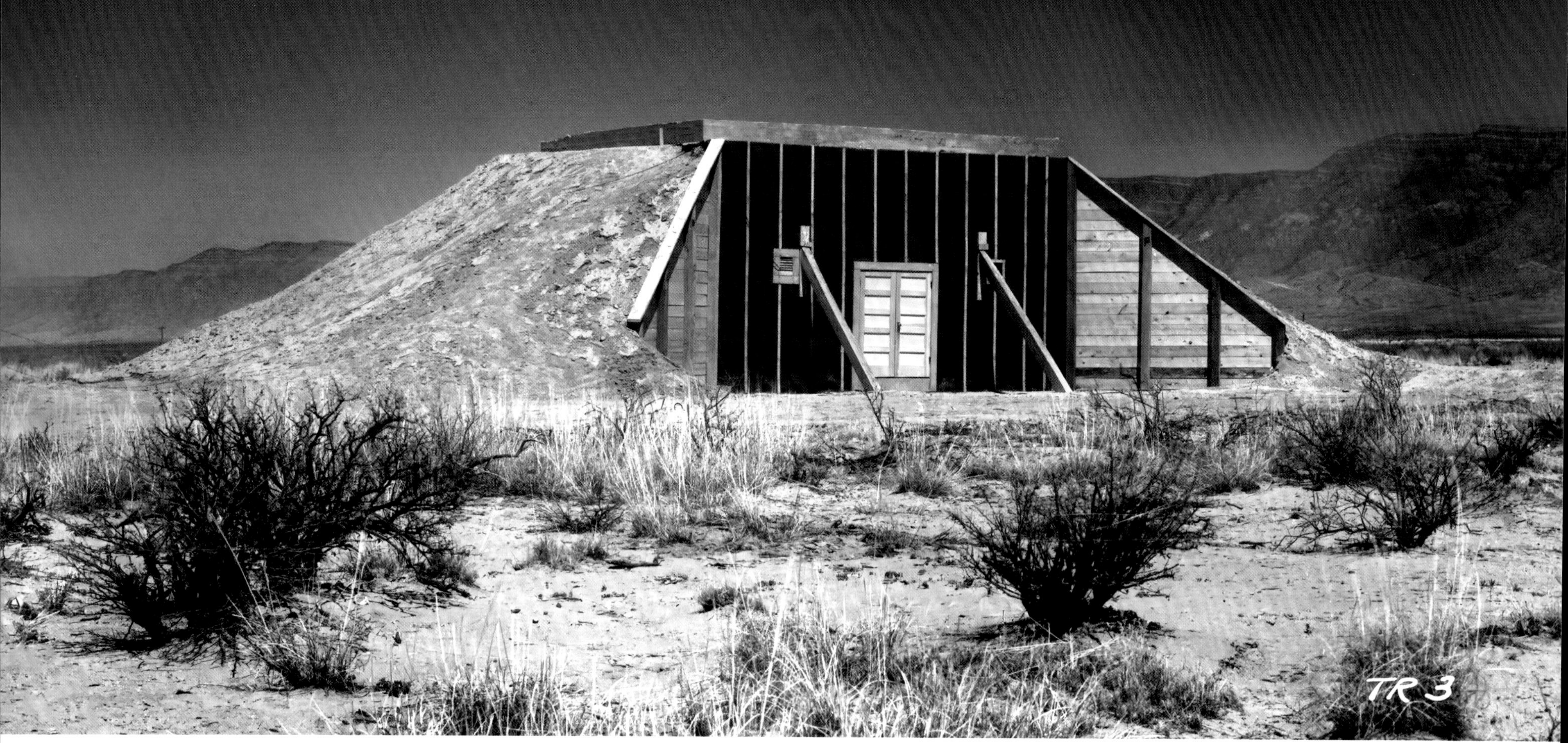

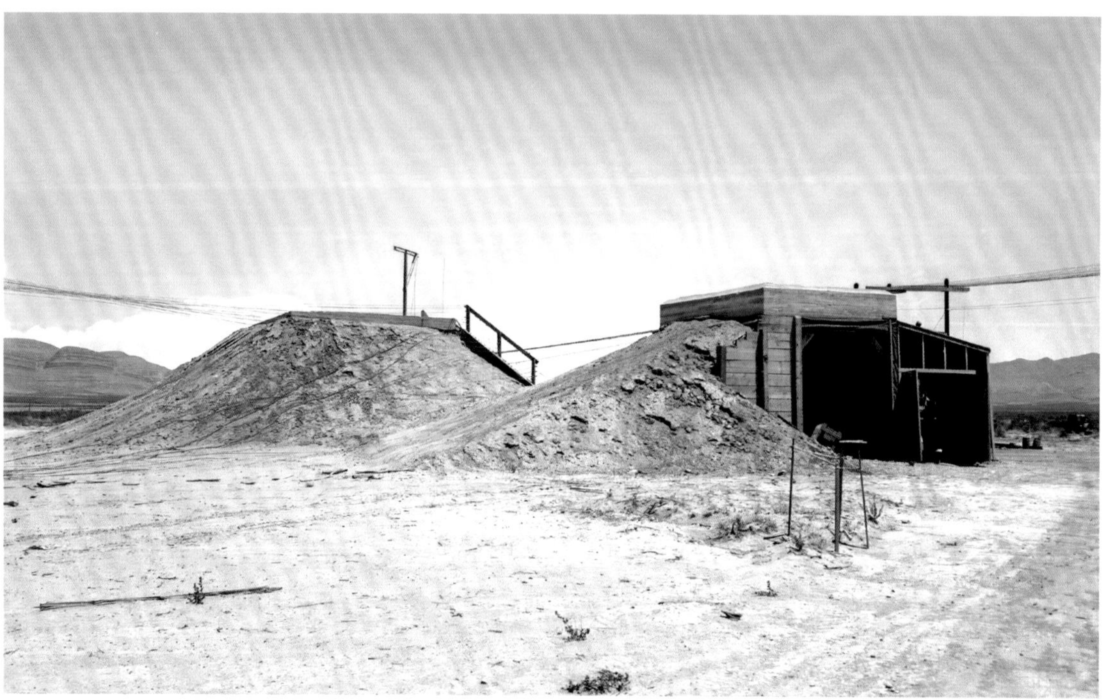

Three main shelters were established 10,000 yards (about five-and-a-half miles) from ground zero—one each to the north, south, and west. These would house diagnostic instruments and personnel during the test shot.

The shelters were fashioned from sturdy timbers supporting concrete slab roofs, with the blast-facing side of each shelter buried under a mound of earth and beams along the opposite wall bracing against the sand fill. The south shelter, shown under construction above, was the control bunker, where the countdown to detonation would be initiated. A small building next to each shelter housed equipment for generating power (*left*).

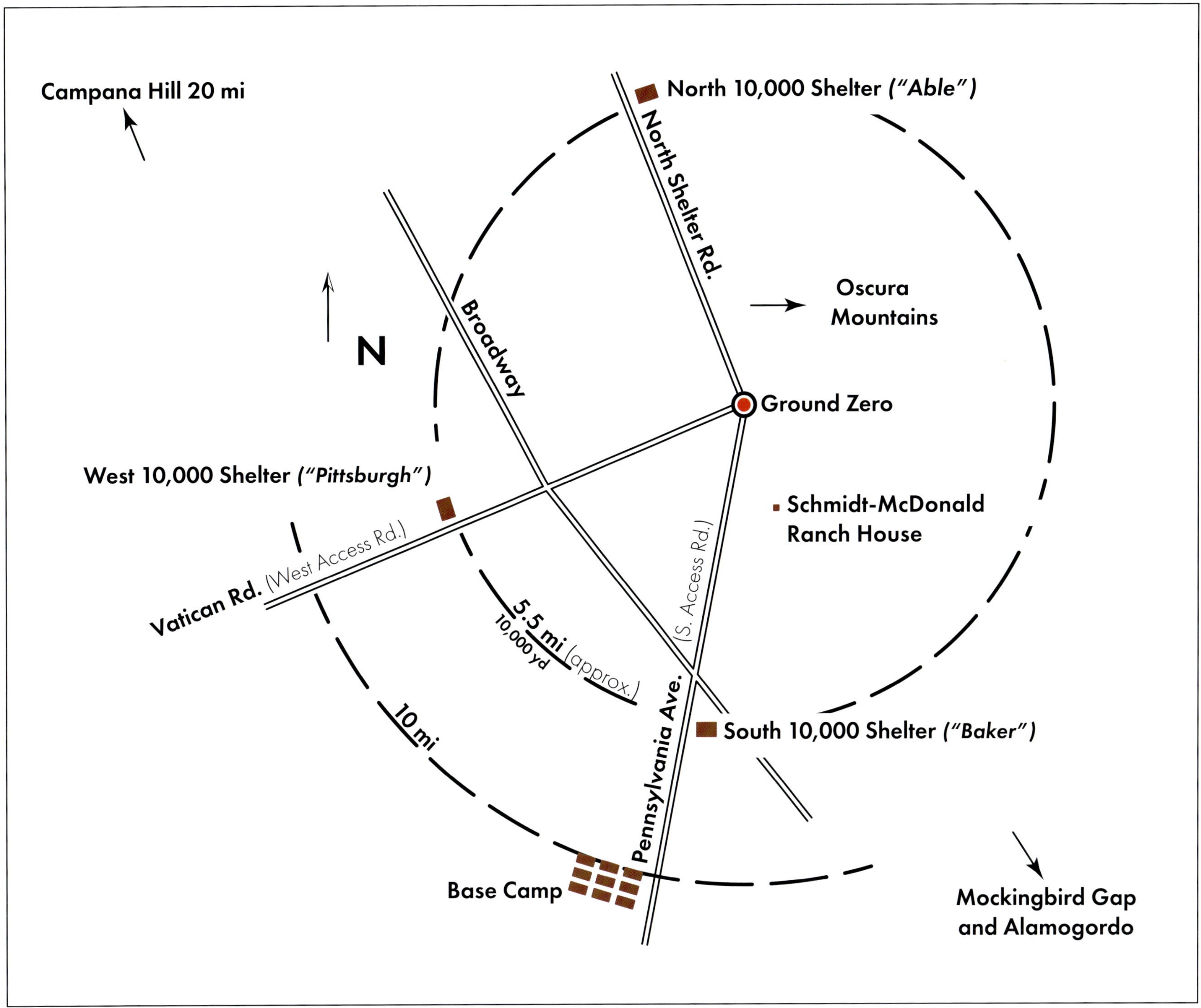

The map above shows the three shelter complexes in relation to ground zero. No facilities were constructed to the east because any airborne radioactive debris was expected to drift in a northeasterly direction. Situated close to the Oscura Mountains—the eastern boundary of the Jornada del Muerto Basin—the location of ground zero allowed for maximum sprawl to the north, south, and west.

A distance of 10,000 yards was chosen for the shelters based on what Jumbino firing tests indicated was outside the range of shrapnel flying away from Jumbo. Though the predicted yield of the weapon—expressed then (and still today) in terms of equivalence to the yield of a given amount of TNT—was 4,000 tons, safety for personnel was insured up to a yield of 200,000 tons.

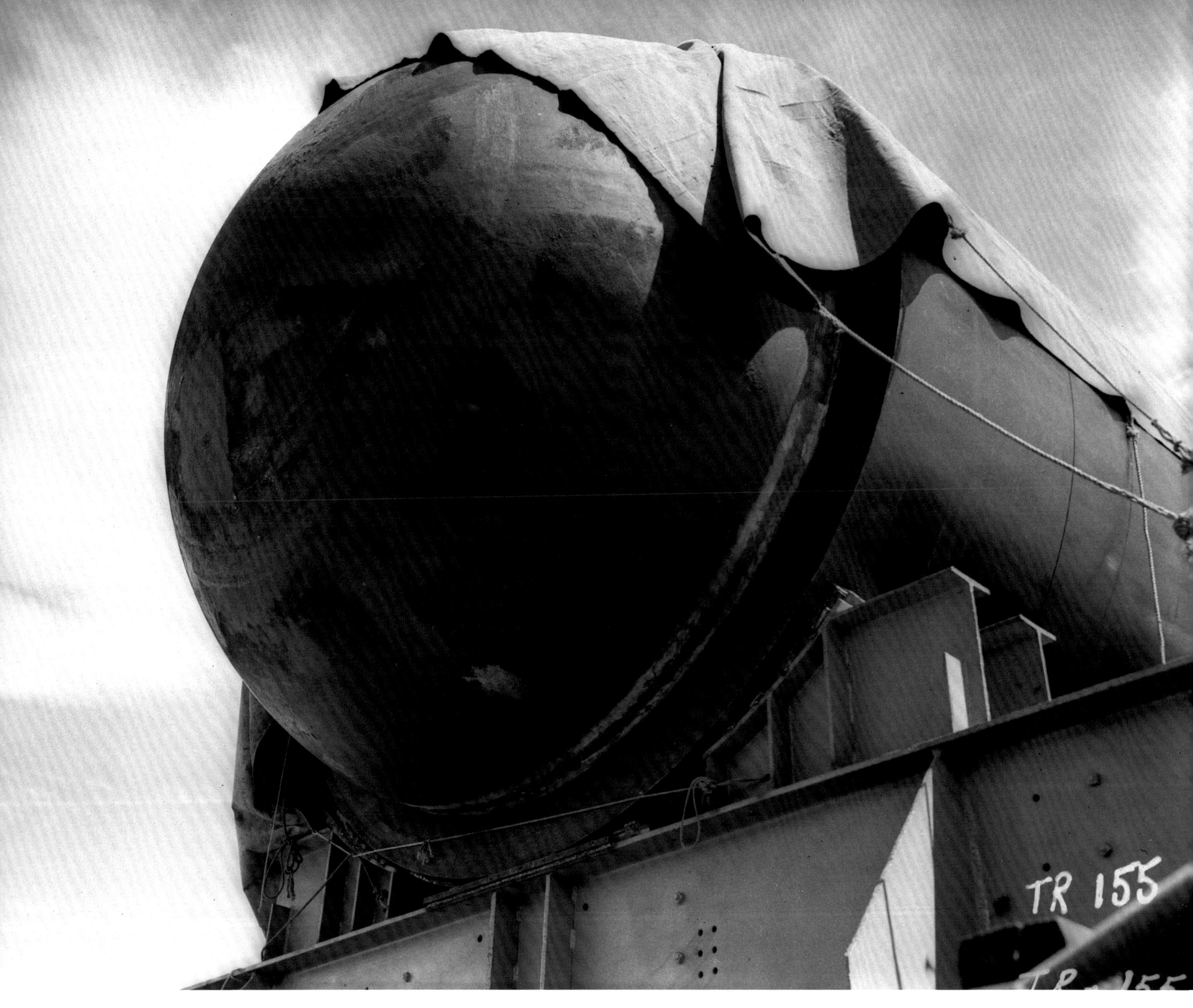

As construction proceeded rapidly in the Jornada del Muerto, Jumbo was slowly taking shape in Barberton, Ohio. A group that included containment team member R. W. Henderson had paid a visit to the manufacturer in December, during construction of Trinity base camp, where they saw the rounded ends of the vessel nearing completion. In light of delays already encountered and those anticipated in the months ahead, B&W projected that the "accumulator" could be shipped out in April of 1945, which would put Jumbo at the Trinity site and ready to house the Gadget by the first week of June.

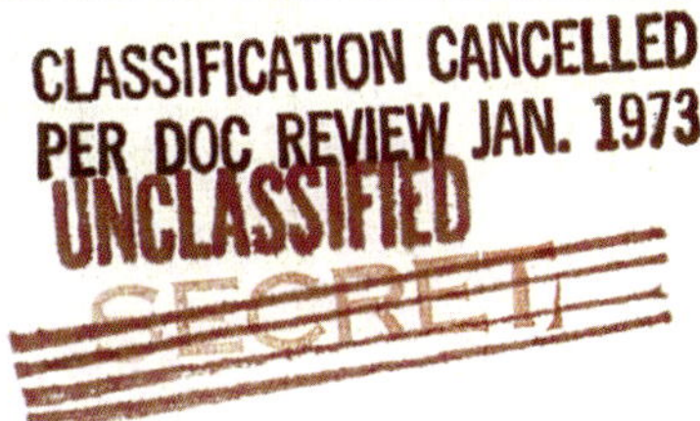

5 Dec. 1944 - P. 3
Status Report of Jumbo
R. W. Henderson

SCHEDULE - Cont'd.

	Operation	Days Required	Date
4.	Stress Relieve and cooling	5	1/27 to 1/31
5.	Transportation of vessel from B. & W. in Barberton, Ohio, to Mesta Machine Wks. in Pittsburgh, Pa.	2*	2/1 to 2/2
6.	Turning outside of vessel preparatory to banding and complete machining of manway closure	28	2/3 to 3/3
7.	Transportation of vessel from Mesta Machine Wks. to B. & W.	2*	3/4 to 3/5
8.	Banding operation to be performed at B. & W.	28	3/6 to 4/2
9.	Hydro-static test of completed vessel at B. & W.	3	4/3 to 4/5
10.	Transportation of completed vessel from B. & W. to Trinity Railroad Siding	21*	4/6 to 4/26
11.	Transportation of vessel overland from Railroad siding to test site at Trinity	21*	4/27 to 5/17
12.	Erection of vessel from horizontal to vertical position	7	5/18 to 5/24
13.	Completion of vessel foundation after vessel is in place	2	5/25 to 5/26
14.	Dismantling of erection scaffolding	2	5/27 to 5/28
15.	Construction and erection of working platforms and scaffolding to be used in the assembly of the gadget in the vessel	7	5/29 to 6/4
16.	Vessel installation complete ready for internal assembly of the gadget		6/5

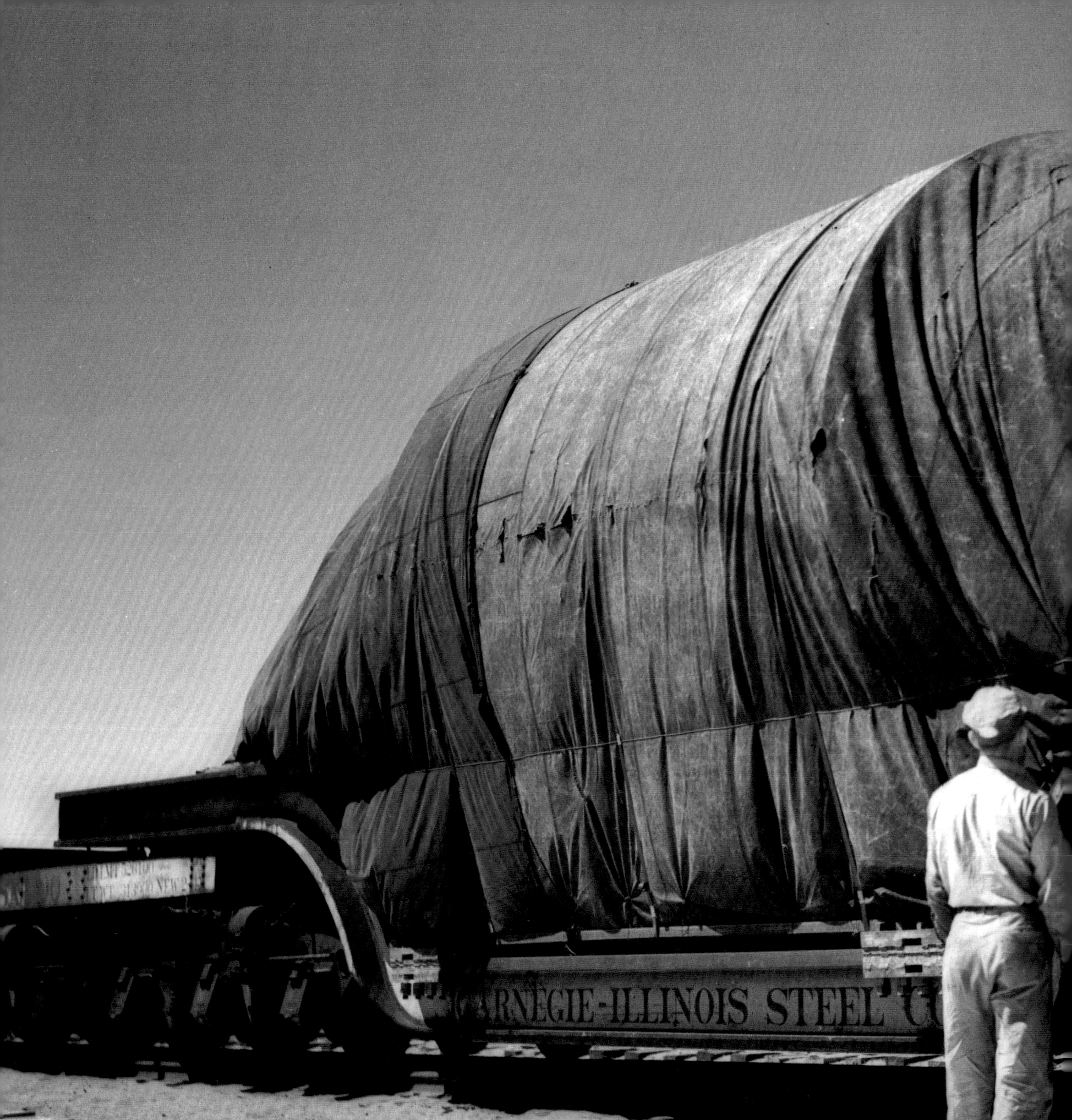
ARNEGIE-ILLINOIS STEEL

After proceeding in parallel for many months, the missions to develop a test site and construct a detonation chamber were set to finally intersect, as a shrouded, shapeless Jumbo began its journey by rail in the early spring of 1945.

Riding on a ninety-foot, heavy-duty train car that normally transported steel ingot molds, the oversized load detoured around tunnels and bridges along a lengthy route that took the cargo as far south as New Orleans before veering west across Texas and into New Mexico.

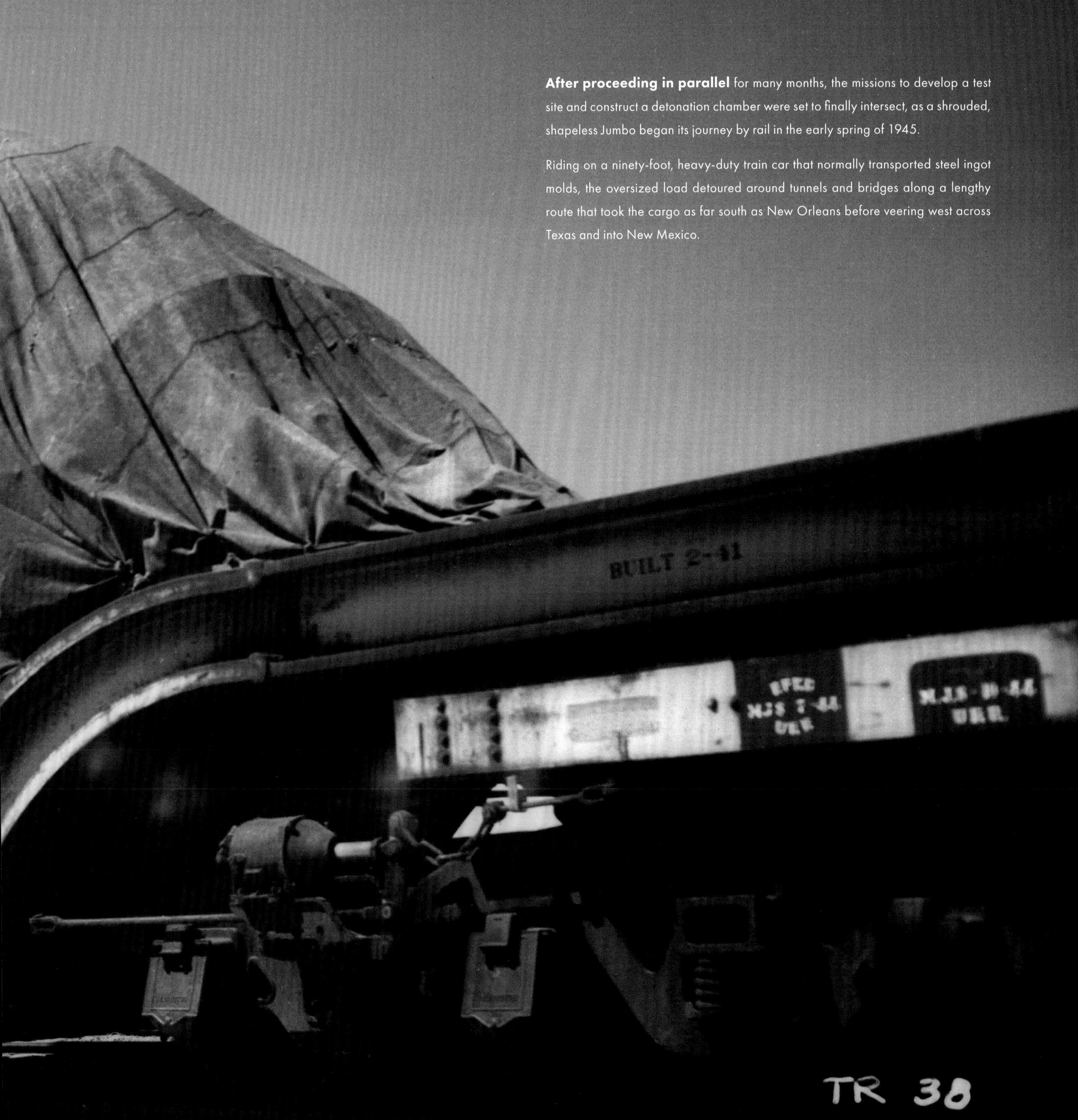

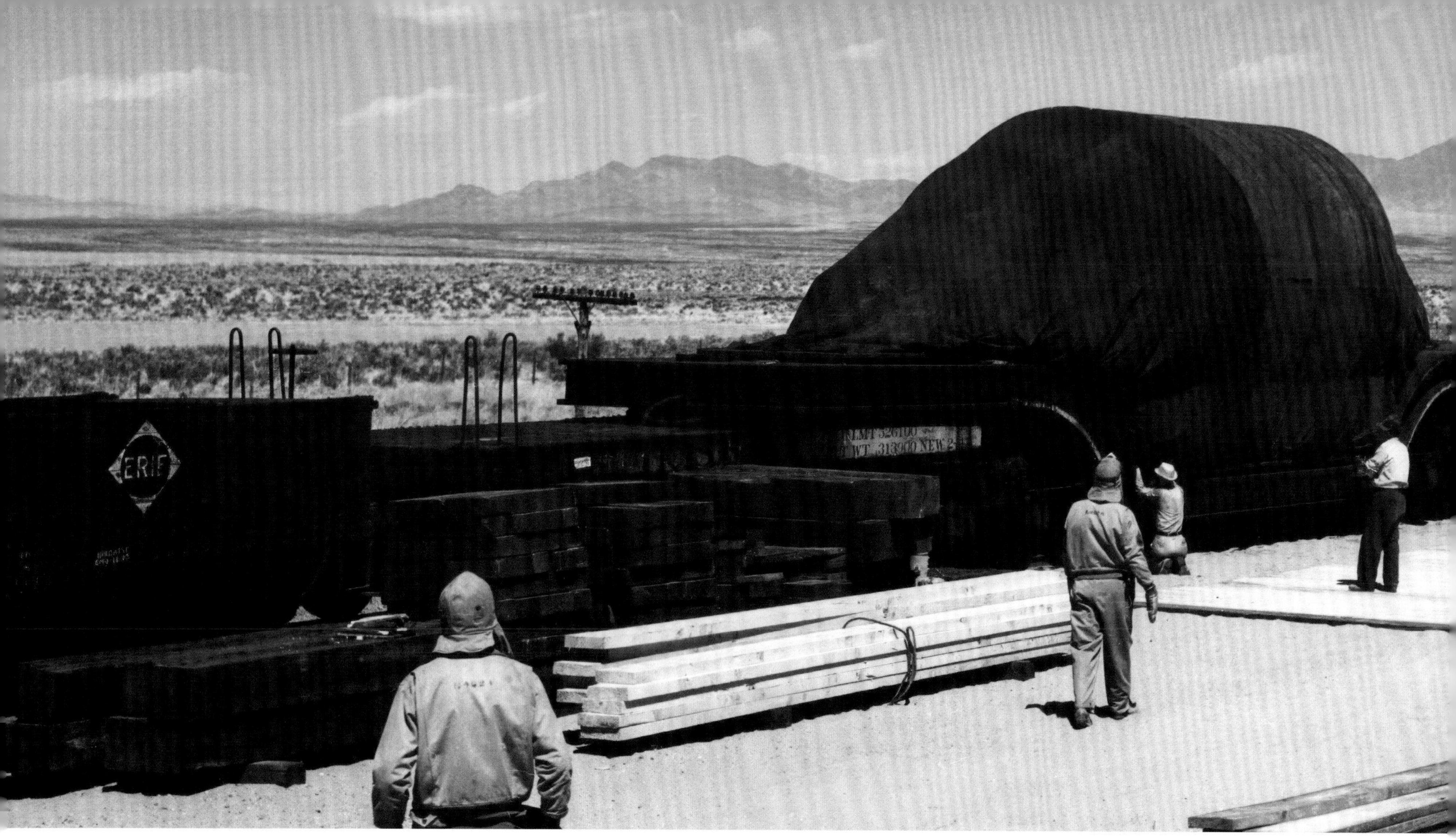

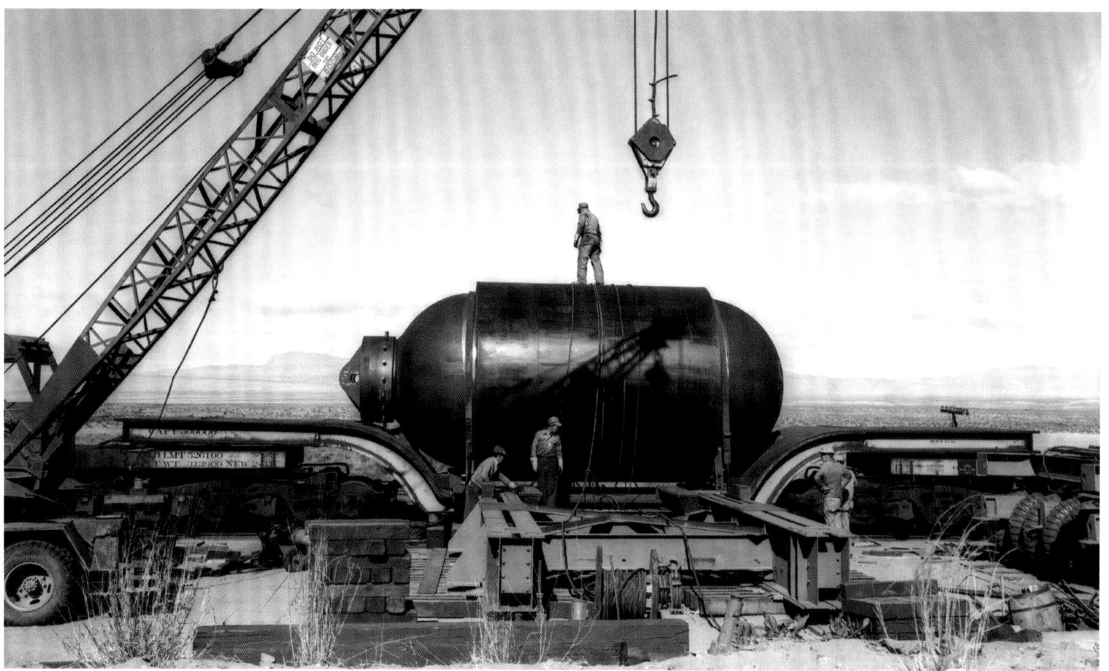

Bainbridge traveled to Pope Siding for the arrival of the long-awaited Jumbo. It was an impressive sight that received commensurate attention from the Spectrographic and Photographic Measurements Group, evident in the many images documenting the vessel's disembarkation from the train car and overland journey to the test site.

Having reached the end of the rail line, the giant bottle would transition onto a specially designed, sixty-four-wheeled trailer (*above*) for the remainder of its travels.

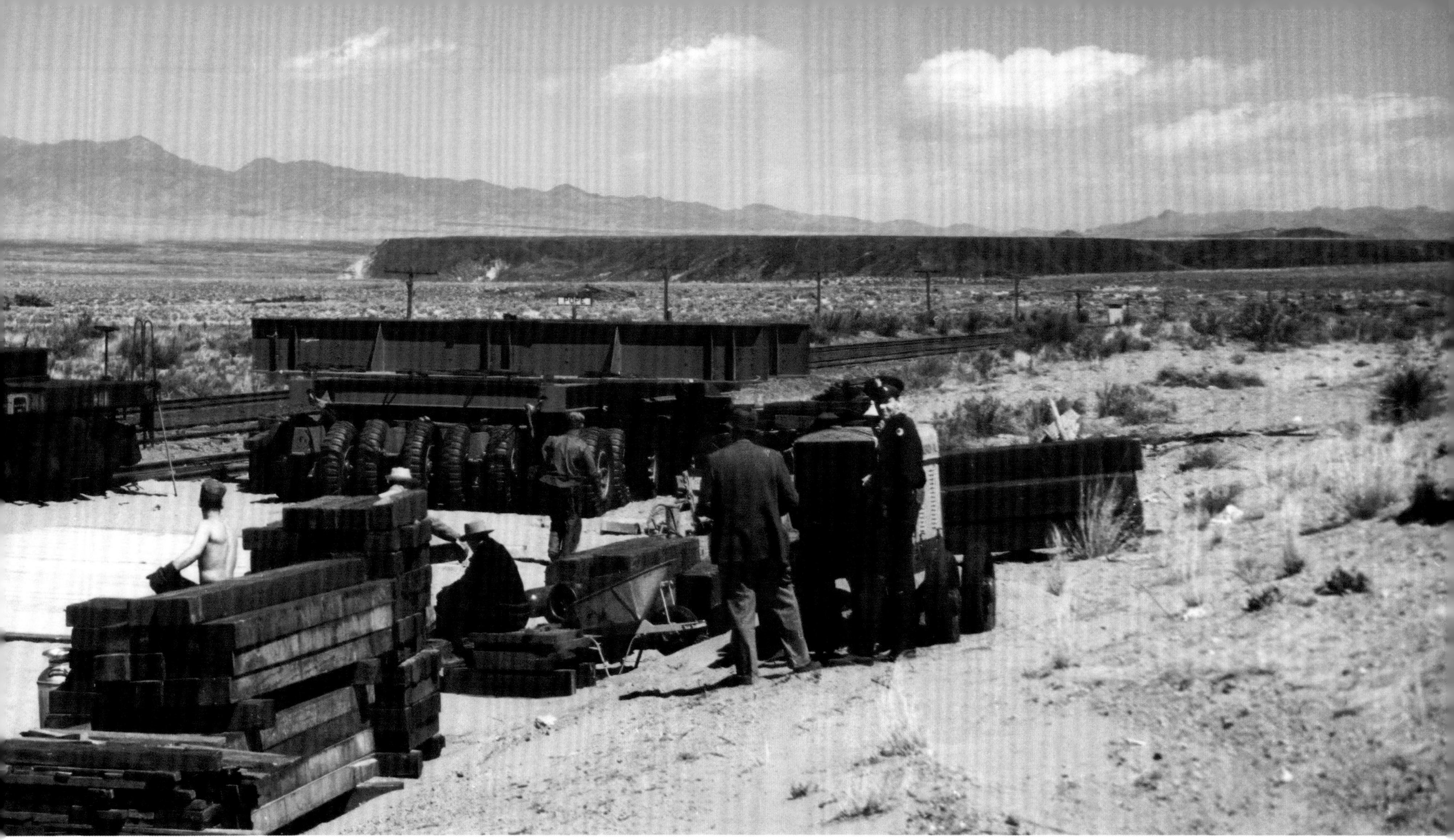

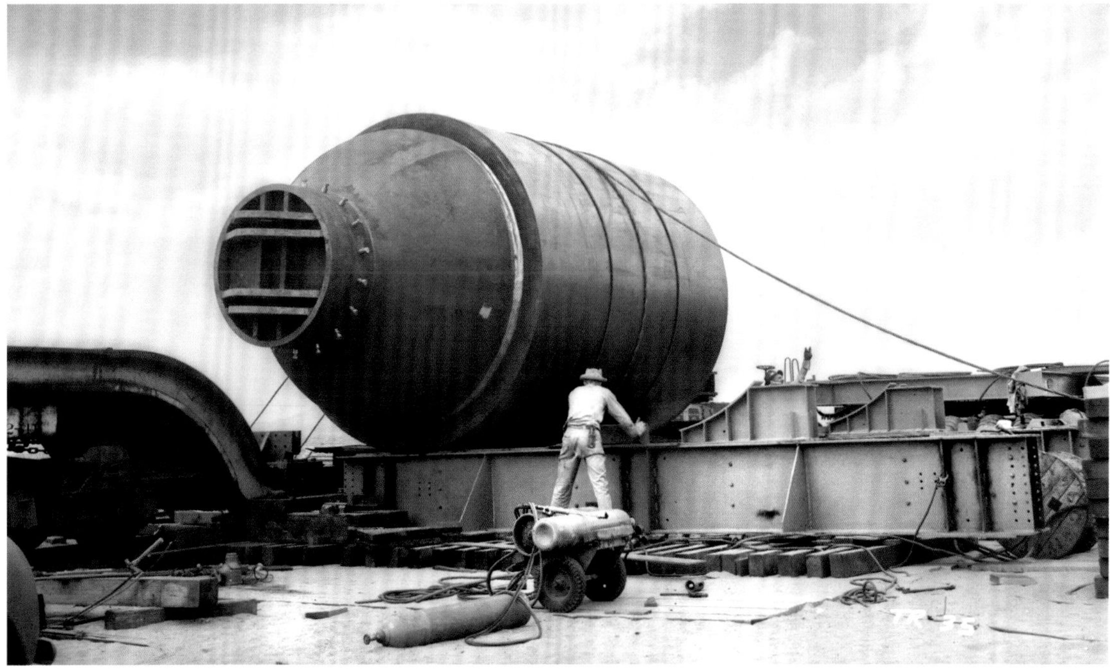

At nearly half a million pounds, however, Jumbo was too heavy to be hoisted by crane. Instead, workers removed the trailer's platform, lowered it to match the height of the railcar, as shown at left, and carefully rolled Jumbo aboard, where curved pieces of steel held the bottle like a giant cradle. The platform was then jacked up so the trailer could scoot underneath it.

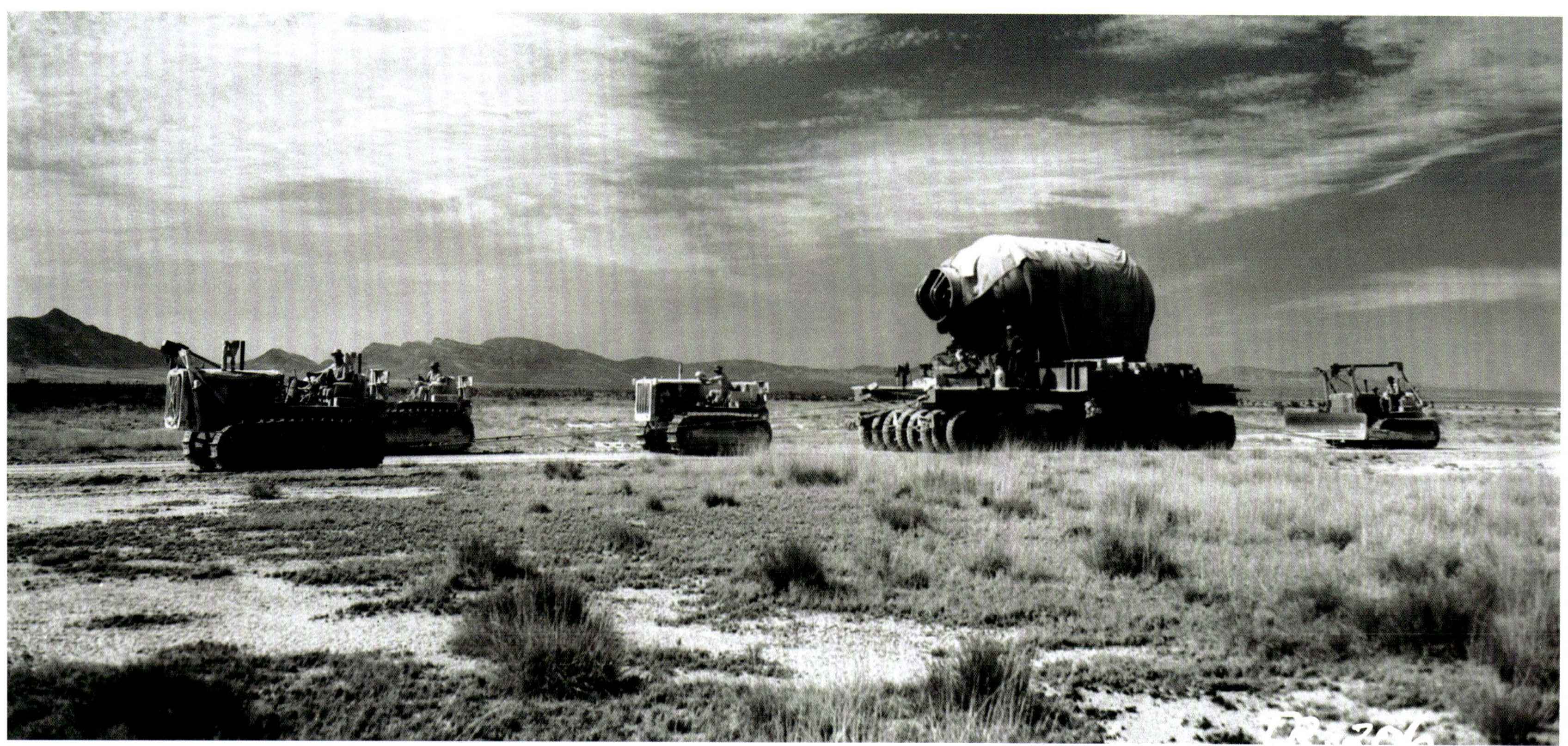

Tugged along by Caterpillar tractors and sporting a more haphazard disguise than on its railway journey, Jumbo plodded across the home stretch to the secret test location. A bulldozer hooked to the rear of the trailer served as brakes, eventually bringing the vessel to a halt at the edge of a large pit.

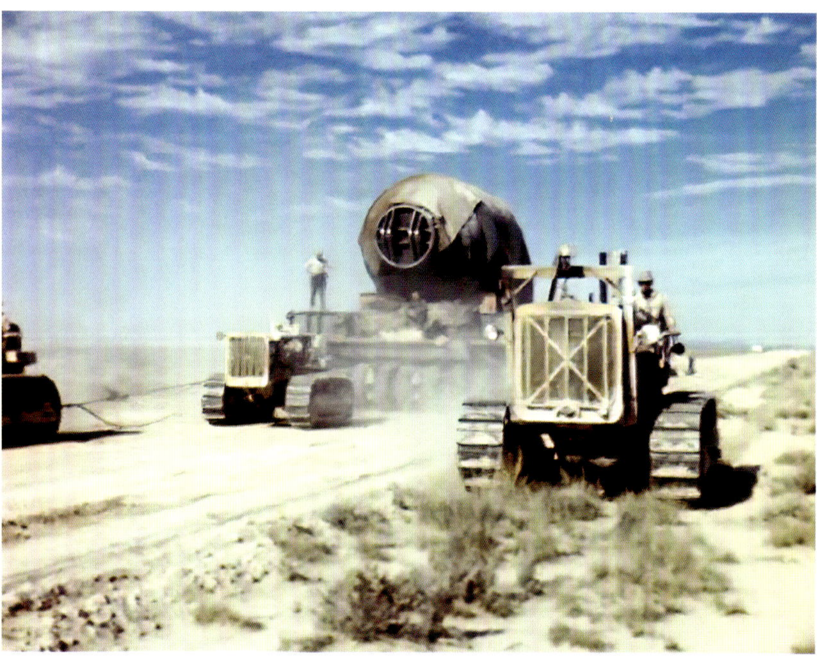

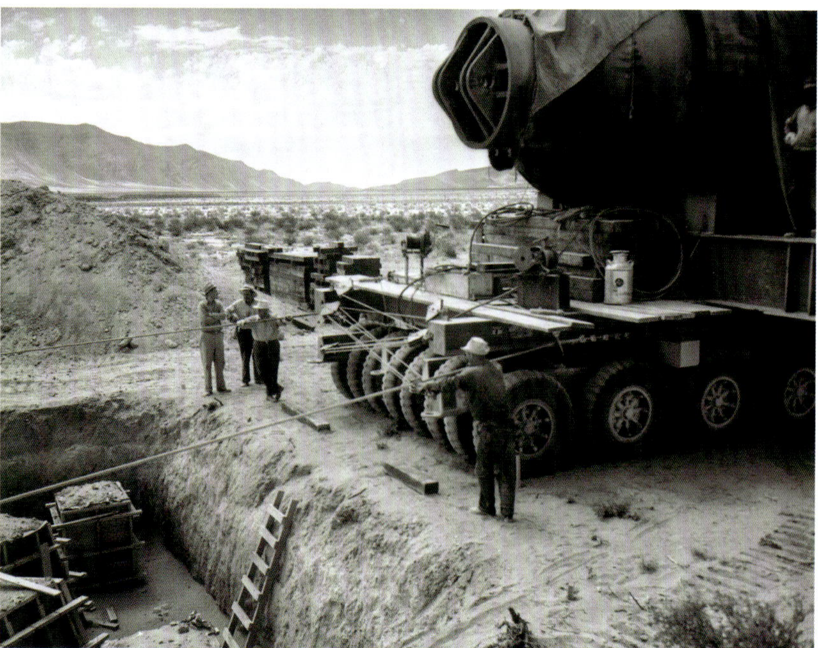

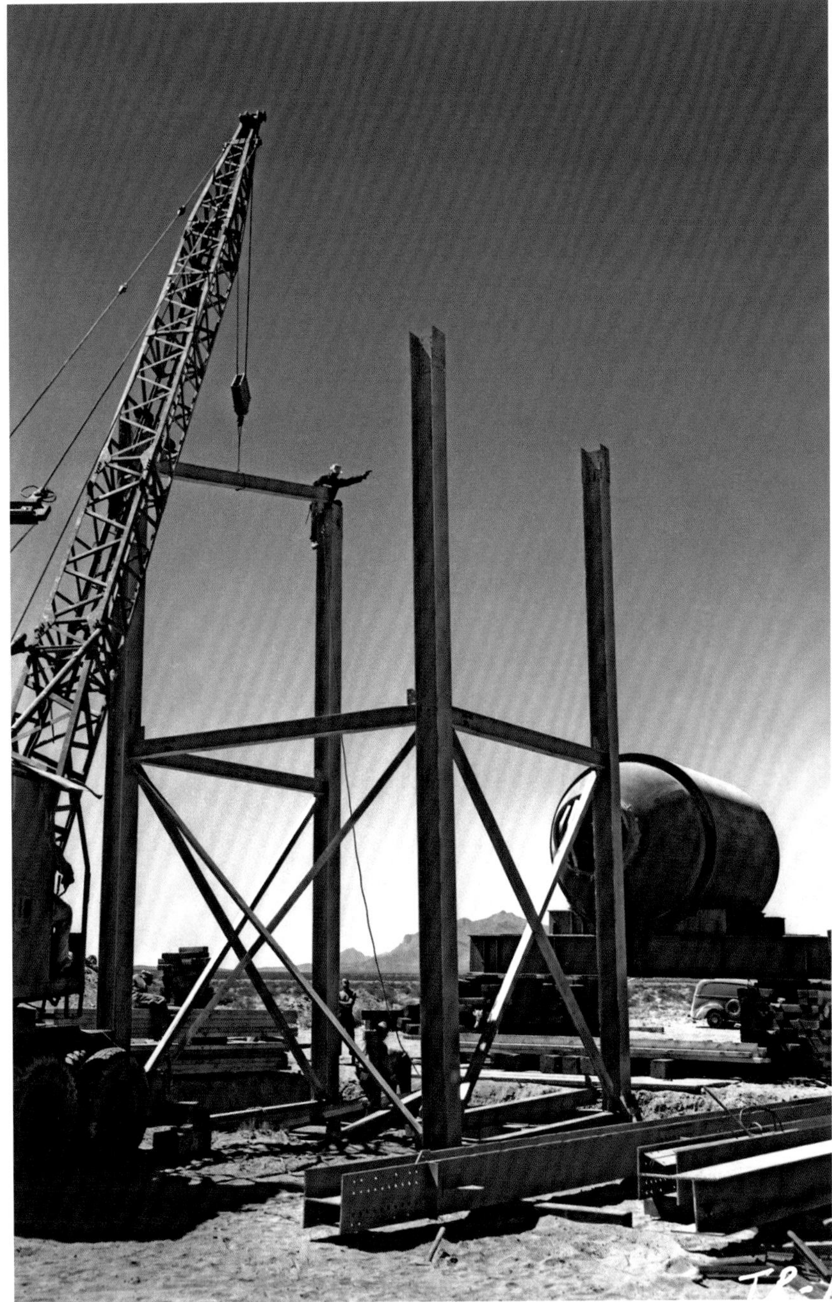

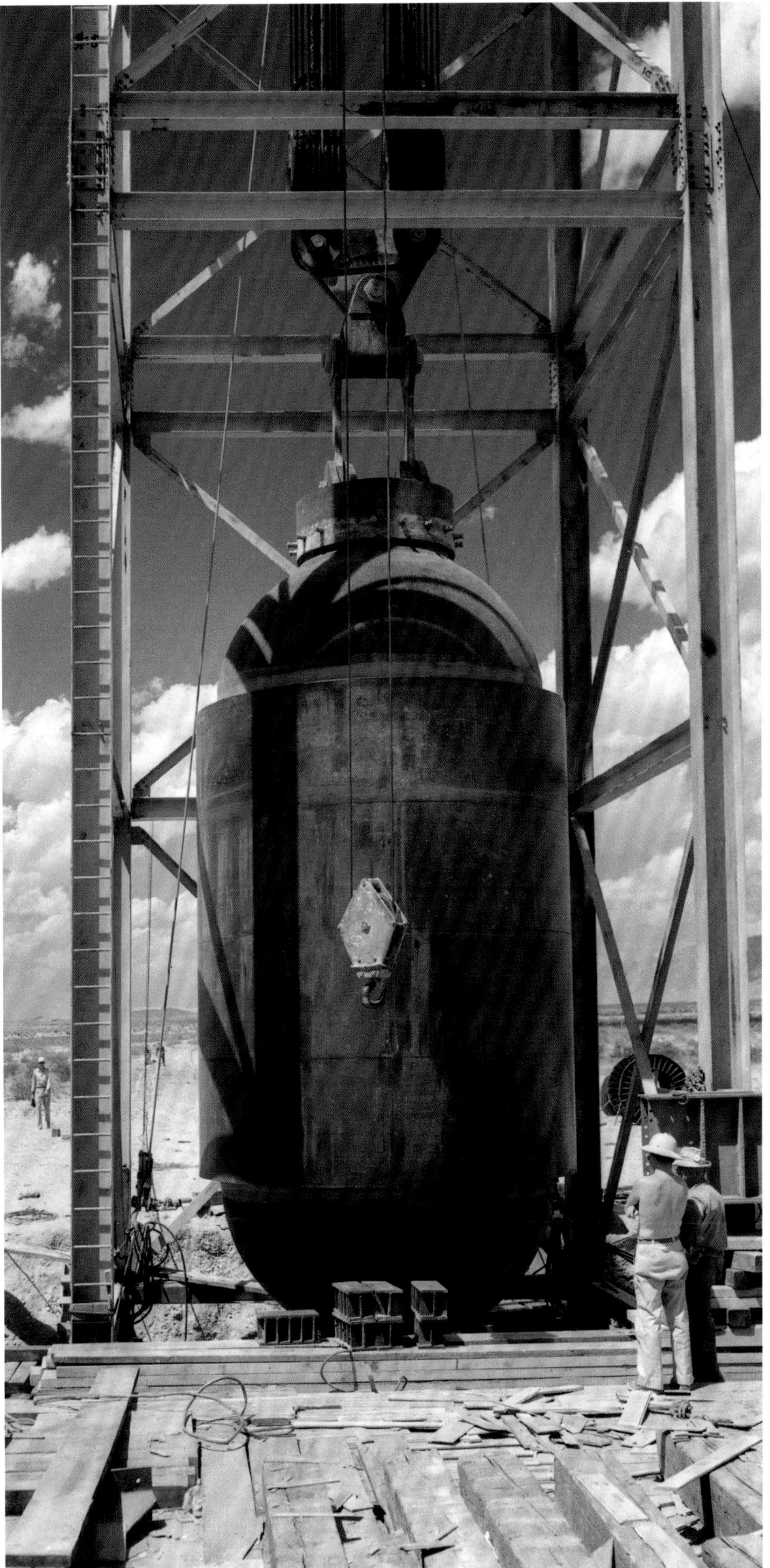

An entire tower was built over the pit to support a heavy-duty pulley system that could lift Jumbo by its top end—referred to as the manway—and lower the container into an upright position. Concrete poured up to the base of its reinforced midsection secured the vessel in place.

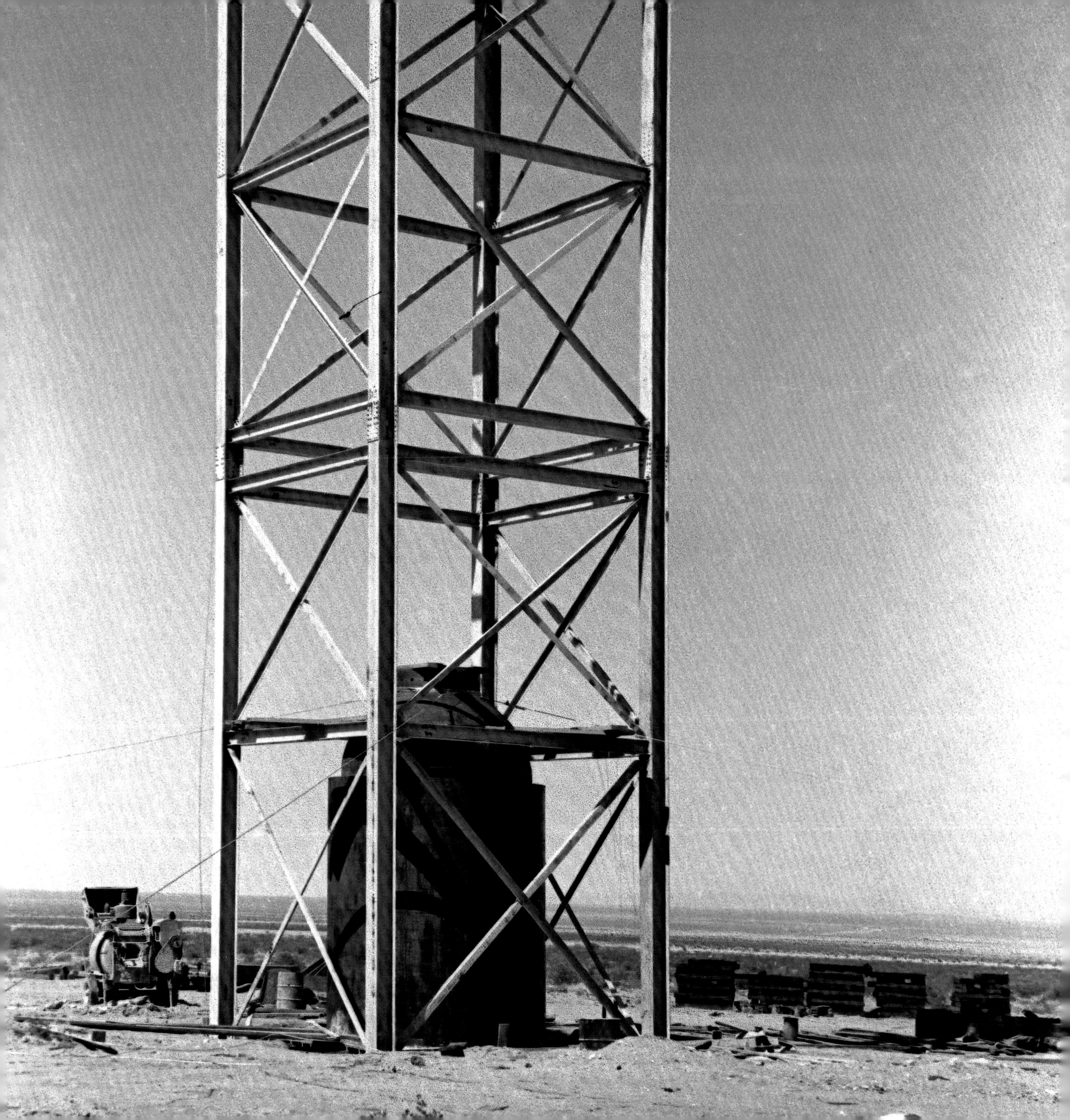

Amid the herculean effort and aggressive scheduling that brought Jumbo to life and then to the desert of New Mexico, the once crucial safety net for the Trinity test was slowly becoming a vestige of more uncertain times. As confidence increased in the implosion method and plutonium production rates improved, interest in the scientific value of the test shot came to supersede the initially paramount concerns about a fizzle. If the Gadget detonated within the thick, steel walls of Jumbo, the behavior of the nuclear explosion, even as it vaporized the containment vessel, would be so skewed as to render the results of any experiments mostly useless.

So it was that by the time of its arrival—and, in fact, its departure from Ohio—Jumbo had been relieved of its role in the Trinity test. Oppenheimer instructed in March that all preparations should assume the detonation chamber would not be used. Though Bainbridge wrote that Jumbo remained "a silent partner in all our plans," the vessel was placed not at ground zero but 800 yards to the northwest, joining a ring of some of the closest-range instrument bunkers. Unless the Gadget encountered a major setback, or it fizzled and a second test was needed, the weighty investment in failure would loom quietly from the sidelines as history unfolded around it.

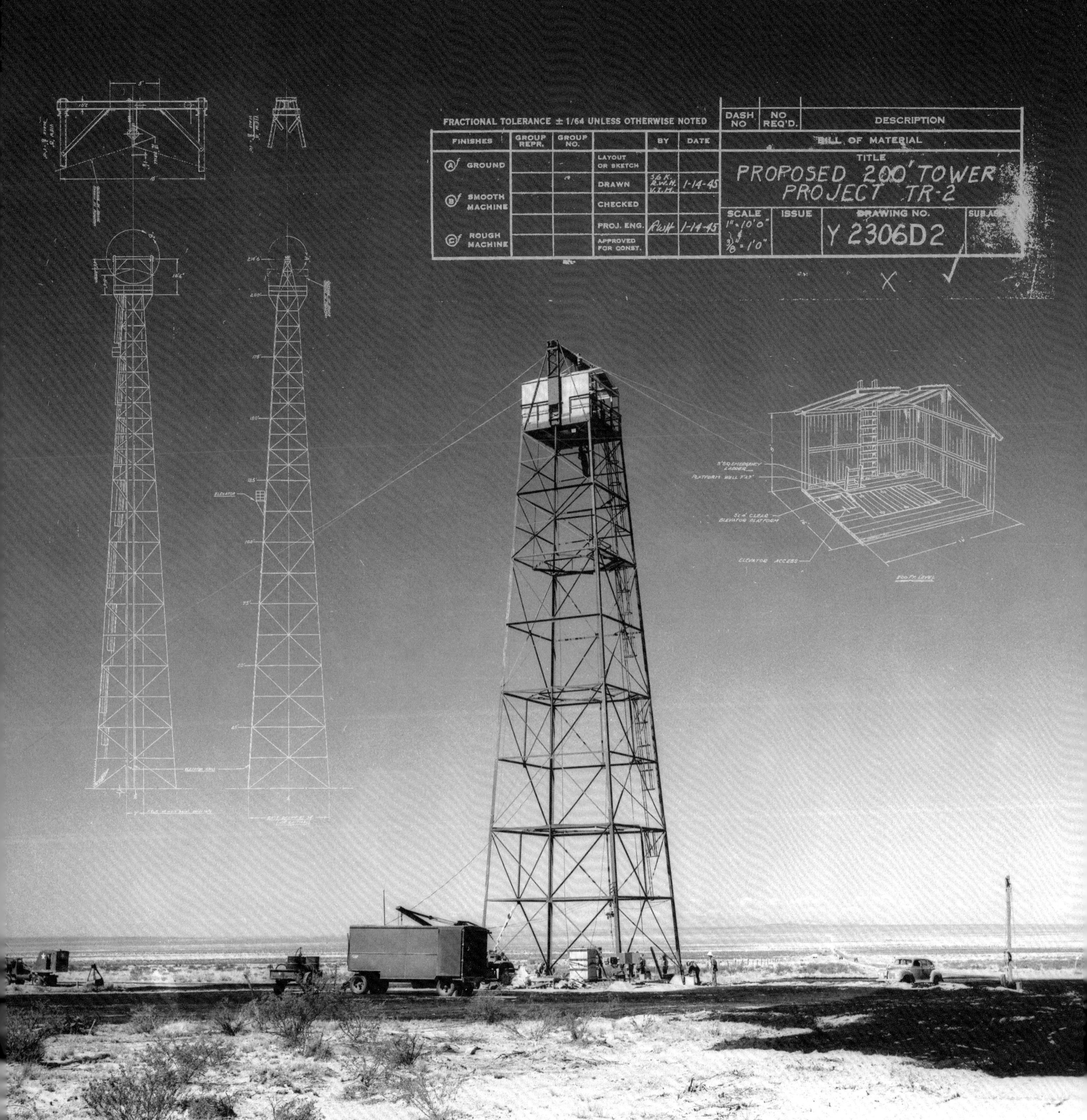
FRACTIONAL TOLERANCE ± 1/64 UNLESS OTHERWISE NOTED
DASH NO
NO REQ'D.
DESCRIPTION
FINISHES
GROUP REPR.
GROUP NO.
BY
DATE
BILL OF MATERIAL
GROUND
SMOOTH MACHINE
ROUGH MACHINE
LAYOUT OR SKETCH
DRAWN
S.G.K. R.W.H. V.I.M.
1-14-45
CHECKED
PROJ. ENG.
RWH
1-14-45
APPROVED FOR CONST.
TITLE
PROPOSED 200' TOWER
PROJECT TR-2
SCALE
1" = 10'0"
3/8" = 1'0"
ISSUE
DRAWING NO.
Y 2306D2

02 Fielding Experiments

The most rudimentary measure of success would, of course, be the eye test: Did the Gadget produce a nuclear explosion? Beyond the answer to this simple question, however, the detonation was also a monumental and complicated science experiment—one that naturally gave rise to as many smaller experiments as could be squeezed, sometimes impractically, into the test plans. Alongside a crowd of Trinity contributors, a robust collection of electronic and mechanical observers was stationed carefully throughout the desert to bear witness to the shot. Although deceptively rugged and nondescript, these devices were highly sophisticated, often employing first-of-their-kind technologies—invented and built by some of the world's preeminent researchers to study a first-of-its-kind event.

After two years of leapfrogging basic nuclear studies in their rush to produce a workable device, the pioneering scientists now "yielded to temptation and conceived experiment after experiment" in anticipation of the Trinity test—much to Kenneth Bainbridge's alarm. In December of 1944, with base camp still under construction at the test site, a selection committee, headed by Bainbridge, had been established to evaluate what was quickly becoming a deluge of ideas. The committee implemented a detailed submission process, requiring scientists to scrupulously outline the personnel and material needs associated with carrying out each idea. Proposed subtests were triaged into three categories: essential experiments, which were fully greenlit no matter the required effort or resources; desirable experiments, which were approved only if they did not interfere with work on the Gadget itself; and unnecessary (i.e., nonessential) experiments, of which only the simplest were allowed.

Completing the Gadget remained the top priority. But while most of the lab continued to work feverishly on readying the test device, a significant battery of physicists, photographers, chemists, engineers, and other specialists began coming and going between Los Alamos and the Trinity site in March, as Jumbo faded into the background. With confidence increasing and a clear emphasis on learning everything possible about the performance of the test device, they were consumed from dawn to dusk with readying cameras and other diagnostic instrumentation to track the effects of the blast from detonation to dissipation.

The greatest unknown was how much energy would be released by the weapon. This value—also known as the yield—would depend on how much of the plutonium fuel underwent fission. The aim was to get as much energy from the explosion as possible. To quantify whatever success or failure awaited, scientists made plans to observe three manifestations of that energy: the amount of radiation emitted from the core (including neutrons, gamma rays, and fission fragments); the pressures and speeds of the air and ground shock waves; and the size and temperature of the fireball, which would indicate the amount of energy released as heat.

With the right combination of approaches and protections, they hoped to record in several ways all effects of the explosion. If the experiments were to have any chance of capturing the various expected phenomena, however, they had to be set up close enough to absorb and record the effects while simultaneously far enough away to withstand them. Difficult decisions about siting and fortification were made harder by the fact that only rough and frequently changing estimates of the Gadget's capacity for destruction were available to serve as guideposts.

Nevertheless, scientists brought the rigor of the laboratory to the desert in dozens of clever ways. Some of the myriad devices relied on communication lines to quickly carry data back to recording instruments in bunkers. Others were programmed to send up visual signals that would be filmed by timestamped cameras placed at safer distances. Still others were purely mechanical, installed dangerously close to the blast but made of resilient materials or buried underground, the plan being to recover them in the aftermath to harvest data. Not knowing which approaches to the balancing act would turn out to be successful, they relied on overlap and redundancy to account for failure.

Although the cameras and instruments ultimately, and not unexpectedly, varied in their performance, the experimental program altogether would succeed in committing to history a comprehensive visual record and a trove of data, built upon the talents and toils of many.

* * * * * * *

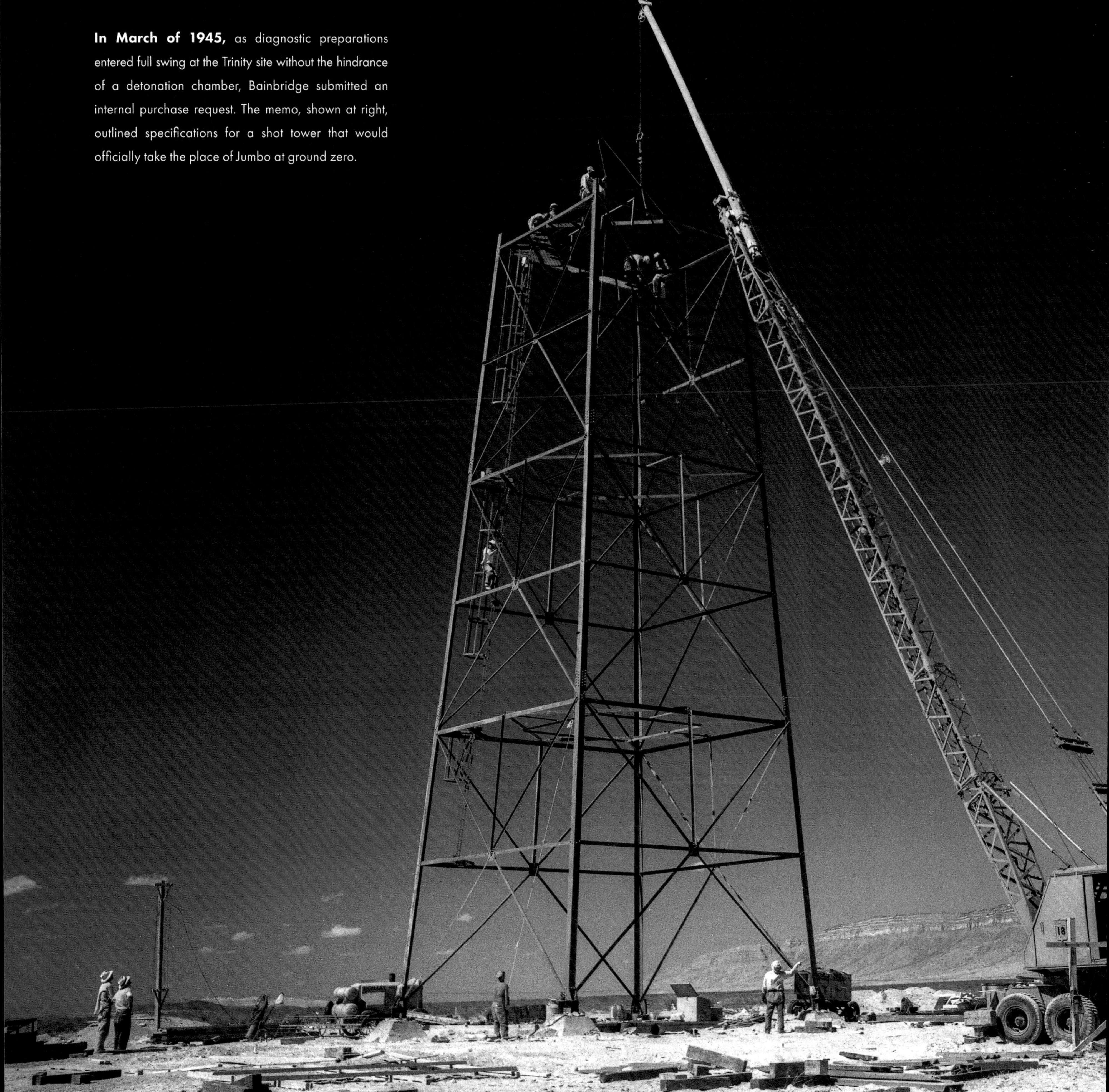

In March of 1945, as diagnostic preparations entered full swing at the Trinity site without the hindrance of a detonation chamber, Bainbridge submitted an internal purchase request. The memo, shown at right, outlined specifications for a shot tower that would officially take the place of Jumbo at ground zero.

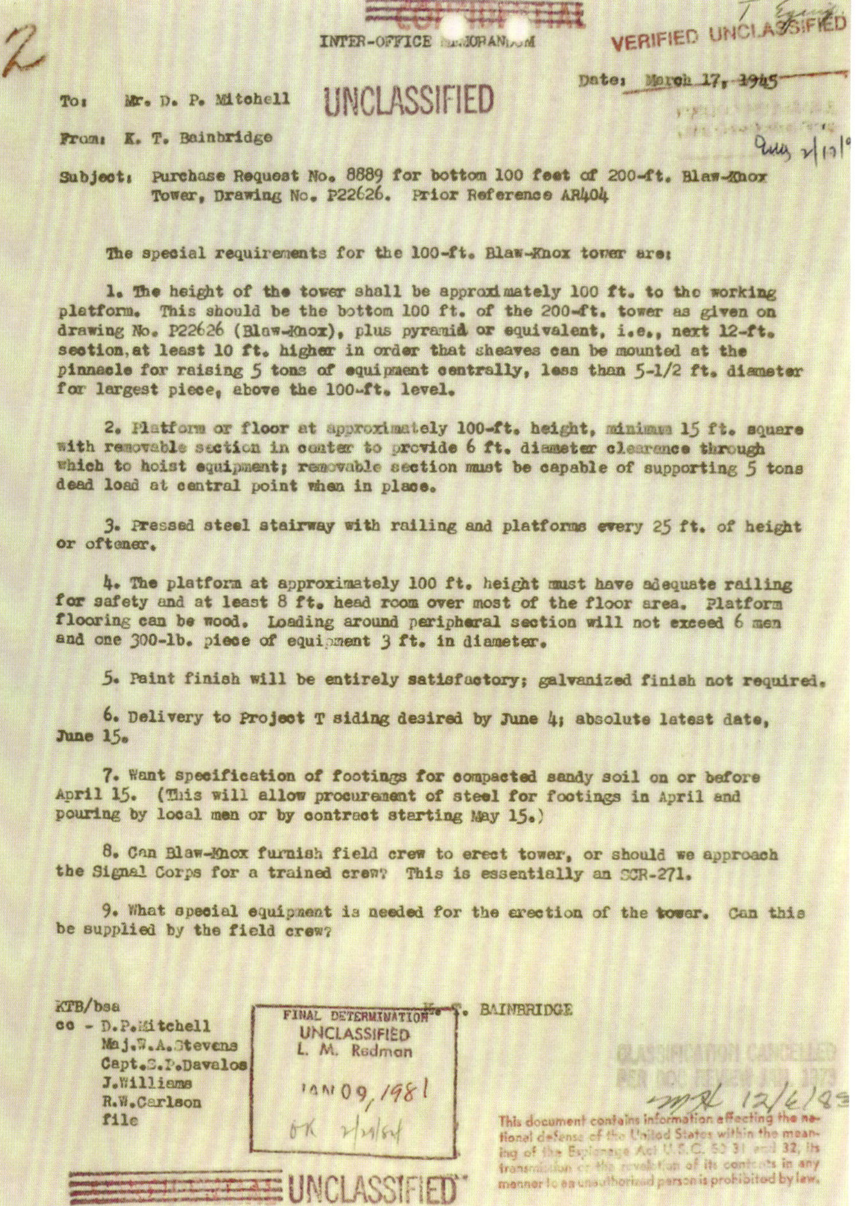

2

INTER-OFFICE MEMORANDUM

VERIFIED UNCLASSIFIED

UNCLASSIFIED

To: Mr. D. P. Mitchell

Date: March 17, 1945

From: K. T. Bainbridge

Subject: Purchase Request No. 8889 for bottom 100 feet of 200-ft. Blaw-Knox Tower, Drawing No. P22626. Prior Reference AR404

The special requirements for the 100-ft. Blaw-Knox tower are:

1. The height of the tower shall be approximately 100 ft. to the working platform. This should be the bottom 100 ft. of the 200-ft. tower as given on drawing No. P22626 (Blaw-Knox), plus pyramid or equivalent, i.e., next 12-ft. section, at least 10 ft. higher in order that sheaves can be mounted at the pinnacle for raising 5 tons of equipment centrally, less than 5-1/2 ft. diameter for largest piece, above the 100-ft. level.

2. Platform or floor at approximately 100-ft. height, minimum 15 ft. square with removable section in center to provide 6 ft. diameter clearance through which to hoist equipment; removable section must be capable of supporting 5 tons dead load at central point when in place.

3. Pressed steel stairway with railing and platforms every 25 ft. of height or oftener.

4. The platform at approximately 100 ft. height must have adequate railing for safety and at least 8 ft. head room over most of the floor area. Platform flooring can be wood. Loading around peripheral section will not exceed 6 men and one 300-lb. piece of equipment 3 ft. in diameter.

5. Paint finish will be entirely satisfactory; galvanized finish not required.

6. Delivery to Project T siding desired by June 4; absolute latest date, June 15.

7. Want specification of footings for compacted sandy soil on or before April 15. (This will allow procurement of steel for footings in April and pouring by local men or by contract starting May 15.)

8. Can Blaw-Knox furnish field crew to erect tower, or should we approach the Signal Corps for a trained crew? This is essentially an SCR-271.

9. What special equipment is needed for the erection of the tower. Can this be supplied by the field crew?

K. T. BAINBRIDGE

KTB/bsa
cc - D.P.Mitchell
Maj.W.A.Stevens
Capt.S.P.Davalos
J.Williams
R.W.Carlson
file

FINAL DETERMINATION
UNCLASSIFIED
L. M. Redman
JAN 09, 1981

This document contains information affecting the national defense of the United States within the meaning of the Espionage Act U.S.C. 50 31 and 32, its transmission or the revelation of its contents in any manner to an unauthorized person is prohibited by law.

UNCLASSIFIED

The new steel structure—the lower half of a standard, 200-foot radar tower manufactured by Blaw-Knox in Pittsburgh—would delay the ground's interference with blast effects and give cameras an improved view of the crucial first moments of the expanding explosion. A foundation was poured in May, and the tower itself was shipped to Pope Siding in pieces, assembled over a two-week period in June.

When construction commenced, however, the crew discovered that anchor bolts on the concrete footings had been spaced incorrectly. New plates were welded to the tops to correct the misalignment and covered with additional concrete, increasing the height of the foundation and the "100-foot" tower by 2 feet. The modified footings can be seen at left, peeking above grade.

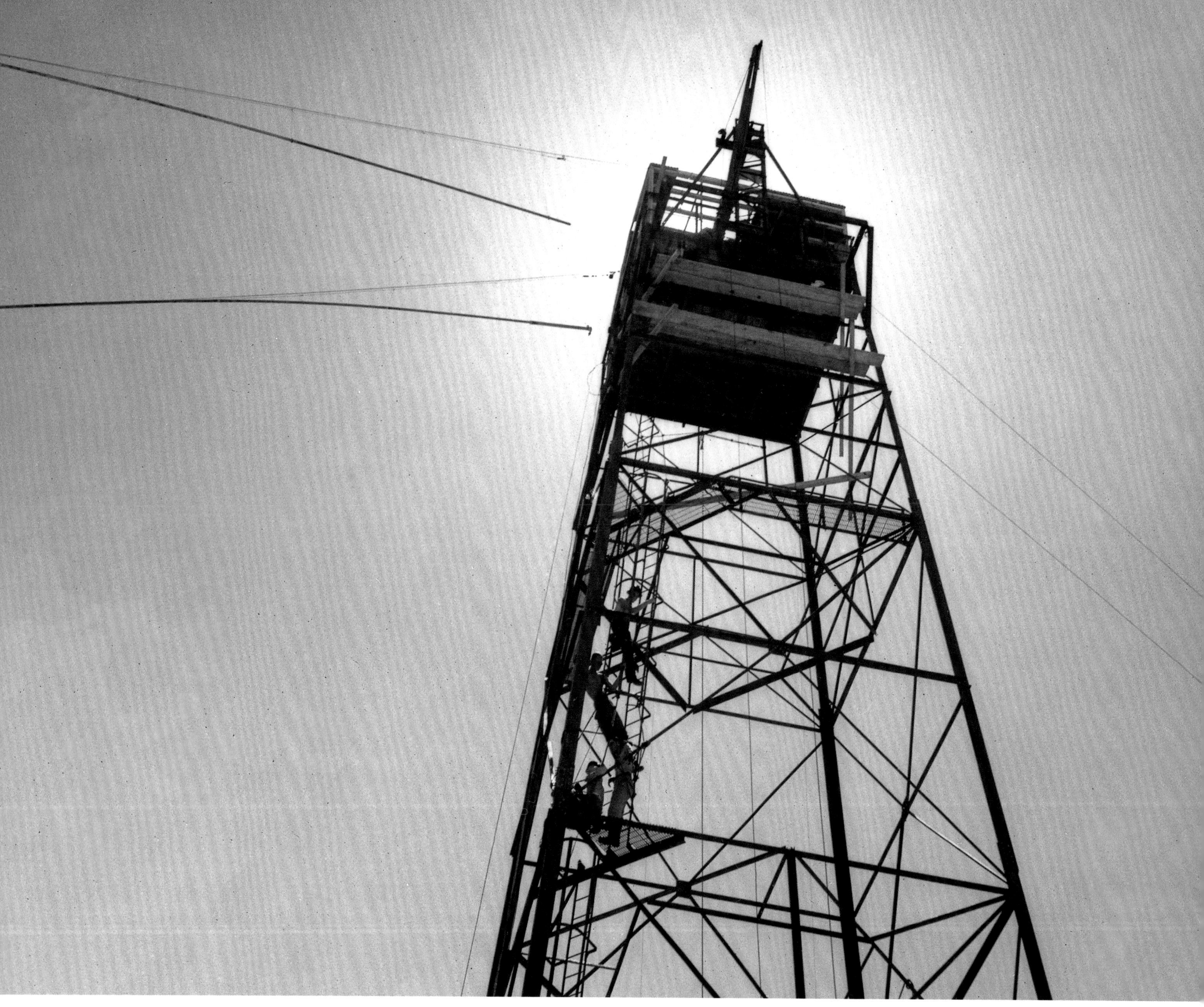

Scientists arranged cameras and instruments both near to and far from the 100-foot tower—underground, in the air, and everywhere in between. Most experiments were set up at safe (or presumed safe) locations, though two, by design, would be within vaporization distance of the Gadget. In their final act, these short-lived devices would transmit data about the timing of the fission chain reaction through heavily-sheathed coaxial cables—the two dark lines running down from the platform at the top of the tower.

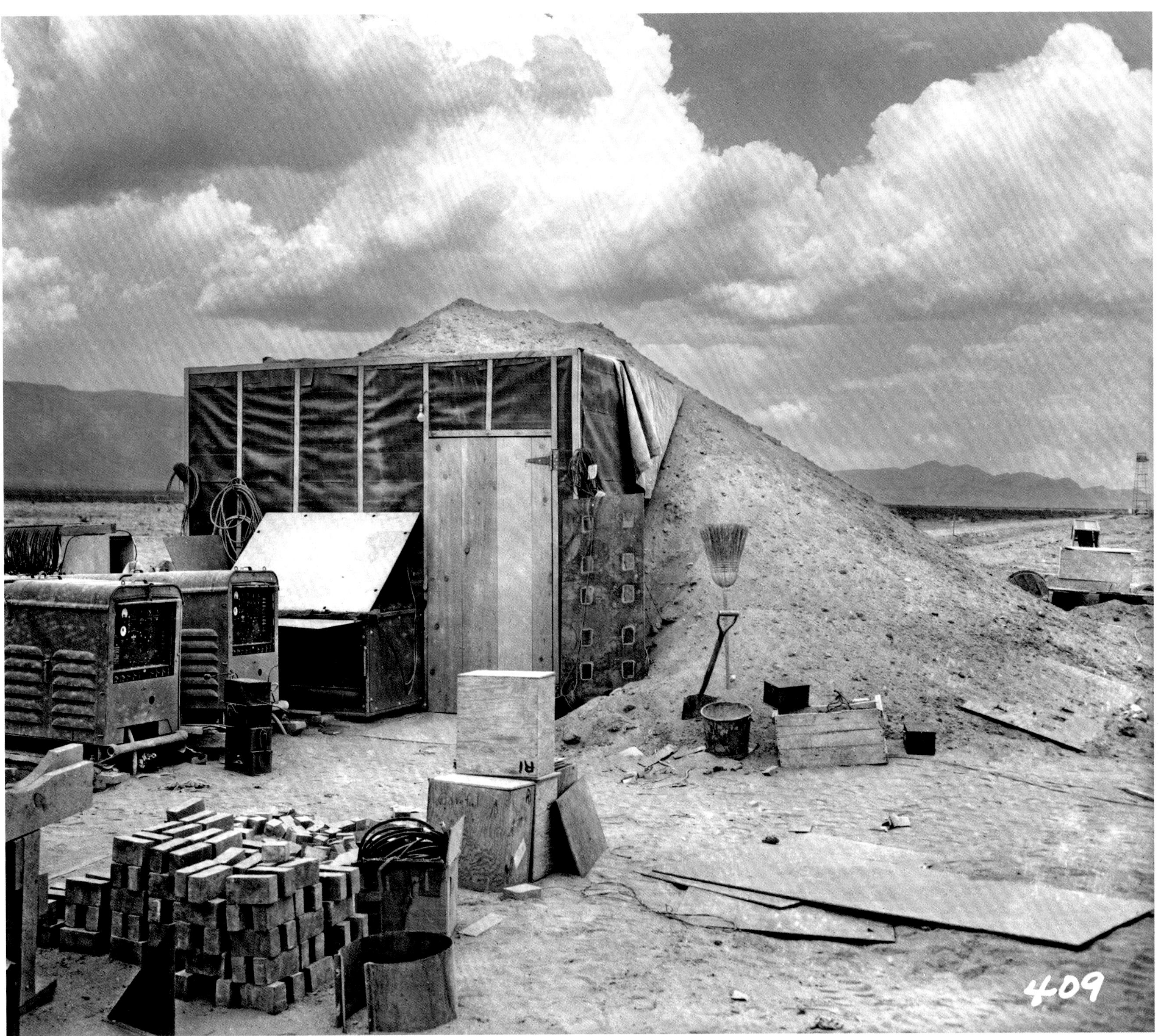

The coaxial cables ran 1,000 yards to the north, buried in trenches ending at the instrument bunker shown above. (Note the tower in the background.) A series of similar bunkers radiating to the north, south, and west housed signal receivers, generators, all sorts of cameras, and other sensitive equipment. The closest was set up just 600 yards northwest of ground zero.

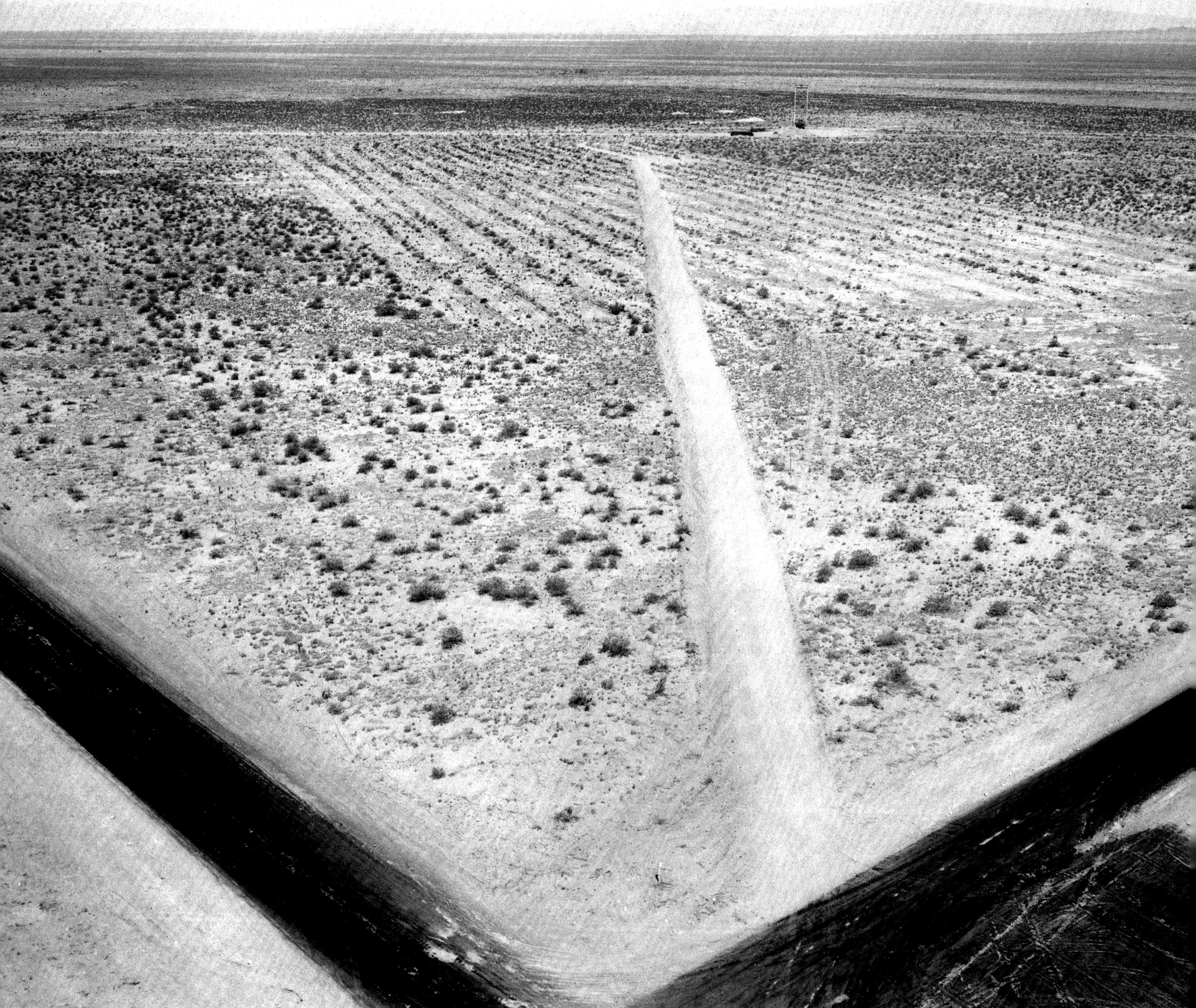

The northwest bunker is just visible along the dirt road in the photograph above, taken from the top of the shot tower. In the same direction, at a distance of 800 yards, a more prominent landmark—Jumbo—rises out of the open desert.

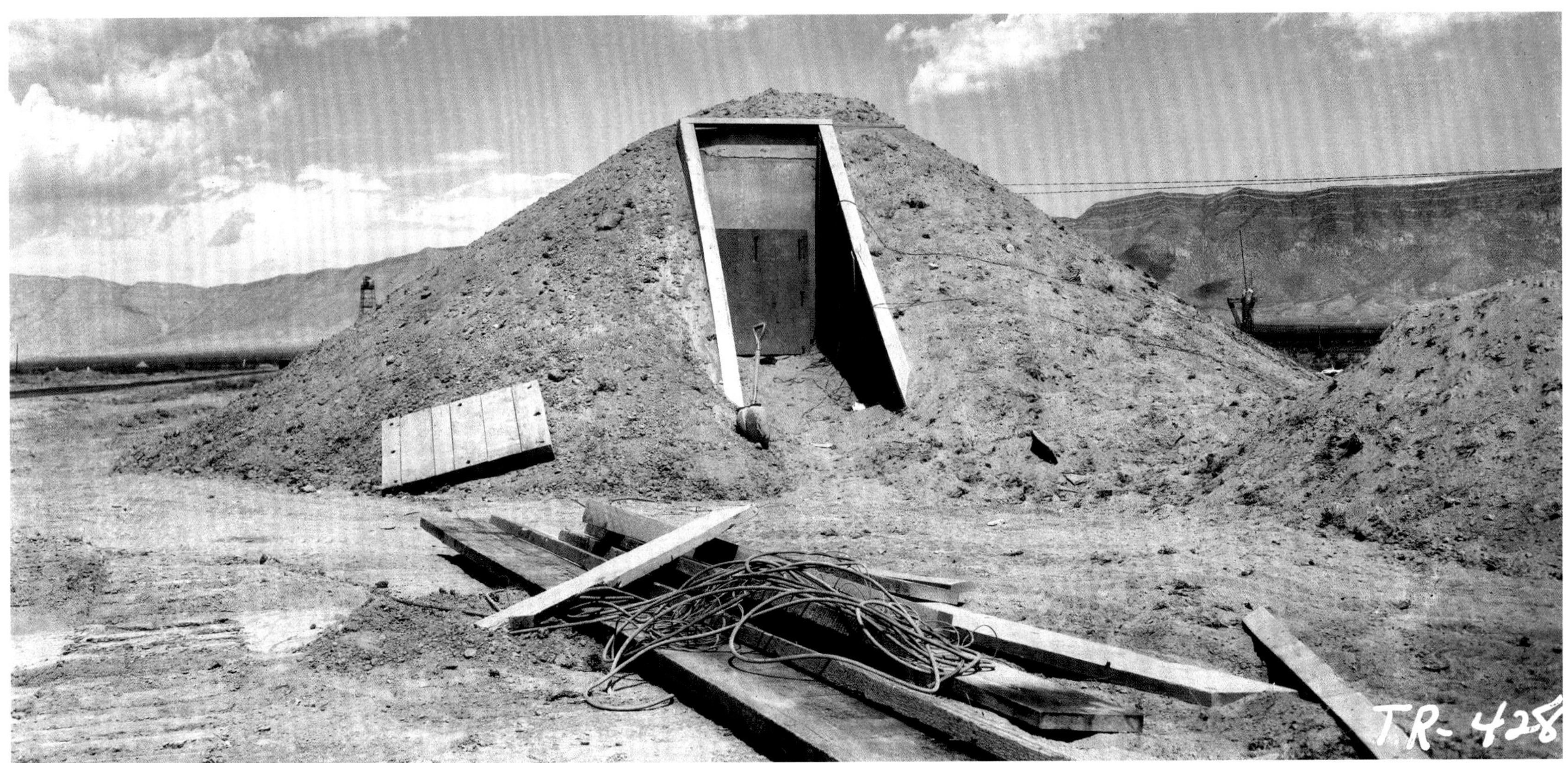

Also at 800 yards—to the west (*above*) and north (*right*) of the tower—were two photography bunkers, small cubes about six feet wide and tall. Like the personnel shelters, each was half buried. The bunkers were similar to the instrument shelters at 600 and 1,000 yards, with the notable addition of three portholes facing ground zero.

The portholes provided lines of sight for high-speed cameras capable of capturing up to 10,000 frames per second. In addition to filming with perfect aim and exposure settings, these special Fastax motion picture cameras had to start rolling precisely when the Gadget fired. Once activated by an electrical signal, they would run through their film supply after only a few hundredths of a second, just long enough to capture the first moments of the nuclear explosion.

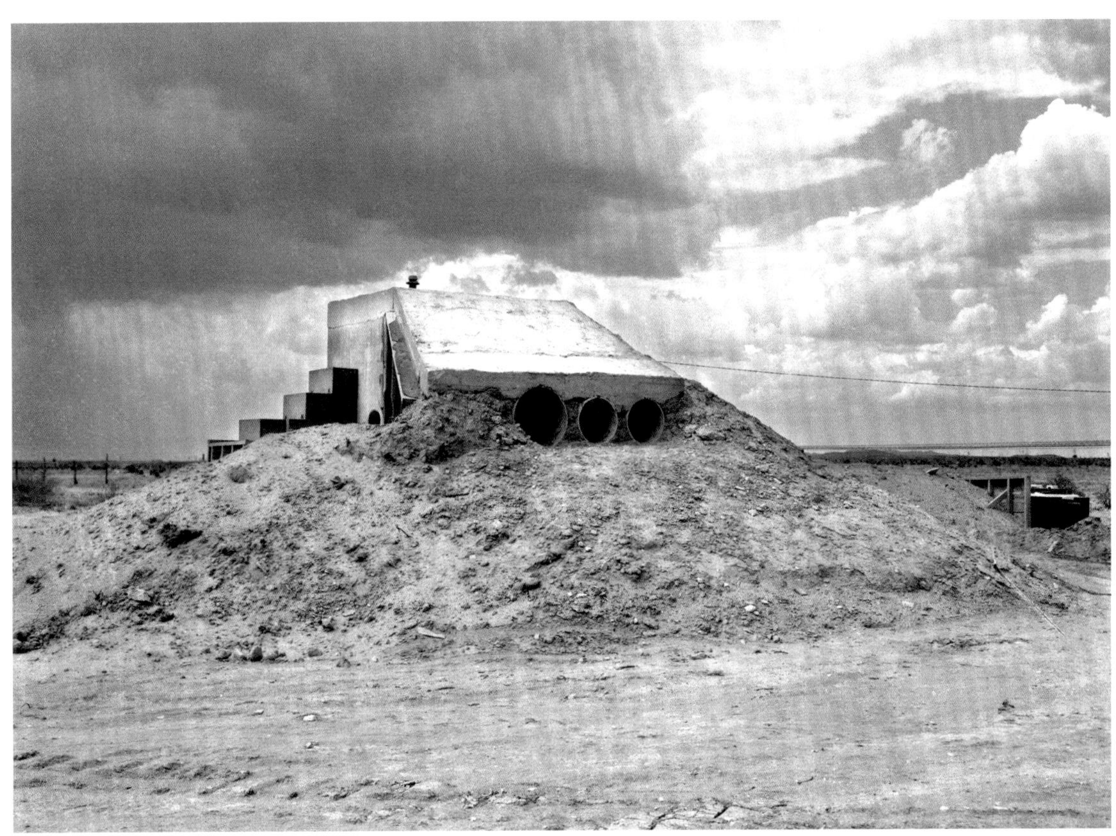

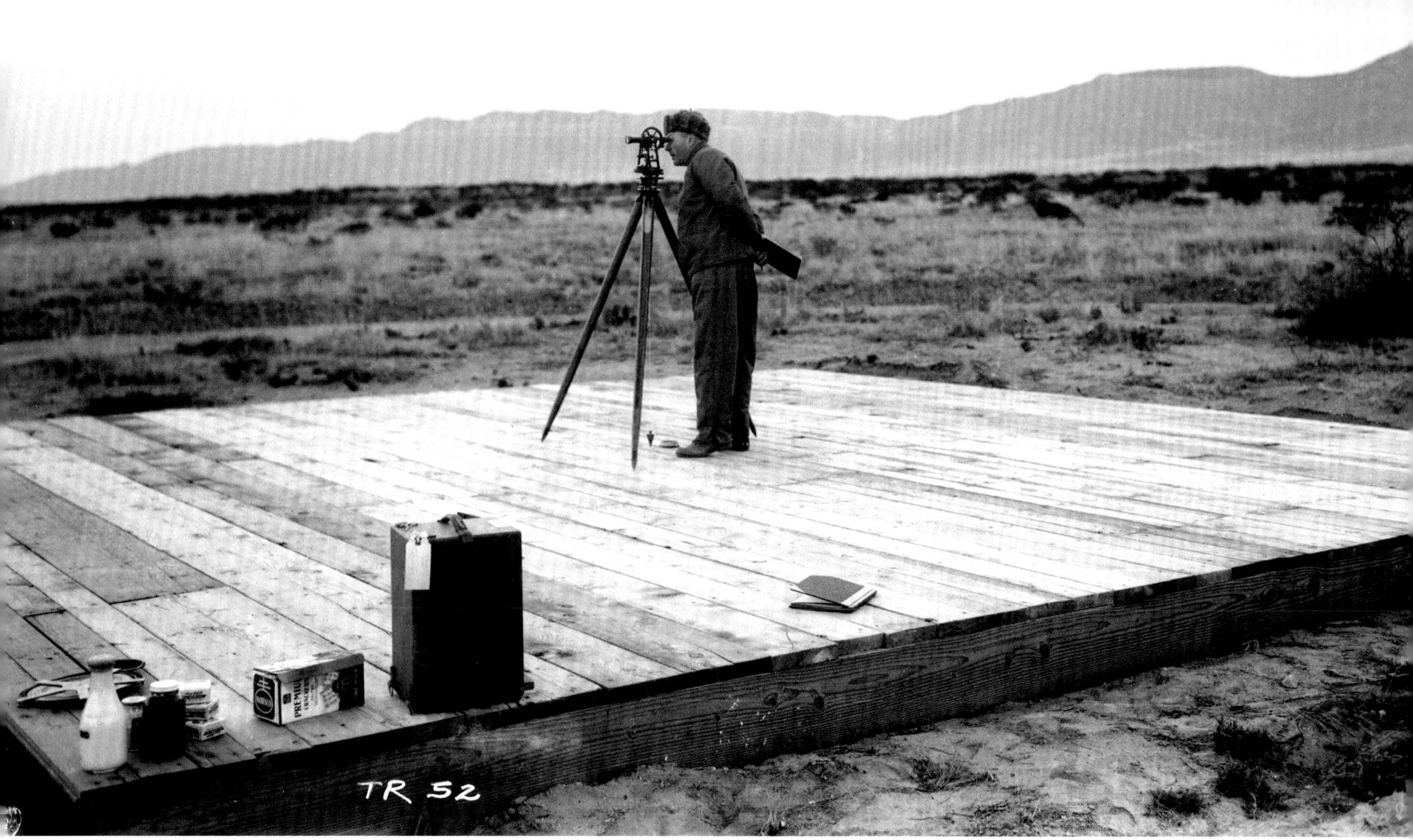

The Fastax was only one of many cameras painstakingly set up to film the test shot. In the spring of 1945, before the 100-foot tower was erected, a small, white flag marked the focal point of the Trinity site. Locating the flag through a camera lens would have been like searching for a needle in a haystack.

Nevertheless, the Spectrographic and Photographic Measurements Group (TR-5), led by Julian Mack, had the task of determining exactly where to aim and how best to configure a total of fifty-two cameras, between 800 and 42,000 yards away from the explosion. Among other tools, such as a surveyor's gadget called a transit (which Mack is peering into, above) the group used a resolution chart to calibrate lenses and check magnification.

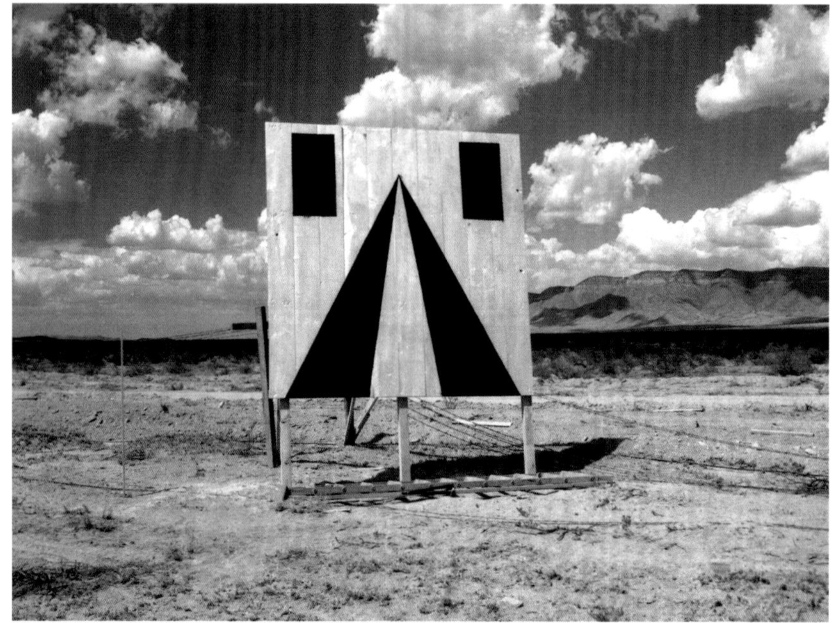

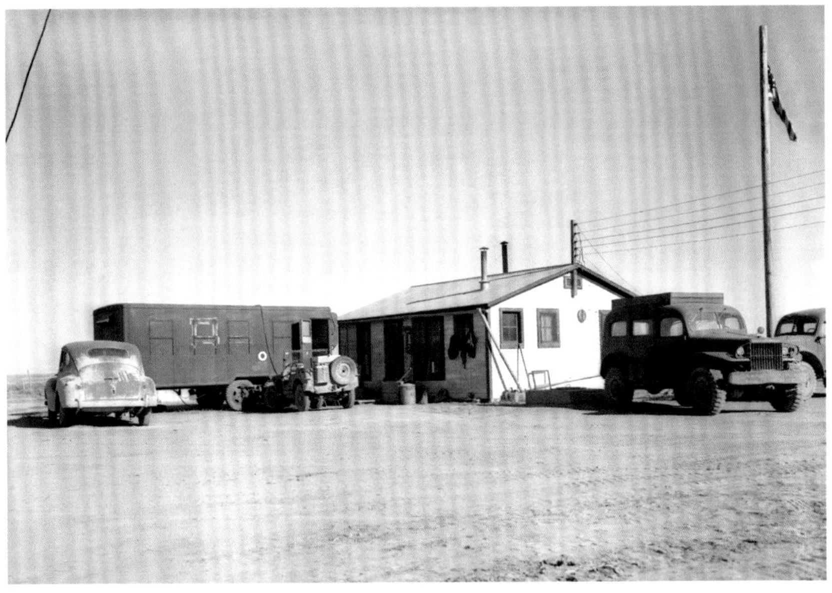

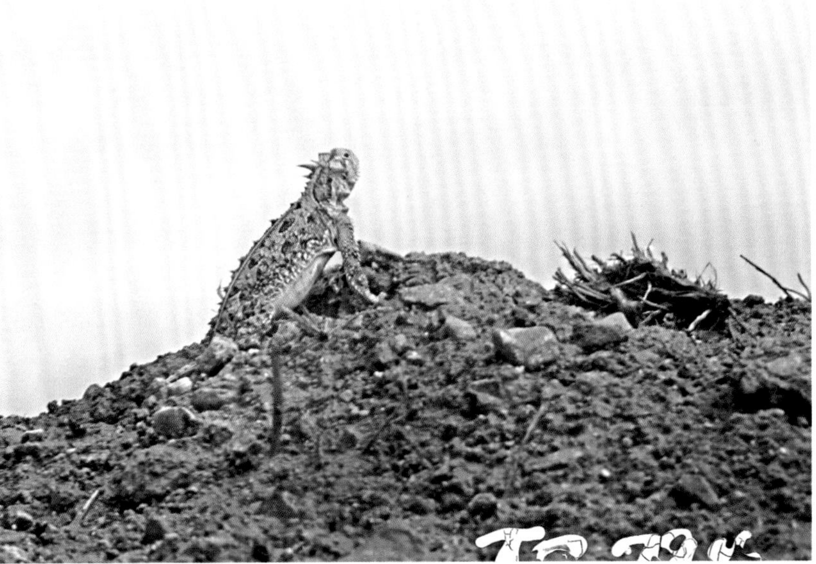

Group TR-5 had a headquarters at base camp—the Ross McDonald Ranch House—for testing equipment and processing film, but much time was spent working in the elements or traveling the empty miles between base camp, ground zero, and the camera bunkers, where it was not uncommon to cross paths with desert creatures, such as the horned lizard above. Below, Mack grabs a bite to eat while in the field with group member Ben Benjamin, a sergeant in the Special Engineer Detachment with a background in camera lens and prism manufacturing.

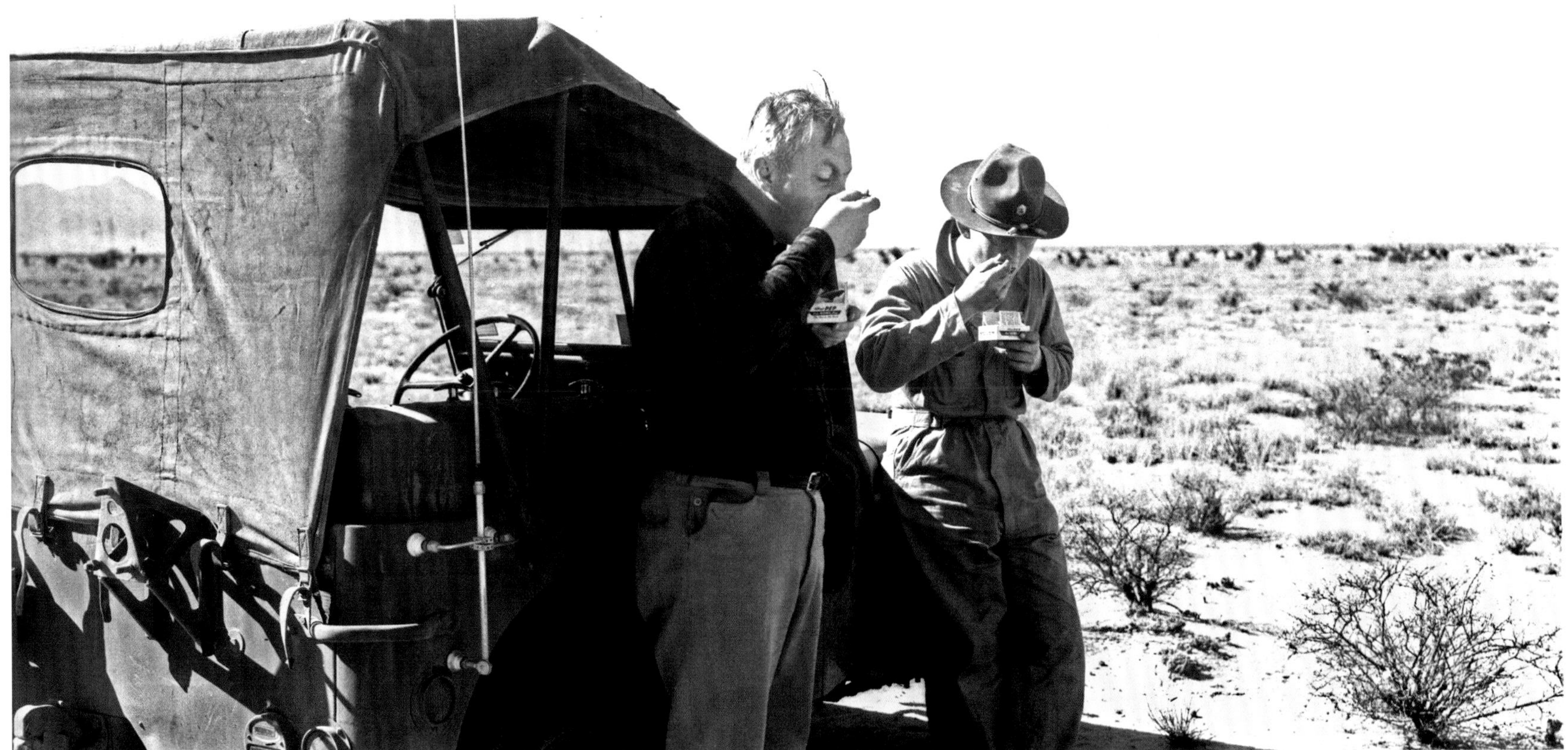

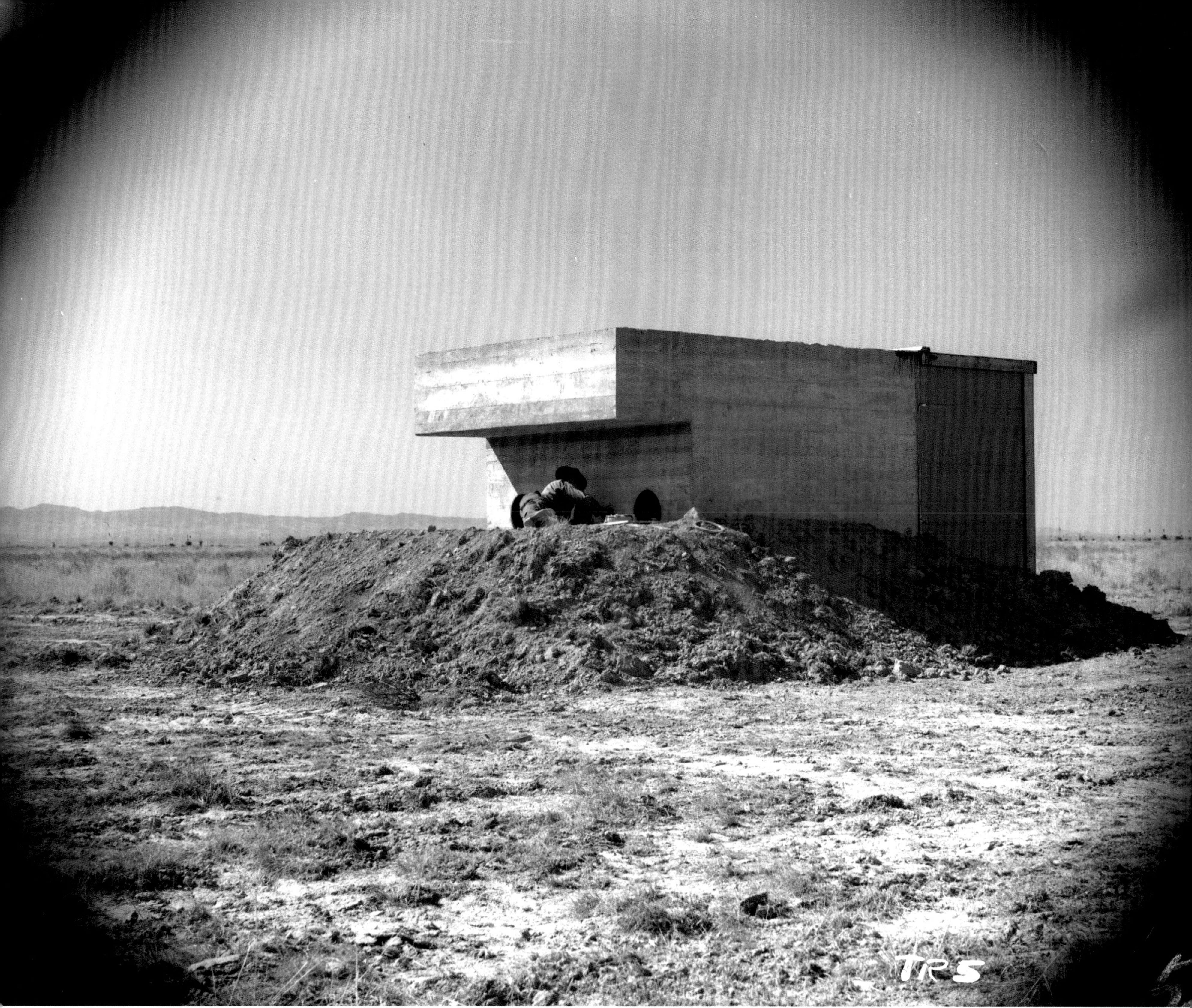

A total of four dedicated photography bunkers were built at the test site, including the unmanned structures at north and west 800 and a pair at north and west 10,000—adjacent to the personnel shelters—that would be staffed by members of Group TR-5 during the test shot. The camera bunker at west 10,000 is shown above. Except for a wooden addition on the rear, built to accommodate instrument overflow, the bunkers were made of concrete.

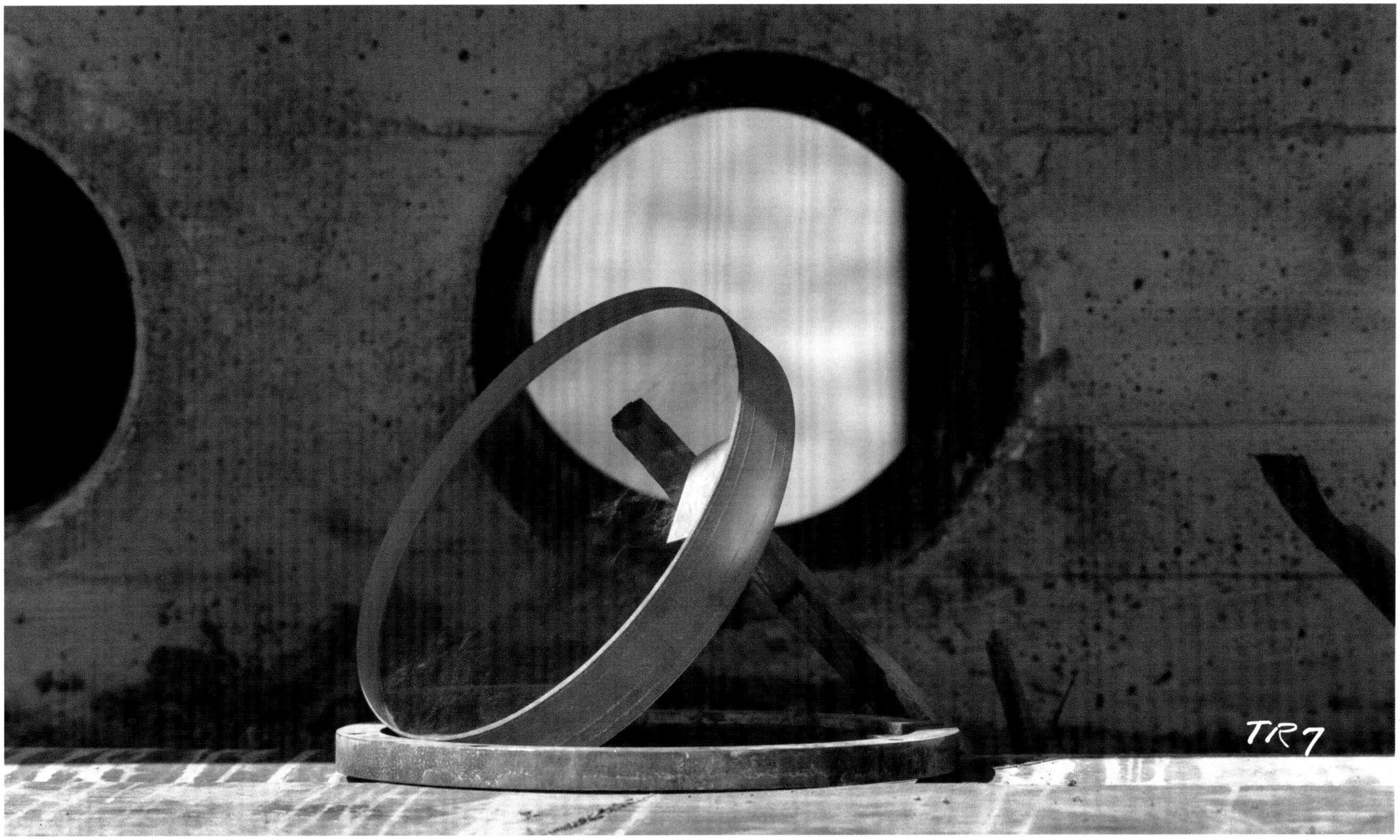

The portholes, made of thick, bulletproof glass, were home to Fastax motion picture cameras and Fairchild aerial cameras. Made to capture the "big picture," the Fairchild K-17B was already in common use by the army for bomb damage assessment. The unit shown at right, inside the north 10,000 camera bunker, was automated to take a photograph every three seconds until its 200 feet of large-format film ran out.

On smaller, sixteen-millimeter film, the Fastax cameras could provide continuous, moment-to-moment imaging, creating a valuable record of the early part of the explosion when it would shine too brightly and expand too quickly to be seen in any detail by the human eye.

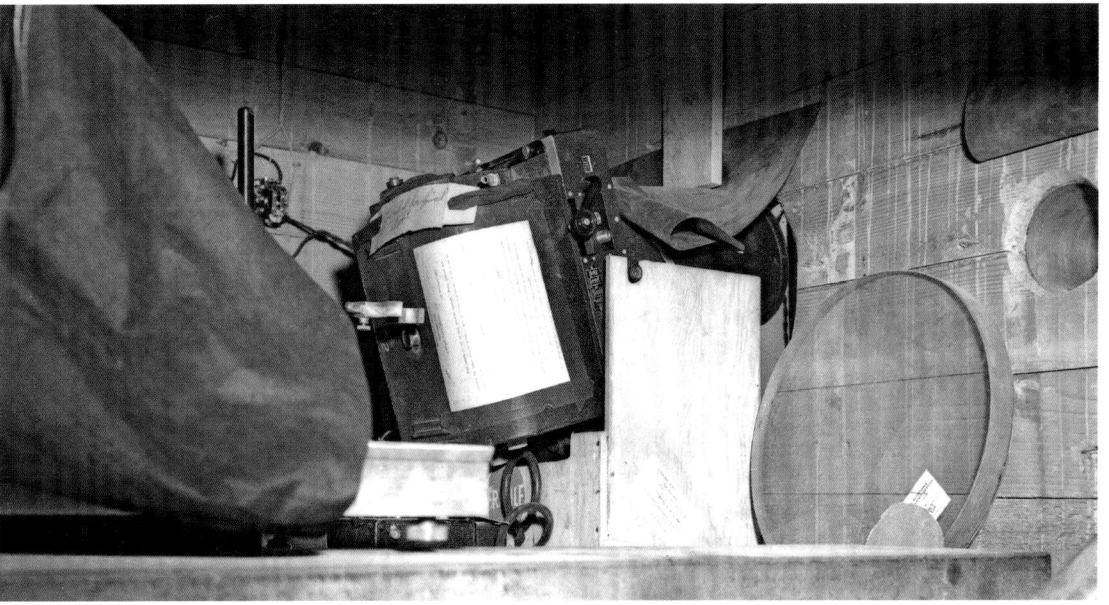

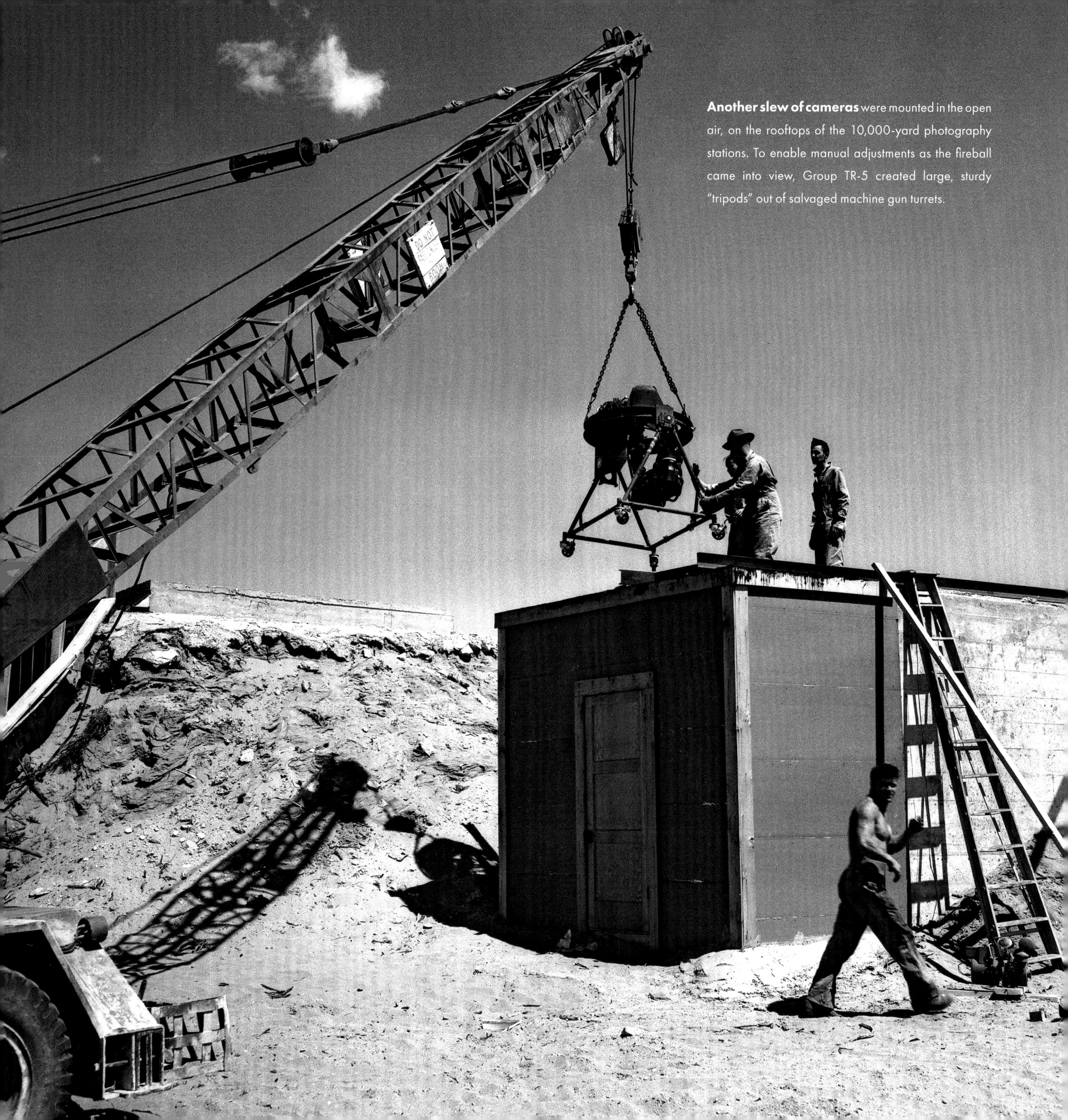

Another slew of cameras were mounted in the open air, on the rooftops of the 10,000-yard photography stations. To enable manual adjustments as the fireball came into view, Group TR-5 created large, sturdy "tripods" out of salvaged machine gun turrets.

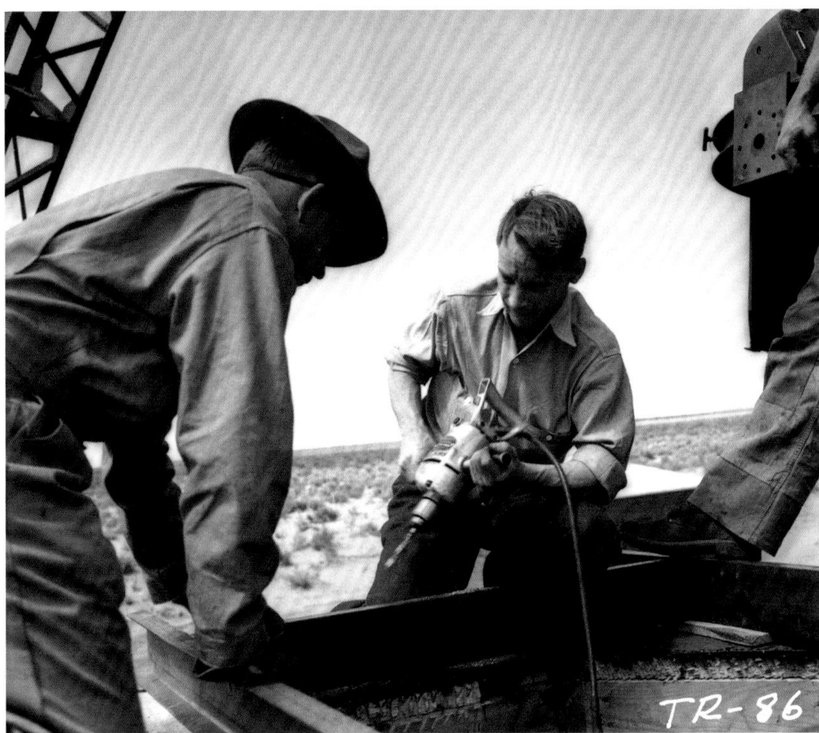

Each 950-pound turret was carefully set by crane onto the rear edge of its respective bunker, where mounting brackets had been preinstalled. Above, Ben Benjamin assists Berlyn Brixner, head of photography, with preparatory work.

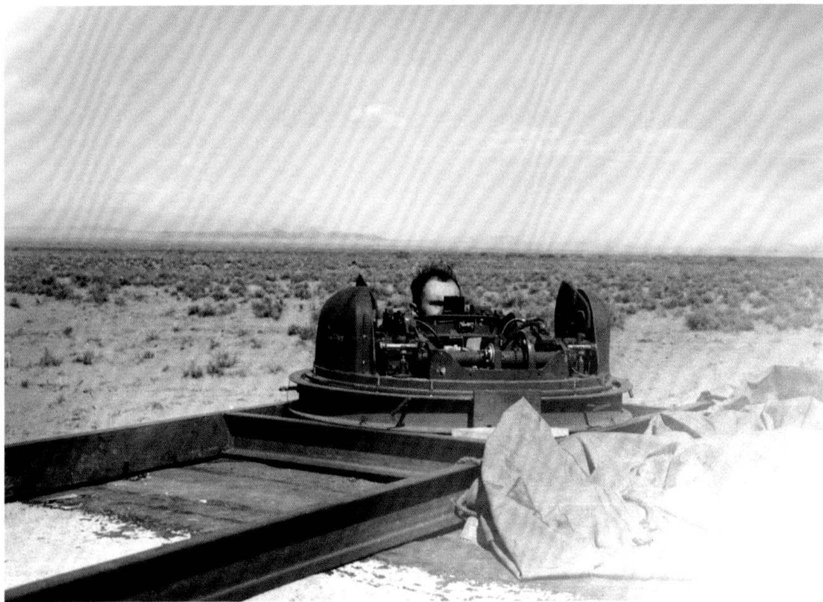

If all went according to plan, the attached cameras would be activated five seconds before the detonation by a timing signal. Brixner and Mack would then control the positions of the turrets by hand.

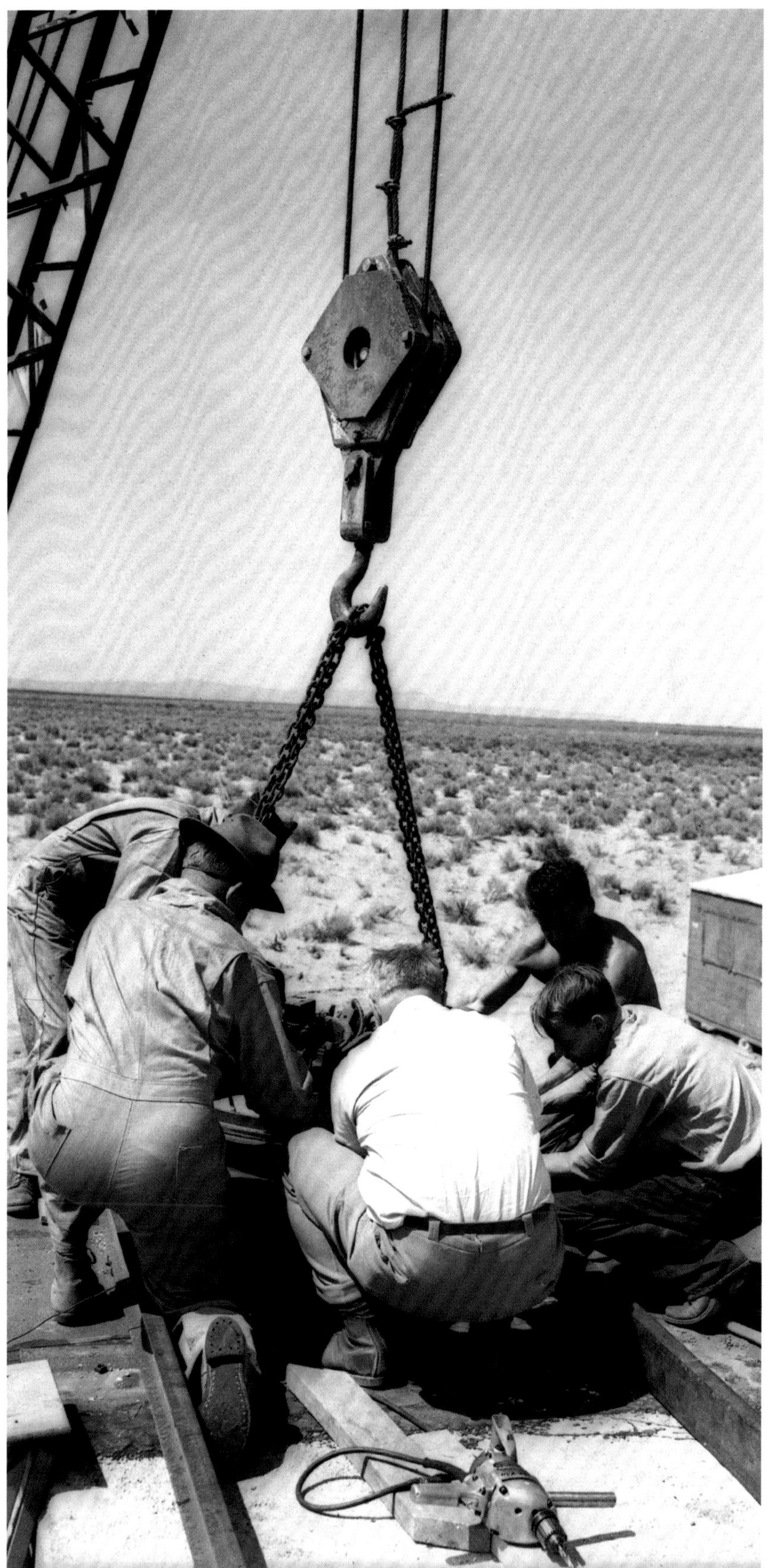

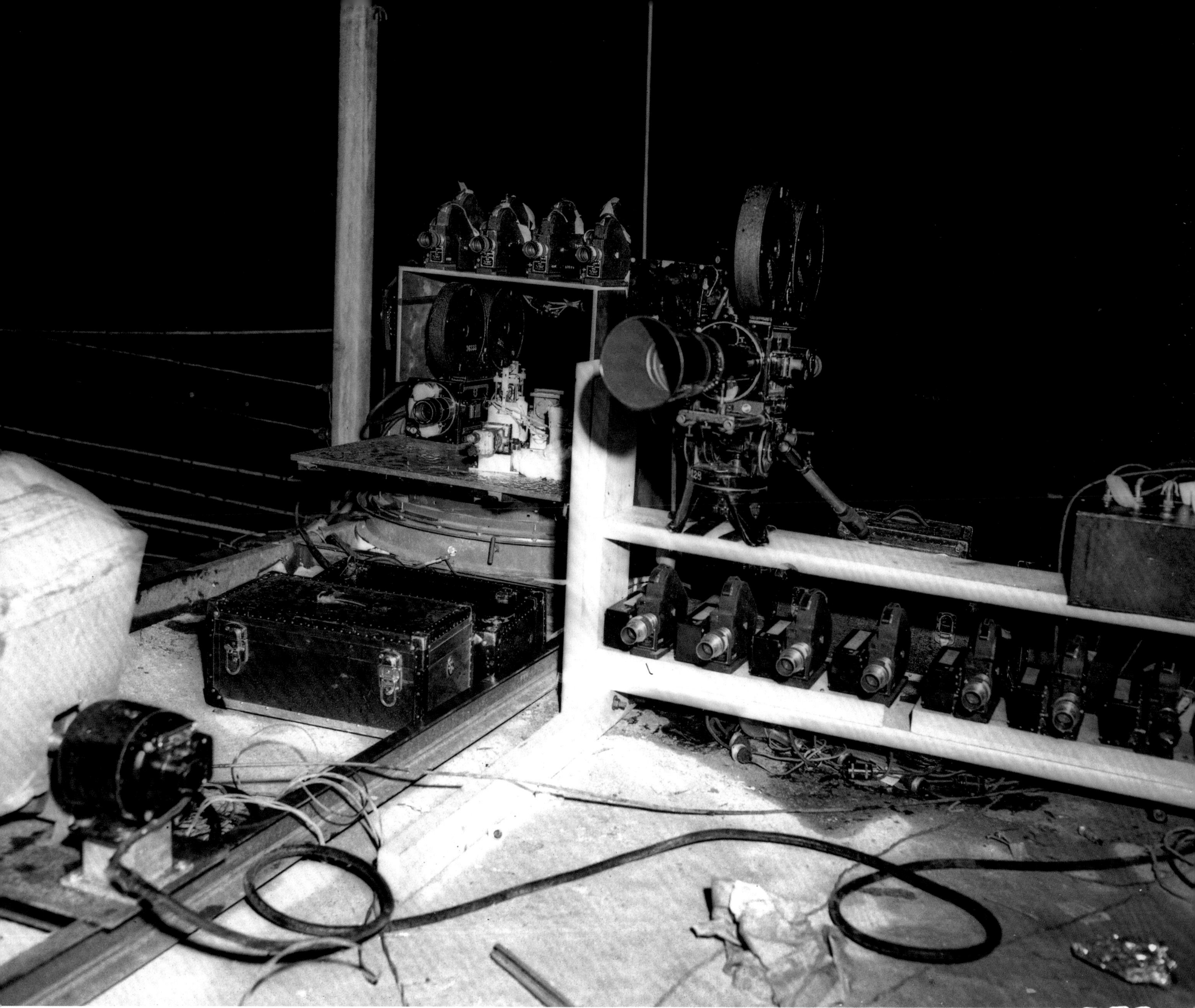

In this nighttime view of the north 10,000 rooftop, the prominent camera with large reels is a Mitchell thirty-five-millimeter—the gold standard for filming Hollywood motion pictures during the 1940s. The smaller cameras on top of the turret and in fixed positions on the low shelves are Ciné-Kodak Model Es, loaded with Kodachrome color film.

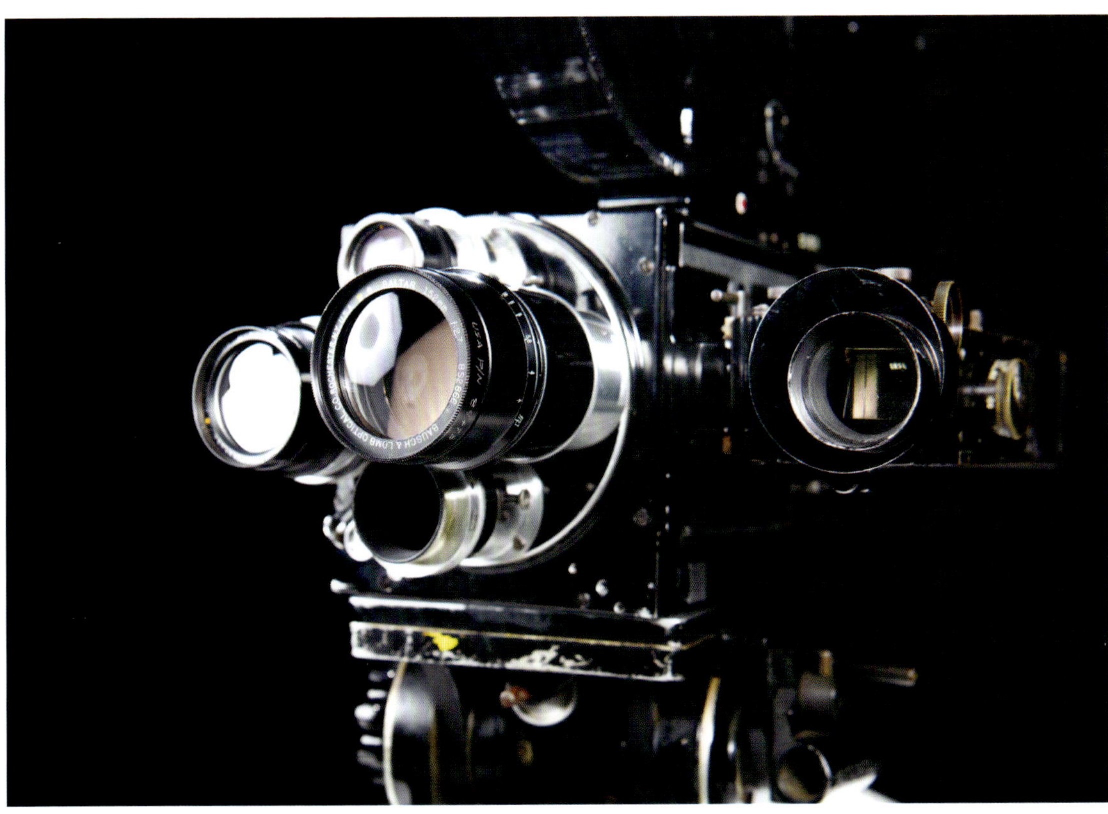

The Mitchells used to shoot the Trinity test were no-frills "government camera" models made for the military. Two were located on the north photography bunker, while another two—one of which was damaged by rainwater and failed to run during the test—were on a matching turret at west 10,000.

The three working Mitchells sported lenses of three different lengths—the longer the lens, the narrower the field of view and the higher the magnification. At north 10,000, the three-inch and eighteen-inch lenses were meant to capture broad and medium views, while the two-foot lens at west 10,000 would take a close-up of the fireball.

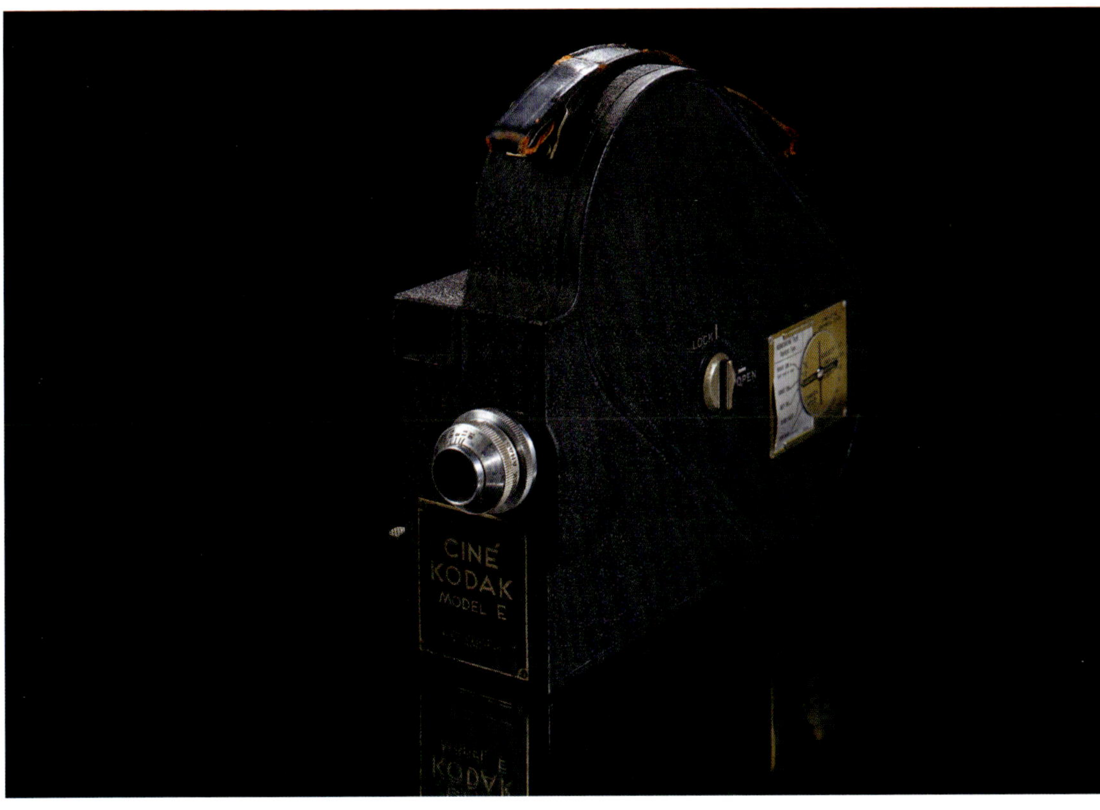

The Model E, with its curved film casing, was designed to accommodate an operator peering into the eyepiece lens while wearing a brimmed hat—a detail Ben Benjamin might have appreciated.

The fifteen Model Es used to record the test reflect an ever-present uncertainty about the effects and yield of the weapon, especially the intensity of the light. Even with the collective input of the world's leading theoretical physicists, decisions as simple as exposure settings for a basic camera were a matter of guesswork and called for redundancy to prepare for a broad spectrum of conditions.

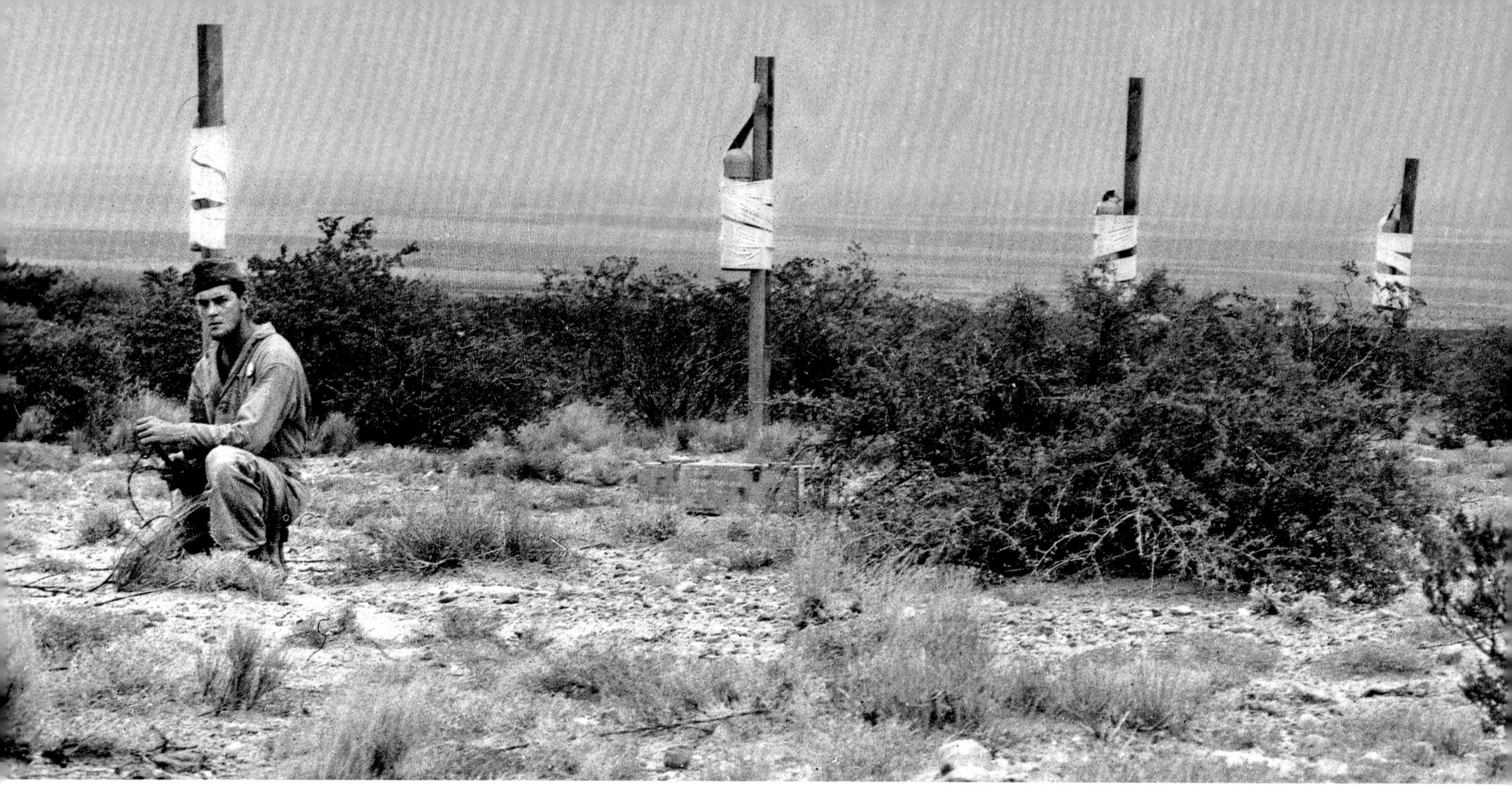

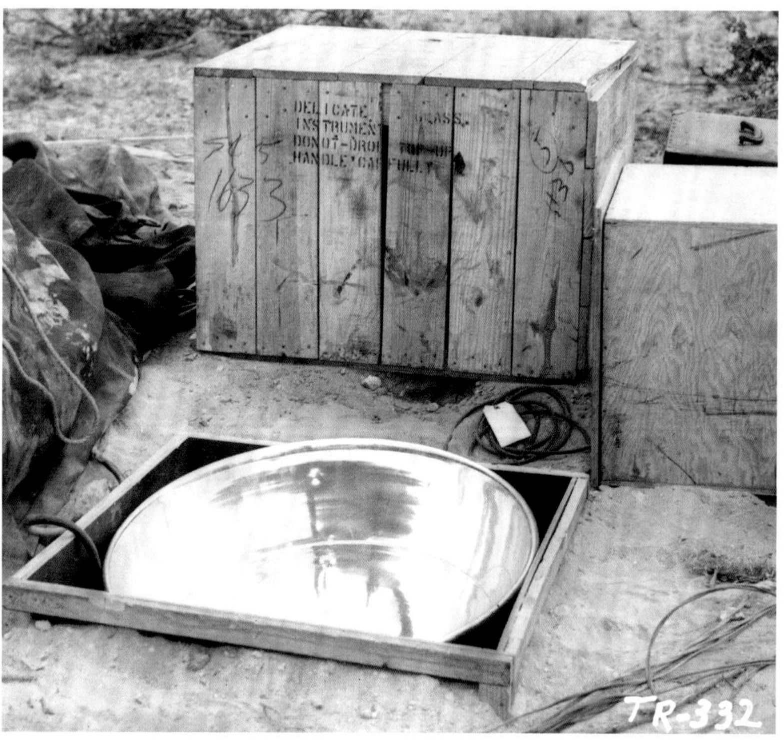

Many difficulties were inherent to filming what might be the brightest and strongest human-made explosion ever seen. And yet, TR-5 faced still another challenge: security measures. The shot was scheduled for four o'clock in the morning in the hope that no more than a few, scattered individuals would be awake to see anything odd in the New Mexico sky. This also meant that as the fireball gave way to a plume of smoke, the explosion's aftermath might fade from view, blending with the backdrop of the dark sky.

To provide a source of light, the group set up 100-pound flash bombs timed to go off every two minutes. The charges were fixed to ground stakes, with flashers (*left*) set up beneath them. A trial run of flash bomb–assisted photography, filmed from west 10,000, is shown at right.

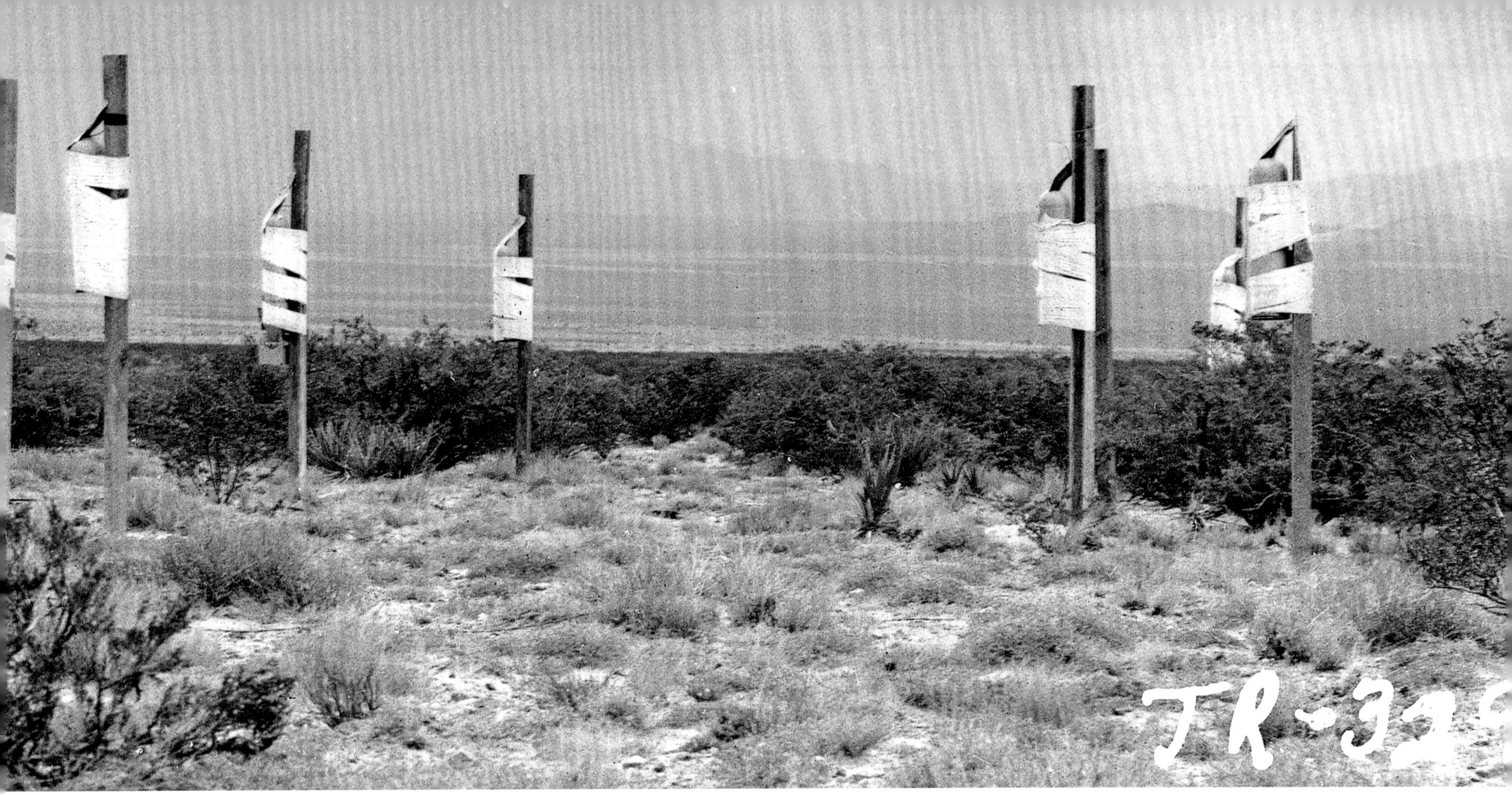

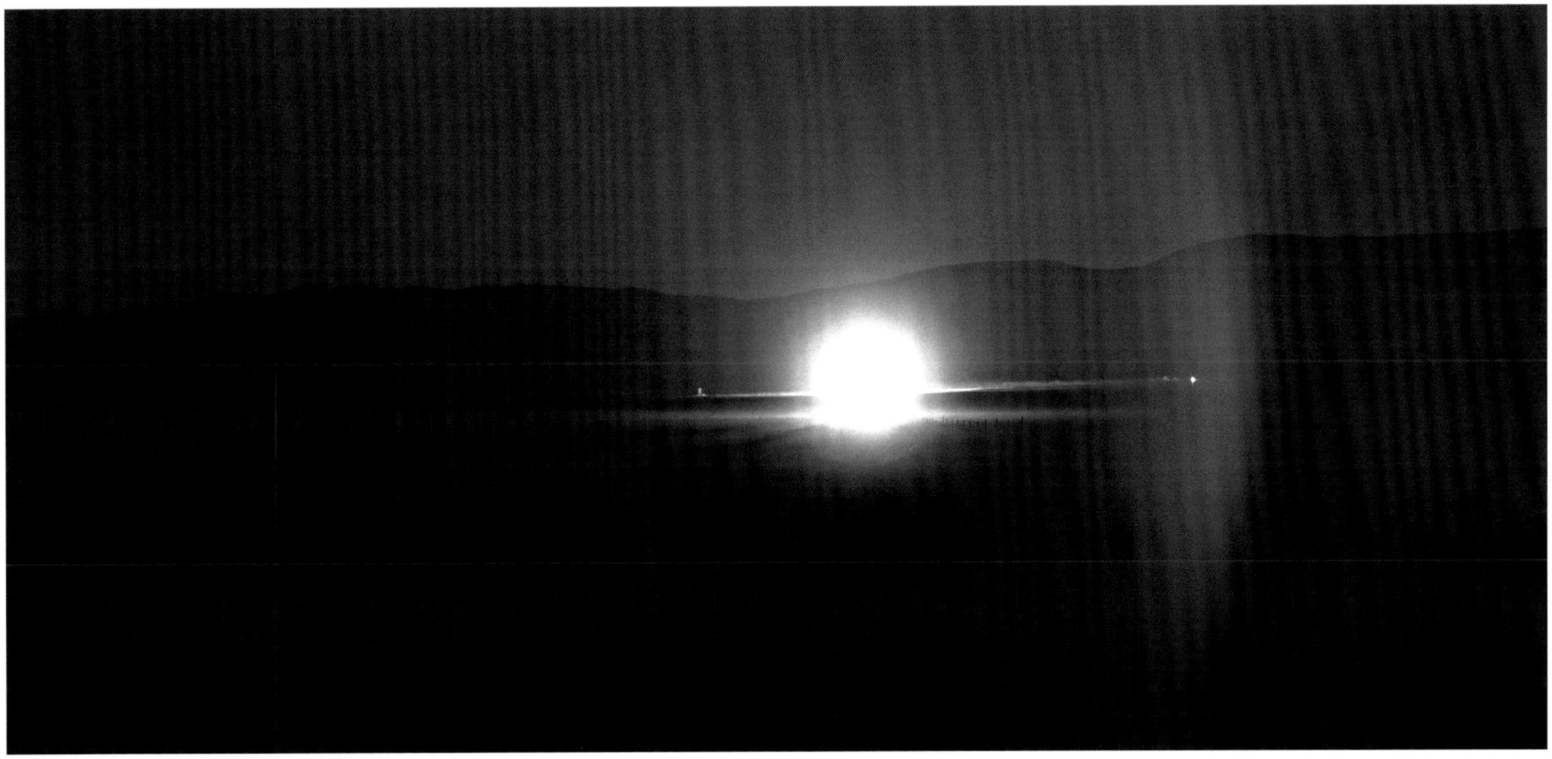

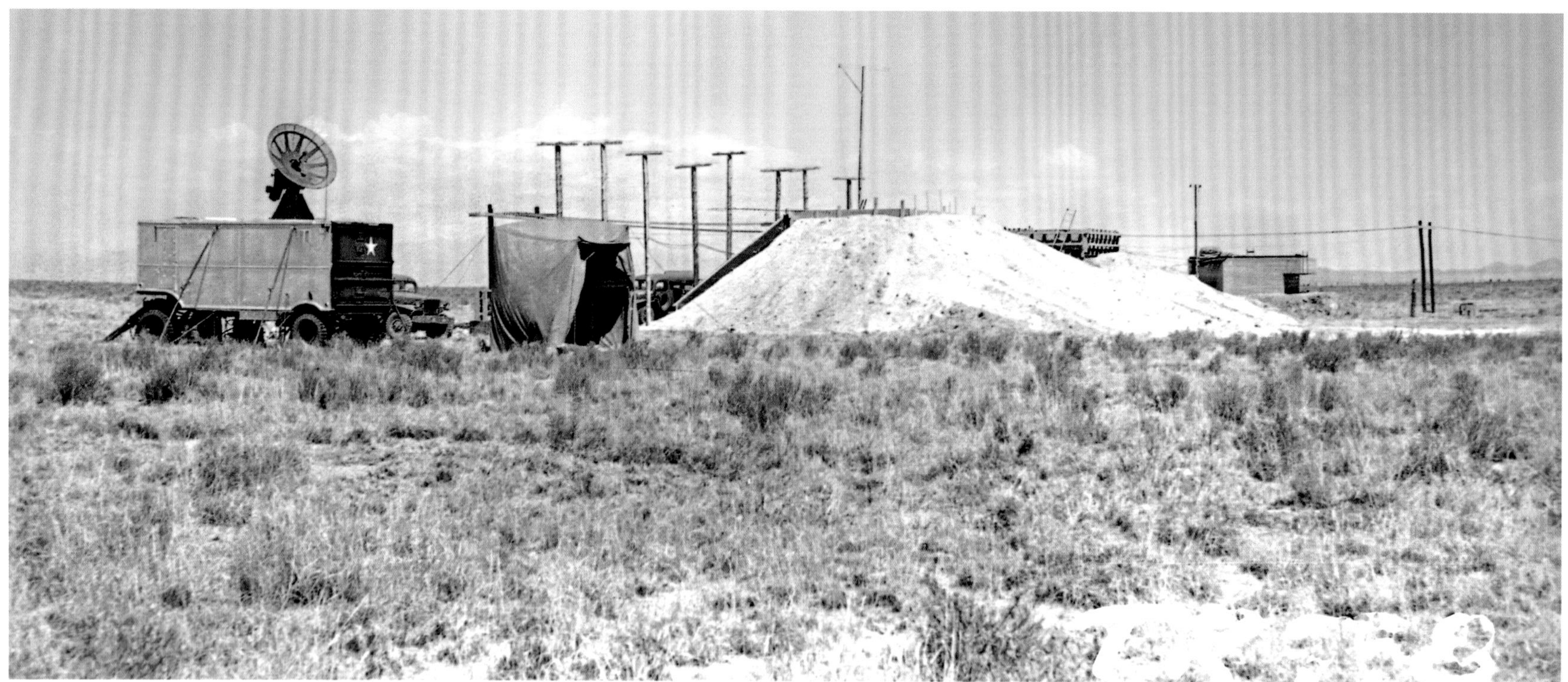

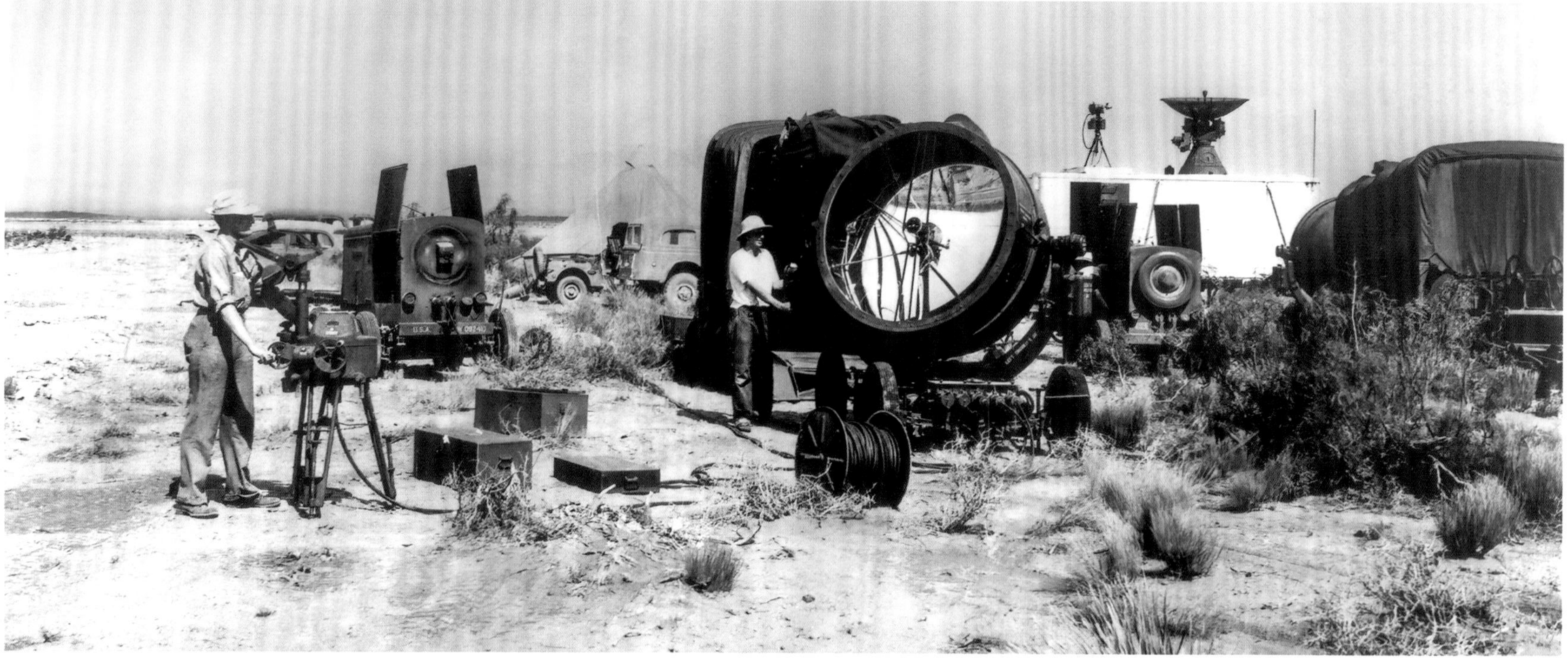

In addition to photographing the fireball and the ensuing cloud, scientists needed to trace the path that the cloud would follow as it drifted away and began to disperse. For this, crews set up SCR-584 radar stations at west 10,000 (*top*) and south 10,000, supplemented by longer-range army searchlights with optical finders to triangulate the cloud's location for as long as possible.

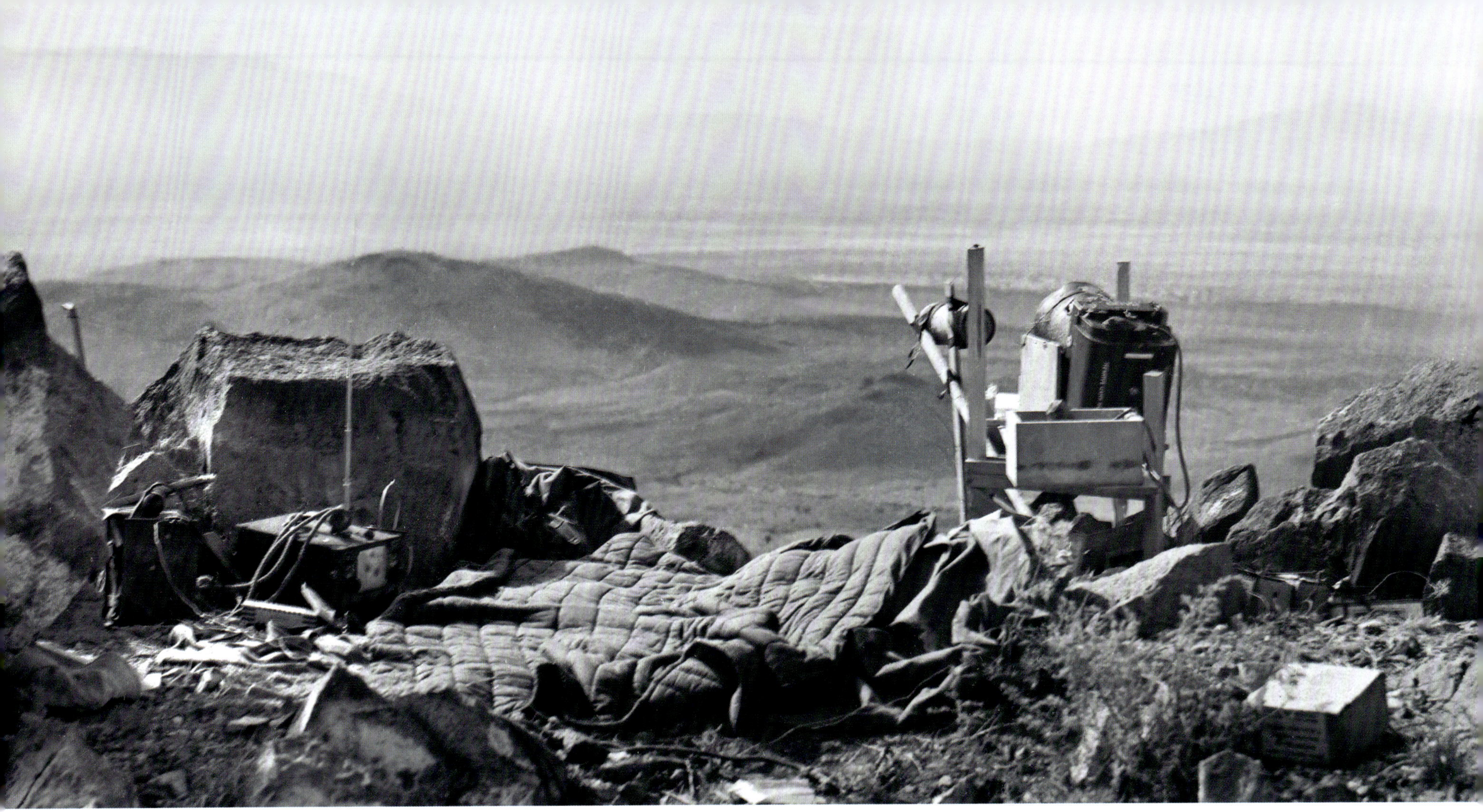

Two pairs of Fairchild K-17Bs would further assist with documenting the fireball and cloud, providing the broadest view from the most distant on-site observation points. One was set up 33,000 yards northwest of ground zero on Campana Hill, at what was called the "distant station" (*above*). Another pair was set up 42,000 yards to the southeast. Operating in synchrony, each of the stereoscopic pairs was programmed to take overlapping aerial photographs, creating the illusion of a three-dimensional image.

The Fairchild cameras were enormous, weighing up to seventy-five pounds depending on the length of the lens cone. Their large-format film is shown at right next to the thirty-five-millimeter film used by the Mitchell cameras. (Photo by Jim Moye, courtesy of Lawrence Livermore National Laboratory.)

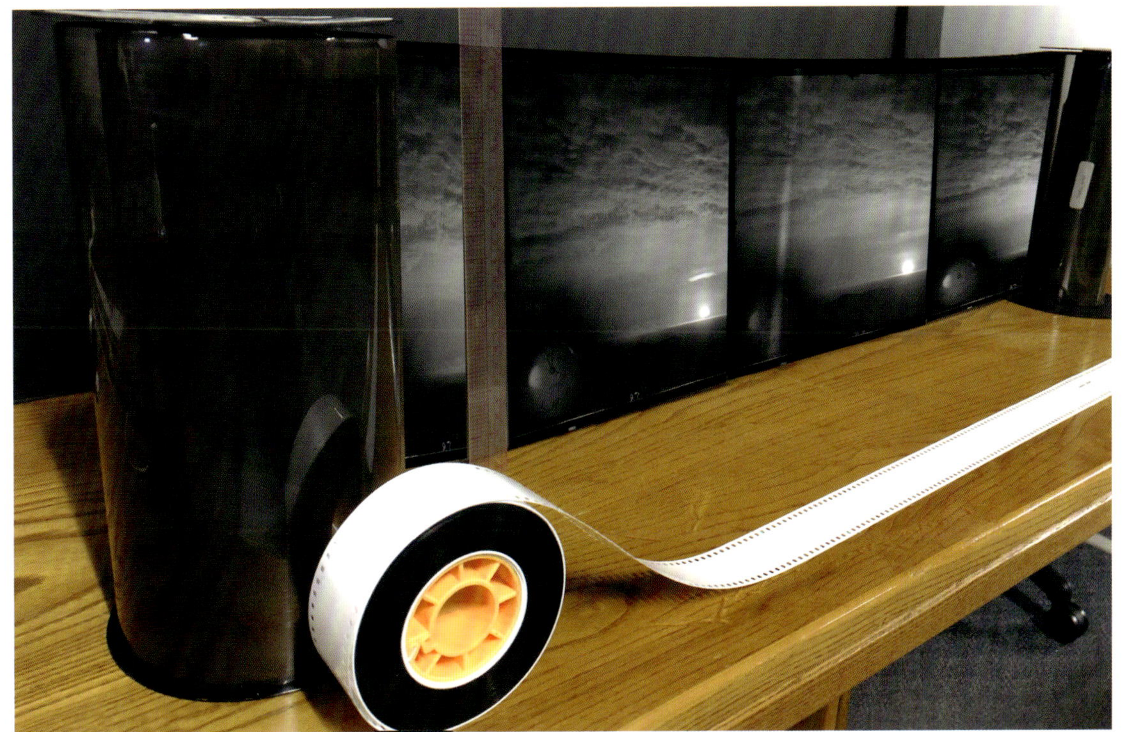

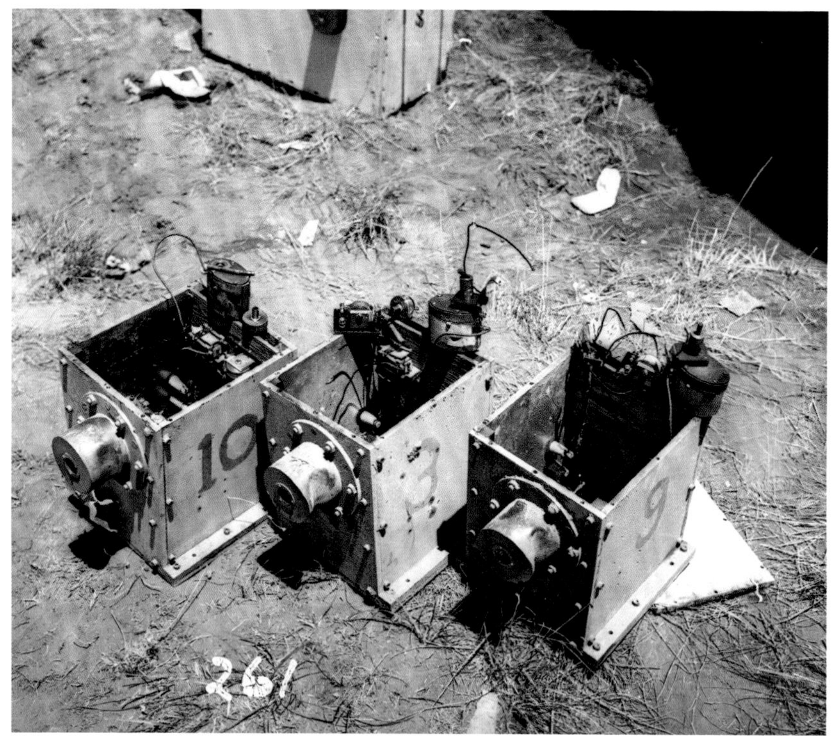

In addition to its task of recording the blast itself, Group TR-5 provided photographic services, upon request, for the other groups at Trinity during setup for the test shot. Photographer Ernest Wallis worked with scientists to visually document their many diagnostic tools. Some of the most important experiments concerned the destructive shock wave that would spread across the basin.

The pressure of the airborne portion of the shock wave was of particular interest because it would help scientists determine for the military the ideal height at which to detonate a nuclear weapon. One device used to obtain this essential measurement was the impulse gauge, configured to squirt a pressure-dependent amount of water out of a piston as the shock wave passed by. Instead of transmitting a signal, which would be vulnerable to interference from radiation, these fully mechanical devices (some placed as little as 350 yards from the blast) etched pressure readings onto a built-in disc spun by a repurposed windshield-wiper motor. Just as cameras were set up to account for different possible lighting conditions, the impulse gauges were set to respond to a wide range of pressures, from 2.5 to 150 pounds per square inch.

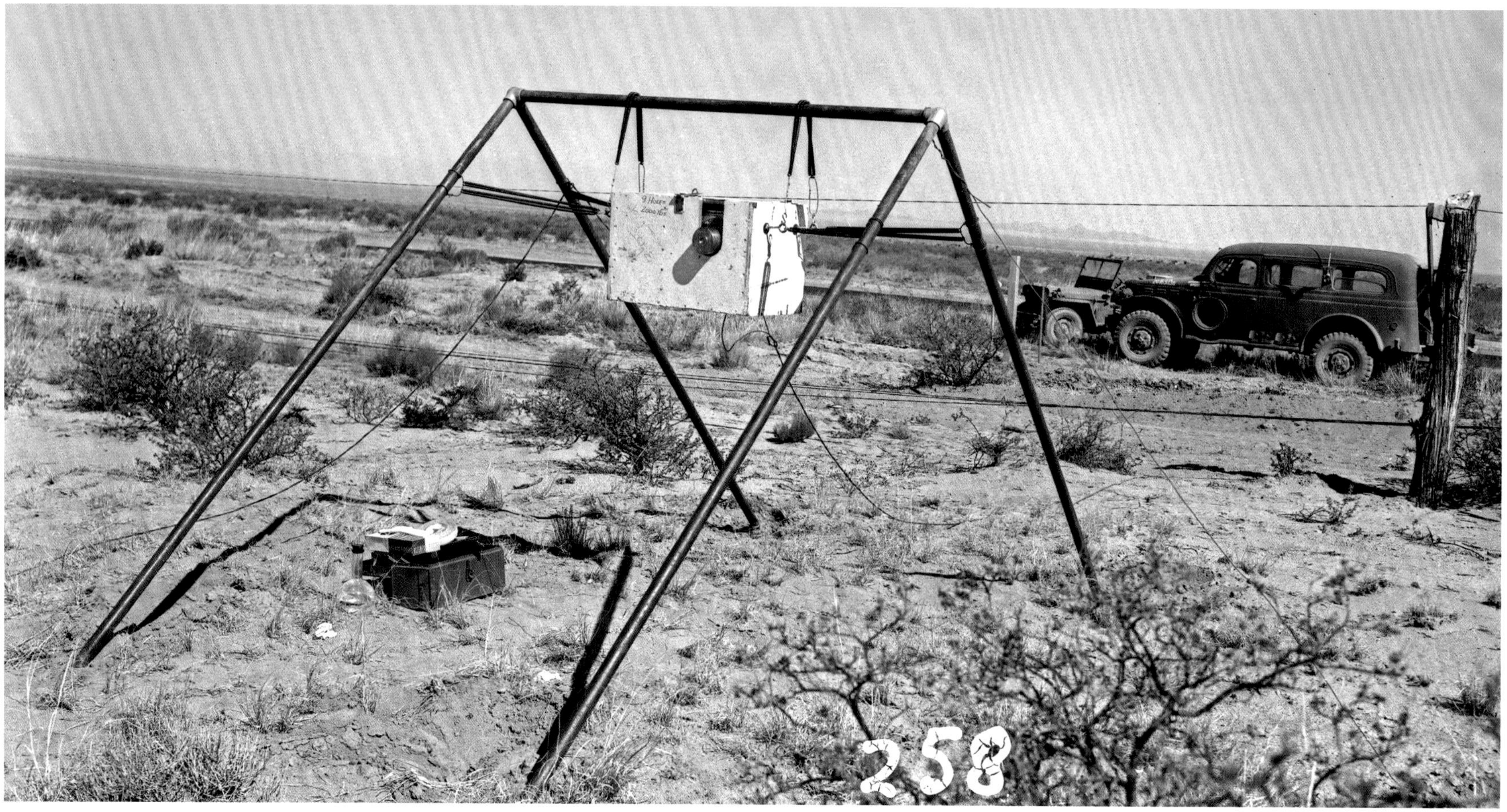

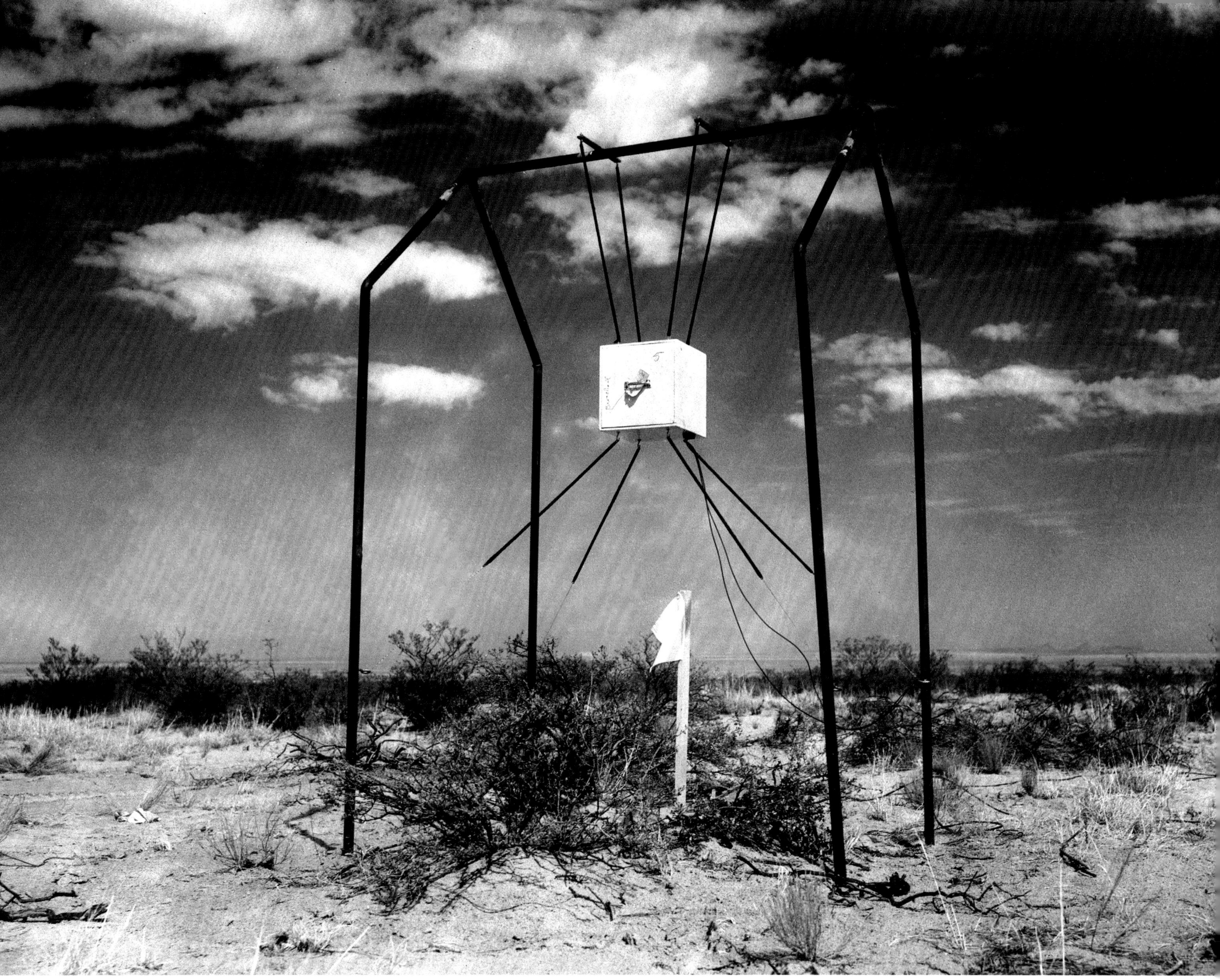

The shock wave's speed was an important marker as well. In simple terms, a shock wave is a sound wave traveling and vibrating faster than the speed of sound. Exactly how much faster depends on the yield of the blast: as yield increases, the speed of the shock wave increases proportionately. To take advantage of this known relationship as a means of calculating the Gadget's energy yield, scientists set up what were called excess velocity experiments. In one elaborate version, an explosive called pentolite—which produces a shock wave of known velocity—was to be detonated near ground zero a few seconds ahead of the Gadget. Dangling plywood boxes containing microphones would register the arrival of the pentolite wave (providing a baseline) and then the subsequent arrival of the Gadget's shock wave—enabling scientists to deduce the unknown velocity by comparison.

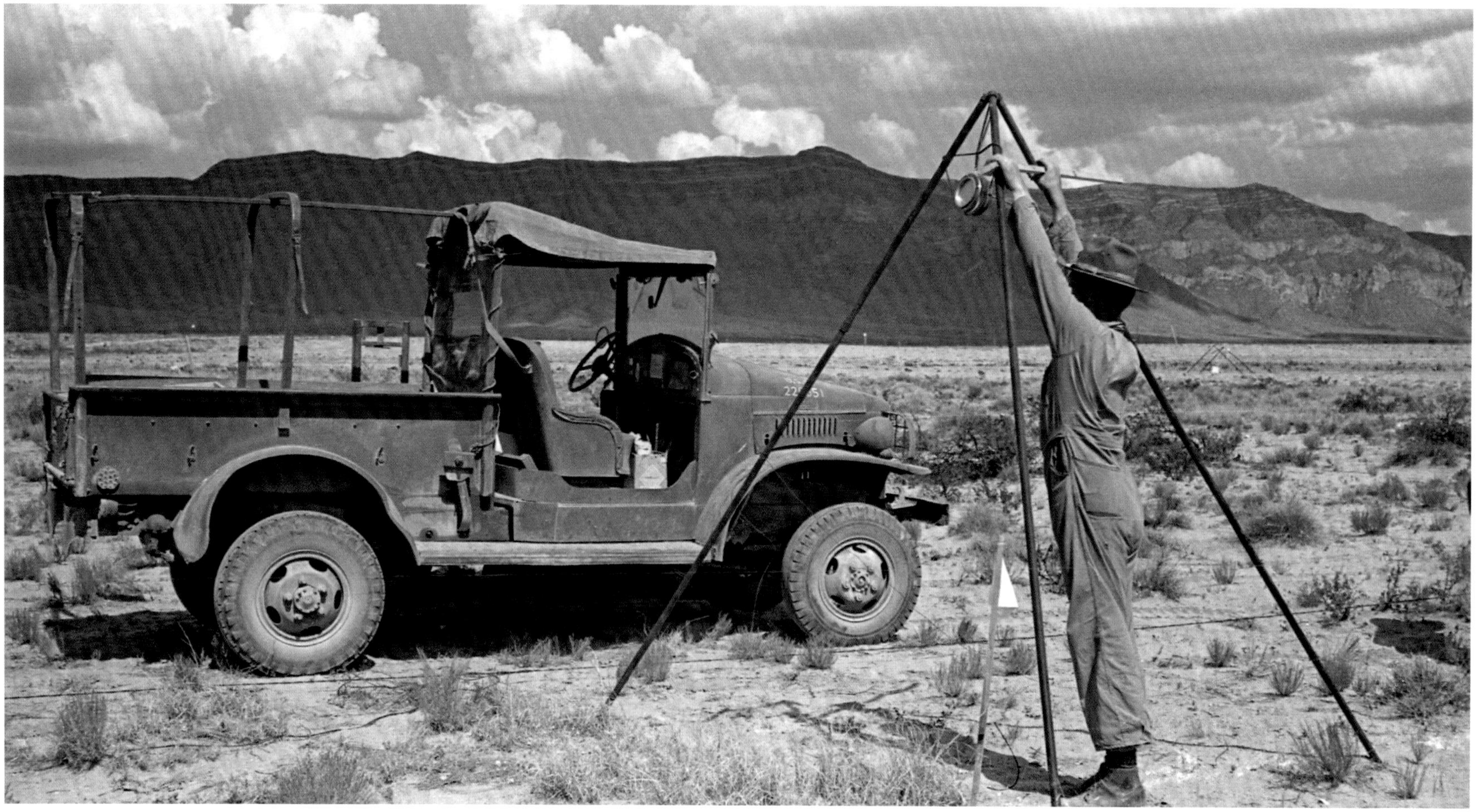

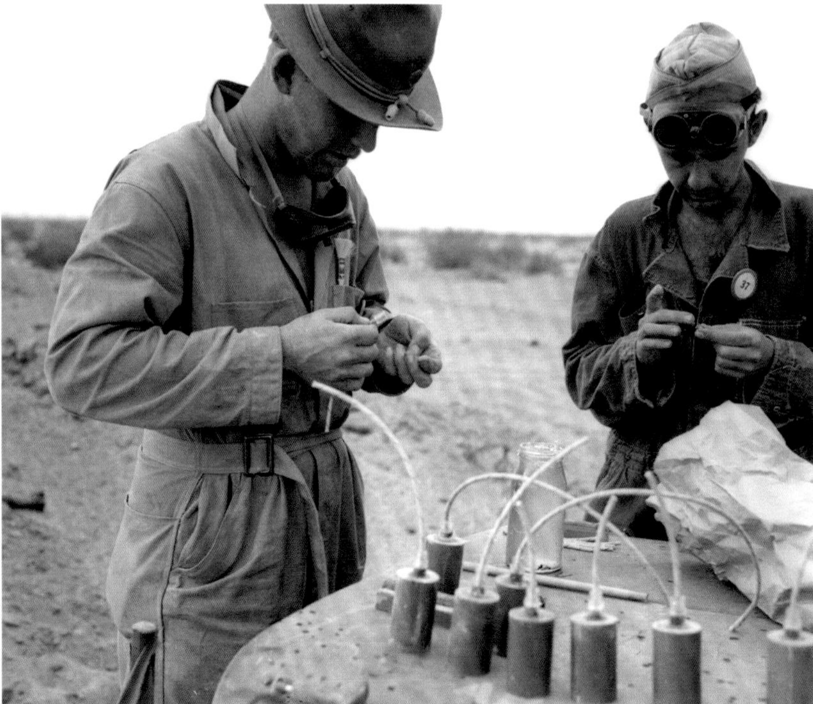

Other methods of tracking the shock wave involved simple devices designed to interact with the wave in a visible way. Two such experiments relied on a Trinity invention called the mechanical shock switch—a trigger consisting of a brass rod and thin piece of copper with a small gap between them. When the shock wave arrived, the copper would recoil and contact the brass rod, closing a circuit tied into a detonator.

One of the shock switch experiments was arranged by Ben Benjamin and the other by George Economou, an optical systems expert and fellow enlisted member of the Special Engineer Detachment. At left, they are shown setting up charges together.

Benjamin's experiment (*above*), like the hanging microphone boxes, was designed to measure the excess velocity of the shock wave, or how much faster than the speed of sound it traveled. His shock switches, hung from tripods, would trigger the detonation of flash bulbs extending in a straight line from ground zero, while a camera at the west 10,000 photography bunker recorded the successively exploding charges to trace the wave's location.

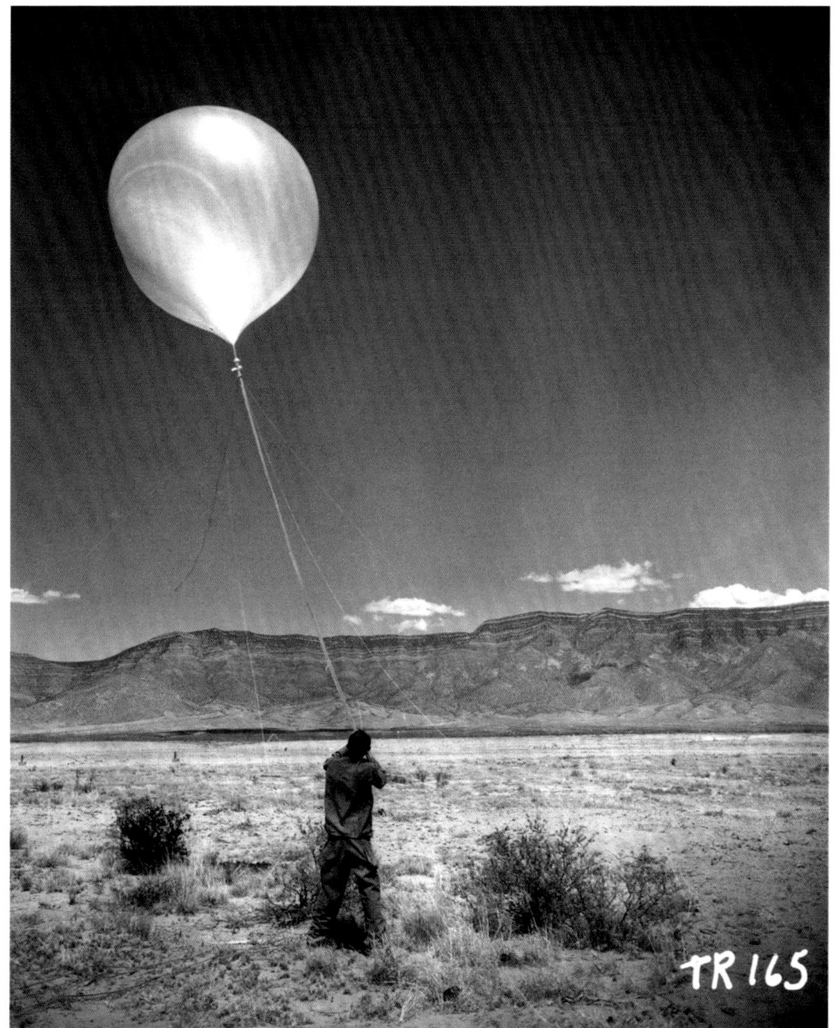

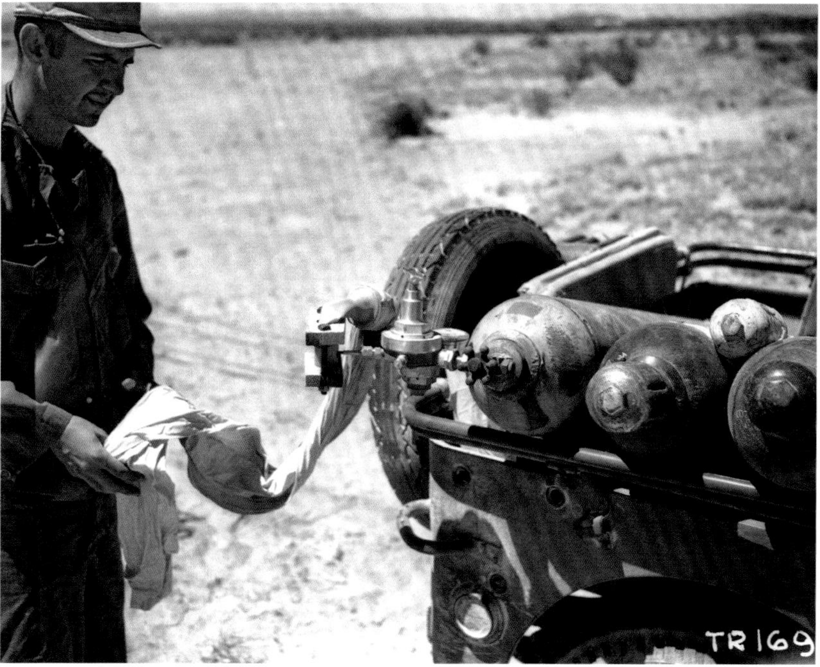

In Economou's experiment, the shock switches were tied into helium balloon–lofted strands of Primacord, a rope-like explosive material that was commonly employed to generate a visual signal with a known time delay. Using a horse stable near base camp as his "laboratory," Economou sprinkled the long strands of Primacord with magnesium flash powder, improving visibility for the Fastax cameras assigned to film the experiment. The arriving shock wave would trip the switches and trigger the flares to ignite.

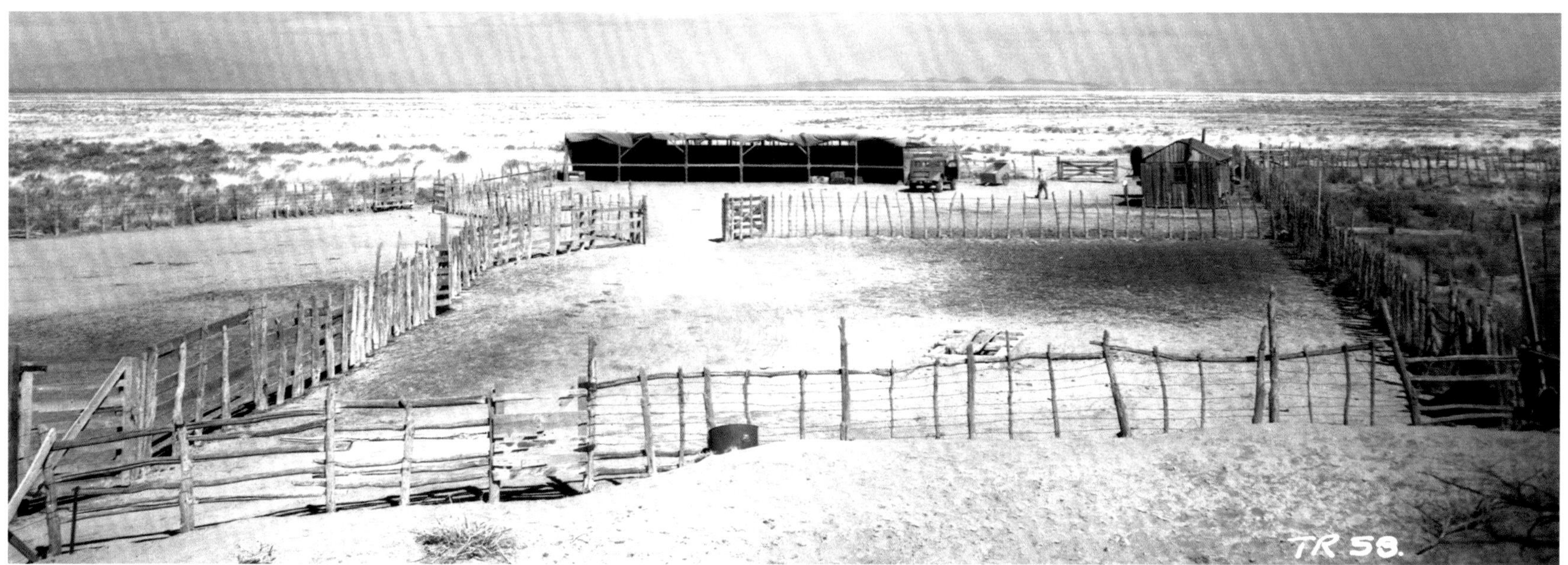

Part of the shock wave would propagate through the ground, instead of the air, and scientists planned to assess its energy by measuring motion of the earth using geophones. Twelve geophones, which transform vibrations into electrical signals, were buried in pits two feet deep at distances of 800, 1,500, and 9,000 yards from ground zero to the north and south.

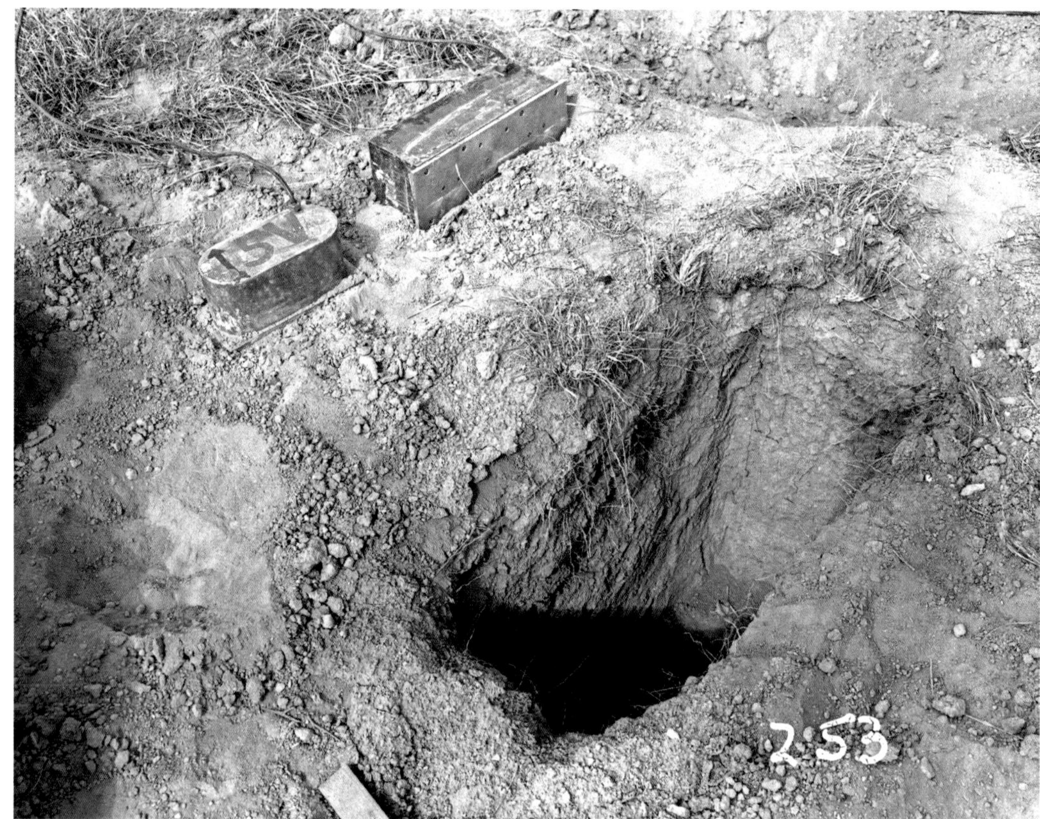

Buried amplifiers, shown at right, would pass along signals from the geophones to the safety of the 10,000-yard personnel shelters.

As a purely mechanical alternative, two-foot-long stakes were driven into the ground at various distances from the blast site to provide a reference point for measuring earth displacement.

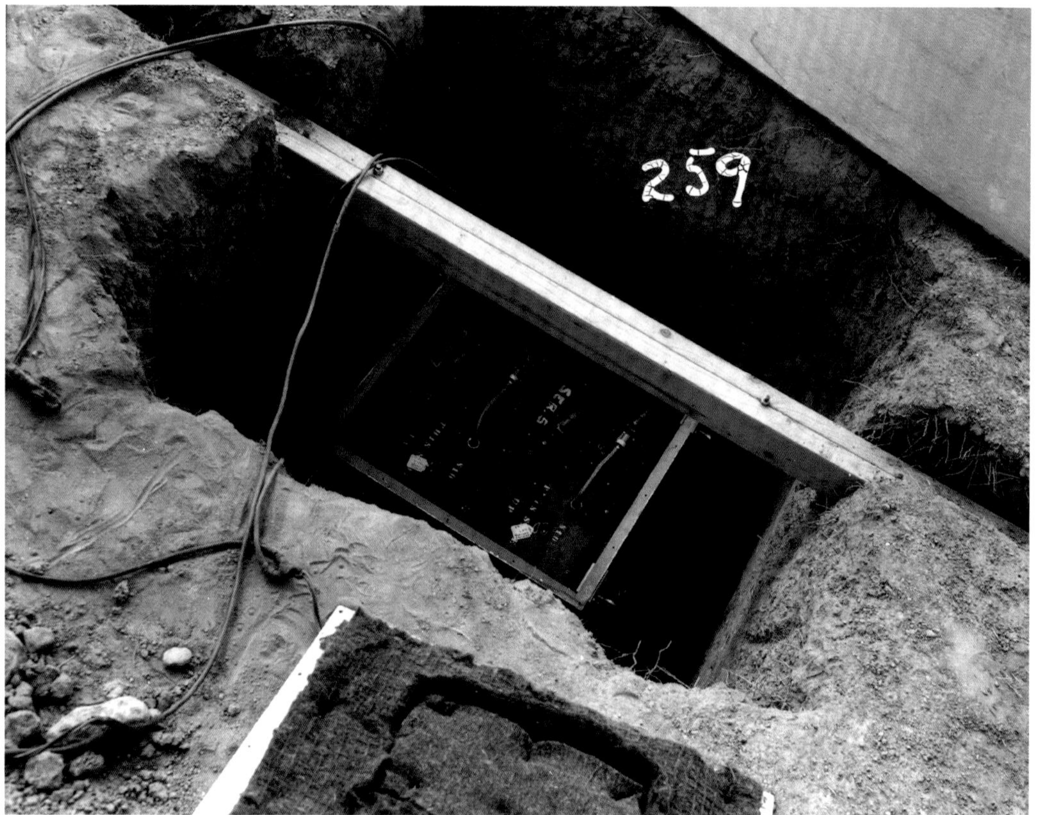

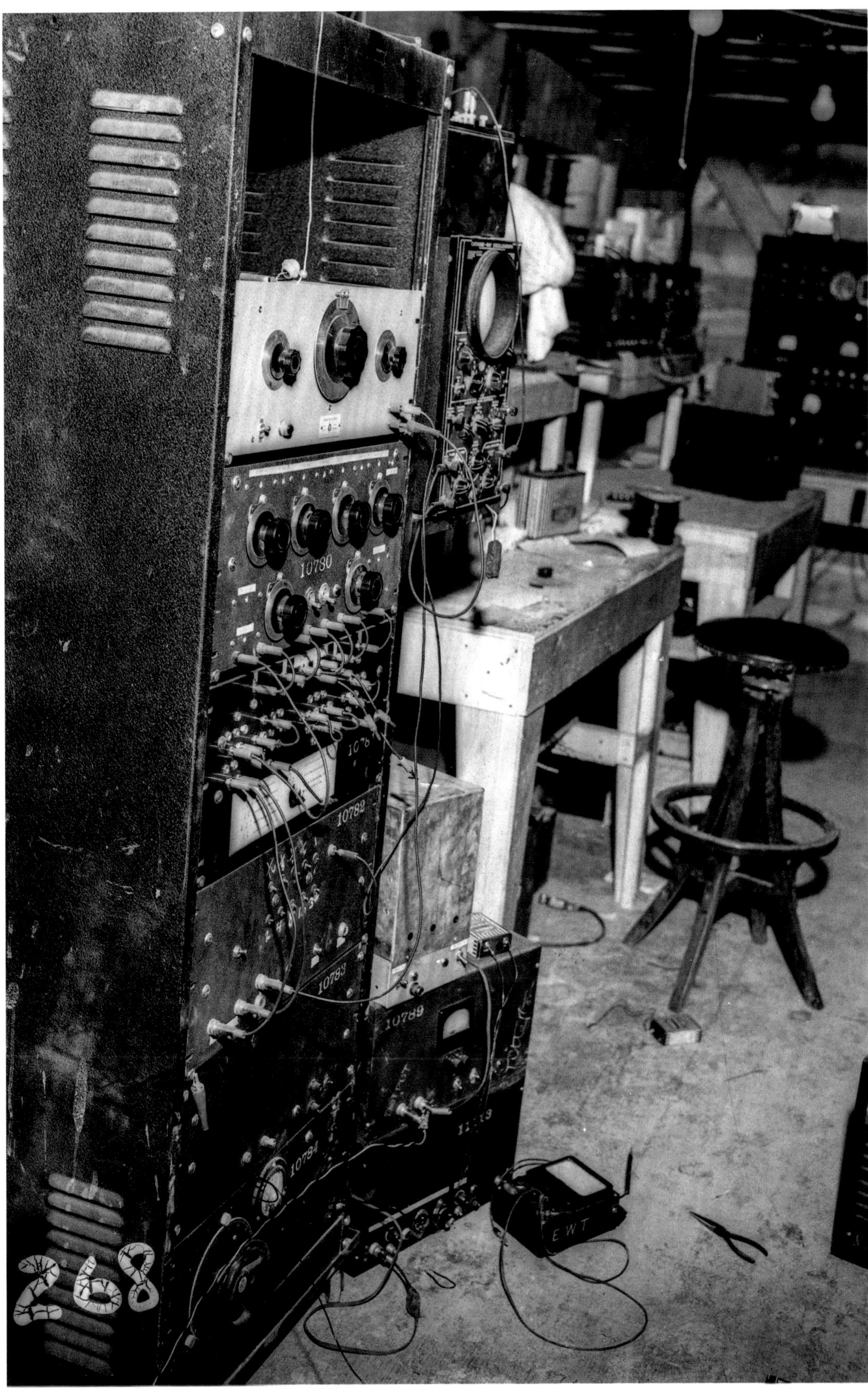

Readings from the geophones and many other instruments would be transmitted in real time during the test to electronic devices called oscilloscopes, which convert electrical signals into readable data, similar in function to a heart monitor.

The image at left shows the inside of the north 10,000 shelter, which housed the geophones' Heiland oscilloscopes; they are set up on the center table.

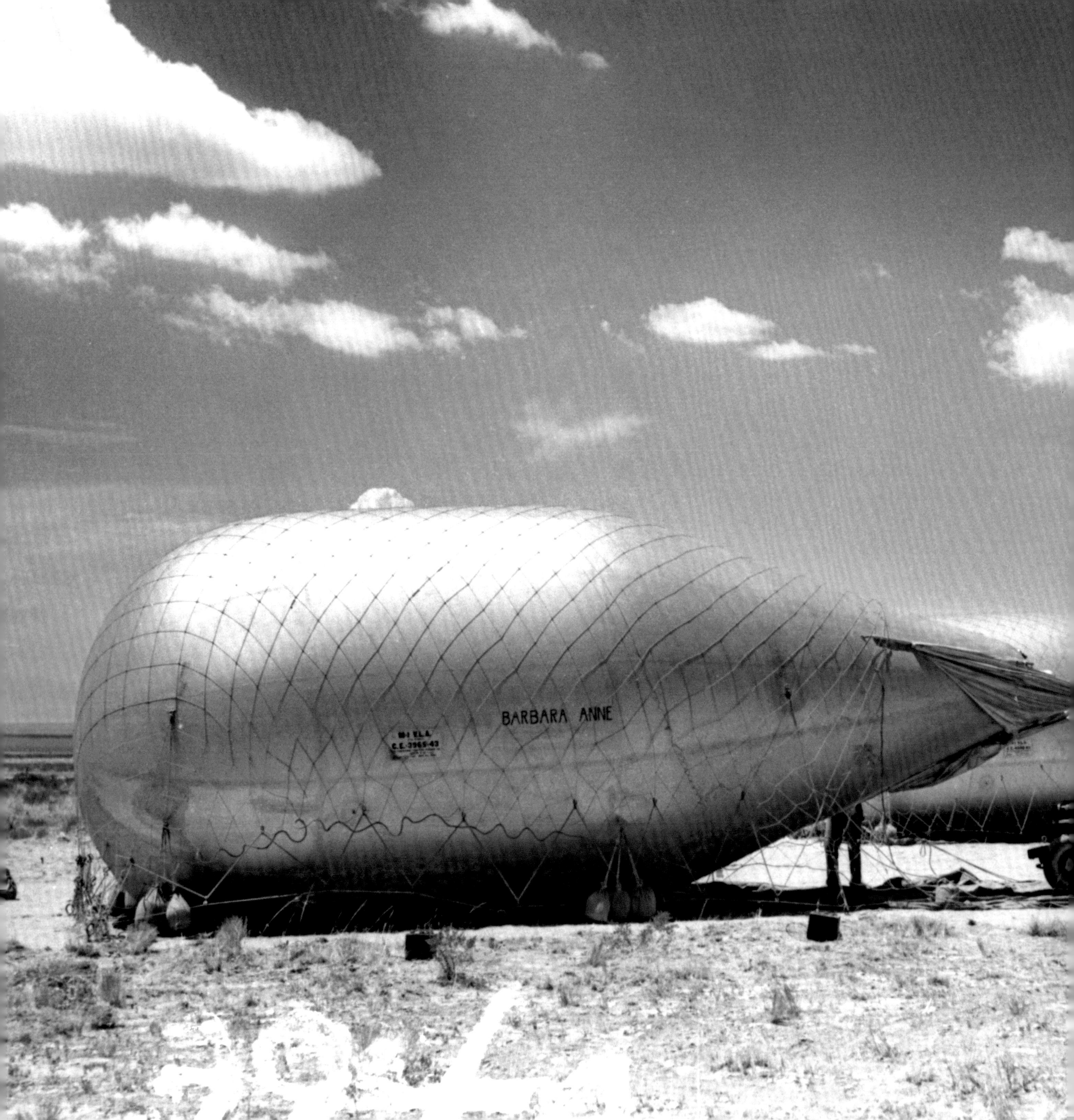
BARBARA ANNE

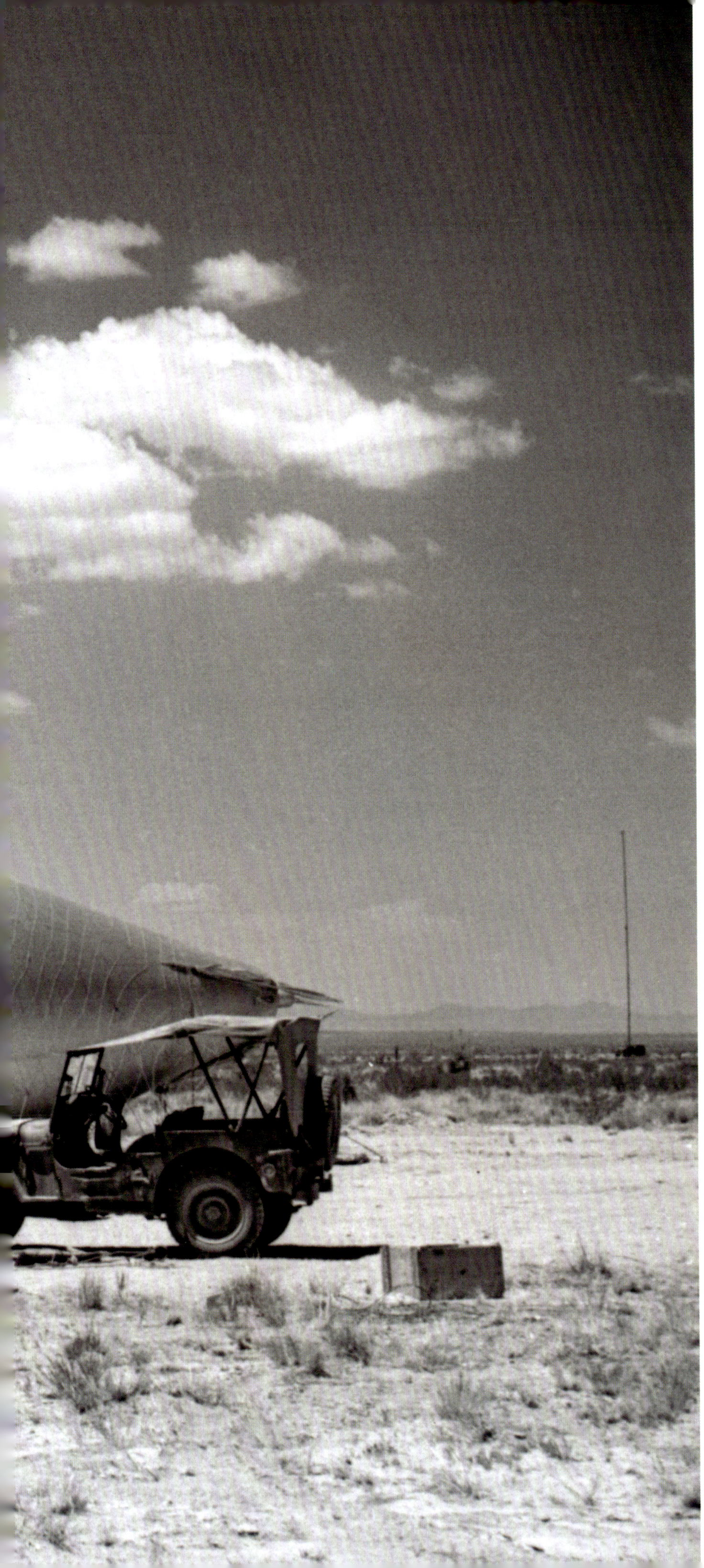

Another measurable product of the nuclear explosion was gamma radiation, an invisible, highly energetic form of light that would emanate from the exploding core. Gamma radiation, though well-known to scientists as a byproduct of nuclear reactions, was expected to reach a peak intensity far above anything ever seen or studied.

Physicist Emilio Segrè designed an experiment to measure gamma ray flux, or the number of gamma rays present over a given period of time—in this case, from 0.01 seconds after the explosion until the shock wave reached his equipment and destroyed it. Gamma ray activity would correlate with the number of fissions that occurred inside the Gadget and, by extension, the amount of energy released during the chain reaction.

The farther away the detection instruments—called ionization chambers—were placed, however, the more gamma rays might be absorbed along the way before reaching them. At the proximity required for accurate measurements, the chambers and attached communication lines would inevitably be destroyed by a successful explosion. There was only a brief window in which data could be transmitted to safety.

Segrè's group determined the best approach to keep their equipment out of harm's way for as long as possible was to suspend it above the ground. For this purpose, they procured a set of barrage balloons, normally used by the military to defend ground forces from air attack.

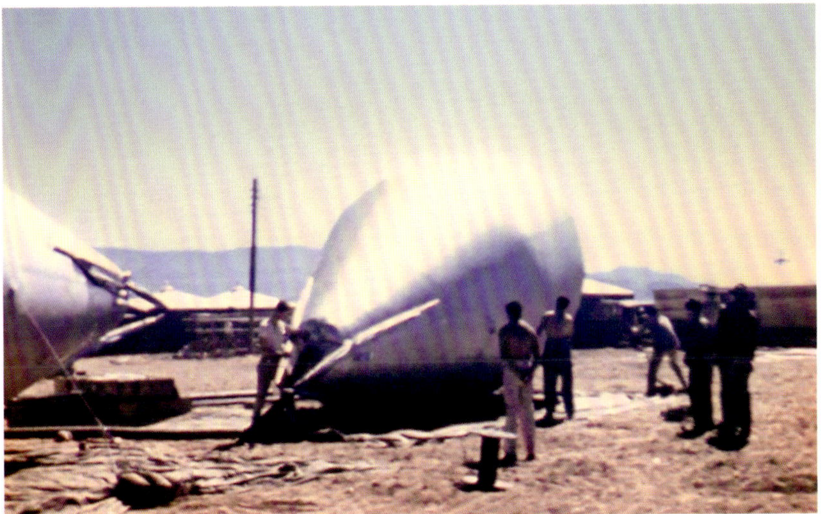

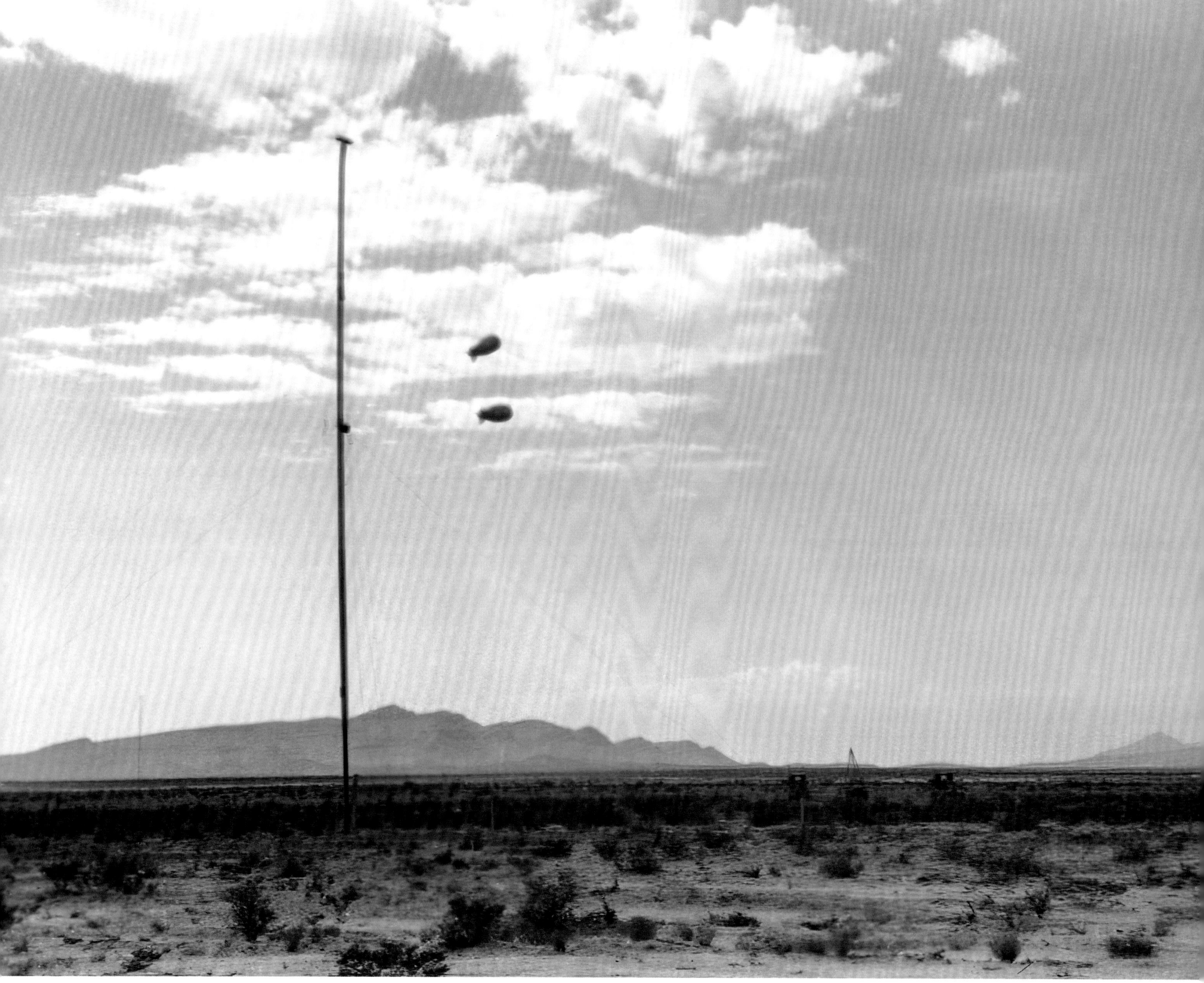

Segrè's barrage balloons were tethered approximately 550 yards from ground zero. The group placed at least one additional ionization chamber on the roof of an instrument bunker at northwest 600—which, ironically, was reported after the blast to have fared better due to protection by the ground. (Photo by Jack Aeby, courtesy of Los Alamos Historical Society Photo Archives.)

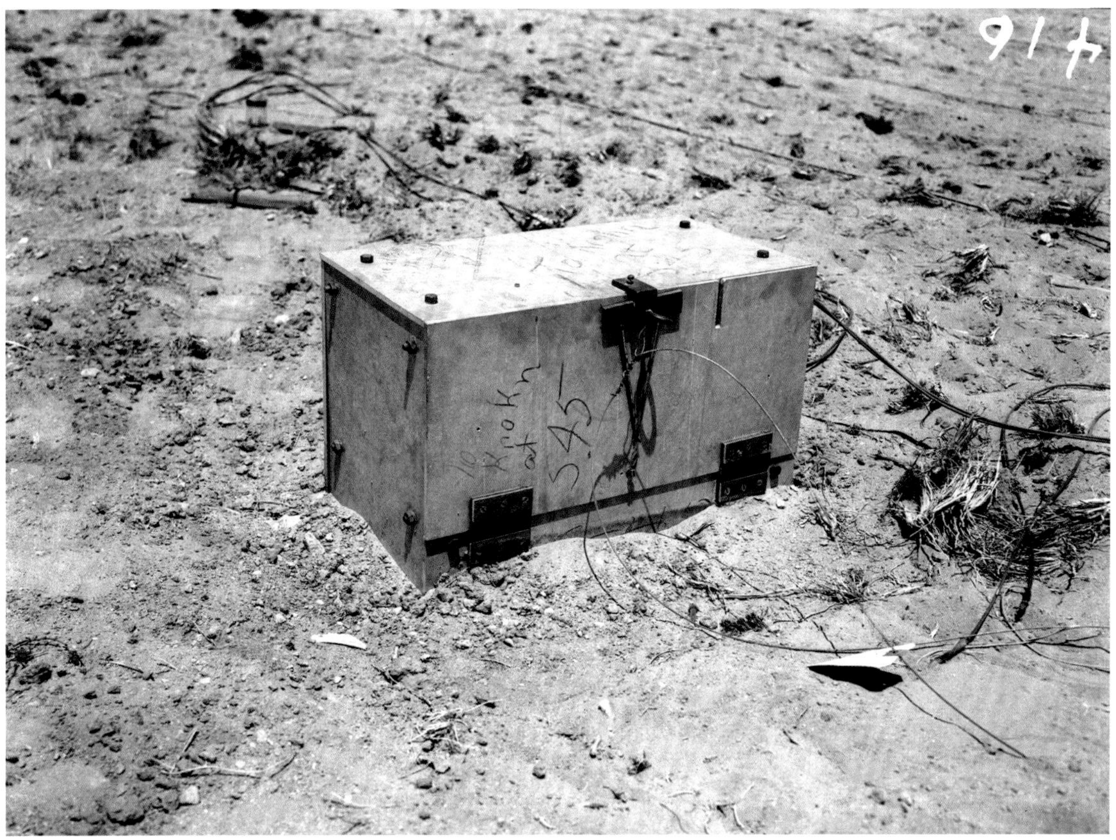

Scientists also used barrage balloons to elevate special cameras designed to count neutrons shooting out of fissioning atoms in the Gadget's core. As the neutrons hurtled outward from ground zero, some would intersect with the special cameras, striking uranium-coated plates inside the devices and causing the uranium atoms to fission. A reel of transparent tape was pulled between the plates, like film, to catch fission fragments, thus creating a timed record of neutron activity.

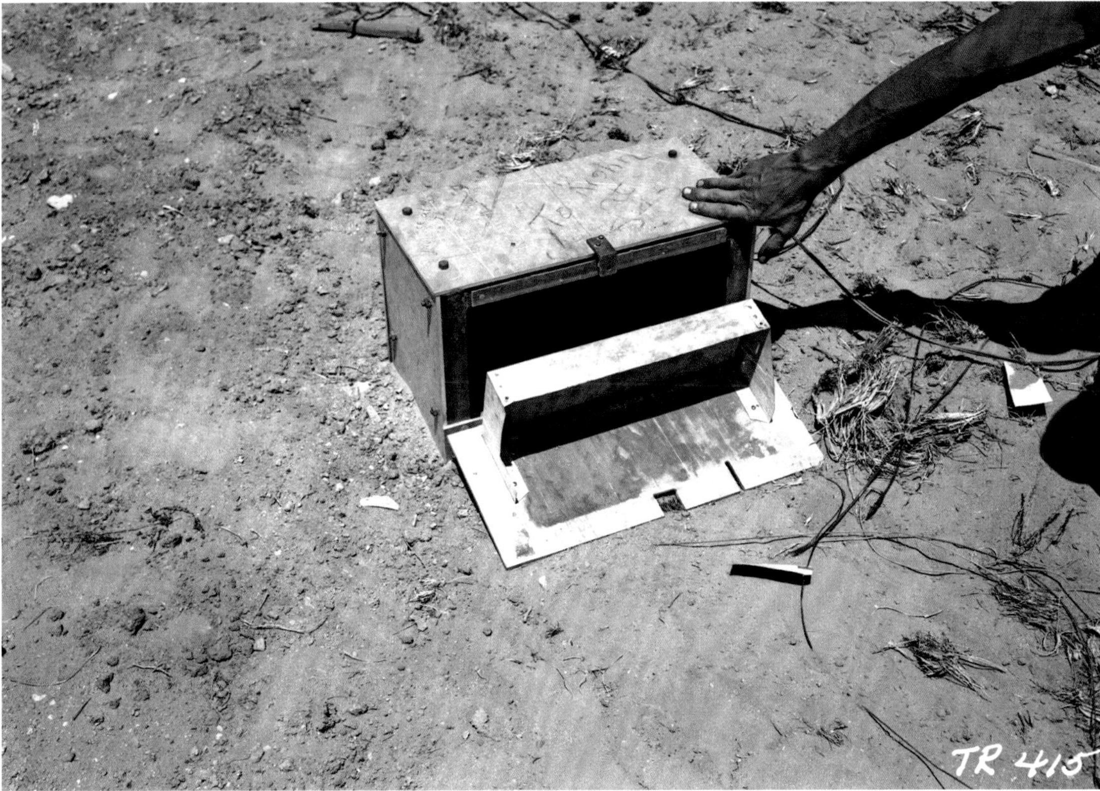

Because neutrons are very readily absorbed by other atoms, such as nitrogen molecules in the air, those freed by fission were at risk of scattering off or being intercepted by other particles on their way to the detection devices. To account for this, the cameras had to be set up very close to the explosion, some only 350 yards away.

Like the gamma ray ionization chambers, some of the neutron cameras were placed at ground level. The interior and exterior of one of the protective boxes is shown at left.

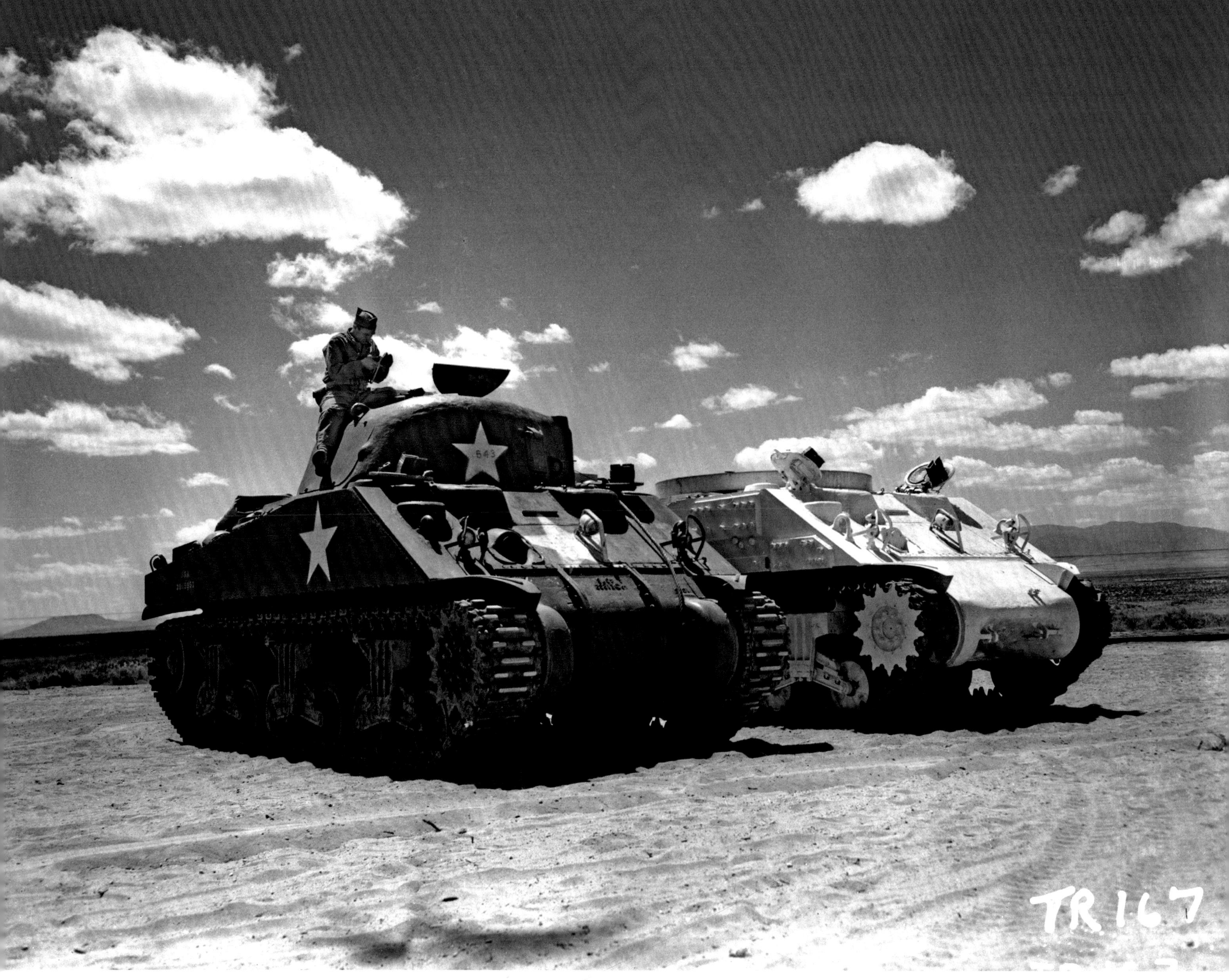

Of the many experiments devised, the one expected to produce the most accurate estimate of the Gadget's yield was also free from any risk of instrument destruction. Led by Herbert L. Anderson, the method involved gathering the radioactive particles that would settle in the soil after the explosion. Instead of counting gamma rays or neutrons, Anderson planned to analyze the products created during fission—isotopes of light elements like strontium, zirconium, barium, and cesium—as well as unfissioned plutonium. Obtaining soil samples to study these unstable particles meant entering the radioactive crater left by the blast as soon as safely possible. An early potential strategy to accomplish this involved use of a blimp to scoop up dirt from above. Instead, two slightly less unwieldy M-4 Sherman tanks, originally procured for implosion testing at the laboratory, were redesignated for the Trinity test.

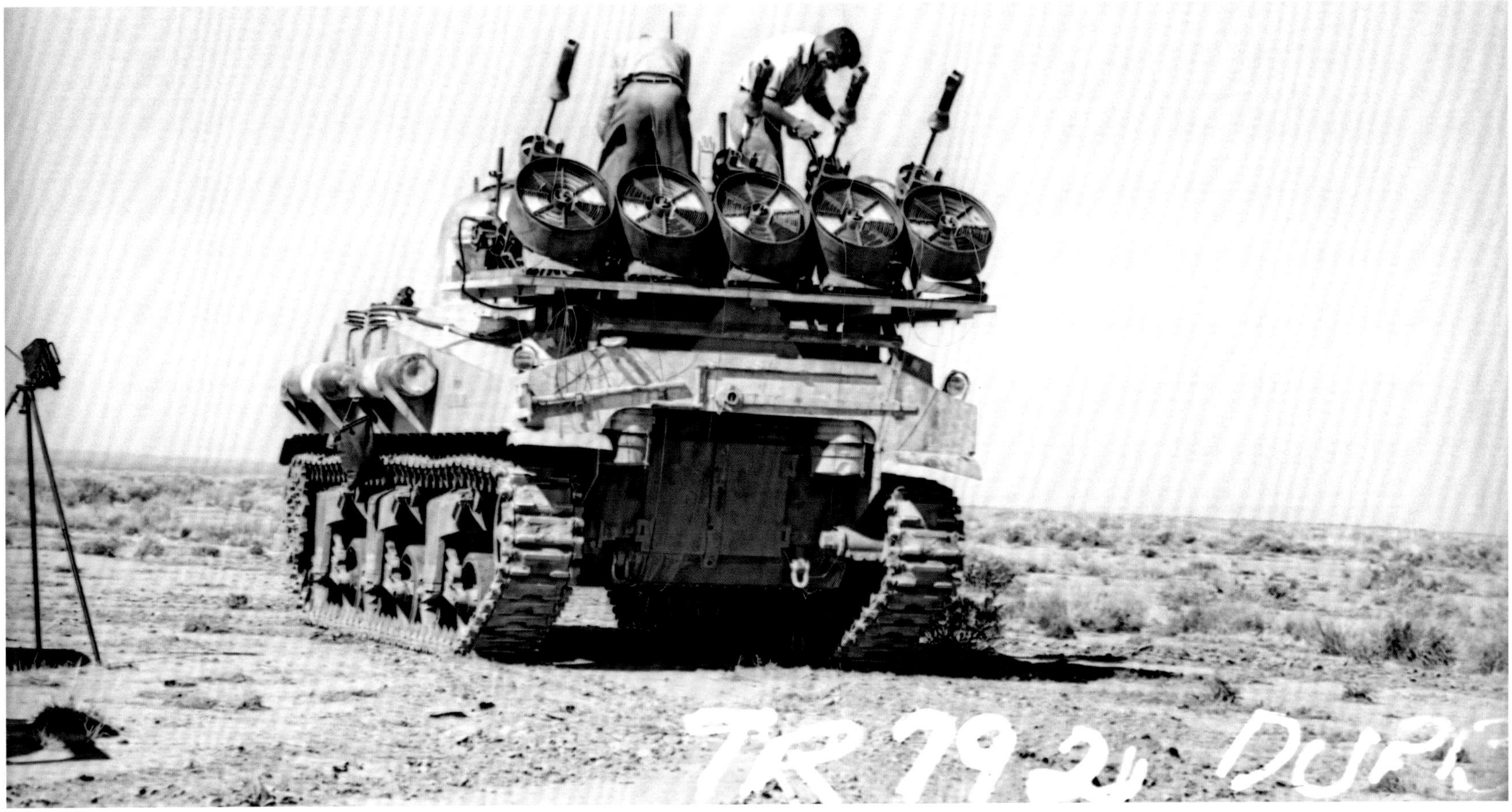

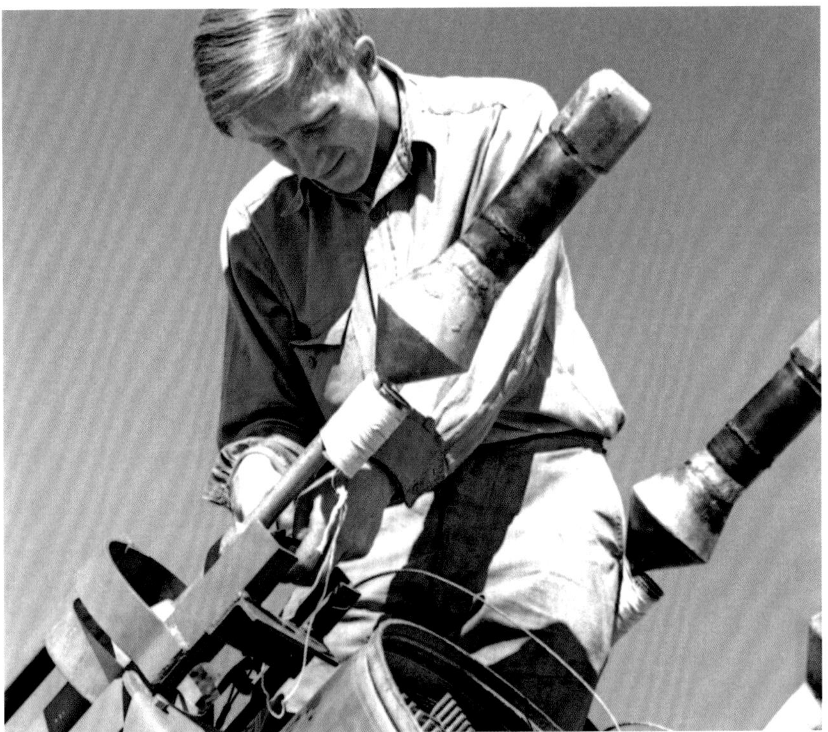

The tanks' thick armor offered some protection at baseline, and one was dressed in 11,000 pounds of lead shielding so that scientists could drive straight into the contaminated area with reduced exposure to radiation. A hatch underneath the shielded tank enabled sample collection through the floor. The second tank, assigned to stay at a safer distance, was retrofitted with special sampling rockets (shown at left, with physicist Hans Staub) that could scoop up 500 grams of soil; after being fired into the crater, the rockets and their payload would be dragged back in by a retractable cable.

The tanks would also help retrieve close-range apparatuses, such as neutron detectors, from highly contaminated areas. Because visibility was limited, rights-of-way had to be marked throughout the Trinity site to keep drivers from rumbling over other experiments.

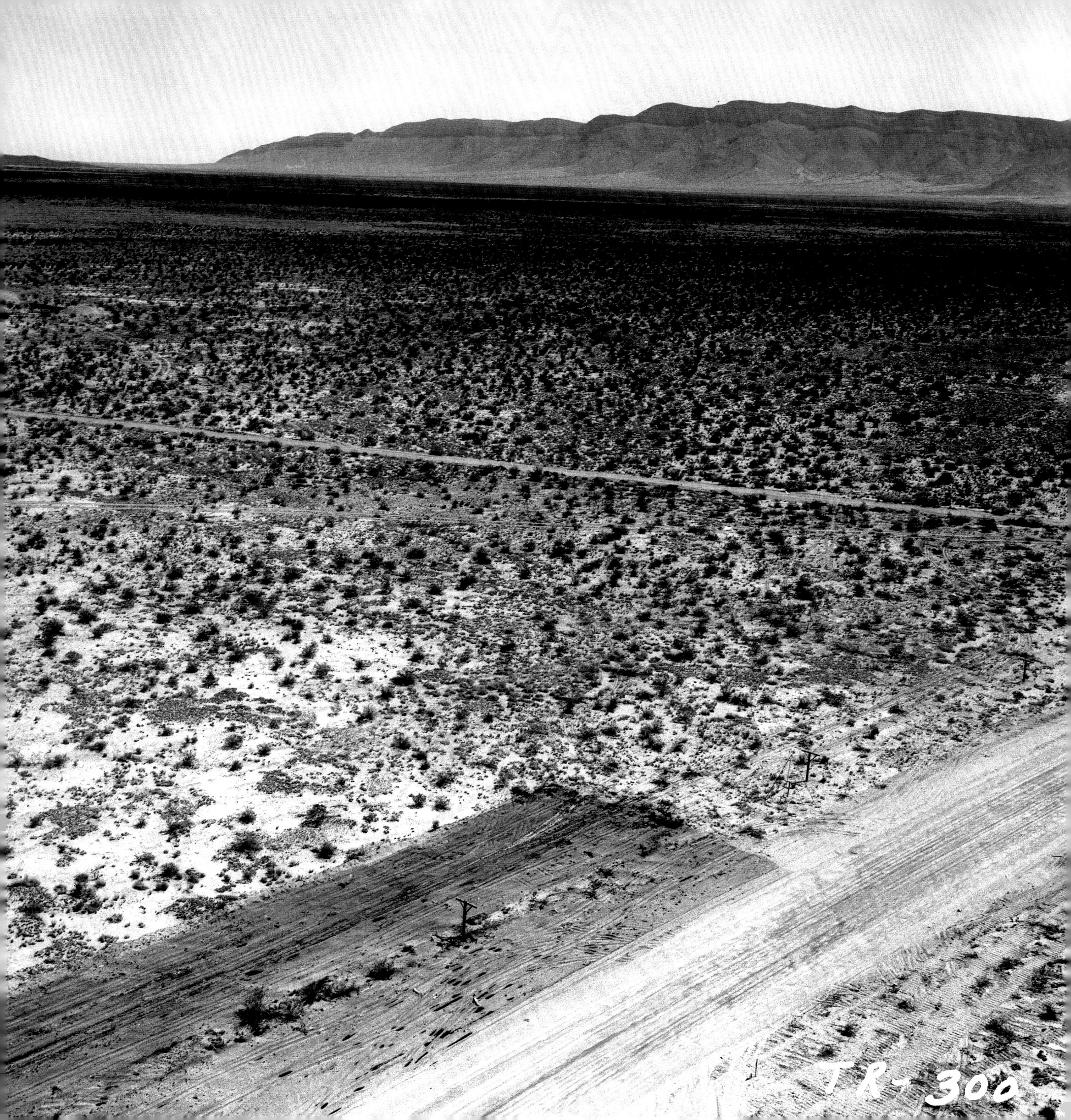
TR-300

The field test in the Jornada del Muerto was a stark departure from the laboratory and academic environment to which most scientists were accustomed. Unwavering in their commitment to the grueling routine of planning, preparing, calibrating, and recalibrating in the desert basin, those scientists and other specialists weathered extreme conditions, learned to operate turrets and tanks, and lived under a long-form countdown for the better part of four months, during which "outdoor air and exercise seemed to serve as a substitute for sleep," as Bainbridge remembered.

Segré later described a typical day at Trinity for the resident physicists, himself included, living in barracks identical to those of military personnel and sharing meals in the mess hall: "We started working intensely at daybreak. The early morning hours were the best; as the sun rose higher in the sky, the heat became oppressive, the light blinding, and we wilted. In the evening, we fell exhausted on our cots, only to start again the next sunrise."

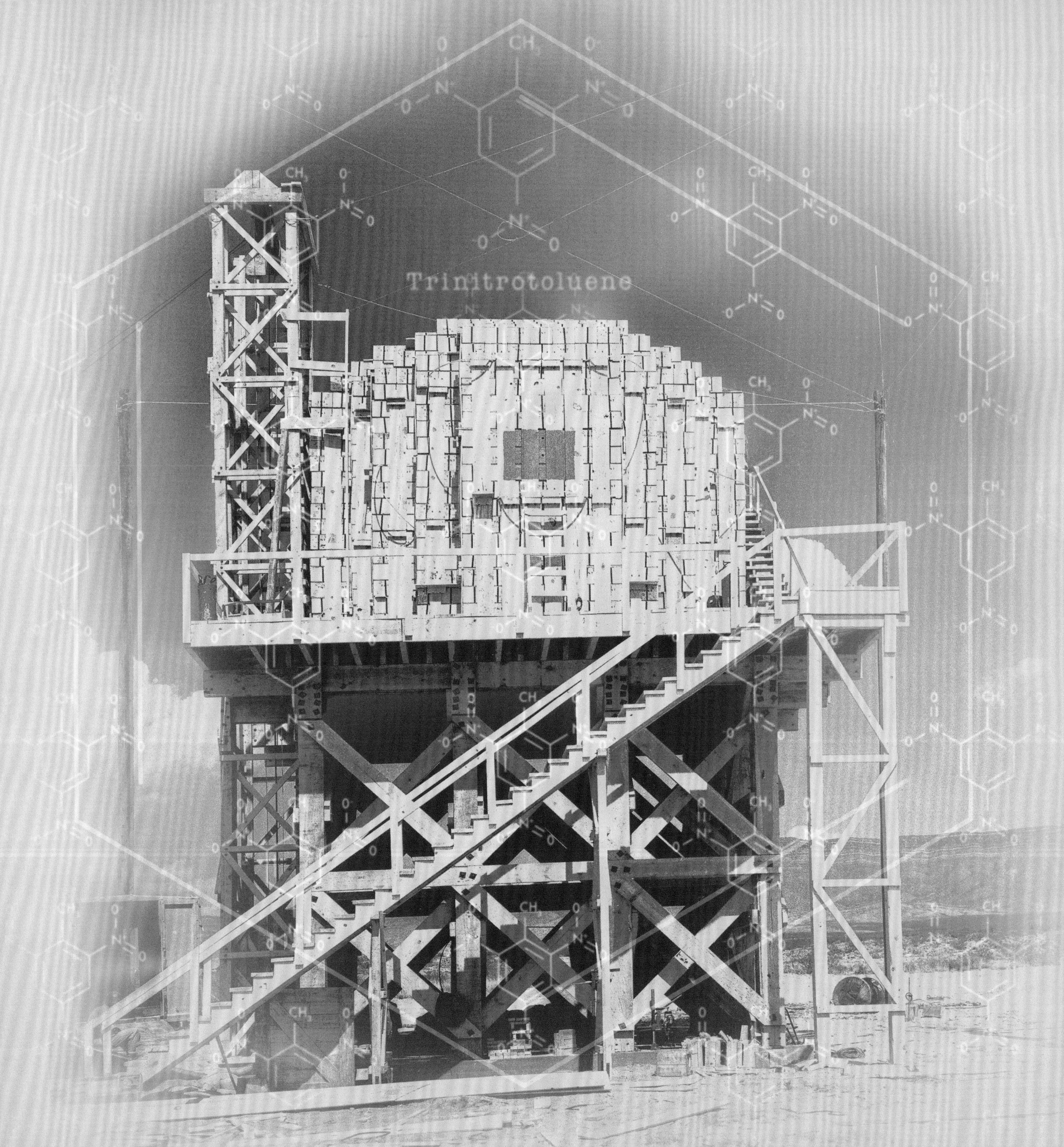
Trinitrotoluene

03 A 100-Ton Dress Rehearsal

The months of intense preparation for the test of the Gadget were punctuated by a rehearsal shot that took place in early May. Kenneth Bainbridge, keenly aware that there was no roadmap for what they were attempting to do in the New Mexico desert, had lobbied Oppenheimer a year before to approve what became known as the 100-ton test—a scaled-down detonation meant to simulate that of the Gadget, using conventional explosives laced with a radioactive tracer.

The major purposes of a dress rehearsal were threefold: One, Bainbridge wanted personnel to have a practice run setting up for detonation, both on the day of the test and in the days and weeks prior, and to go through the countdown procedures all the way to a real T-minus zero. Two, the numerous experiments and devices, most invented just for the Trinity test and many proofed in a laboratory setting, needed to be properly calibrated and field tested. These instruments had been designed based on extrapolations of available data, which existed only for explosions of up to 66 tons of TNT—nothing close to the expected yield for the Gadget, thought most likely to be around 4,000 tons. The scaled-down dress rehearsal was a way of bridging the gap. And three, scientists needed to check their predictions about the behavior of the radioactive material that would explode out of the Gadget; this would inform both the safety plan overseen by physician Louis Hempelmann and the radiochemistry-based yield analysis spearheaded by Herbert Anderson.

The practice test held many possibilities, but interested parties would only be able to gain from the dress rehearsal what they were able to invest in it. Time was the greatest limiting factor. On-site setup of experiments for the test of the Gadget had only really taken off in the second half of March, and yet the 100-ton test would necessitate a whole other set of preparations, to be completed in parallel throughout the month of April. Many experiments were still in the materials procurement phase at this point or had not yet been conceived or introduced into the test plan. Amid supply chain delays and last-minute scrambling as new apparatuses continued to enter the picture, Bainbridge was forced to delay the practice shot until May 7, two days after its originally scheduled date.

Facing many requests to push the date back even further, Bainbridge made the determination that further delays would harm the more fixed schedule for the test of the Gadget and put "intolerable burdens on the whole group." The remaining preparations that depended on results from the trial run—including laying new wiring, building close-range bunkers, and improving roadways—would have to start immediately to keep the schedule for the July test within the realm of possibility. The practice shot could wait no longer.

A few days after the 100-ton test, with dust settled around a small crater to the southeast of ground zero, Bainbridge convened a Trinity town hall, inviting personnel to air any concerns that had either been created or given credence by the scaled rehearsal. Scientists were already taking the needed actions to fix flaws in their experiments that came to light, but the occasion brought out a full suite of more systemic grievances, from living conditions to radio interference to unreasonable safety protocols. Broadly, the test brought into focus the many potential missteps that could affect the schedule, the detonation itself, and the success of diagnostic instruments.

A particularly troubling result concerned the dispersion of radioactive debris. Scientists involved in evaluating the possible danger of the test had, up to this point, believed most of the hazardous material would remain near ground zero. The 100-ton test indicated a new reality: while the immediate area would certainly be contaminated, a large percentage of radioactive material would likely be blasted into the air, where it would gradually fall back to the earth downwind of the test site. This was a revelation, and not a welcome one—necessitating a significantly more elaborate approach to the safety plan.

Failures and surprises were not ideal at this point, but for those experiments and procedures that did not meet expectations, the May practice shot left a manageable amount of time to redesign or recalibrate in preparation for the all-important test of the Gadget. A little breathing room was granted when the laboratory's Cowpuncher Committee—a supervisory body tasked with "riding herd" to keep the implosion program on track—agreed to the first postponement of the July 4 test, setting a new date of July 13 for the main event.

* * * * * * *

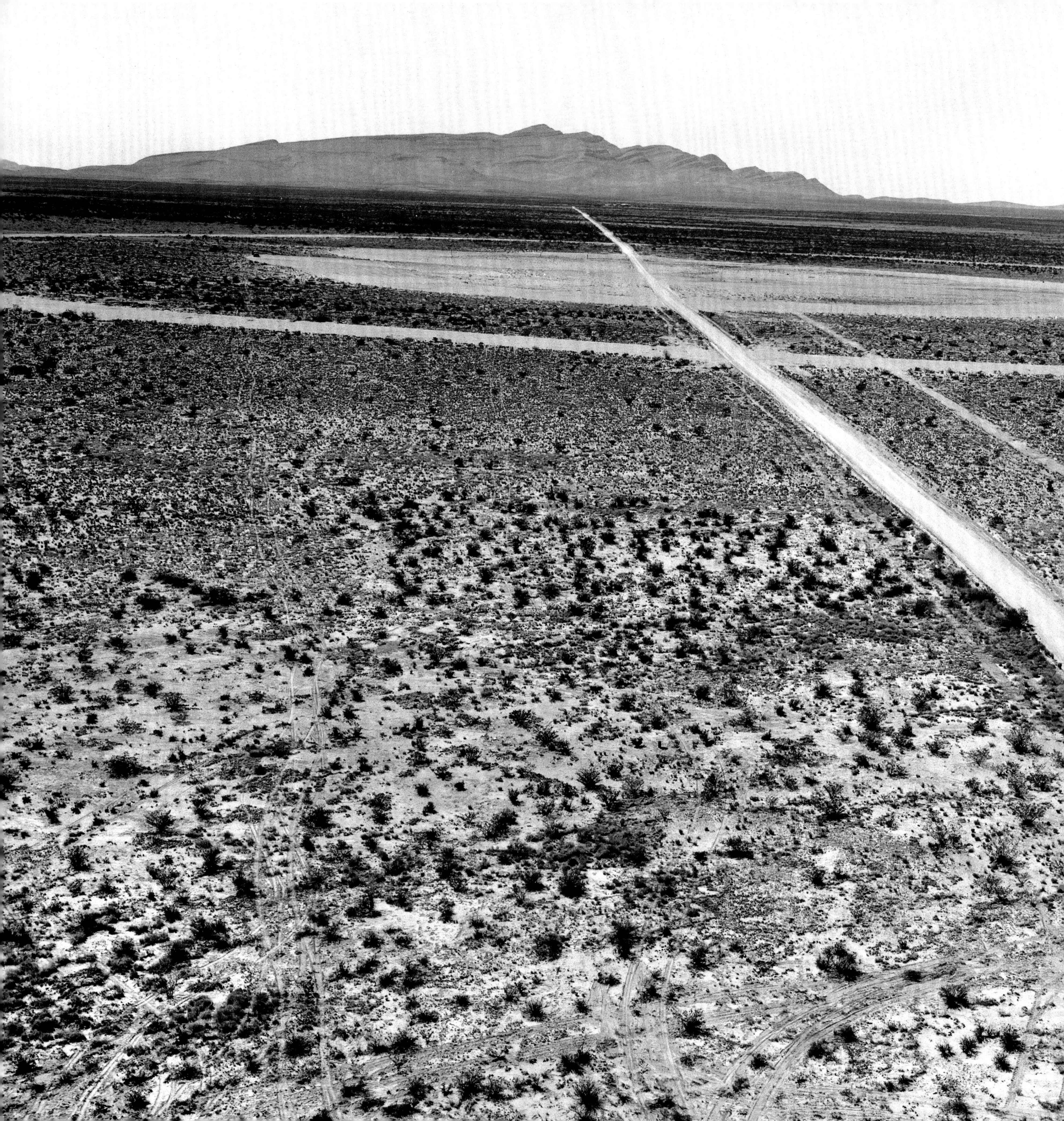

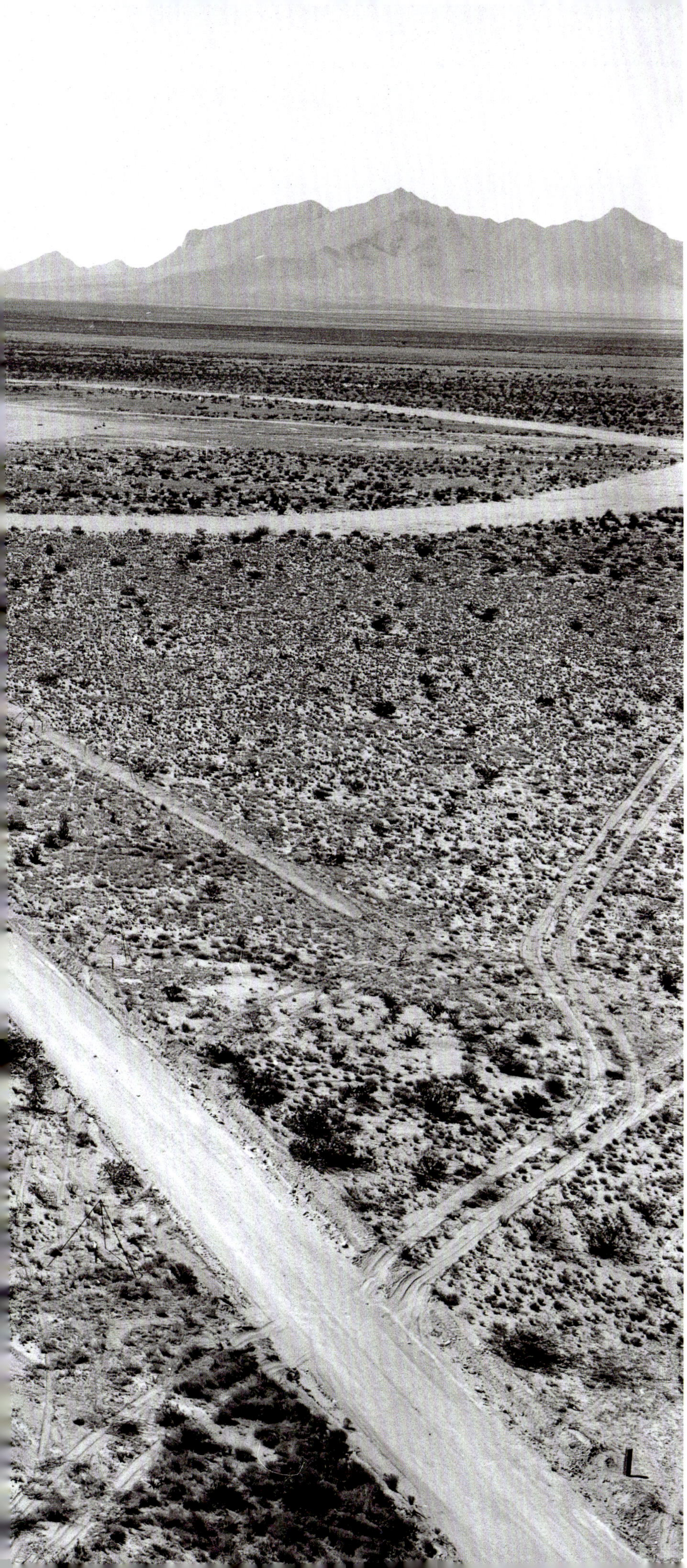

On the final day of March 1945, John Williams—Bainbridge's second in command and the supervisor of field preparations at Trinity—issued orders to clear away brush and scrape out a wide firebreak around a location 800 yards southeast of ground zero. In the center of the clearing, shown at left, a sturdy wooden tower would become host to a tremendous explosion, one that for two months would stand as the largest ever to be scientifically measured. Serving as an opportunity to uncover flaws and firm up plans, this rehearsal detonation paved the way for the blast that would ultimately supersede it—the full-scale test of the first nuclear device.

Codenamed the Z test but referred to ubiquitously as the 100-ton test, the rehearsal explosion actually consisted of just under 90 tons of TNT plus a mixture of TNT and RDX known as Composition B, one of the chemical formulas used in the Gadget. The 14.91 tons of Composition B was equivalent to the explosive energy of 17.9 tons of TNT, bringing the total charge of the practice shot to approximately 108 tons. The explosives and their arrangement were precisely chosen to correspond with the predicted yield and spherical shape of the real test device. While the pre-test did not itself rely on a nuclear reaction, it did incorporate radioactive material, infused throughout the stack of conventional explosives (on the tower, below), to model dispersion of fission products from the Gadget.

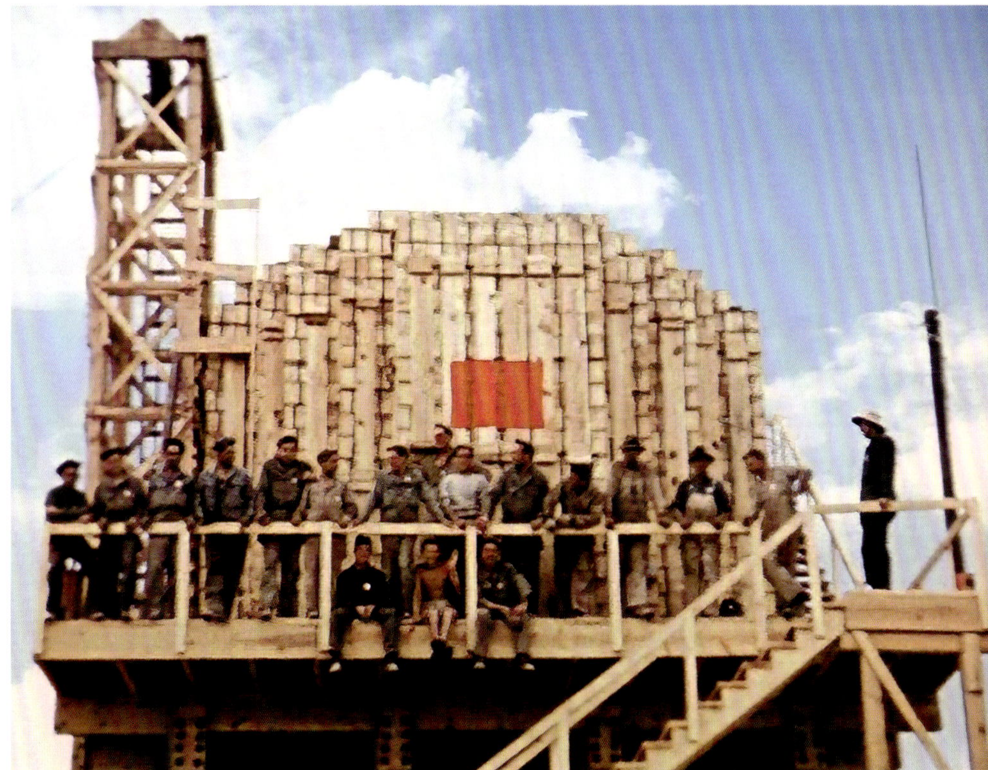

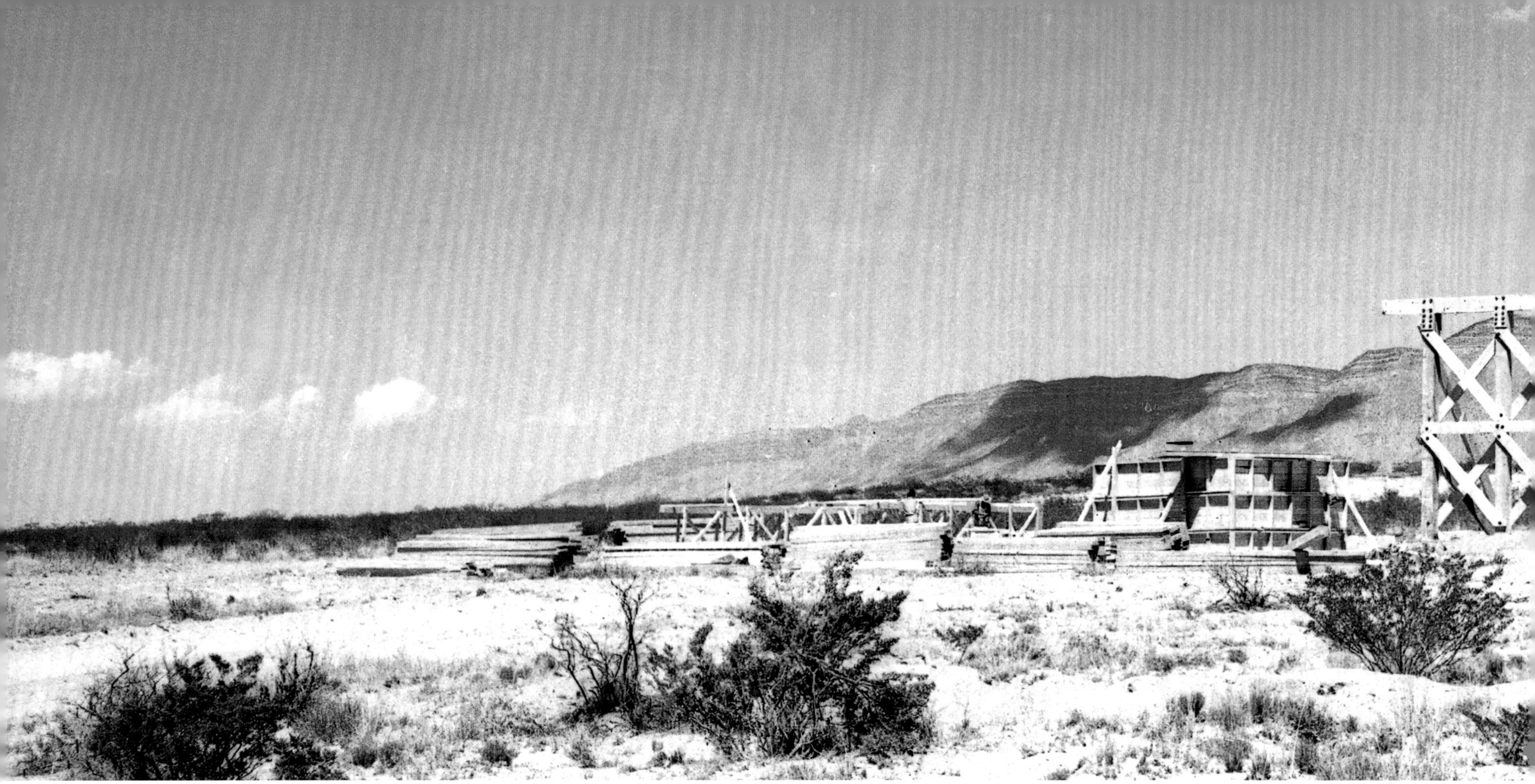

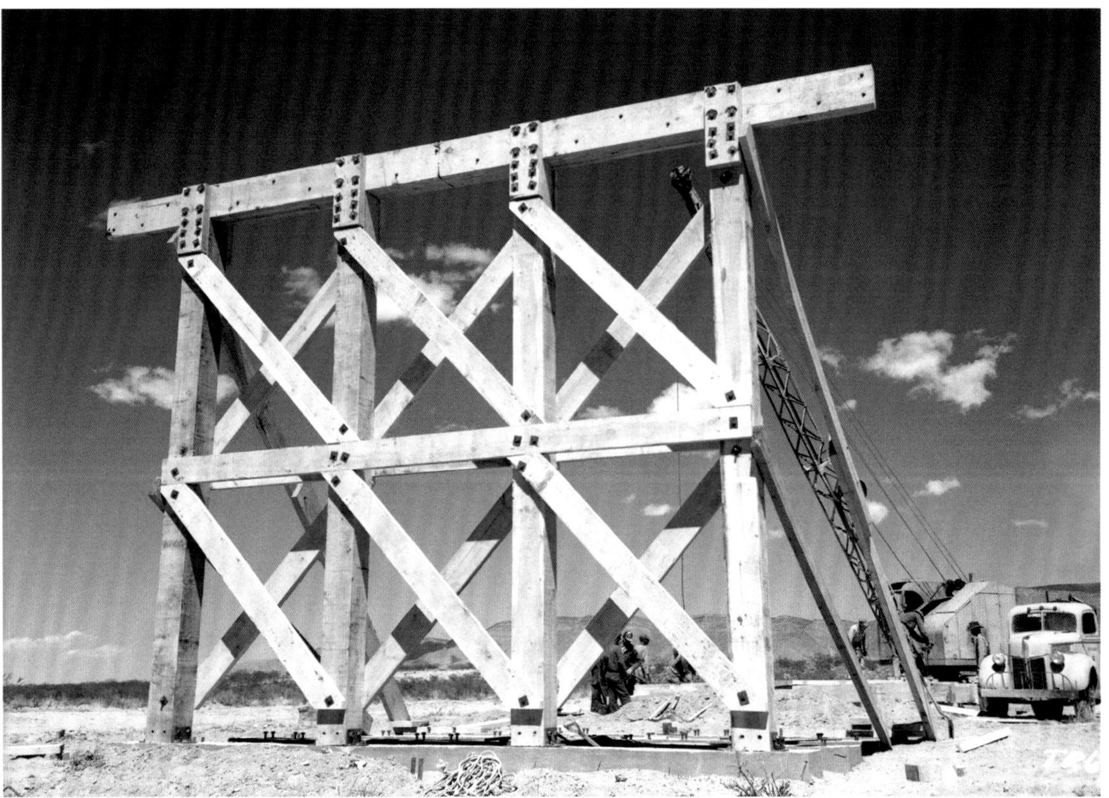

As a crew of contractors assembled the huge timbers of the tower—its purpose, supporting 200,000 pounds of explosives, unbeknownst to them—scientists began adapting their various experiments to measure the effects of the scaled-down explosion at proportionately scaled-in distances.

This calibration was necessary for instruments designed to measure phenomena like shock wave pressure, ground disturbance, and dispersal of radioactive material, which would be negligible or even imperceptible at the stations mapped out for the Gadget's much larger explosion. Photography and communication equipment, not dependent on receiving physical input from the blast, was set up at the full distances planned for the real test, with cameras aimed at red target cloths hung on the north and west faces of the completed stack.

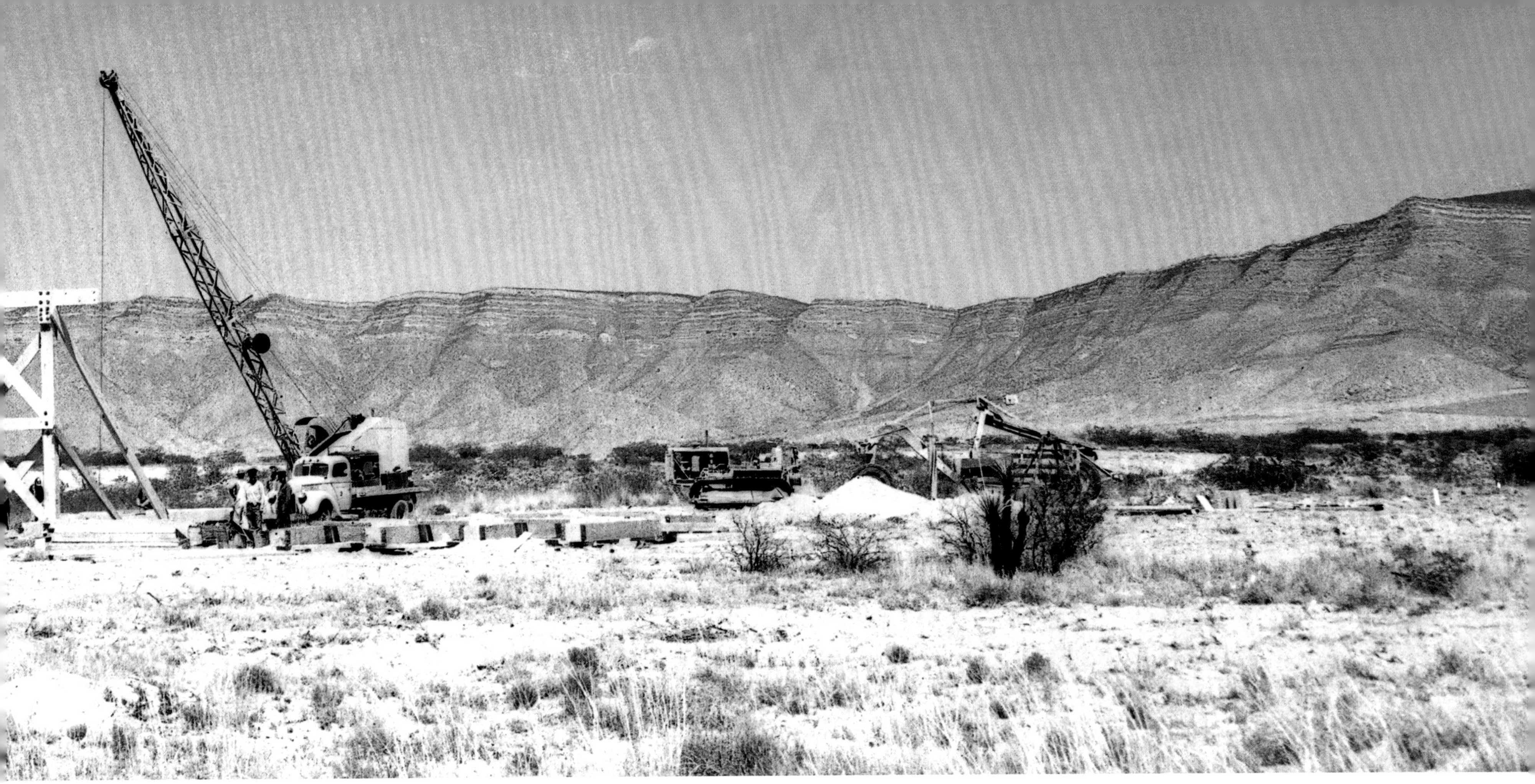

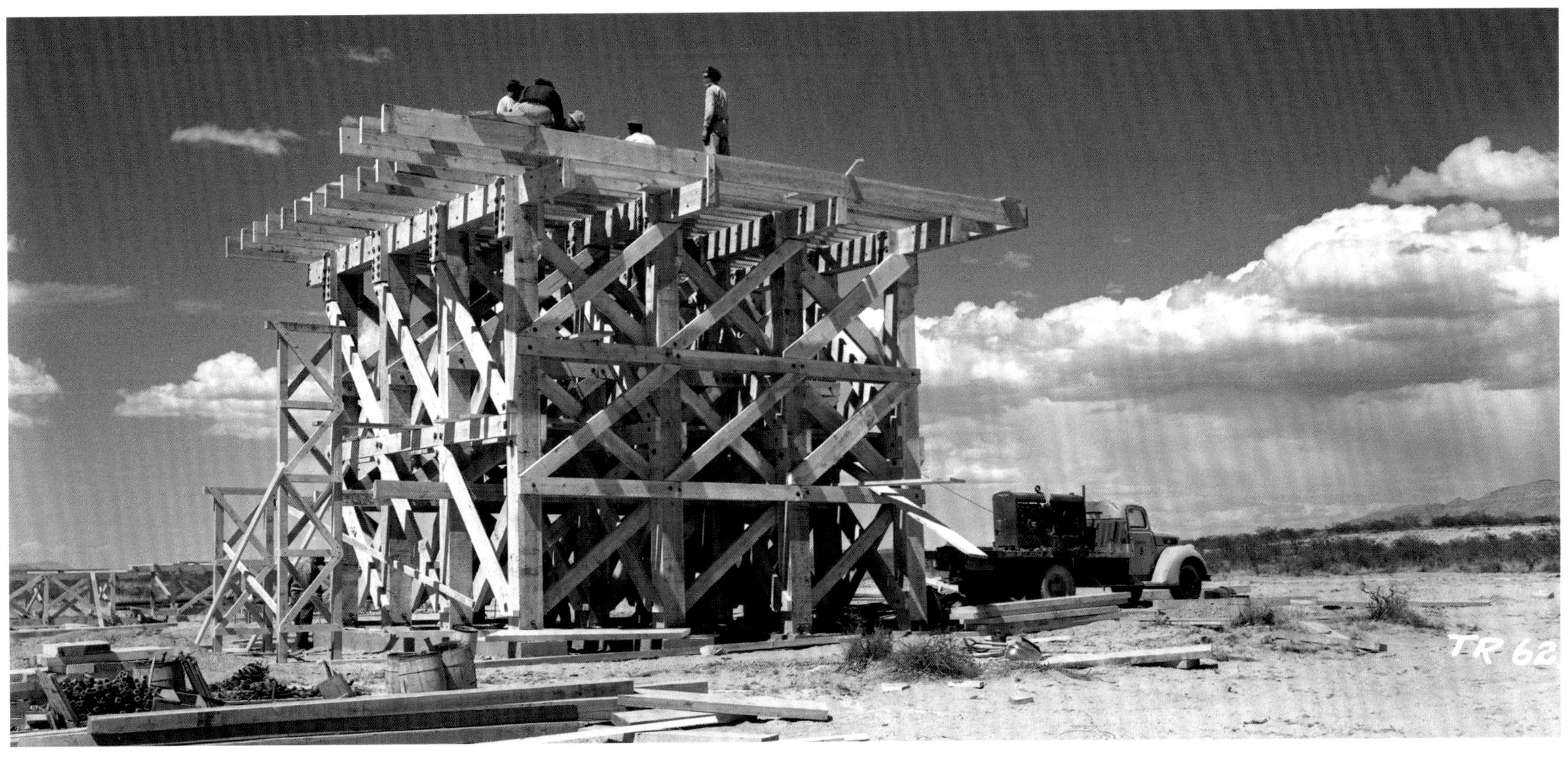
TR 62

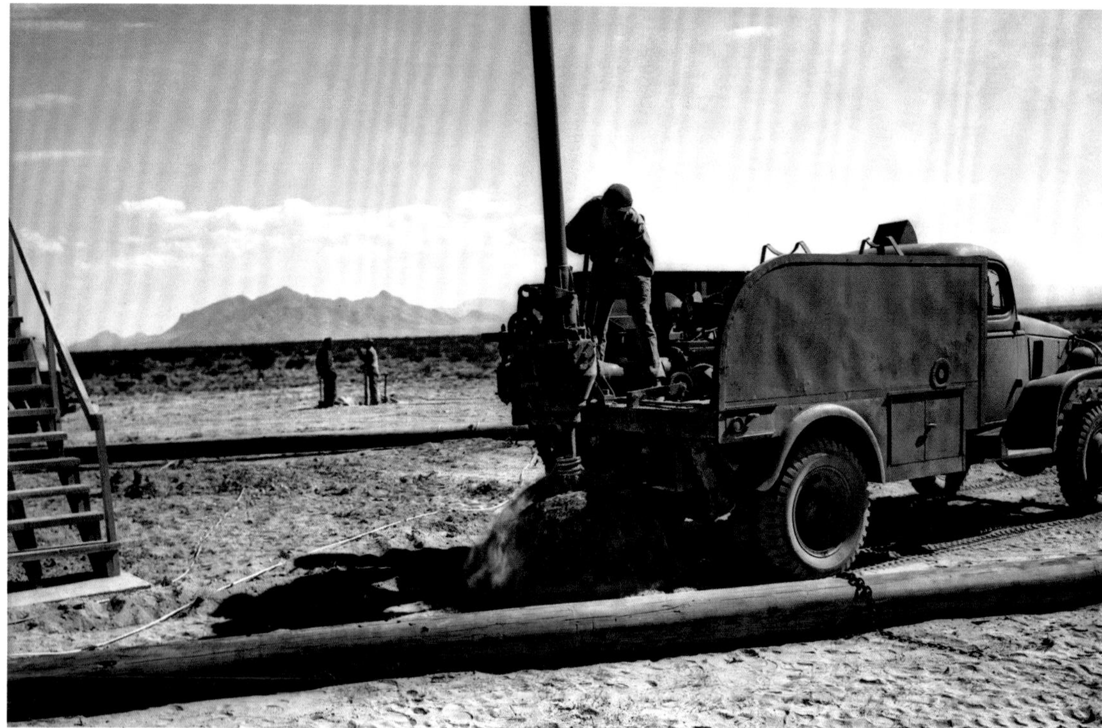

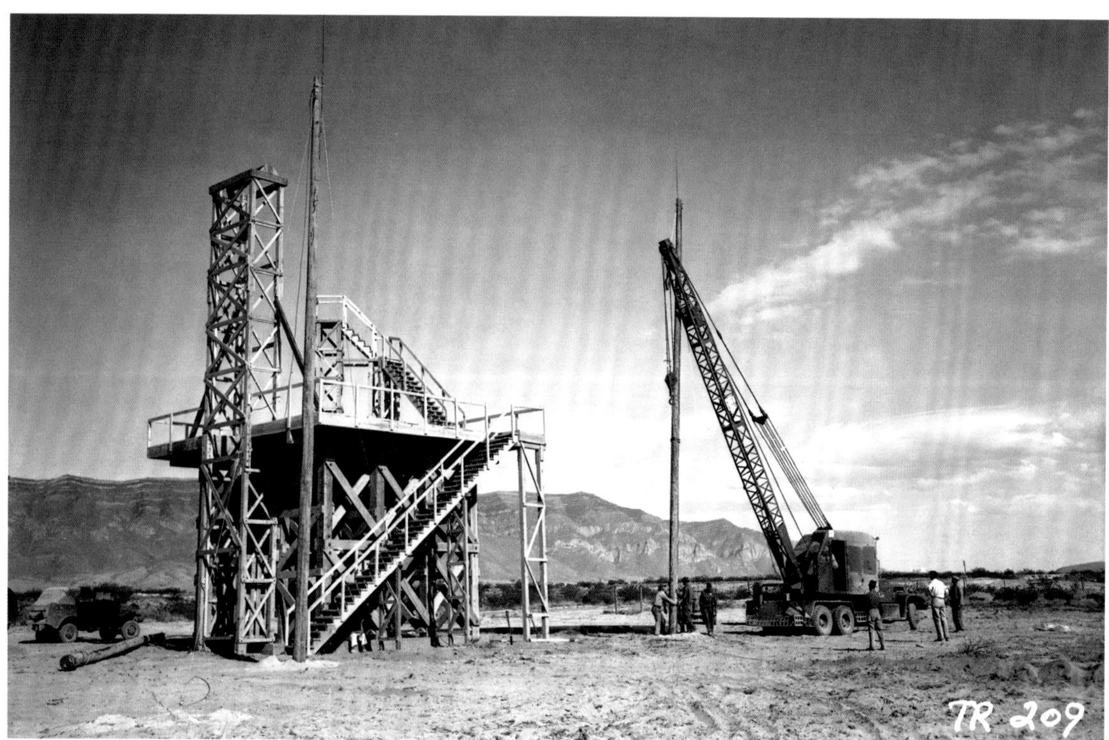

Before loading the platform with explosives, crews planted four telephone poles topped with steel flag poles around it. The makeshift lightning rods were tied to each other and the earth by copper conductors, forming a safeguard against accidental detonation that might be caused by a late spring lightning storm.

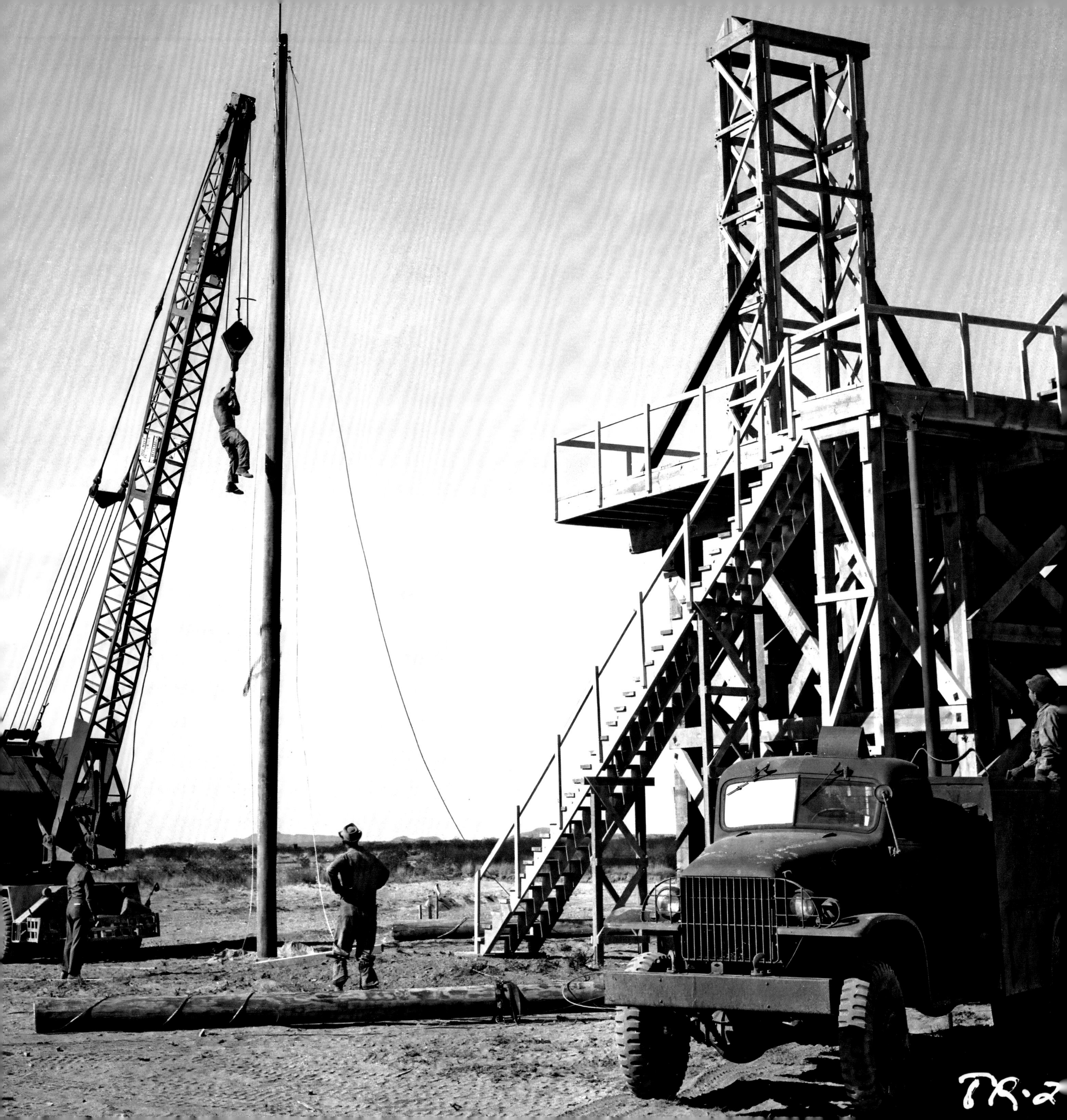
TR-2

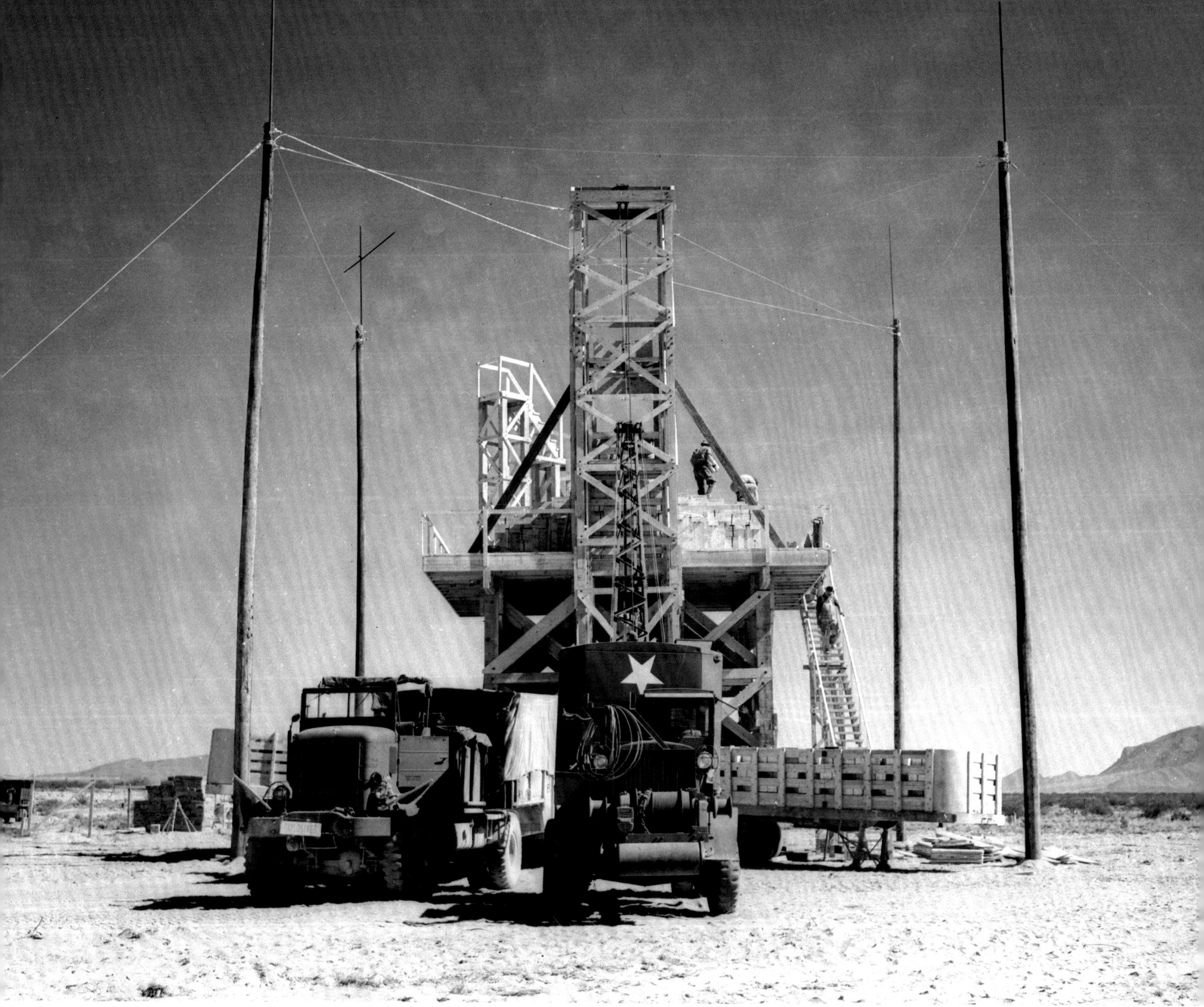

The tallest component of the tower was the elevator shaft, used to lift more than 4,000 crates onto the twenty-foot-high platform. A stairway opposite the shaft allowed workers to climb onto the pile as it grew to a final height of eighteen feet.

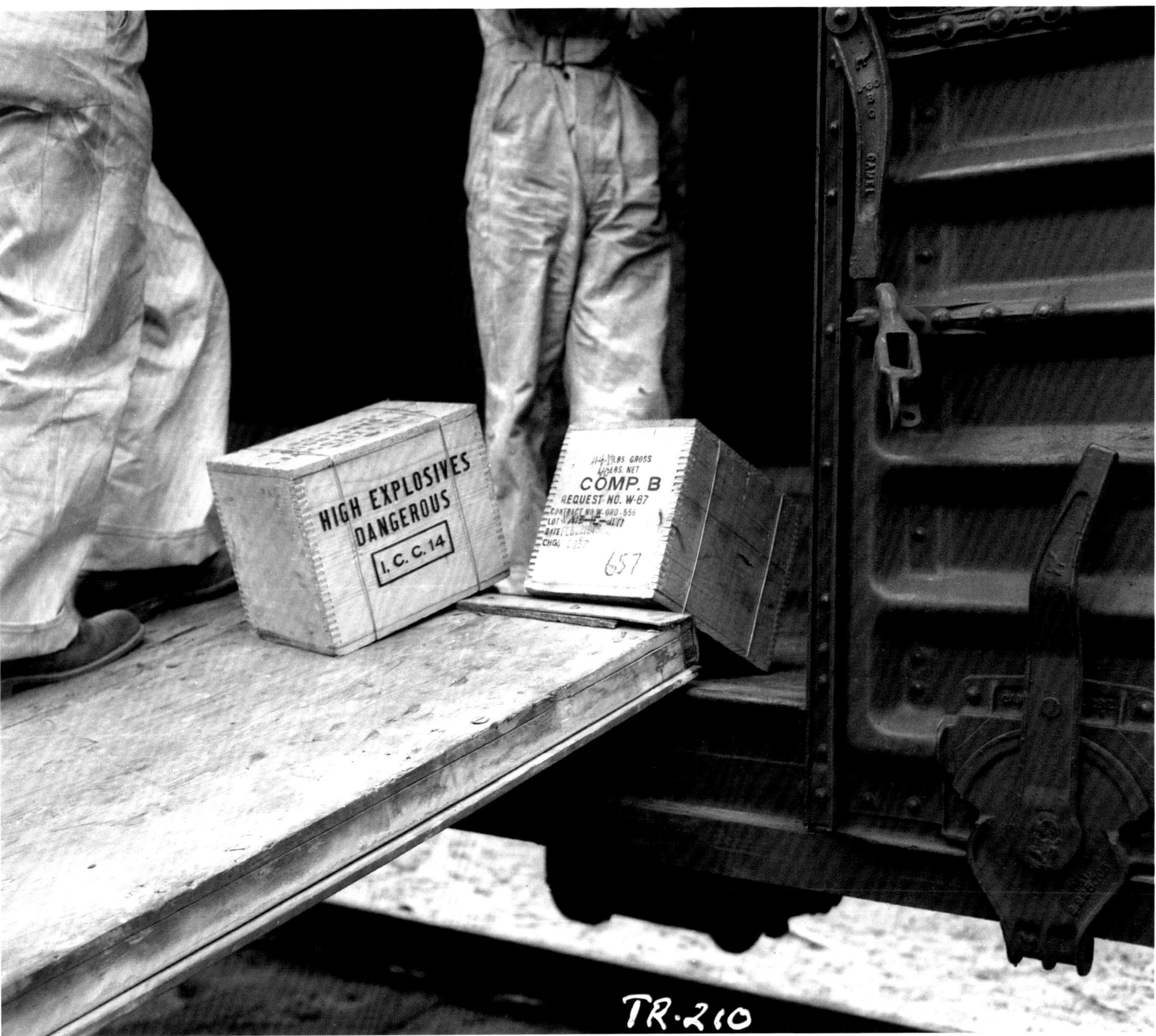

George Kistiakowsky, head of the Explosives Division, had assured Bainbridge that the crated explosives would not be sensitive to mechanical shock. This was demonstrated twice over—first, when some crates fell off a truck on their way down Pope Road to the Trinity site, and again when several more fell off the tower elevator.

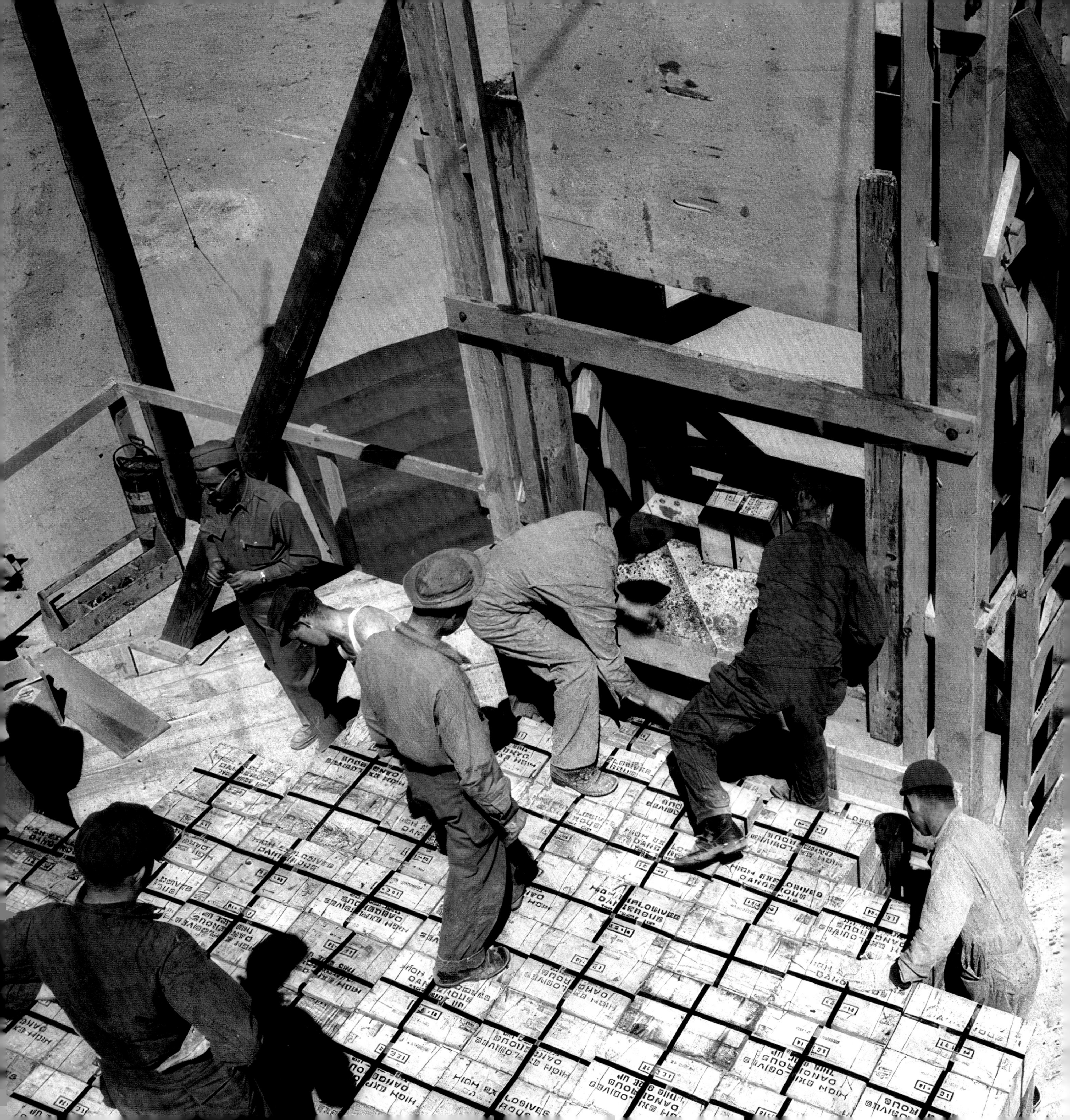

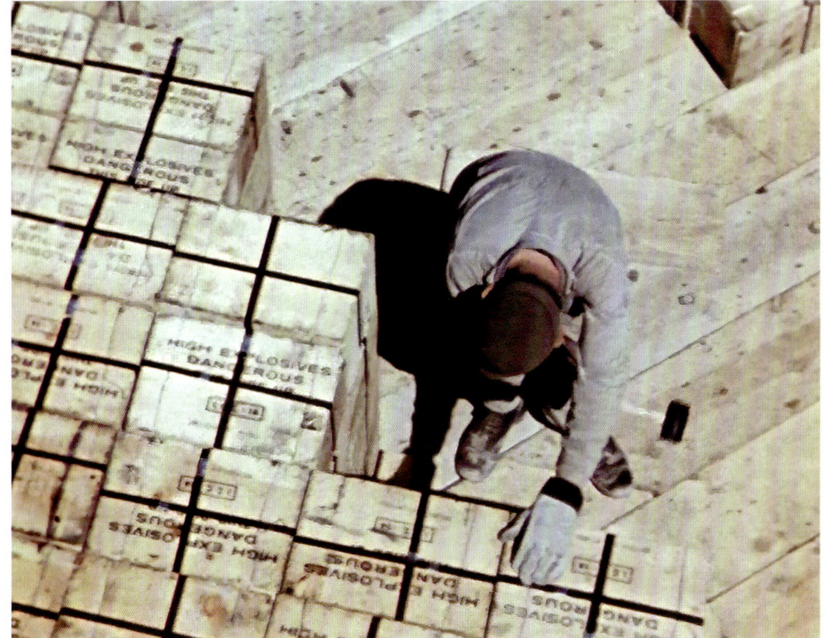

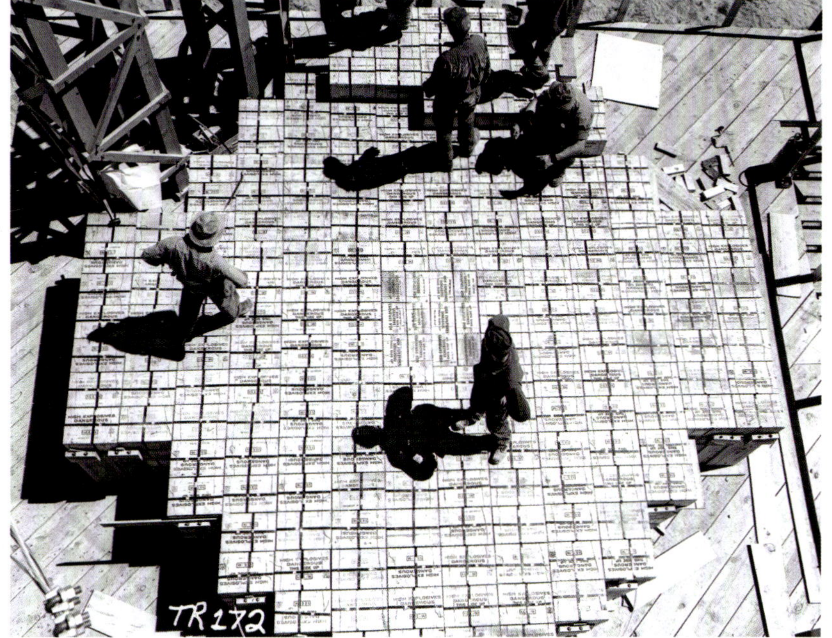

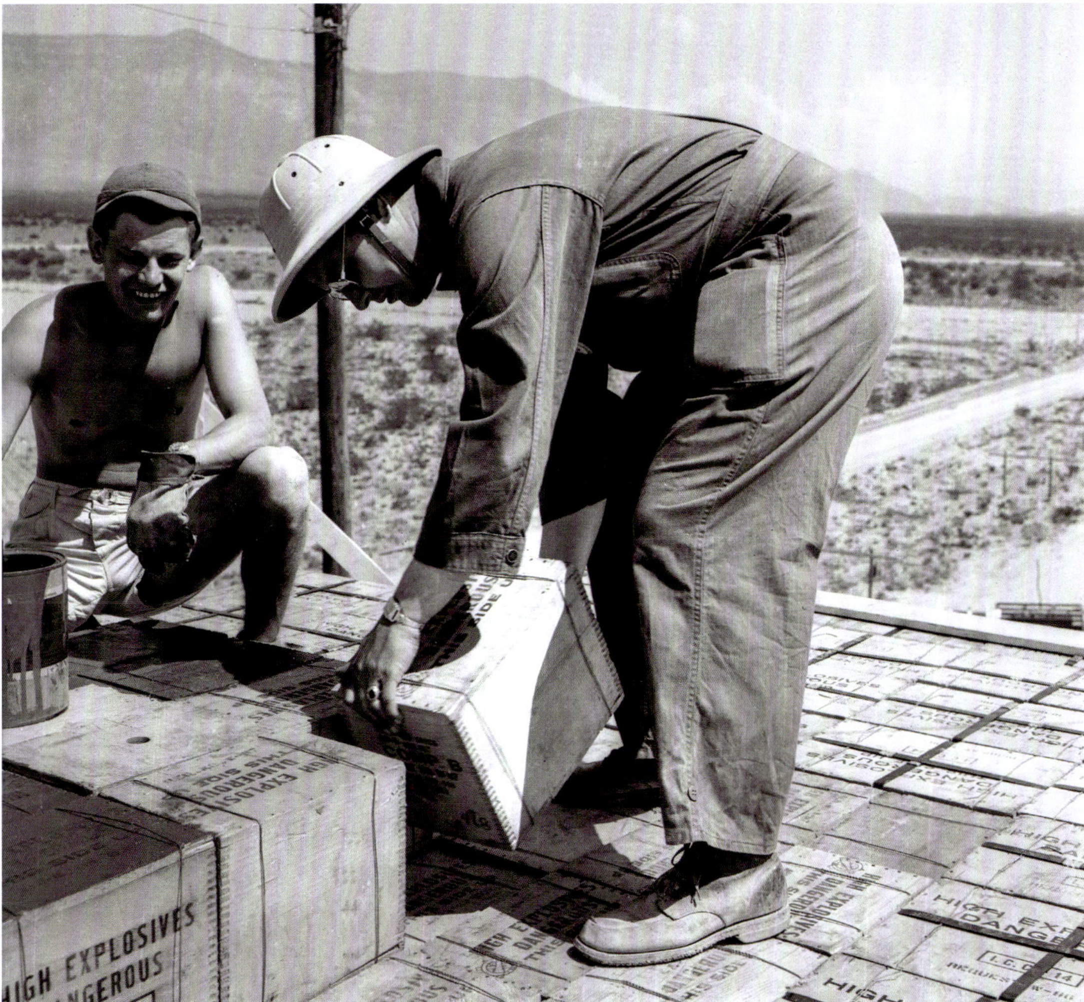

While the tower was built by civilian contractors, the explosives were handled by servicemen from the Special Engineer Detachment and the Military Police. Under the intense heat of the New Mexico sun, the workers retrieved crate after crate from the elevator and stacked them on the platform, placing trust in Kistiakowsky as they swung rubber mallets to pack the high explosives snugly into place.

The operation was overseen by Capt. Wilbur F. Schaffer, shown at left setting the final crate.

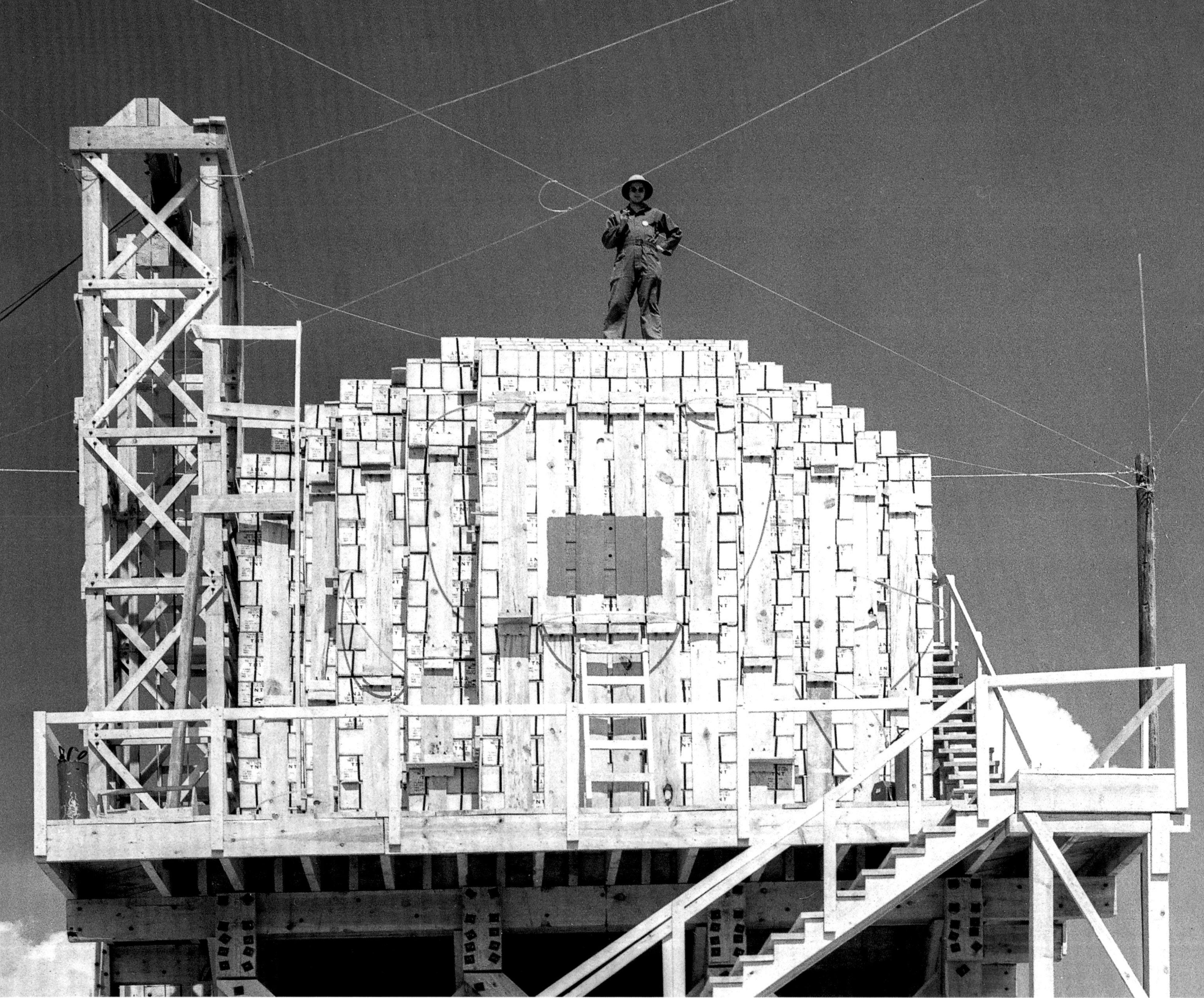

An important goal of the dress rehearsal was to observe how a large explosion dispersed radioactive debris. The energy released by the 100-ton test was derived only from conventional explosives—not a nuclear process—but nuclear material was integrated into the practice test by running a network of tubes throughout the stack of crates. One-inch cleated planks laid between the layers of crates, as shown at right, served the dual purpose of stabilizing the arrangement and creating channels for the plastic tubing, which is visible looping across the front of the stack, above.

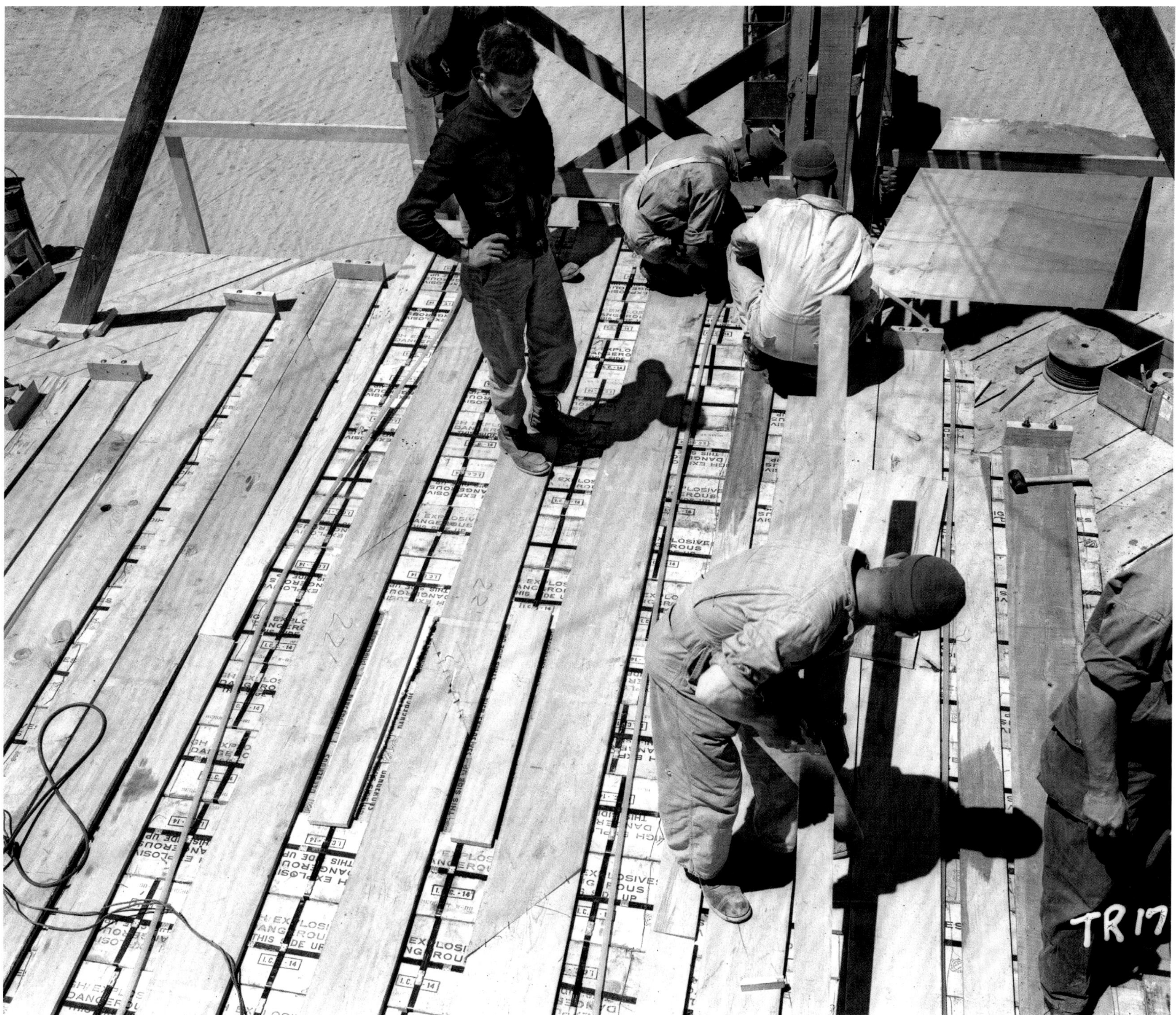

The nuclear material that would be pumped into the tubing originated in Hanford, Washington, where plutonium for the Gadget's core was being slowly created in nuclear reactors. One atom at a time, a rounded block of uranium called a slug—the reactor's fuel—absorbed neutrons and decayed into plutonium and other radioactive particles. In the days before the May 7 practice test, one of these slugs, after spending 100 days in the reactor, was rerouted to Trinity.

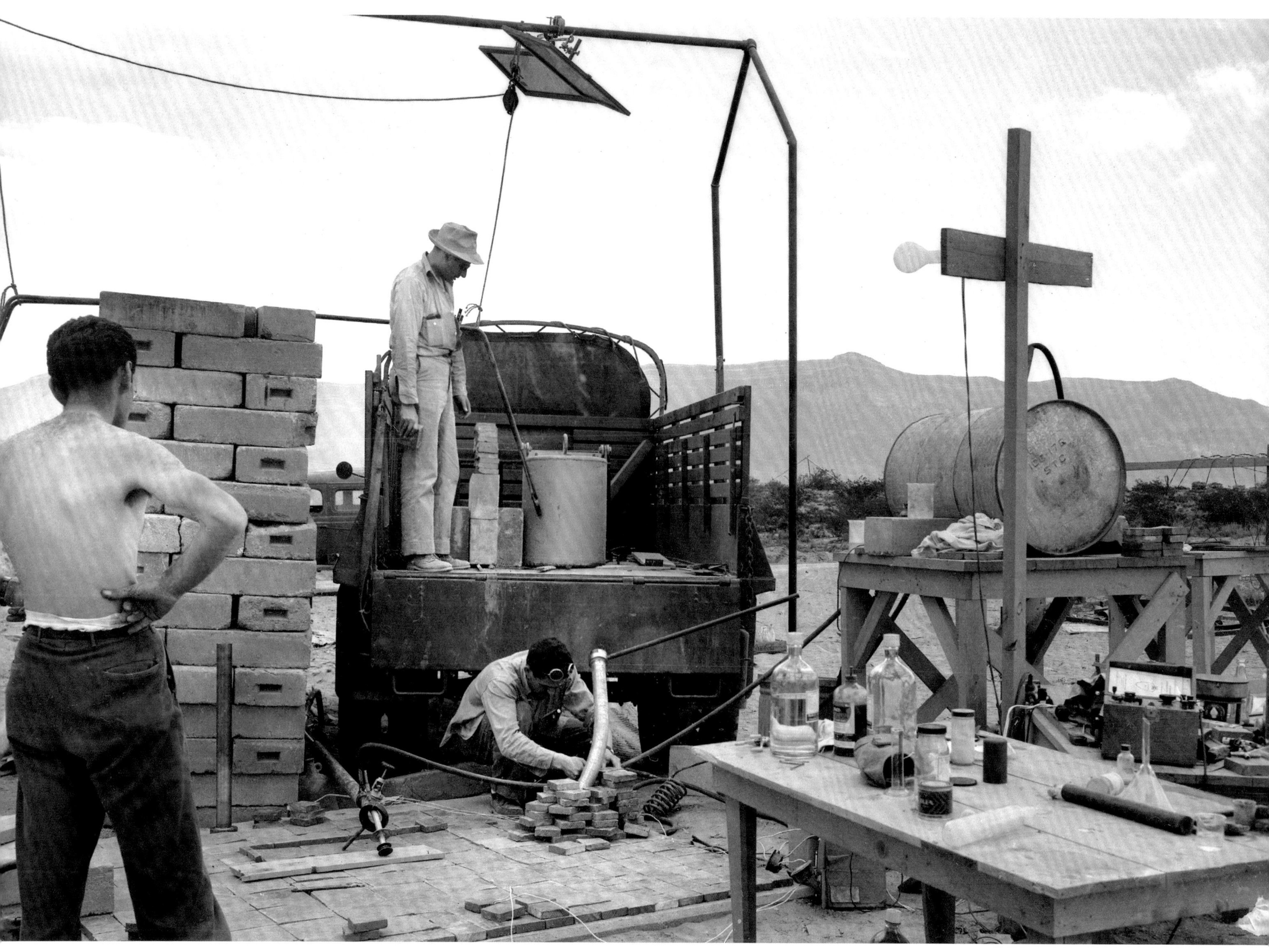

At the Hanford plant, technicians used robotic tools to harvest plutonium from the irradiated slugs, which remained sealed off in a special chemical separation building to prevent any exposure to radiation emanating from the highly dangerous material. To safely work with the slug at the Trinity site, scientists likewise took many precautions, as shown in the image above. The slug was delivered inside a lead container (the cylinder on the truck bed) kept behind a thick stack of concrete blocks. The taller stack was supplemented by another, smaller stack of shielding right next to the container. Mirrors positioned over the truck bed were angled so that workers could maneuver the slug while standing behind the protective wall.

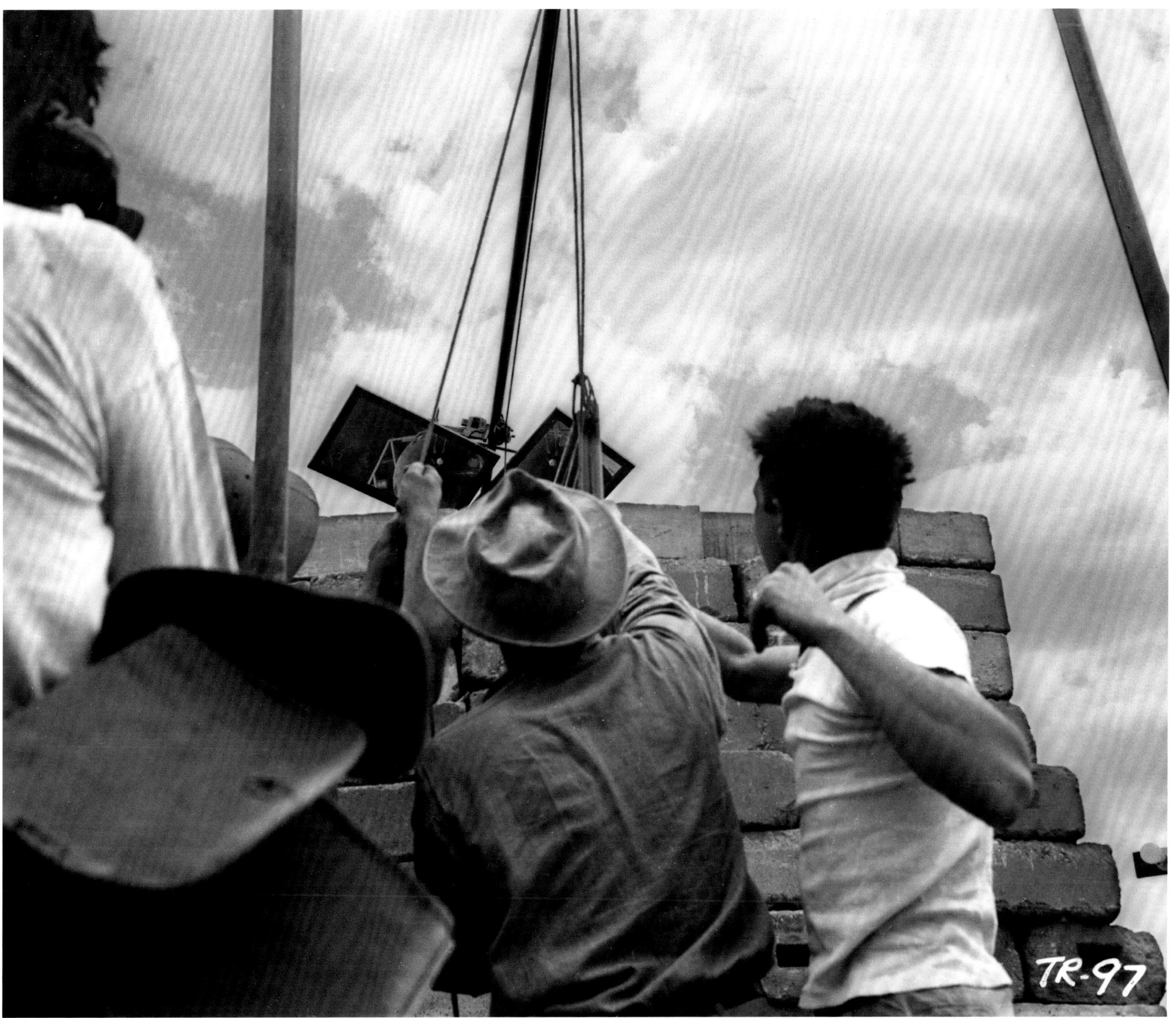

After a crane removed the heavy, lead top from the container to expose the slug, the workers used a pair of long tongs to move the radioactive material through a lead funnel and into an underground, concrete dissolving tank layered over with lead bricks.

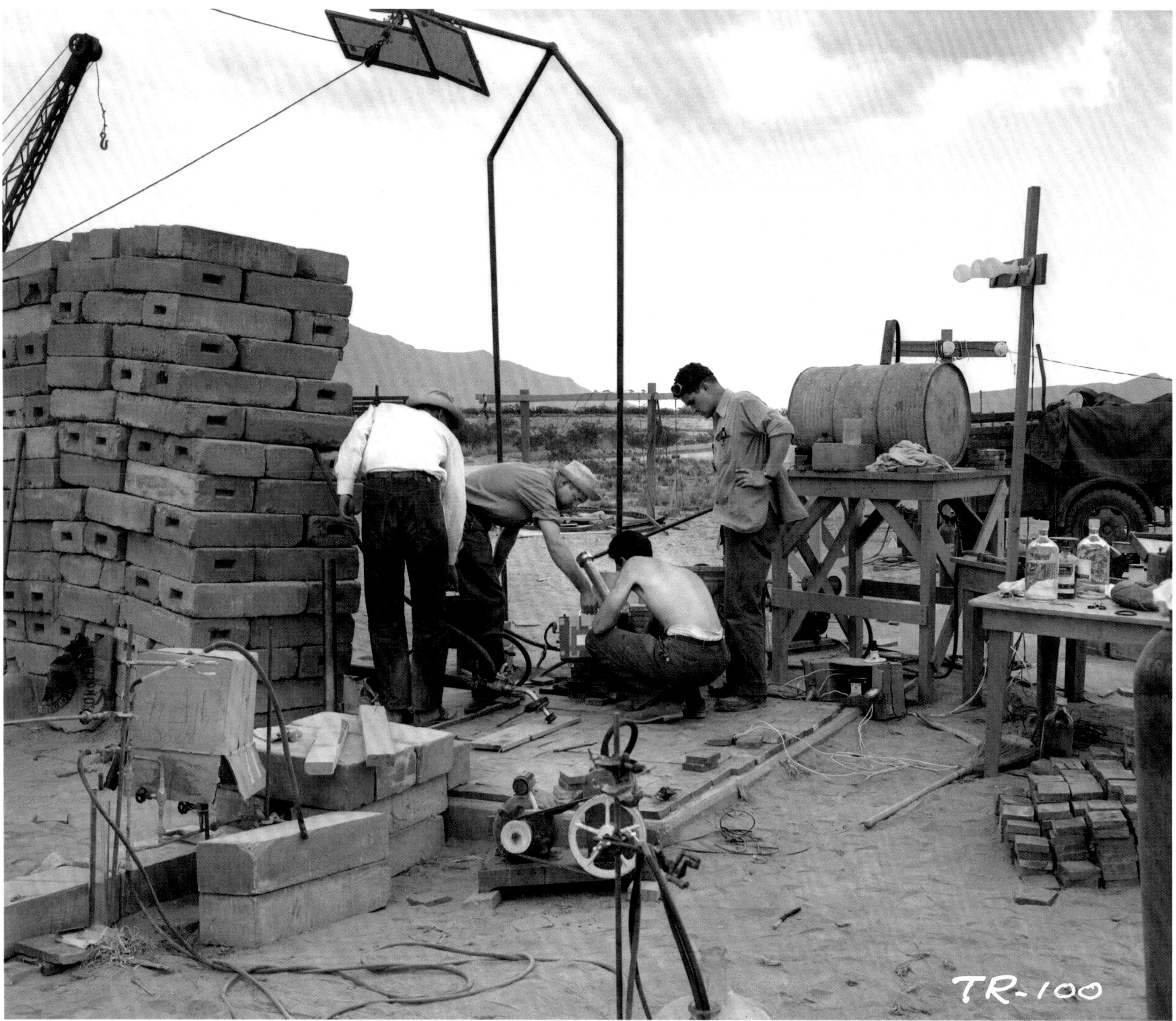

Inside the dissolving tank, the slug would be broken down by acid into a slurry suitable for pumping into the tubes amid the high explosives. Nitric acid—stored in the barrel on the workbench, above—was fed down an inlet tube into the dissolving chamber to begin the process.

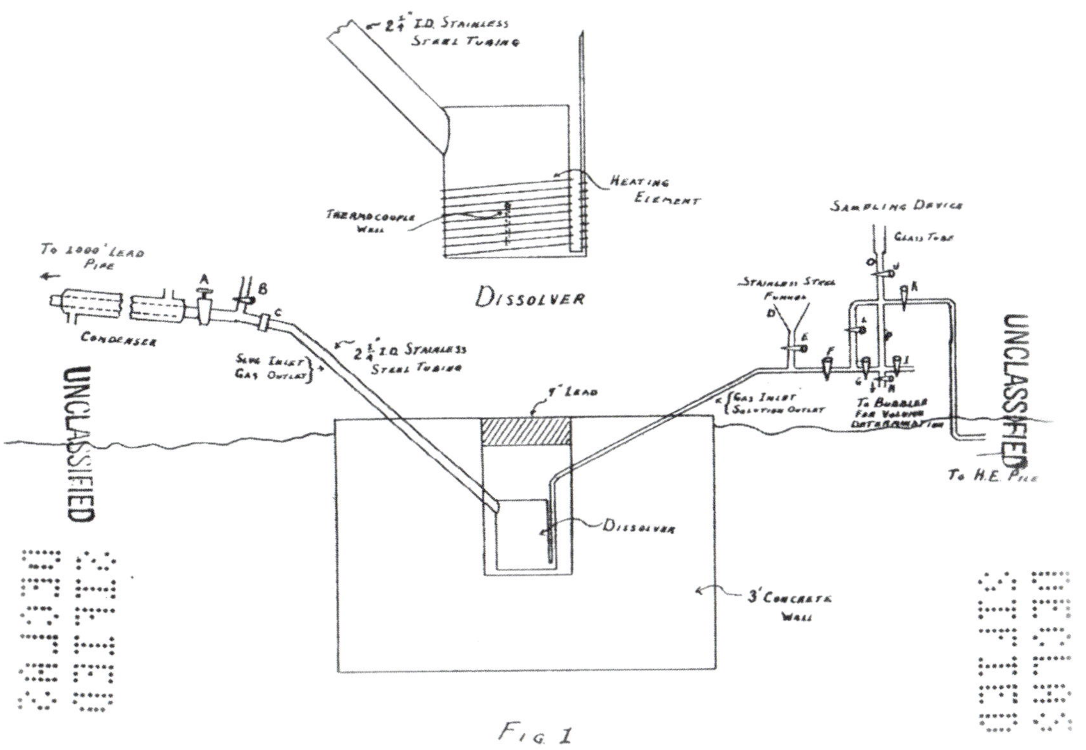

A 1,000-foot exhaust pipe vented away the gaseous byproducts of the reaction, while the slug slowly liquefied underground. Scientists monitored reddish-brown nitrogen dioxide wafting from the end of the pipe with a telescope, waiting for the emissions to stop and signal that the slurry was ready.

As the final step, they added a weaker acid to raise the pH of the mixture, which would prevent it from igniting the high explosives in the event of a leak.

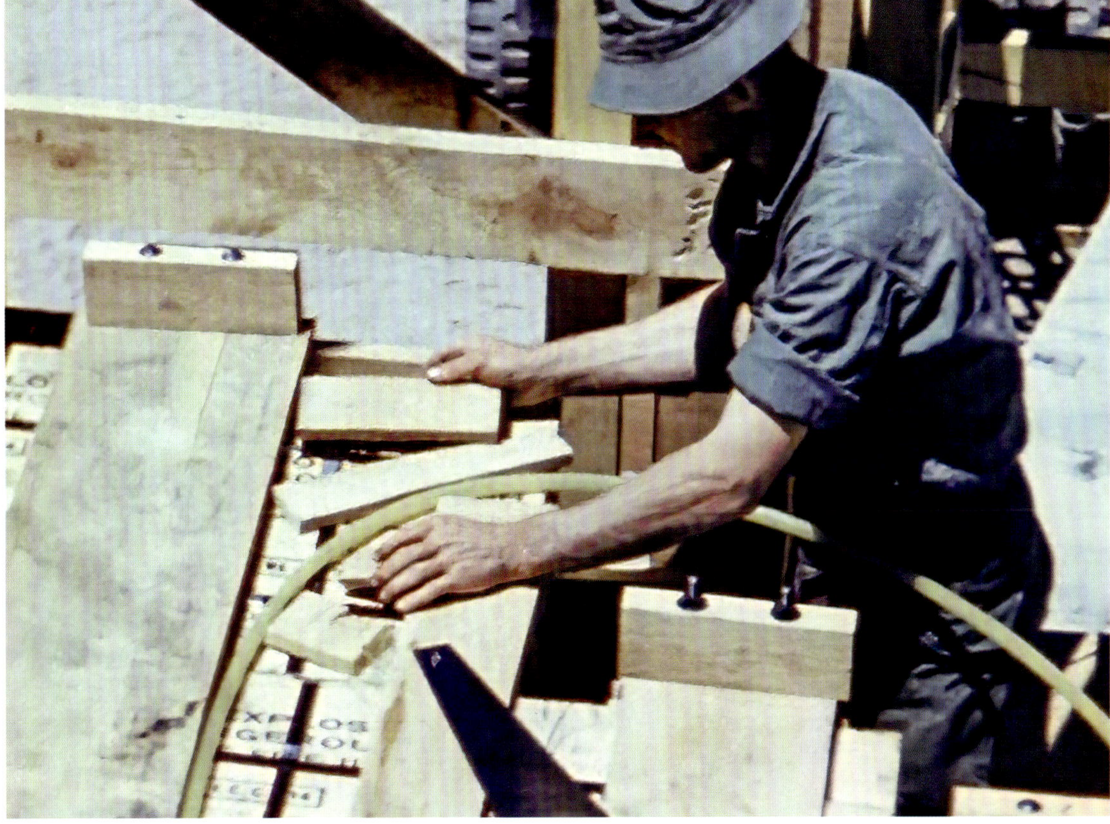

On the afternoon of May 6, sixteen gallons of slurry were pumped from the dissolving tank into the tubing system of the stack, about seventy-five feet away. Since the tubing was made of plastic and its radioactive contents were thus unshielded, access to the area near the 100-ton test became limited to only those personnel with remaining responsibilities for test preparation.

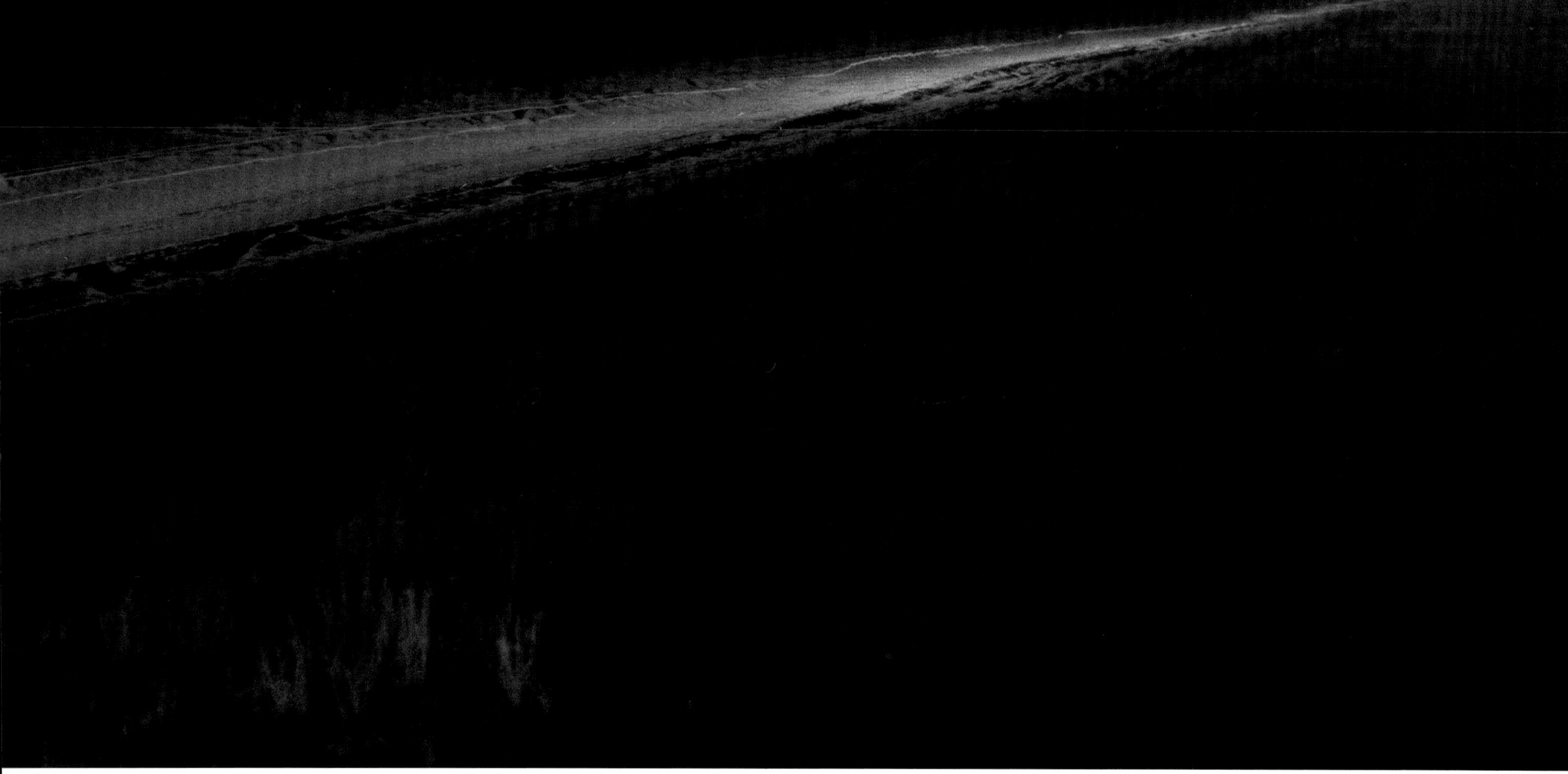

Fully readied for countdown, the tower stood in waiting overnight. A searchlight shines from its location, above, in a photograph taken from west 10,000. The explosion would be detonated from the control bunker (south 10,000) at the same predawn time as the real Trinity test—at least, according to the schedule. As the hour approached, firing was delayed while an observation plane carrying blast gauges navigated into position.

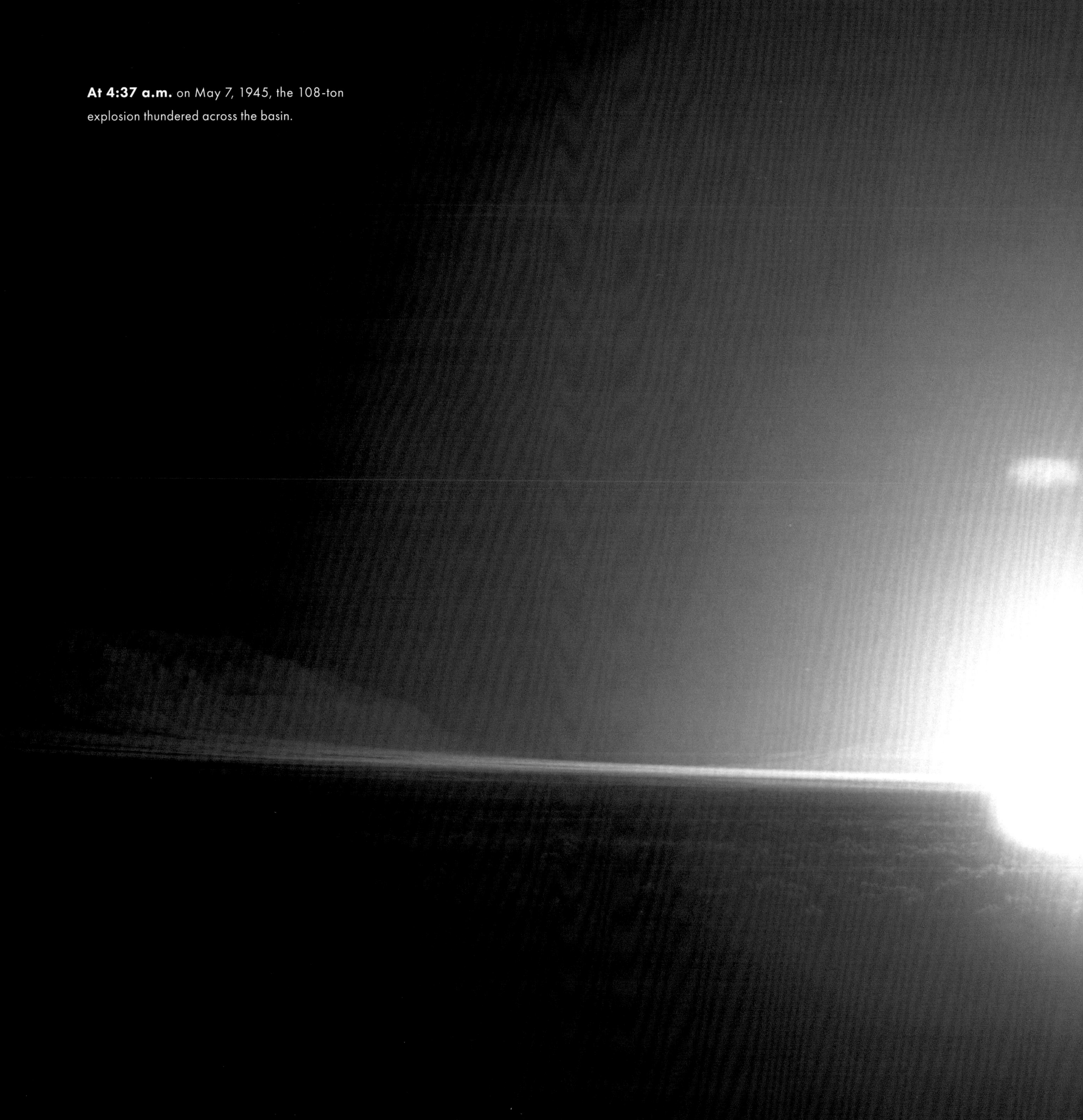

At 4:37 a.m. on May 7, 1945, the 108-ton explosion thundered across the basin.

What began as a brightly glowing orb erupted into a spectacular fireball, beginning to rise and then fading to black as the reaction concluded and visible light raced away.

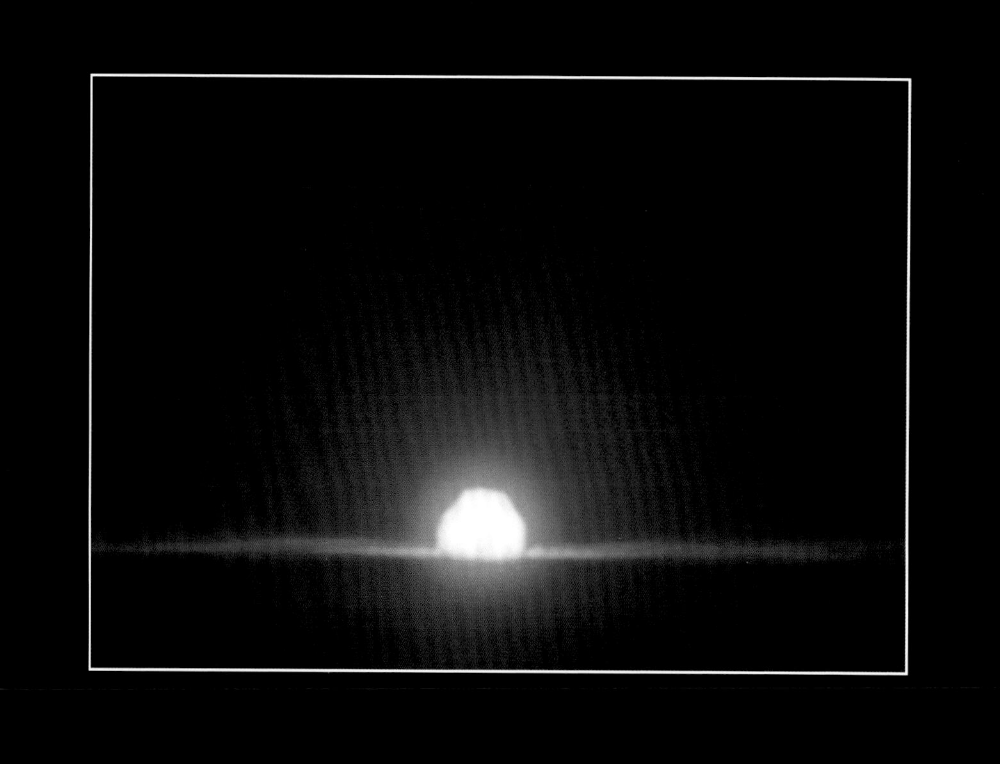

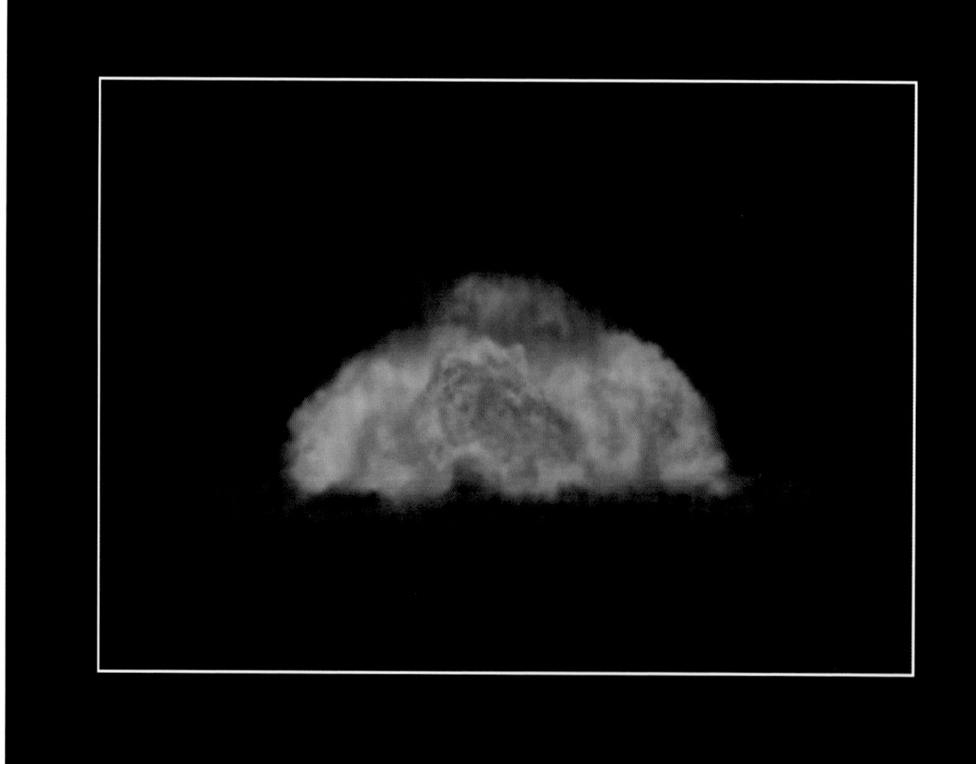

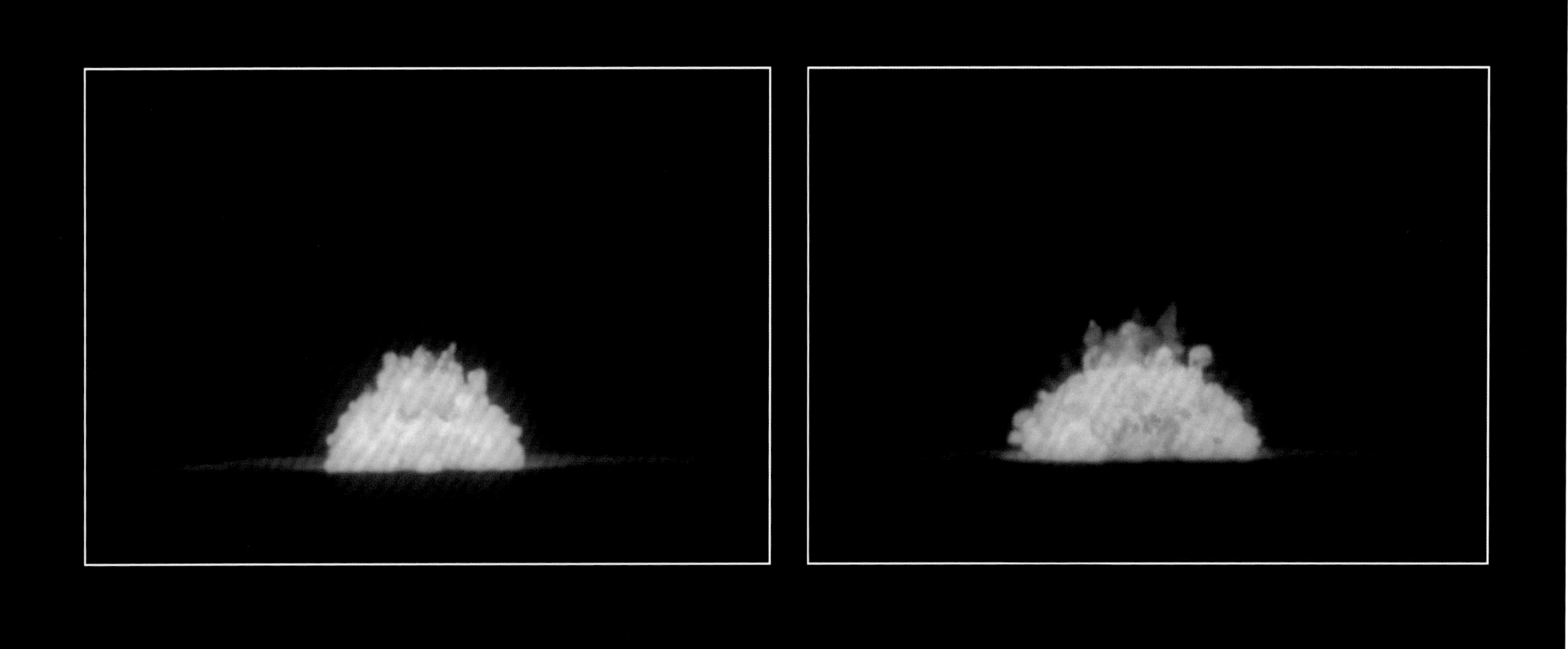

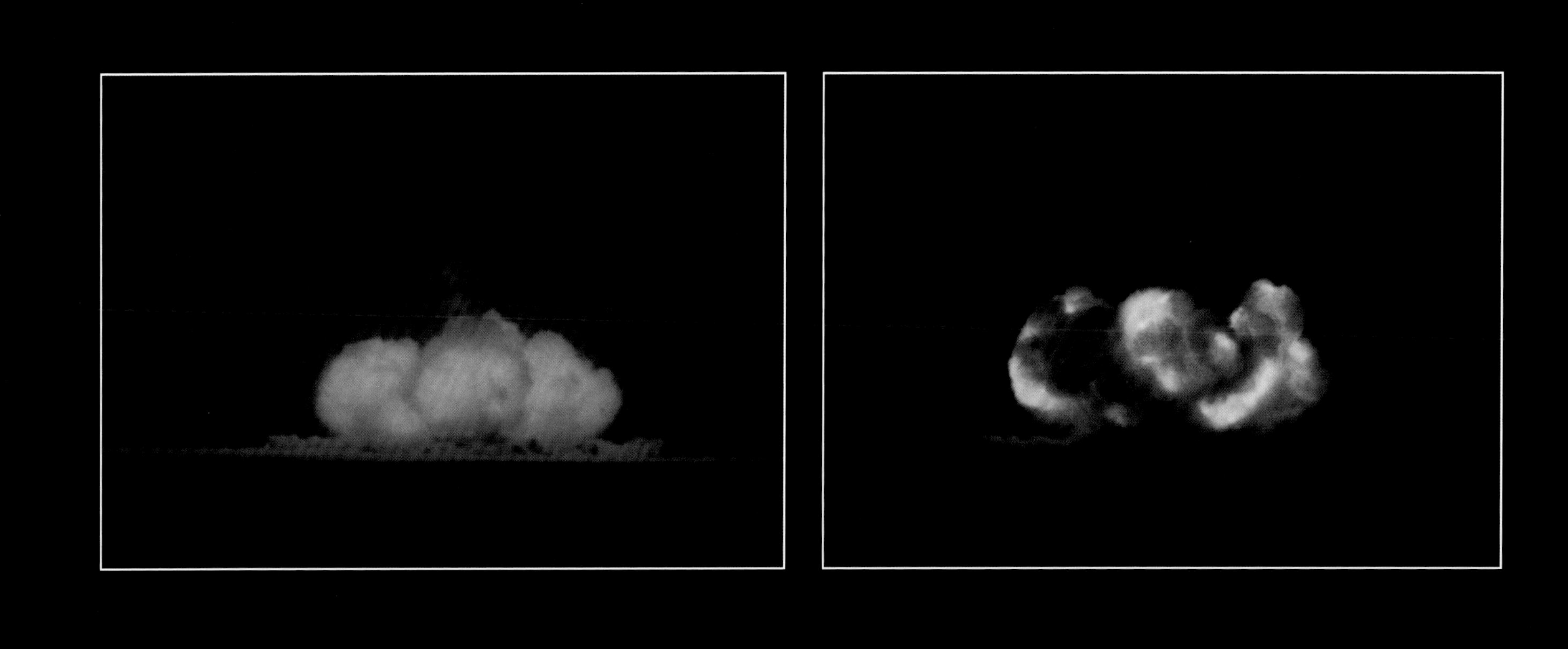

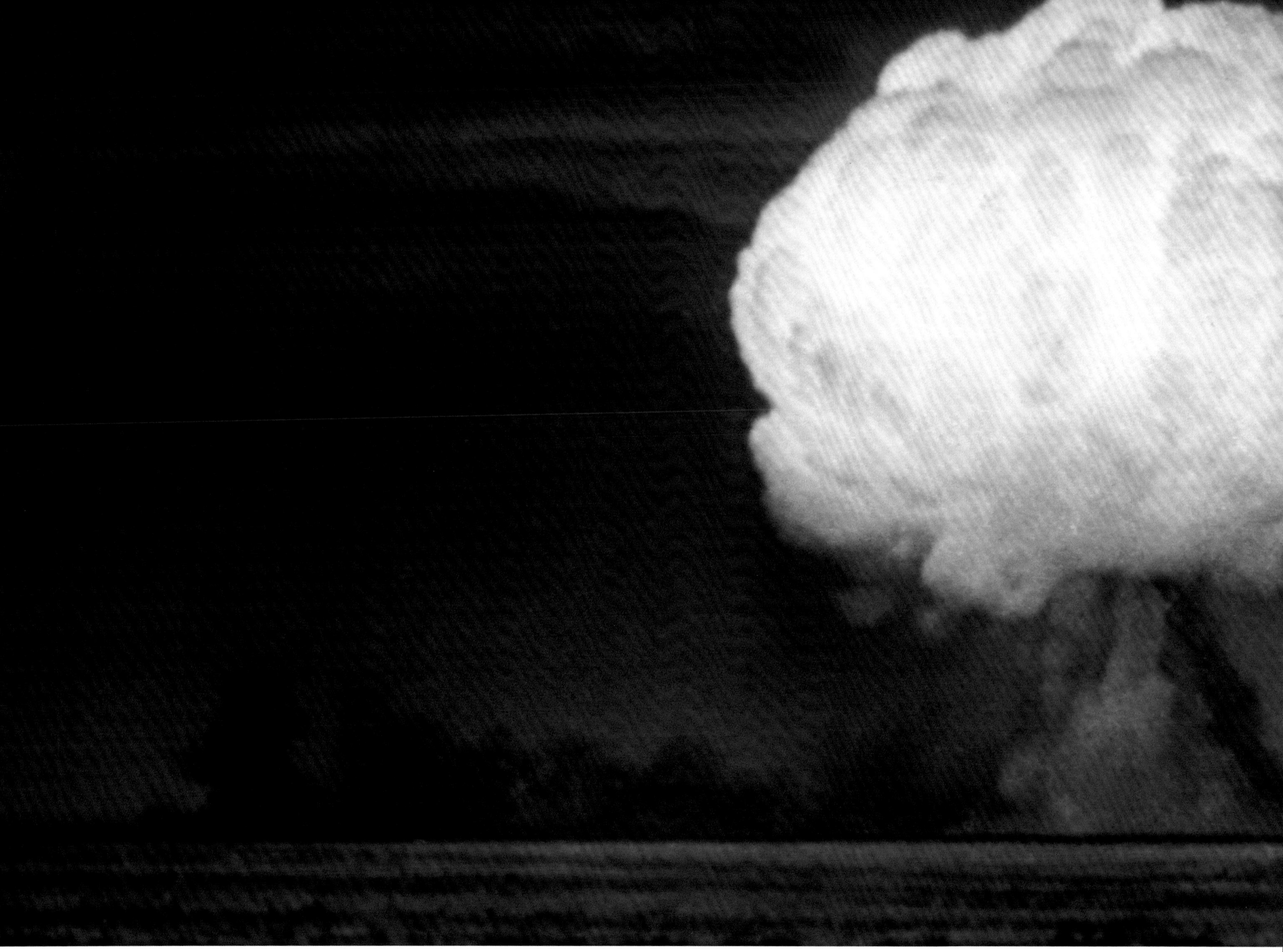

Another camera captured a later stage of the fireball as it morphed into a distinctive cloud, expanding outward at a point 15,000 feet above the desert landscape. The now-familiar mushroom formation, although associated with atomic blasts, can occur after any sufficiently large explosion where extremely hot gases rise quickly into the atmosphere. (Photo courtesy of Defense Threat Reduction Information Analysis Center.)

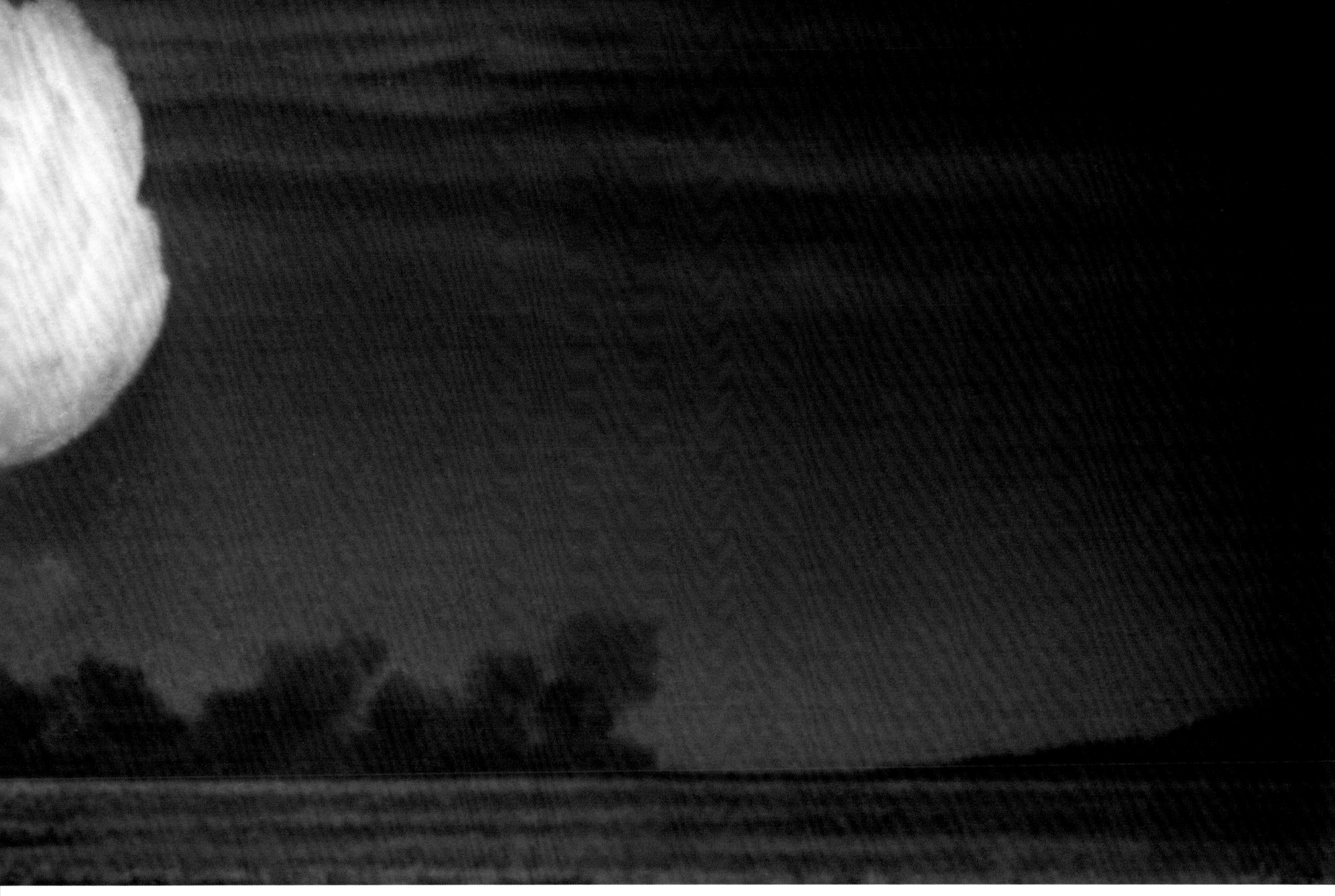

A designated watchperson in the control tower at Alamogordo Army Air Field, sixty miles to the southeast, was able to see and hear the explosion. The cloud drifted away to the east and was reportedly still visible hours later and at least 160 miles away. Despite the impressive audible and visual effects, ground disturbance was imperceptible at base camp (about nine miles away) and even at the 10,000-yard shelters.

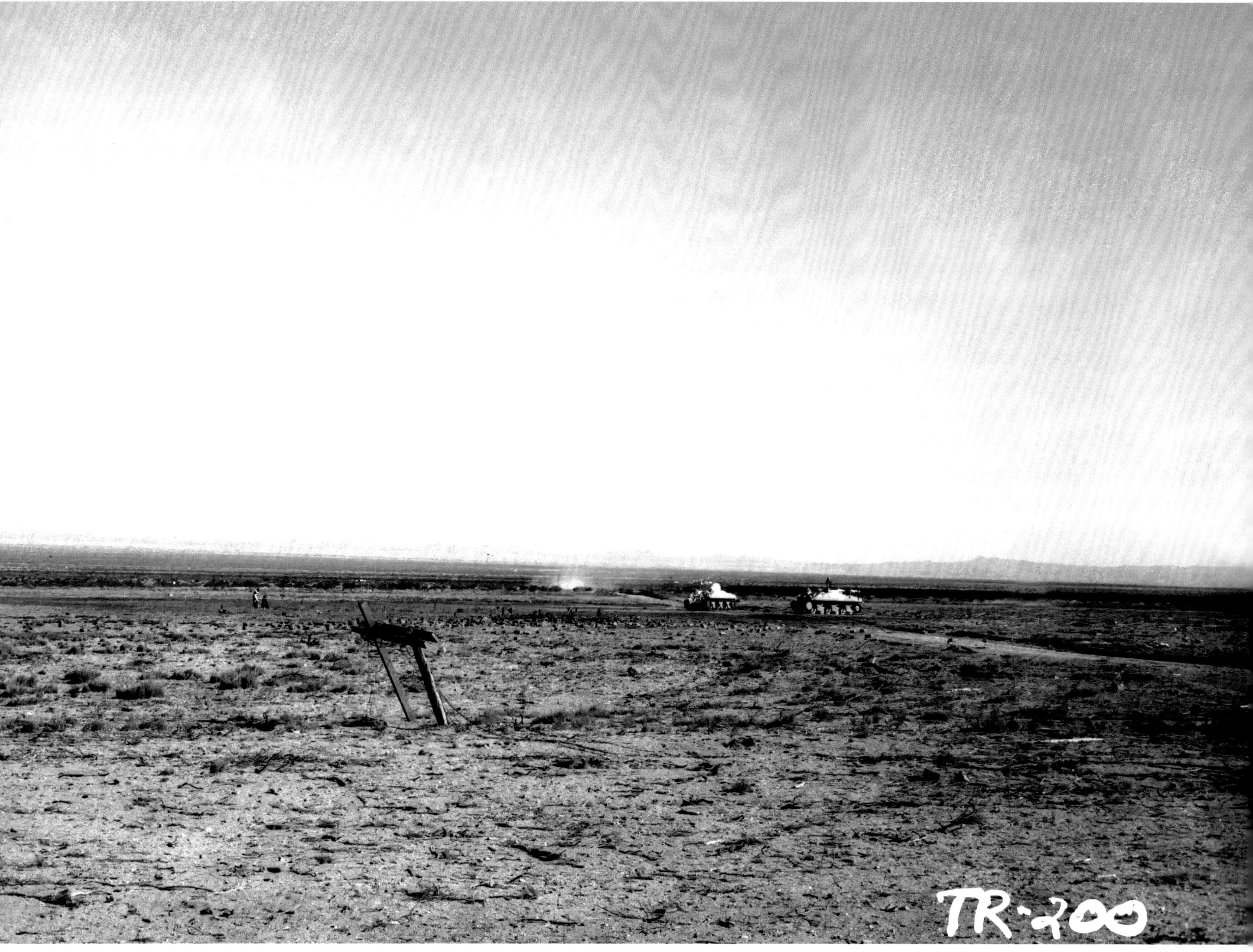

After the shot, Herbert Anderson's Sherman tanks made their way toward the site of the explosion, where scientists were surprised to find a crater only about four feet deep and thirty-six feet across. Construction crews returning to the Trinity site for their next assignment would later be surprised to find nothing remaining of the twenty-foot platform.

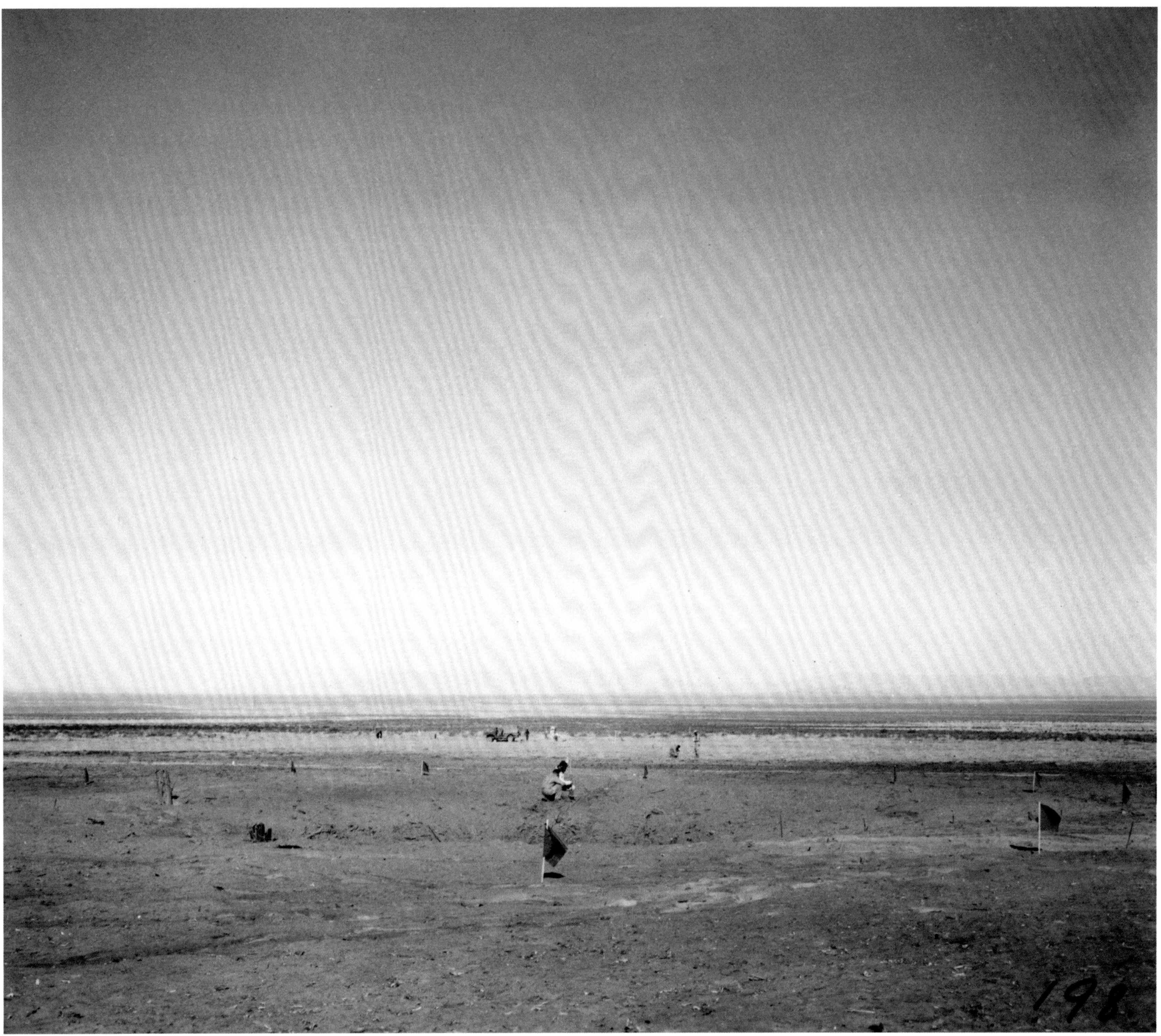

The tanks were able to navigate the rubble and sample the brush-cleared area successfully, collecting soil that would hopefully contain remnants of the radioactive slurry for analysis. Meanwhile, members of the Health Group marked an area around the crater with flags as they assessed the level of contamination, finding another surprise: Only a very small percentage of the radioactive material from the tubing had been deposited near the shot. Most seemed to have been dispersed, quite unexpectedly, into the atmosphere. This was a troubling indication of how radioactive fallout might behave after the nuclear explosion.

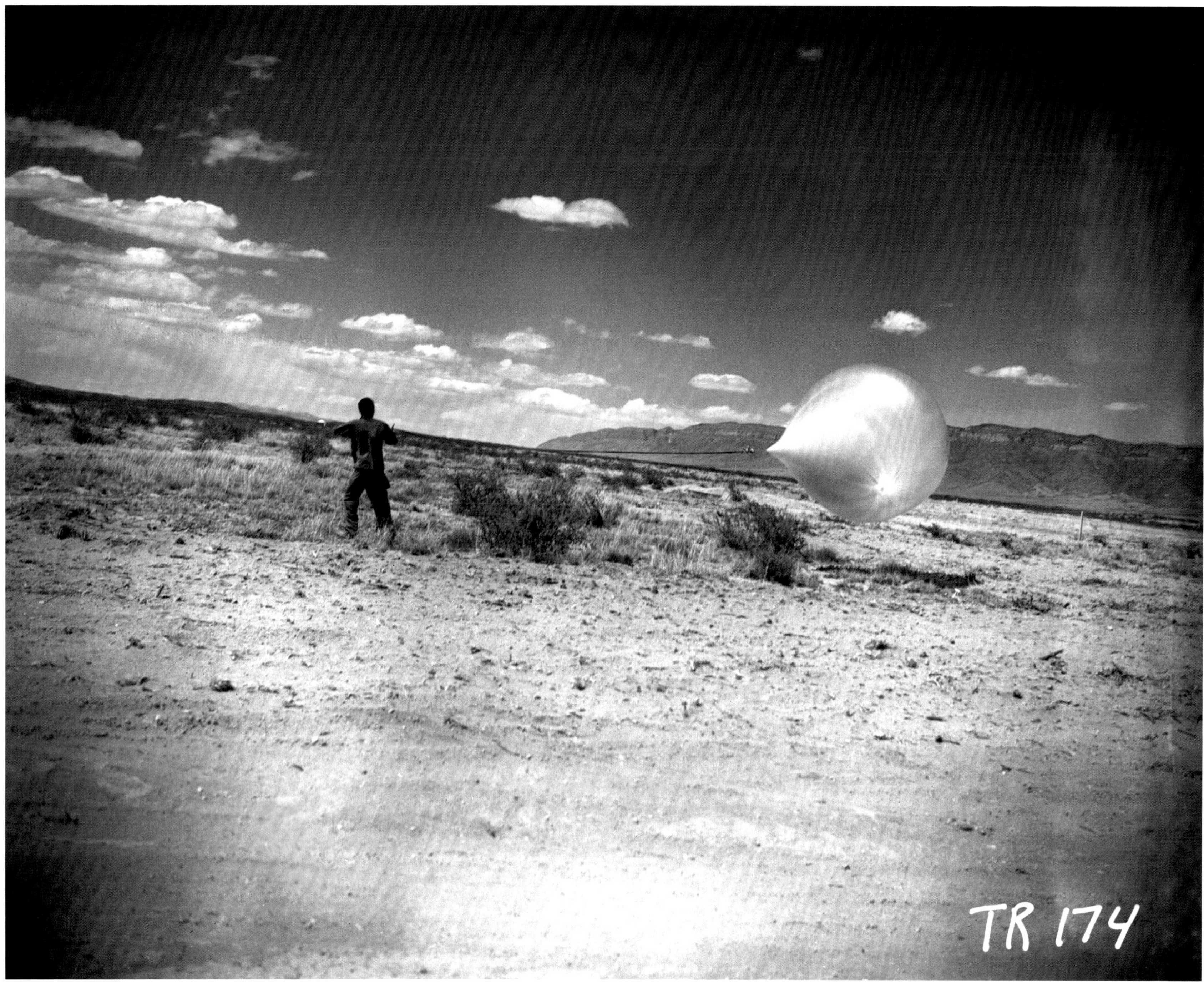

On the diagnostics front, results of the trial run were exciting. Anderson's team affirmed that soil samples collected by their tanks should allow an accurate measurement of the Gadget's yield. Various gauges correctly registered the 108-ton yield of the blast as well. In general, the instruments set up for the practice test performed as designed—although there were some notable failures. The air-dropped gauges that had delayed the shot did not provide useful results, nor did the attempt to elevate strands of Primacord with helium-filled balloons, which could not provide sufficient loft for the material. Most troubling among the various mishaps was a timing malfunction that caused the high explosives to detonate sooner than intended. After running over half an hour behind, the blast was ultimately a quarter of a second early. This meant that instruments designed to record the first milliseconds of the rehearsal shot activated a moment too late. Losing those early data during the real Trinity test would have been disastrous.

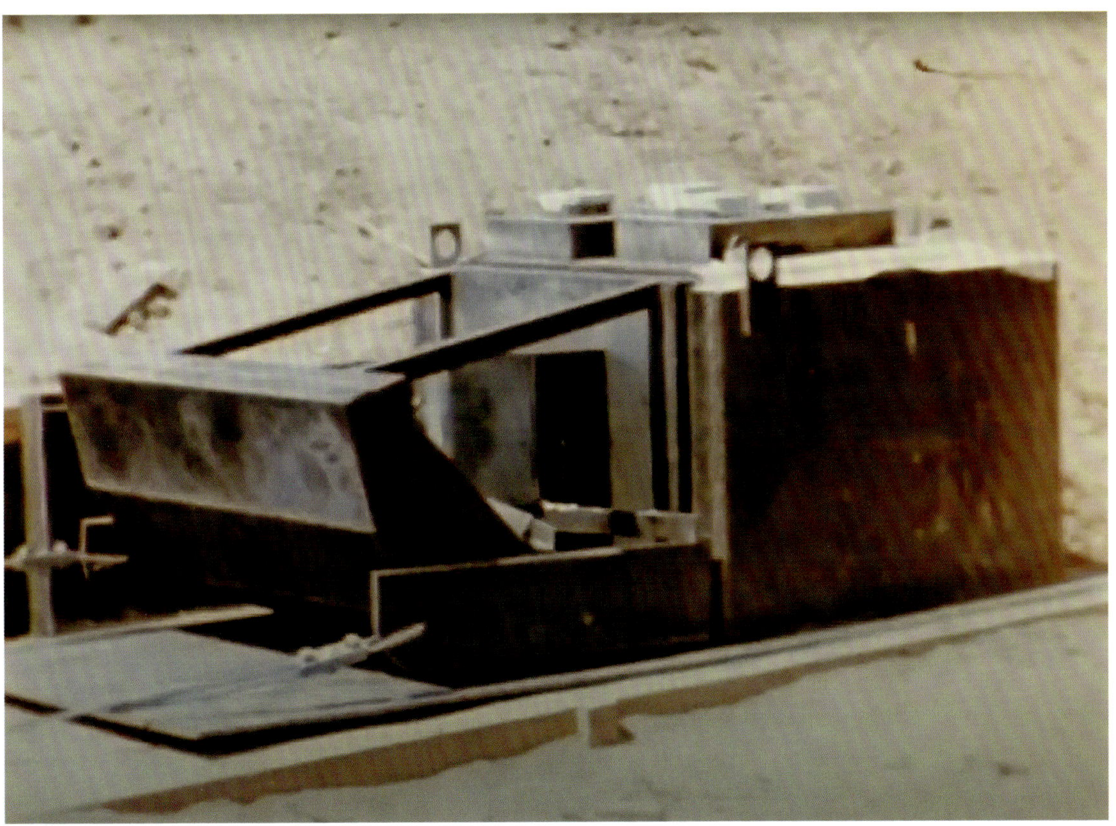

Despite the explosives detonating ahead of cue, timing signal devices were shown to communicate quite effectively with the various instrument stations, including the 800-yard Fastax bunkers.

There was, however, another problem with the close-range Fastax cameras. Ongoing studies were making clear that at only 800 yards, gamma radiation from the Gadget would penetrate the thick portholes of the bunkers, spoiling camera film. After the 100-ton test, the Spectrographic and Photographic Measurements Group decided to relocate the cameras into new, smaller protective boxes made of lead (*left*).

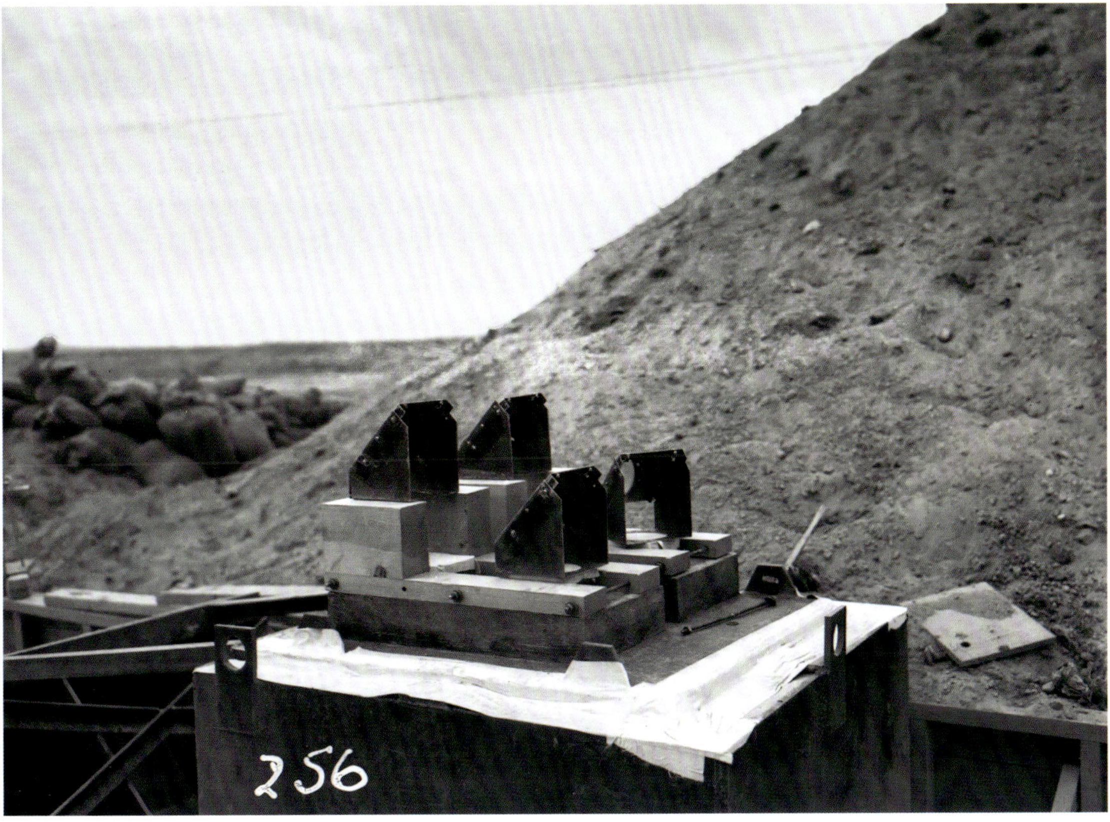

The four cameras in each new bunker were angled straight up toward thickly leaded glass windows, facing the sky instead of the nuclear explosion. Mirrors above the four windows were affixed at forty-five-degree angles so that the cameras could film the detonation indirectly: radiation would pass straight through the mirrors, while light would reflect down into the camera lenses.

As an added benefit over the now-abandoned stationary bunkers, the Fastax bunkers were light enough to be towed out of the contaminated area from 1,000 feet away via steel cables—a new responsibility assigned to the drivers of the tanks.

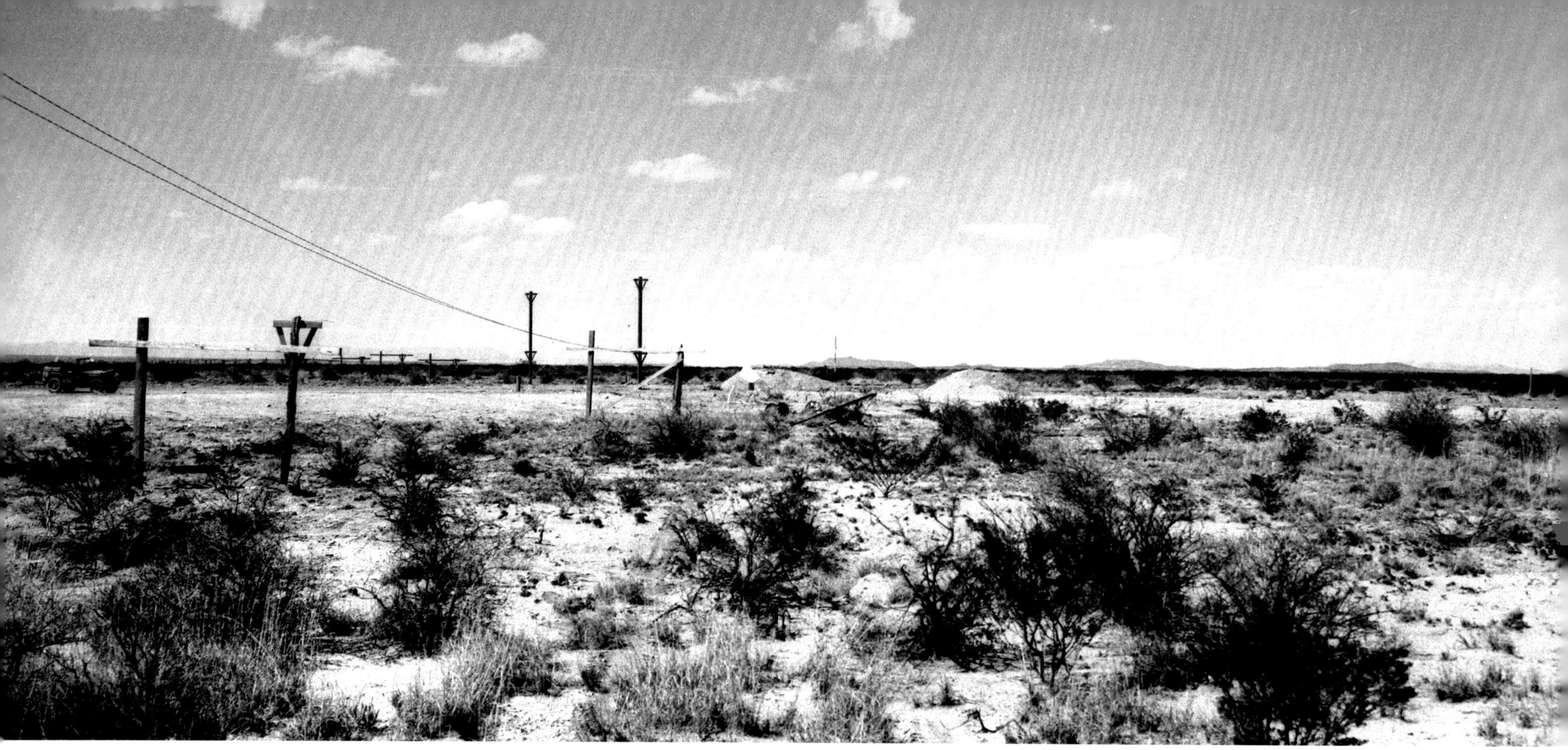

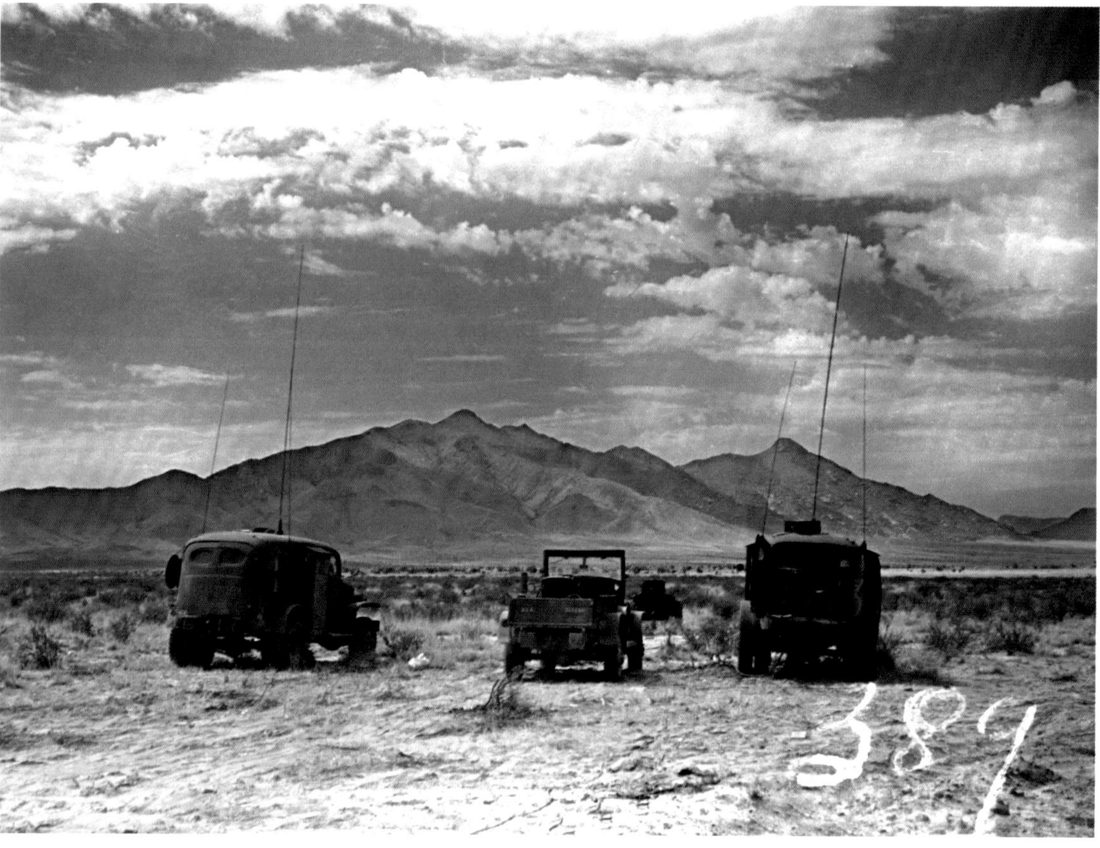

While calibrating experiments was a critical part of the 100-ton test, many of the problems revealed on May 7 were logistical in nature. With this in mind, Bainbridge later remarked that the rehearsal "accomplished what we had hoped for, or feared."

That is, long-standing problems could no longer be ignored, with their impact to operations put on full display. Hundreds of miles of power lines, for instance, had originally been hung from low poles to allow for easy repairs and layout changes. The idea fully backfired by increasing the frequency with which repairs were needed, including last-minute emergency fixes that sent workers into restricted areas after the test site had already been cleared for firing.

Following the 100-ton test, a set of high poles was installed every 100 or 200 yards so drivers could pass underneath with ample clearance; as shown at left, some vehicles were quite tall with radio equipment installed.

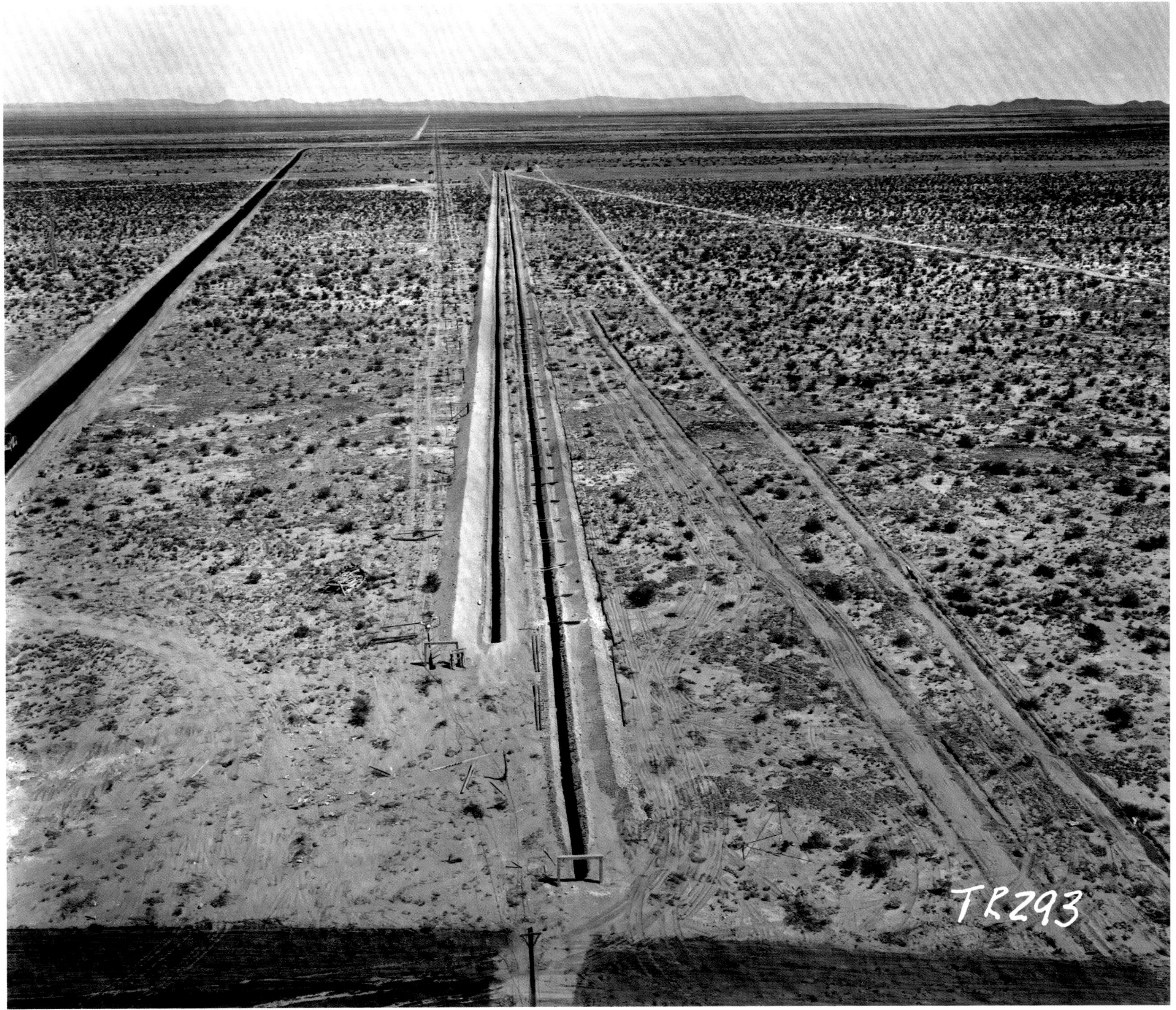

Upgrades were also approved for the problematic dirt roads that crisscrossed the basin. Agreeing to a thin, waterproof seal over the adobe and caliche, Groves authorized the blacktopping of twenty-five miles of roadway, including an area surrounding the shot tower (visible at the bottom of the photo above). In addition to controlling dust, the improvements would help reduce repair work needed to keep vehicles operable. Bainbridge later characterized the new roads as "a superlative example of planned obsolescence. . . . Built to last three months, they started to break up 92 days after completion."

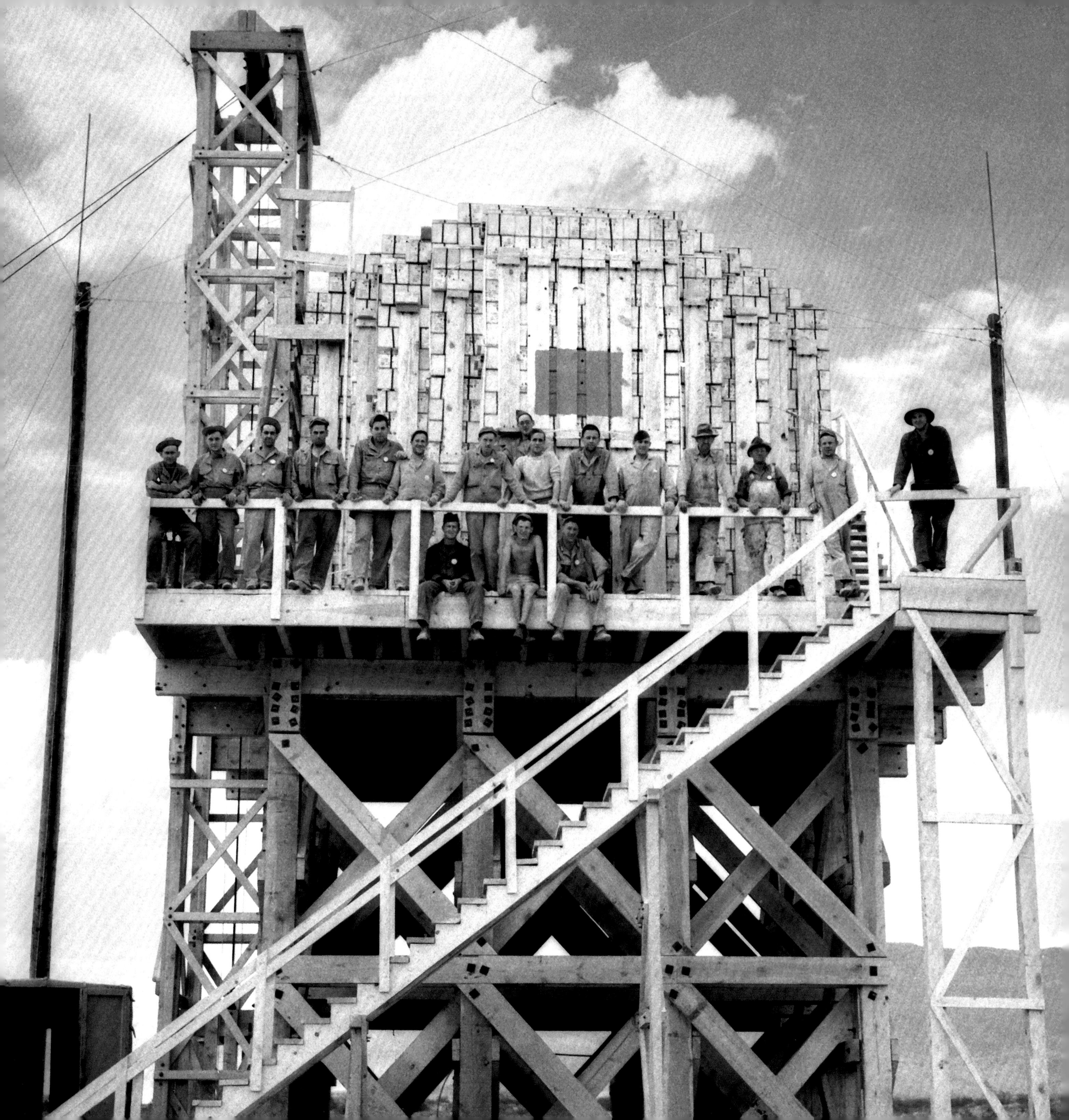

For all the problems it revealed, the 100-ton test also bolstered confidence in the upcoming Trinity test and was viewed afterward by Bainbridge and others as an indispensable component of preparation. It was also a tangible representation of the military value of nuclear weapons. The Gadget, small enough by design to fit inside an aircraft, was predicted to explode with a force equivalent to 4,000 tons—or another thirty-nine stacks—of TNT.

Armed with well-earned new information, the Trinity workers entered the home stretch of preparation, beginning to assemble the final pieces of the puzzle as their counterparts at Los Alamos used every minute and resource to finalize the test device itself.

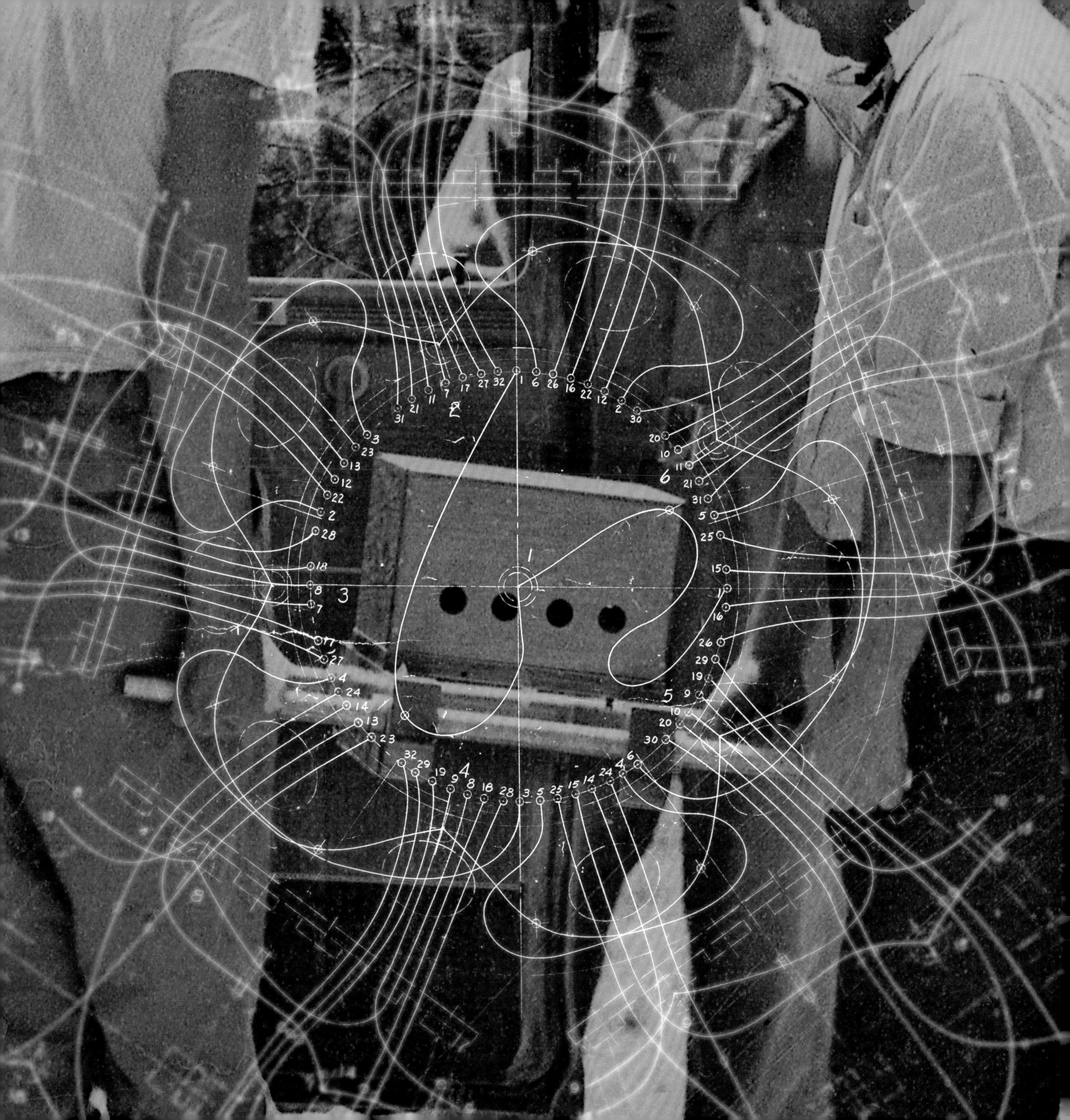

04 Gadget Complete

Just over two years after the laboratory at Los Alamos opened its doors, the many parts of the Manhattan Project were at last coming together. Though the urgency and necessity of the project's success were initially fueled by the threat of a German nuclear bomb, the plan was never to use a working weapon against a specific German target. This was in part because of the risks associated with a failed device falling into German scientists' hands. Instead, the earliest known plan, discussed among Groves and a small group of advisers in May of 1943, was to detonate the first bomb above the Japanese fleet stationed at Truk Harbor; if the weapon was a dud, it would ideally be lost in the deep ocean, and if it was recovered, it would not be the Germans who obtained nuclear secrets for their own use. Such precautionary thinking dissolved by mid-1944—as the tide of war turned in the Allies' favor—into the expectation that Hitler's regime would unravel before either a German or American atomic bomb became available.

Indeed, when the Soviet army took Berlin in April of 1945, the German dictator took his own life; a week later—on the day of the 100-ton test—Germany surrendered unconditionally. Despite the fall of Japan's only remaining ally and the end of fighting in Europe, the lengthy conflict in the Pacific Theater wore on. Hundreds of American bombers, launching air raids from captured islands, continued to bring the full might of conventional firepower to major cities across Japan; the devastating attacks killed and injured hundreds of thousands of civilians. At the same time, a naval blockade hindered the Japanese government's efforts to transport food, reinforcements, and other supplies. An eventual invasion of the mainland, though costly to both sides, would inevitably end in an Allied victory if carried out. And yet, Japanese leaders—for numerous reasons, including the desire to retain the nation's Imperial system of government—were unwilling to accept the unconditional surrender demanded by the United States.

As the bloody, nearly three-month Battle of Okinawa was unfolding in the Pacific (through late June of 1945) and as development of the Gadget progressed on the home front, scientists and government officials alike began turning attention to the prospect of a successful test at Trinity. President Harry S. Truman, inaugurated on April 12 following the sudden death of Roosevelt, had inherited his predecessor's commitment to the Manhattan Project and a fast-approaching decision as commander in chief that had been discussed abstractly but was now an imminent reality—how to leverage a working atomic bomb. With Truman's approval, Secretary of War Henry Stimson convened a civilian advisory group called the Interim Committee on Nuclear Power, which considered throughout May and into June the long-term implications and policy decisions attached to nuclear technology as well as the more immediate question of how it could be used to secure Japan's surrender.

A Scientific Panel that included Oppenheimer and Fermi represented the views of the weapon's creators to the Interim Committee. Scientists' opinions ranged broadly: Some advocated for a technical demonstration, fearing that military use would prejudice the United States' ability to negotiate an international ban on nuclear arsenals. Others supported military use, emphasizing the opportunity to hasten the end of the war and the possibility that the existence of nuclear arsenals could prevent future wars altogether. In a June 16 letter, the Scientific Panel endorsed this latter view, expressing an "obligation to our nation to use the weapons to help save American lives." They concluded that a technical demonstration was unlikely to make an impact strong enough to induce surrender.

The Interim Committee concurred, reaffirming its earlier recommendation that the weapon should be used as soon as possible, no warning should be given, and the strike should be against a dual target: a war plant surrounded by workers' homes. It also suggested that President Truman reveal the atomic bomb's existence at a forthcoming meeting in Potsdam, Germany, between the leaders of the Big Three—the United States, Great Britain (already involved in the Manhattan Project), and the Soviet Union. Among other advantages, alerting Soviet leader Joseph Stalin—a close ally during the war but an ideological adversary—to the bomb's existence might help soften the Soviets' bargaining posture and increase the likelihood of future cooperation in establishing international controls for nuclear weapons.

Going into policy negotiations, Truman needed to know if the United States possessed a nuclear monopoly. With the Potsdam Conference set to begin on July 17, the laboratory's Cowpuncher Committee, responsible for keeping the Trinity schedule on track, asked managers on June 30 to provide the earliest possible date by which their respective groups, including those scrambling to prepare the Gadget, would be ready for the test. A new target date of July 16 emerged, aligning just barely with political needs. But project meteorologist Jack Hubbard—tasked with identifying feasible conditions for the test shot—predicted a problem that might jeopardize the operation: stormy skies.

Despite potential interference from the weather, a detailed schedule for pre-test dry runs and the eventual "TR Hot Run"—the full-scale test shot—was issued on July 1. While Hubbard closely tracked the forecast, a rush of final construction and calibration at Trinity came to a close as the test device began arriving in pieces on July 12. Near the end of the Hot Run schedule, concluding the day of July 14, two underlined words announced that scientists would soon cede control over their creation to the natural process of fission and then—if their efforts were successful—to the waiting US military: Gadget complete.

* * * * * * *

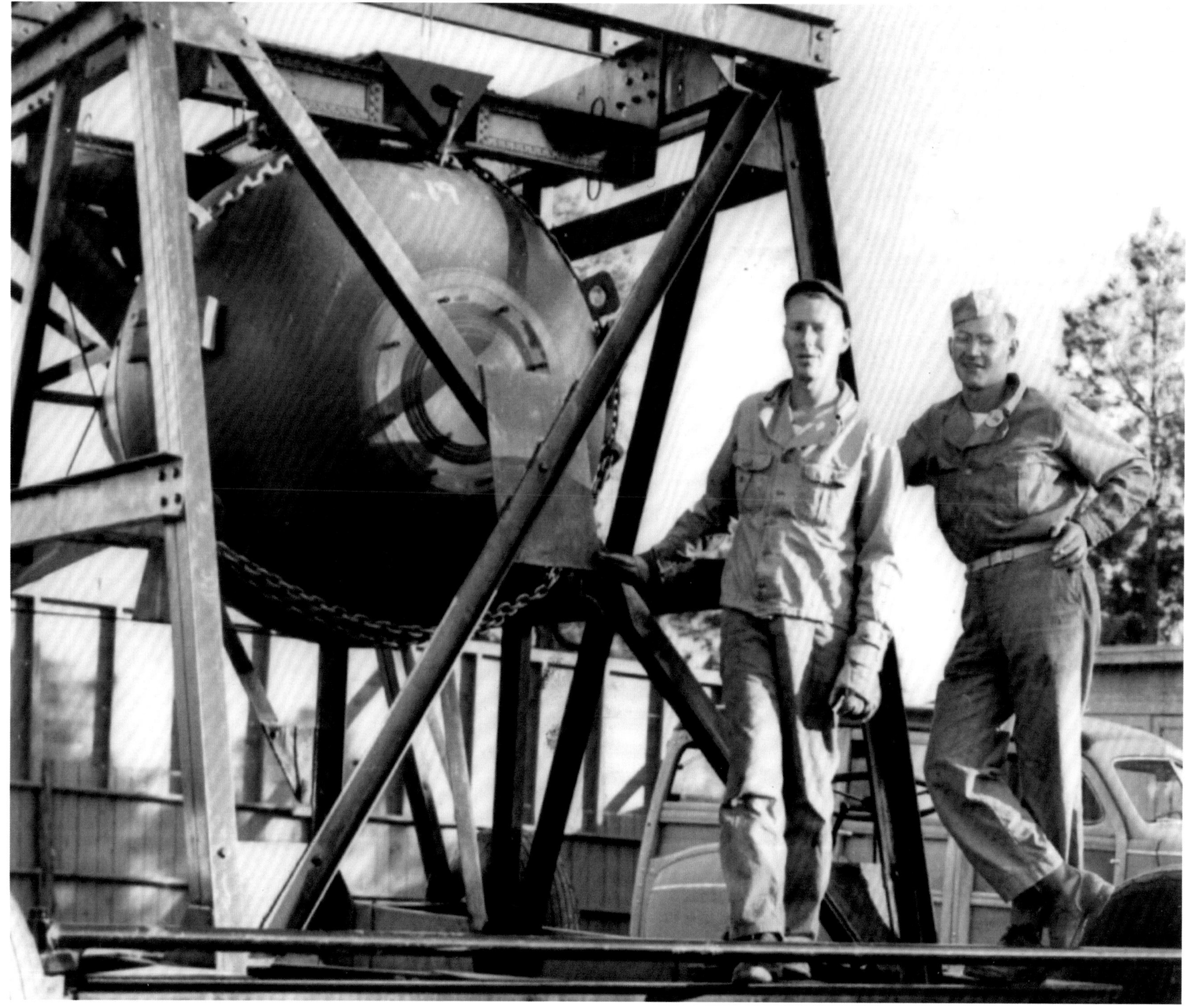

As final preparations were unfolding at the Trinity site, the Explosives Division was perfecting protocols for moving the Gadget across the rough roadways to Trinity and, once there, to ground zero. Many so-called shake tests had been conducted at Los Alamos's V-Site —a high-explosives handling and assembly facility—beginning in 1944, often in metal contraptions (like that above) meant to replicate the vibrations and in some cases even temperature conditions of a bomb bay. These tests proved useful, too, for planning ground transportation.

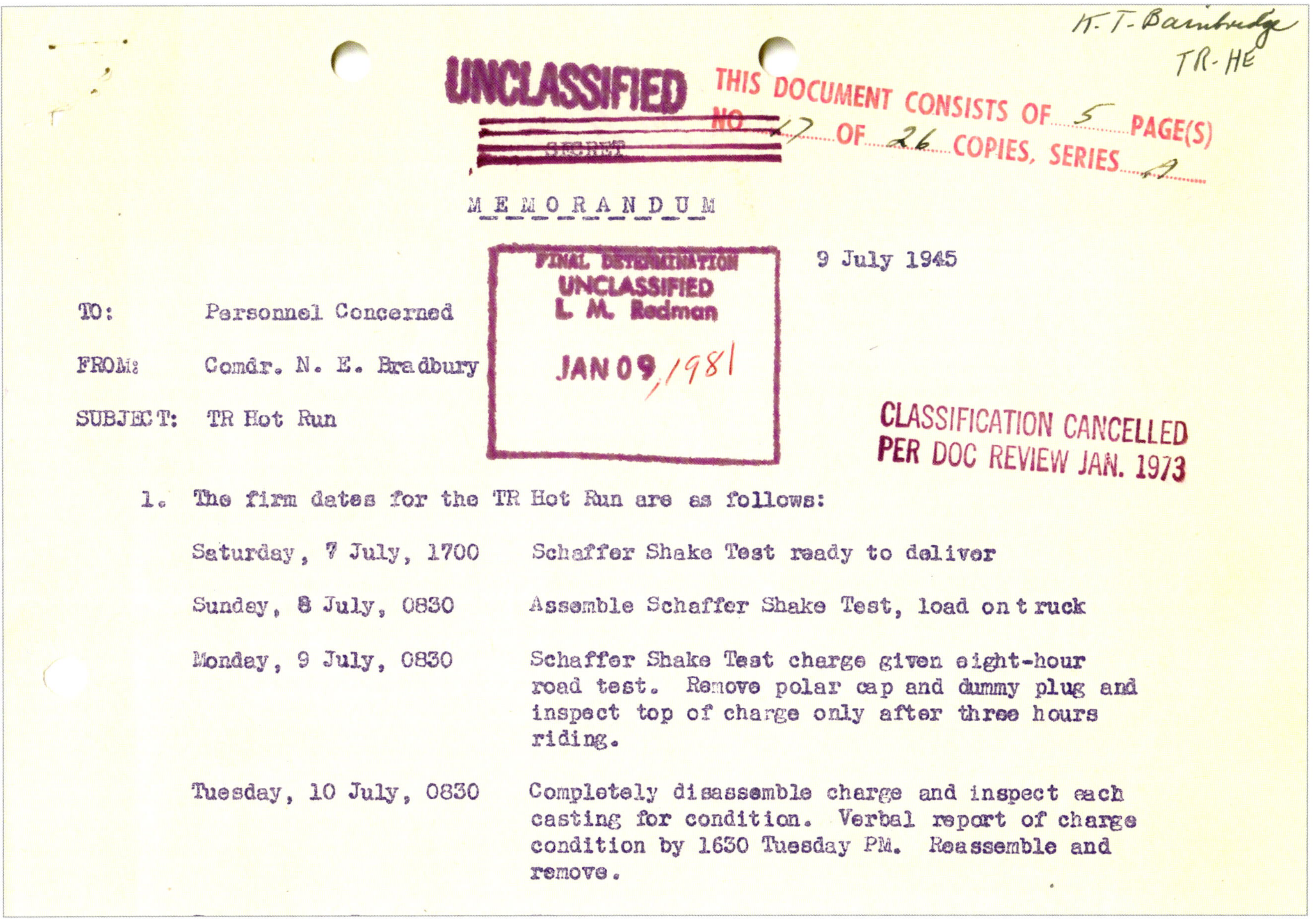

K. T. Bainbridge
TR-HE

UNCLASSIFIED THIS DOCUMENT CONSISTS OF 5 PAGE(S)
NO 17 OF 26 COPIES, SERIES A

SECRET

MEMORANDUM

FINAL DETERMINATION
UNCLASSIFIED
L. M. Redman
JAN 09 1981

9 July 1945

TO: Personnel Concerned

FROM: Comdr. N. E. Bradbury

SUBJECT: TR Hot Run

CLASSIFICATION CANCELLED
PER DOC REVIEW JAN. 1973

1. The firm dates for the TR Hot Run are as follows:

Saturday, 7 July, 1700 — Schaffer Shake Test ready to deliver

Sunday, 8 July, 0830 — Assemble Schaffer Shake Test, load on truck

Monday, 9 July, 0830 — Schaffer Shake Test charge given eight-hour road test. Remove polar cap and dummy plug and inspect top of charge only after three hours riding.

Tuesday, 10 July, 0830 — Completely disassemble charge and inspect each casting for condition. Verbal report of charge condition by 1630 Tuesday PM. Reassemble and remove.

On July 8, with the Trinity test just over a week away, the best practices developed earlier at V-Site were put to use. A full-scale mockup of the test device—with four live, high-explosive lenses—underwent an eight-hour, on-road shake test designed to draw out any effects that transportation might have on the Gadget. If there was any movement among the three-dimensional lenses of the real test device, everything had to come apart and the all-day process of assembling the lenses would begin again, creating a major delay. Capt. Wilbur F. Schaffer drove over rough terrain with the mockup strapped carefully to a truck bed, checking it briefly after three hours and then fully disassembling the device for a thorough inspection the following day. The Schaffer Shake Test was a success: all charges remained in perfect alignment.

2293 1265

Tuesday, 10 July, 1730	TR and Creutz charges ready for delivery. Start papering. Arrange for night shift if necessary to paper charges. Additional personnel will be furnished as required.
Wednesday, 11 July, 0830	Information will be furnished as to which charges will be used in TR shot and which in Creutz shot. Separate charges. Complete papering of TR charges. Complete papering of Creutz charges (use separate groups--request additional personnel as needed).
Wednesday, 11 July, 1730	Both charges completely papered. Work another night shift if necessary to complete this job. Request personnel as necessary. This job must be done so that assembly of both charges can start on Thursday, 12 July, at 0830.
Thursday, 12 July, 0830	Use two groups--one at V Site to assemble TR charge (Lt. Schaffer supervise), and one at Pajarito for Creutz charge (Mr. North supervise). Tamper needed by 1000 at V Site.
Thursday, 12 July, 1500	Assembly of TR charge complete. Notify interested personnel that it is ready for inspection if desired.
1600	Seal up all holes in case; wrap with scotch wrap (time not available for strippable plastic), start loading on truck. Tie down to truck body.

This document contains information affecting the national defense of the United States within the meaning of the Espionage Act U.S.C. 50 31 and 32, its transmission or the revelation of its contents in any manner to an unauthorized person is prohibited by law.

~~SECRET~~

UNCLASSIFIED

Transportation procedures were settled, but even a week before the scheduled test day, a full suite of high-explosive lenses was not yet available for use in the Gadget. The lenses encountered delay after delay during manufacturing, as builders attempted to cast them into perfect shapes that would create the perfect compression needed around the plutonium core. When the first batch was finally ready, George Kistiakowsky and Norris Bradbury, who authored the Hot Run procedures excerpted above, picked out the best-crafted lenses for use in the Trinity device. However, even some of the finest castings were rife with air pockets, which scaled testing had shown would distort the shock wave from the high explosives and possibly cause a fizzle. It was July 10, and there was no time to procure new lenses; daily dress rehearsals were already beginning at the test site. Kistiakowsky later recalled "spending most of one night, the week before the Trinity test, drilling holes in some faulty castings [with a dentist's drill] so as to reach the air cavities indicated on our x-ray inspection films." He filled the cavities with molten explosive slurry, salvaging enough of the lenses to build two complete spheres.

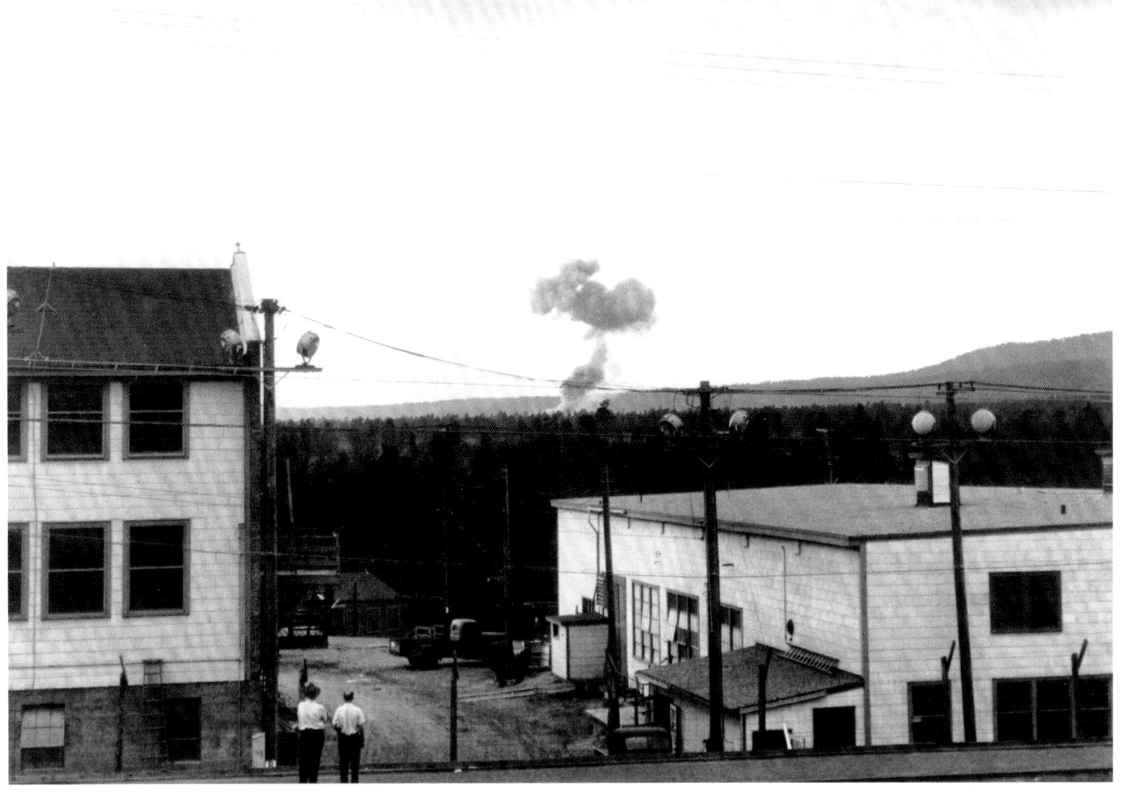

The precise configuration of the implosion lenses had evolved over many months as scientists conducted thousands of tests using prototypes. The photo at left, taken from the town of Los Alamos, shows a test in the distance at a high-explosives firing site.

In the course of their work, various groups at the laboratory invented no fewer than seven methods of studying implosion. Due to the delay in lens production, however, none of these methods had yet been used to observe a detonation of the entire, full-scale explosive lens assembly that would soon be set off inside the Gadget.

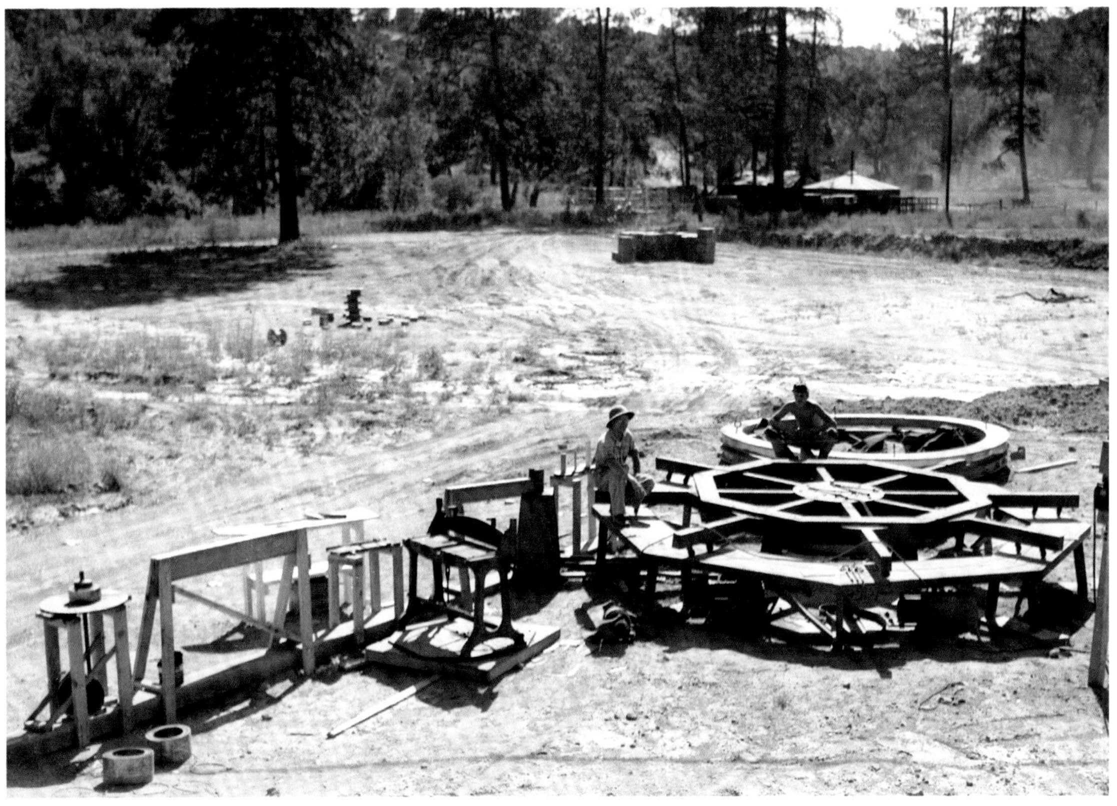

With only days remaining until the first nuclear test, as the Gadget was being assembled at V-Site, a Gadget-like device—consisting of a complete implosion lens system but lacking the plutonium core—was simultaneously assembled at Los Alamos's Pajarito Site by a group under physicist Edward Creutz. The doppelganger was to be fired two days before the Trinity Gadget in an octagonal testing apparatus (shown at left in a staging area). What became known as the Creutz test would give scientists their first opportunity to analyze the compression produced by a full set of lenses—and hopefully, at the last minute, validate the implosion design.

The Gadget's exterior shell was made of duraluminum (an aluminum-copper alloy) and consisted of seven segments held together with bolts: two polar caps (at top and bottom) around five curved pieces that formed the midsection.

Had the test device been prepped as if for military deployment instead of for an experiment, a steel ballistic case—the outermost layer in the blueprint—would have been layered over the duraluminum cover.

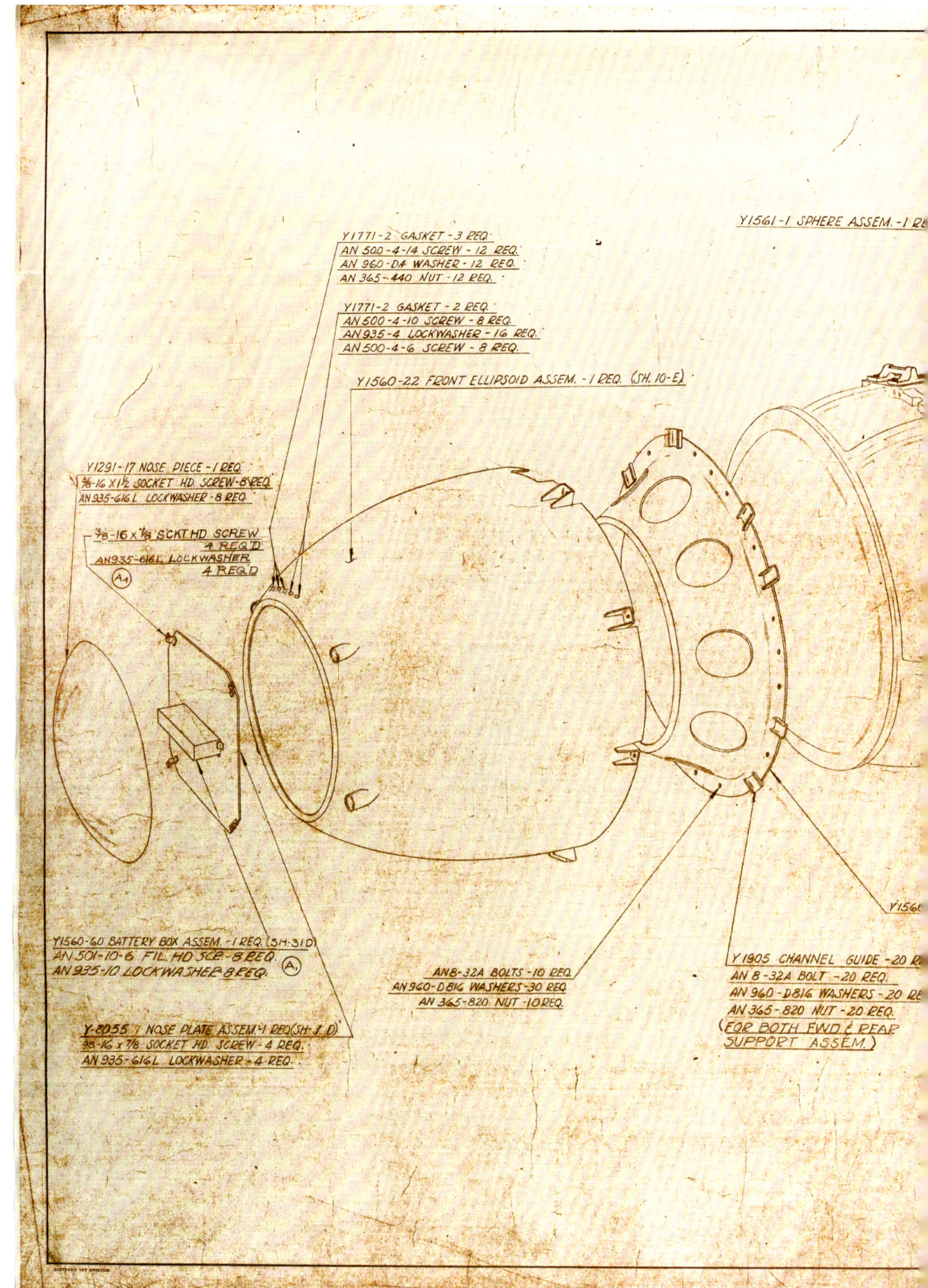

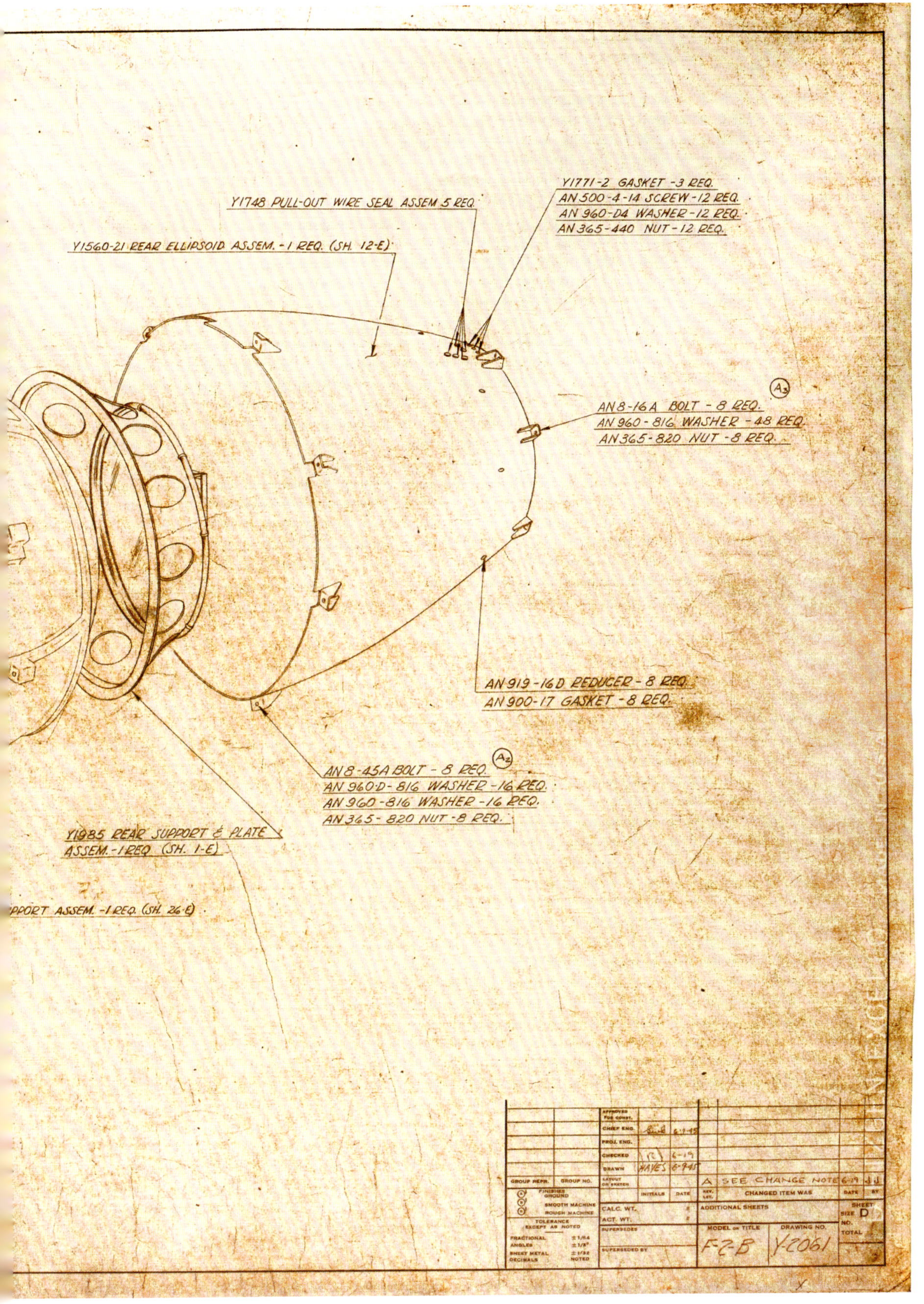

Beneath the outer shell, concentric layers of high explosives and other components encircled a small gap at the Gadget's center. The device was designed with an access channel to streamline preparation. This "trap door" design would allow complete lens assemblies to be put together with the fissile component inserted last, just before deployment.

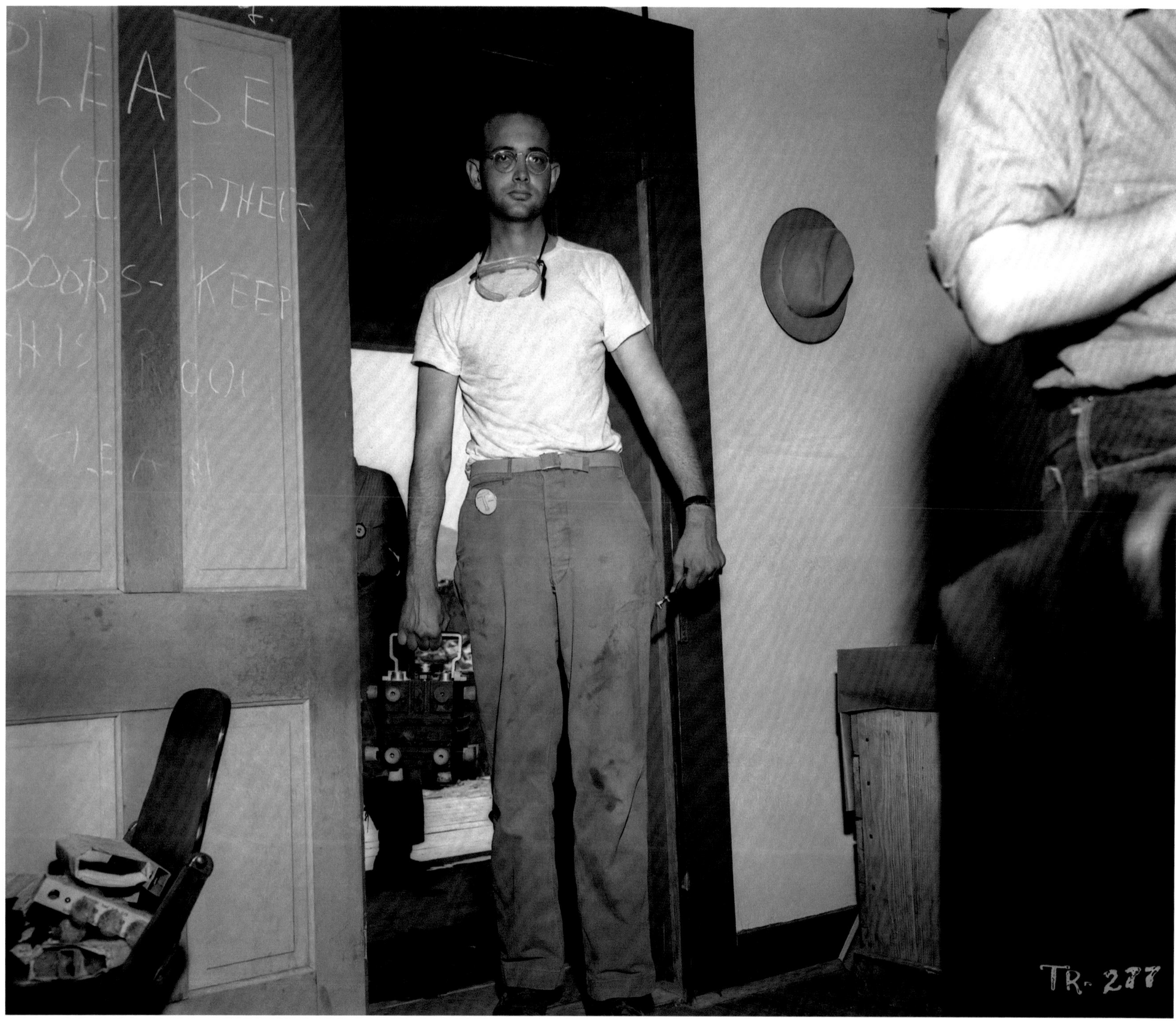

On July 12, 1945, as the lenses of the Gadget and Creutz device were being assembled at Los Alamos, the plutonium core began its journey to the Trinity site. Escorted by physicist Phillip Morrison and Health Group member Paul Aebersold, the two hemispheres made their way in an army sedan from the laboratory to the Jornada del Muerto. Kenneth Bainbridge greeted the convoy and directed the lieutenant in charge to a ranch house two miles from the shot tower, where Groves's deputy, Brig. Gen. Thomas Farrell, awaited. Above, Herbert Lehr, a member of the Special Engineer Detachment, holds the core inside its carrying case in the doorway to a designated clean room, or assembly room, where it spent the night under armed guard. Lehr is wearing a Trinity access badge on his hip.

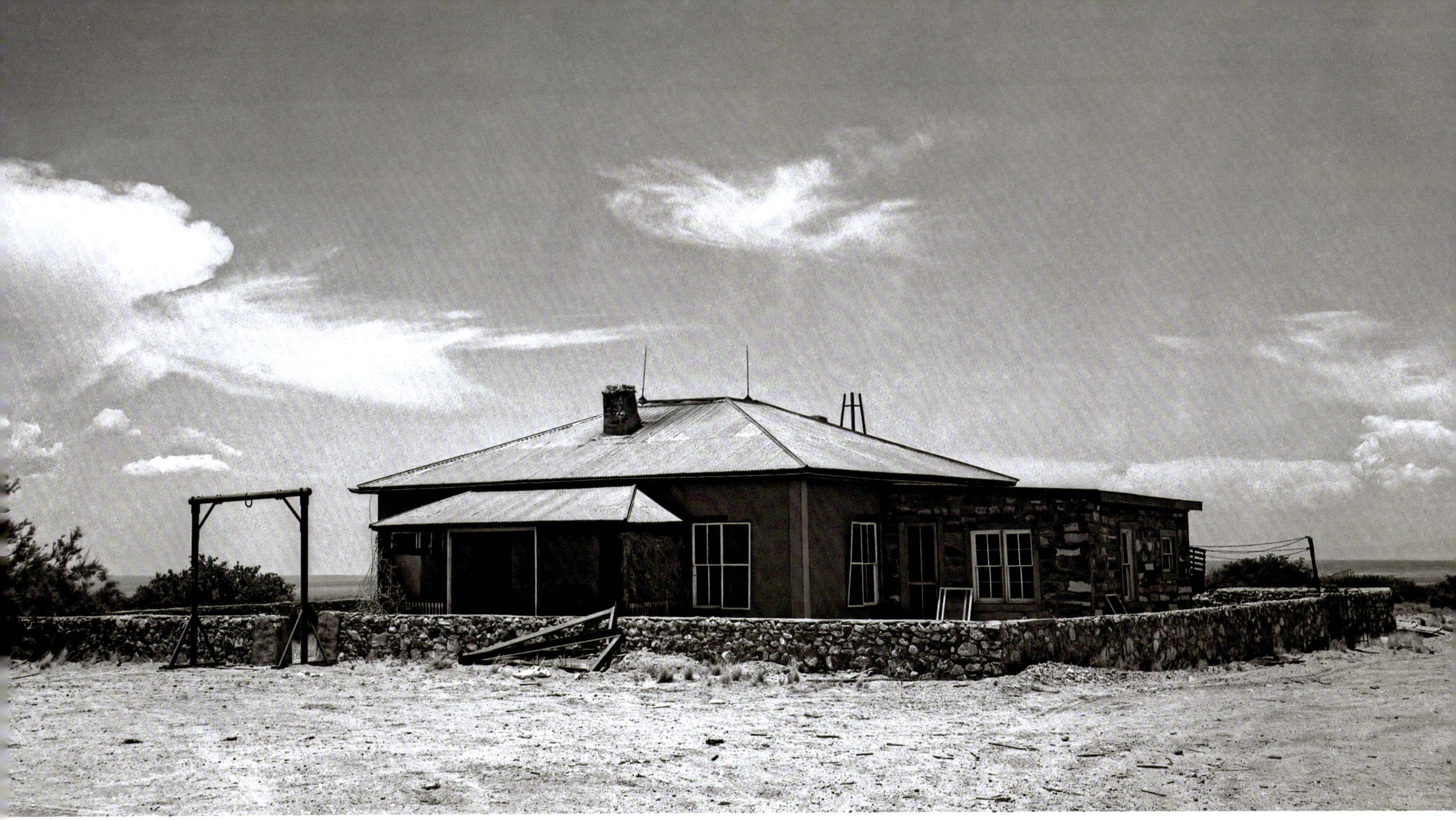

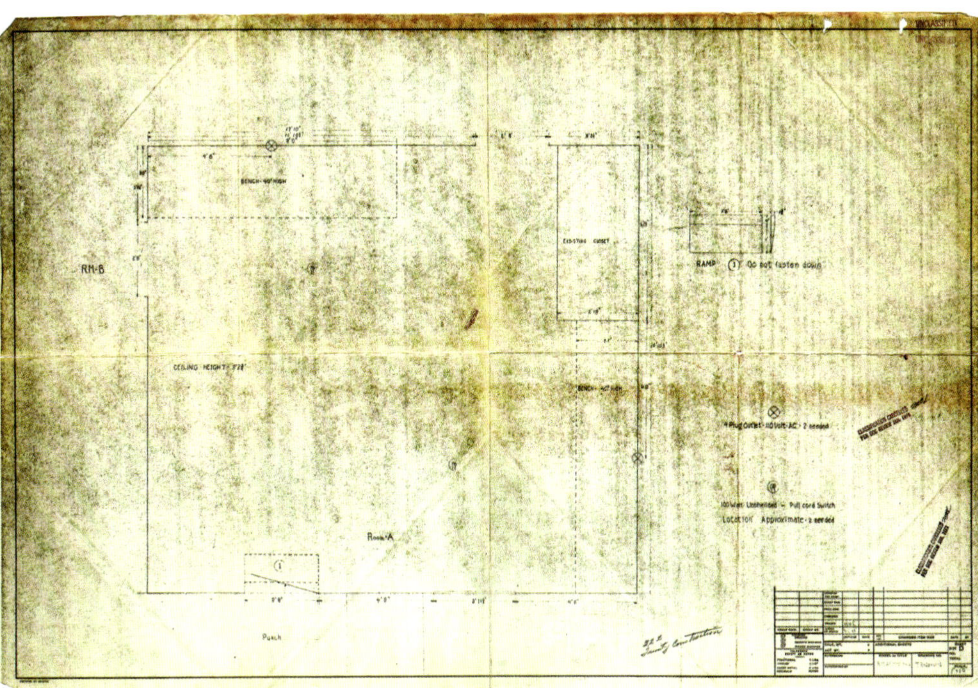

Earlier that year, the Franz Schmidt–George McDonald Ranch House had been modified to serve as a workspace for the core assembly team. The remodeling plans are shown at left.

In the main bedroom, which became the assembly room, workers taped plastic over windows, doors, and any cracks in the plaster walls to minimize dust and sand intrusion. They added a support frame and ramp out front for unloading cargo and moving equipment inside.

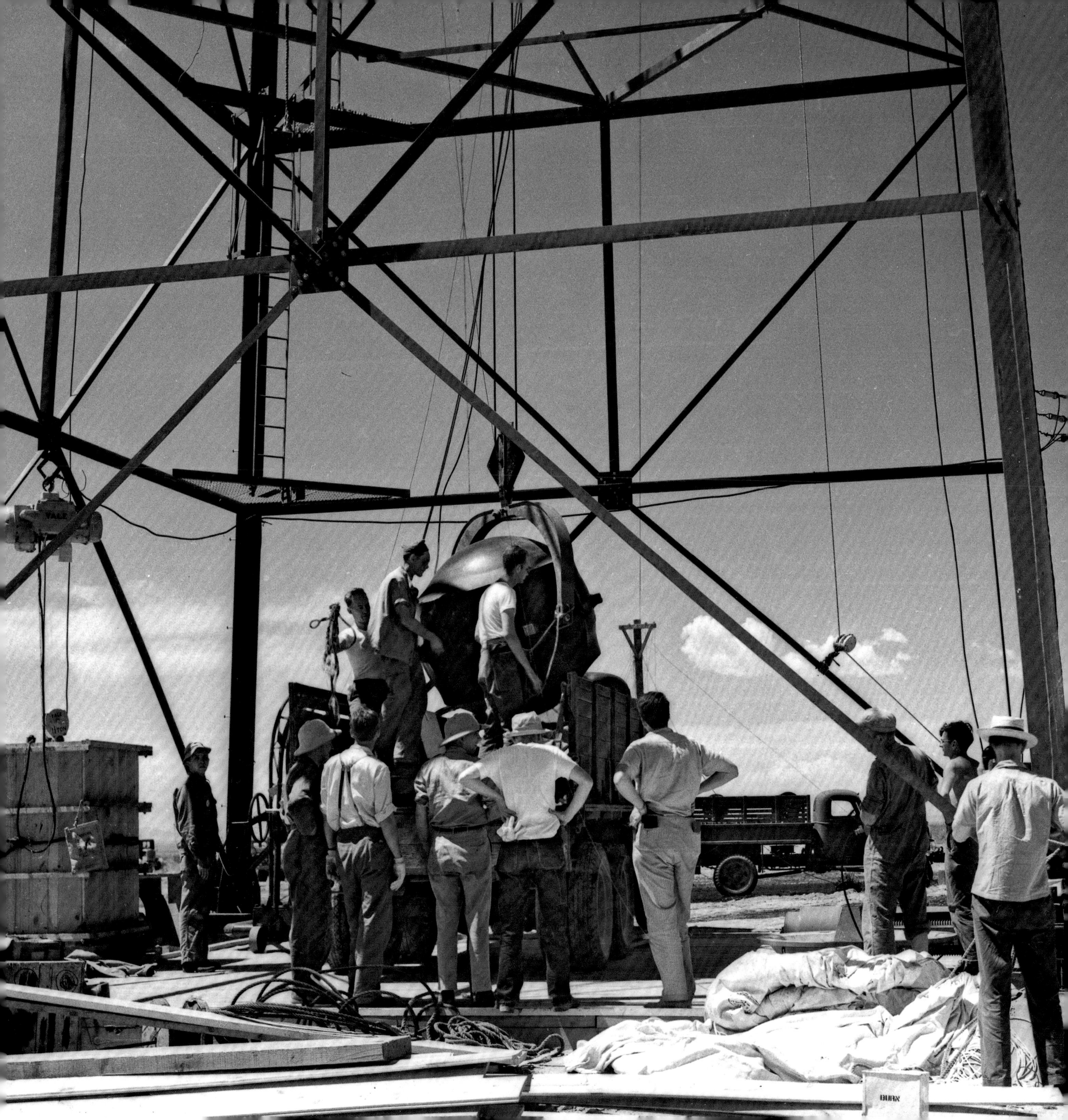
YALE
BURN

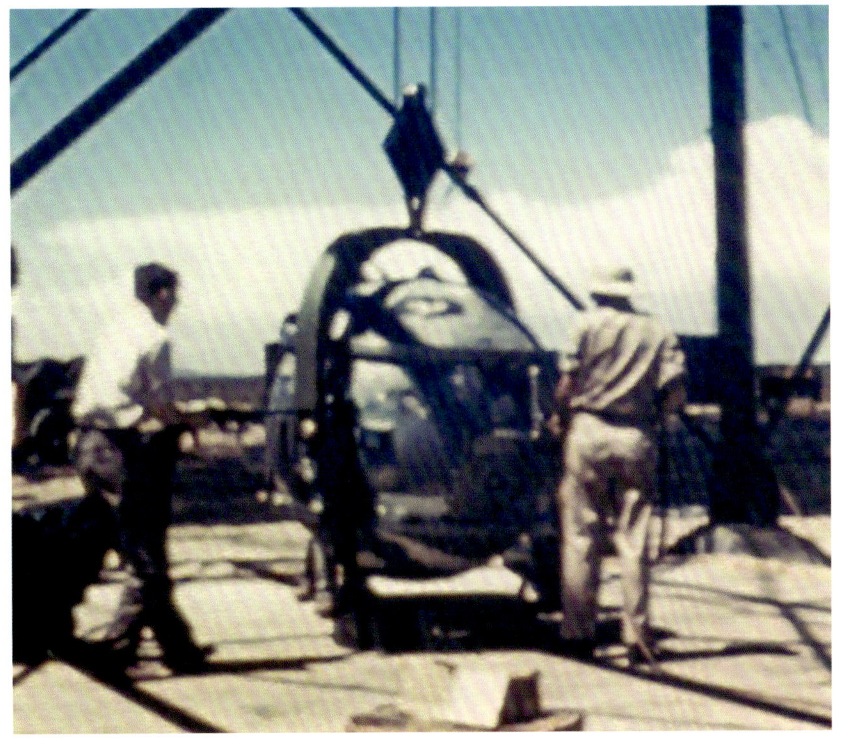

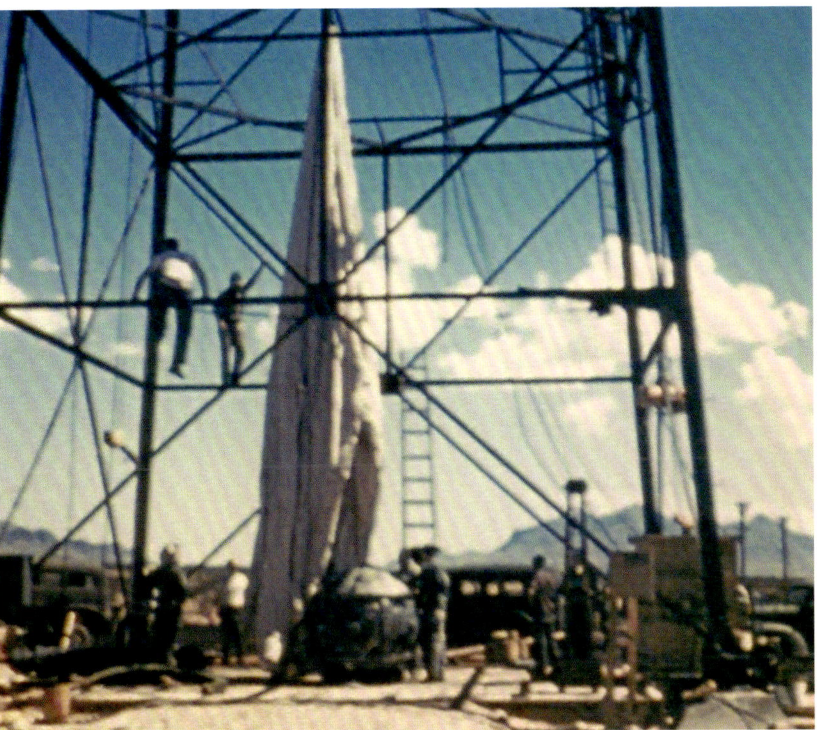

As the clock struck midnight on Friday the thirteenth, another convoy struck out from Los Alamos carrying the non-nuclear components of the Gadget—wrapped in waterproof material and strapped to the bed of a truck surrounded by escort vehicles. The group arrived at ground zero around midday.

At the base of the tower, workers attached the Gadget to a set of lifting tongs and then pitched a canvas tent to make a "clean enough" room for assembly. It was here that the high-explosive lenses and the plutonium core would be brought together that afternoon—completing, for the first time, a fully active implosion device.

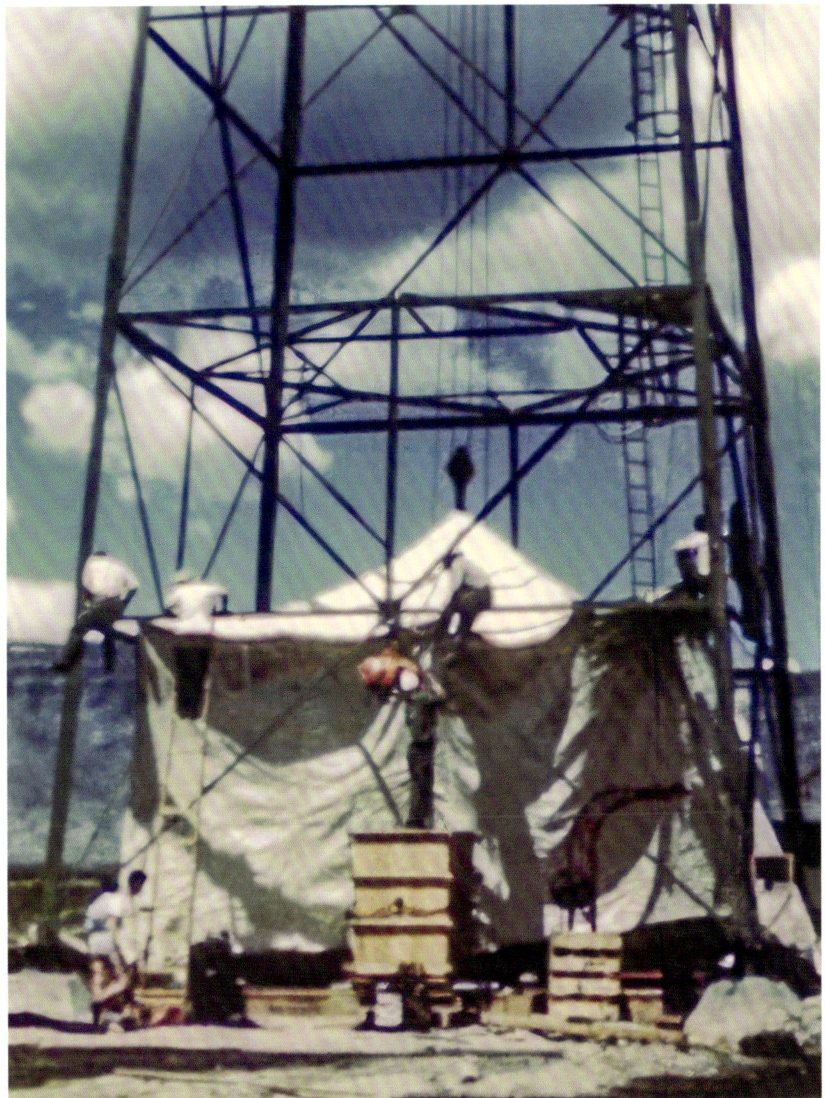

As the Gadget stood by under the assembly tent, its plutonium core was prepared at the Schmidt-McDonald Ranch House (just out of frame, at right).

If a sufficient amount of plutonium is amassed in one location, it can become critical, meaning a fission chain reaction can begin. The core did not contain enough plutonium to become critical in isolation. But, the amount of mass required to reach that threshold can be reduced by increasing the plutonium's density—as was the mechanism of the implosion Gadget—or by reflecting neutrons emitted from naturally occurring fission in the core back toward the plutonium. There would be no explosion if the core became critical during assembly, but a lethal dose of radiation could be delivered to everyone in the room within a single moment. The procedure at the ranch house unfolded without incident, but two criticality accidents would lead within a year to the tragic deaths of Trinity veterans Louis Slotin and Harry Daghlian.

When the assembly team had completed their work, two men (including Daghlian, on the right) carried the core to a waiting vehicle, which would take it to the base of the tower to meet the Gadget. With Oppenheimer looking on (in the blue shirt, below), the assembly team lowered the core through the trap door and into the center of the test device. As they closed up the Gadget once and for all, a metal wire was inserted to monitor neutron activity until zero time and confirm the radioactive core remained stable.

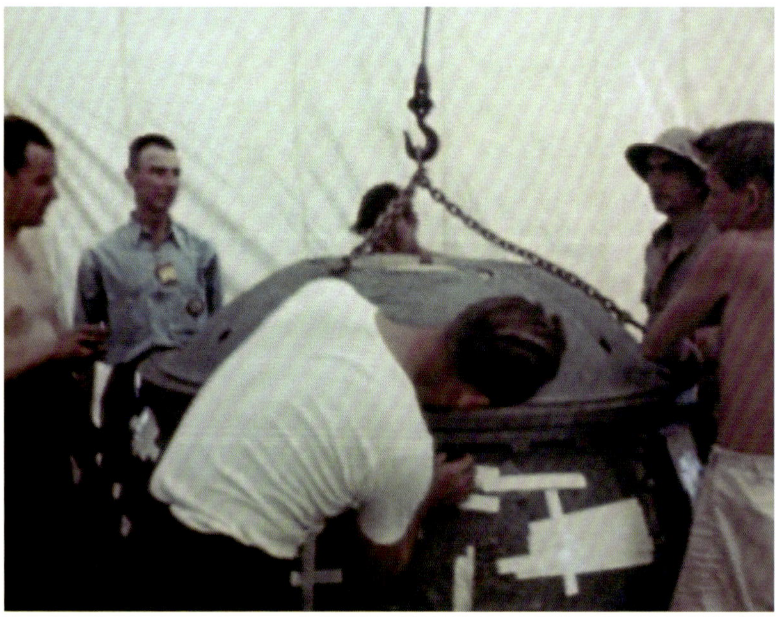

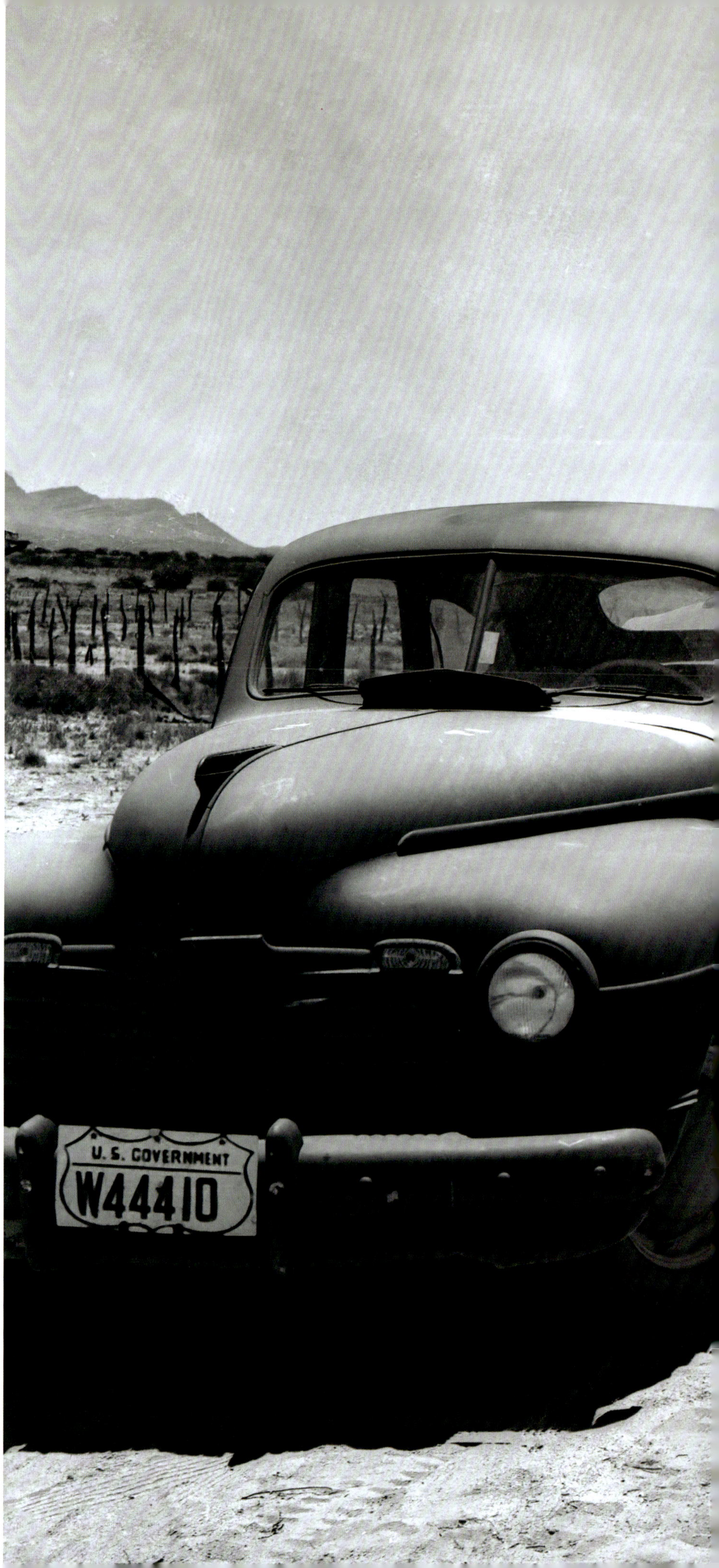

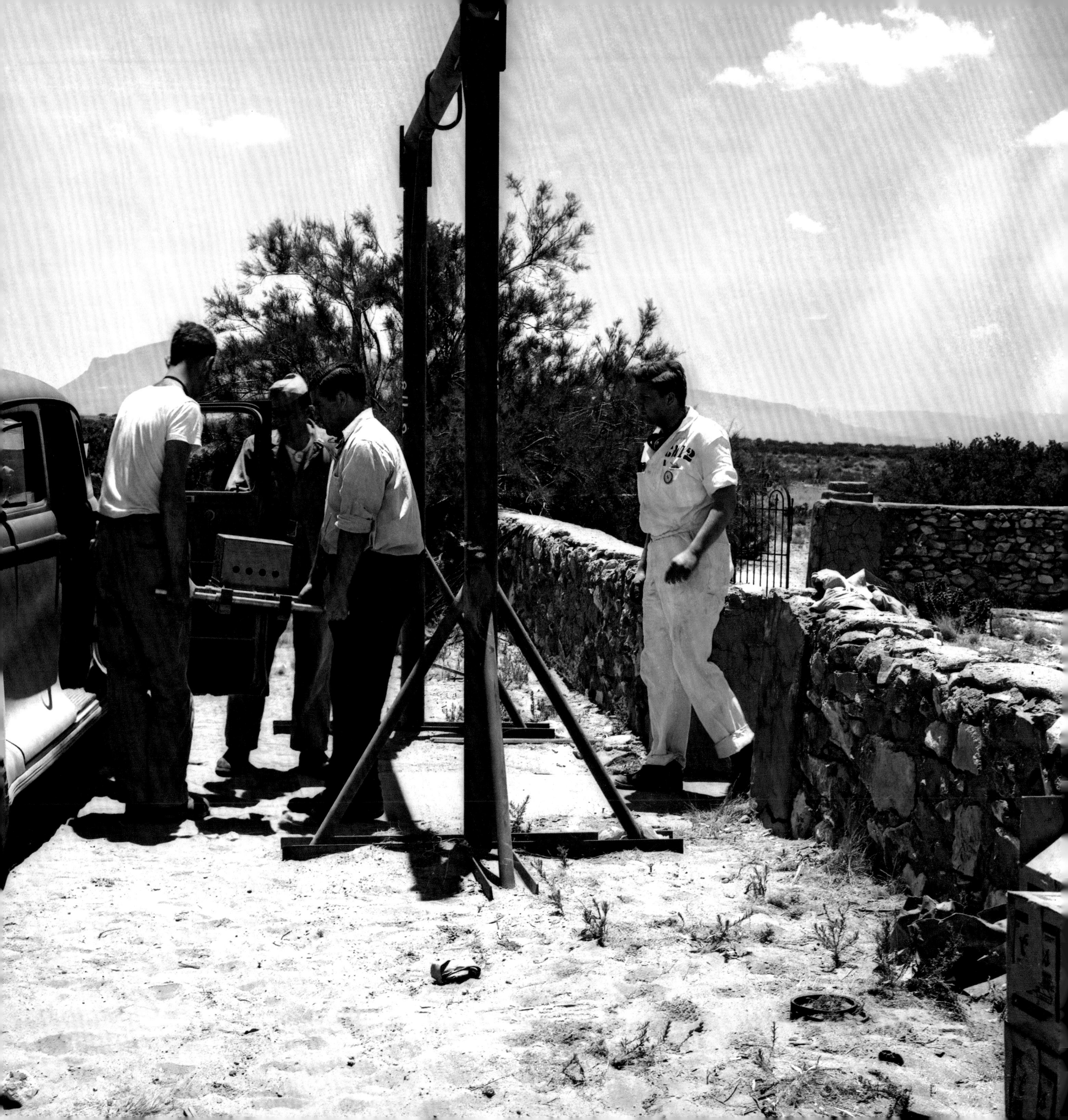

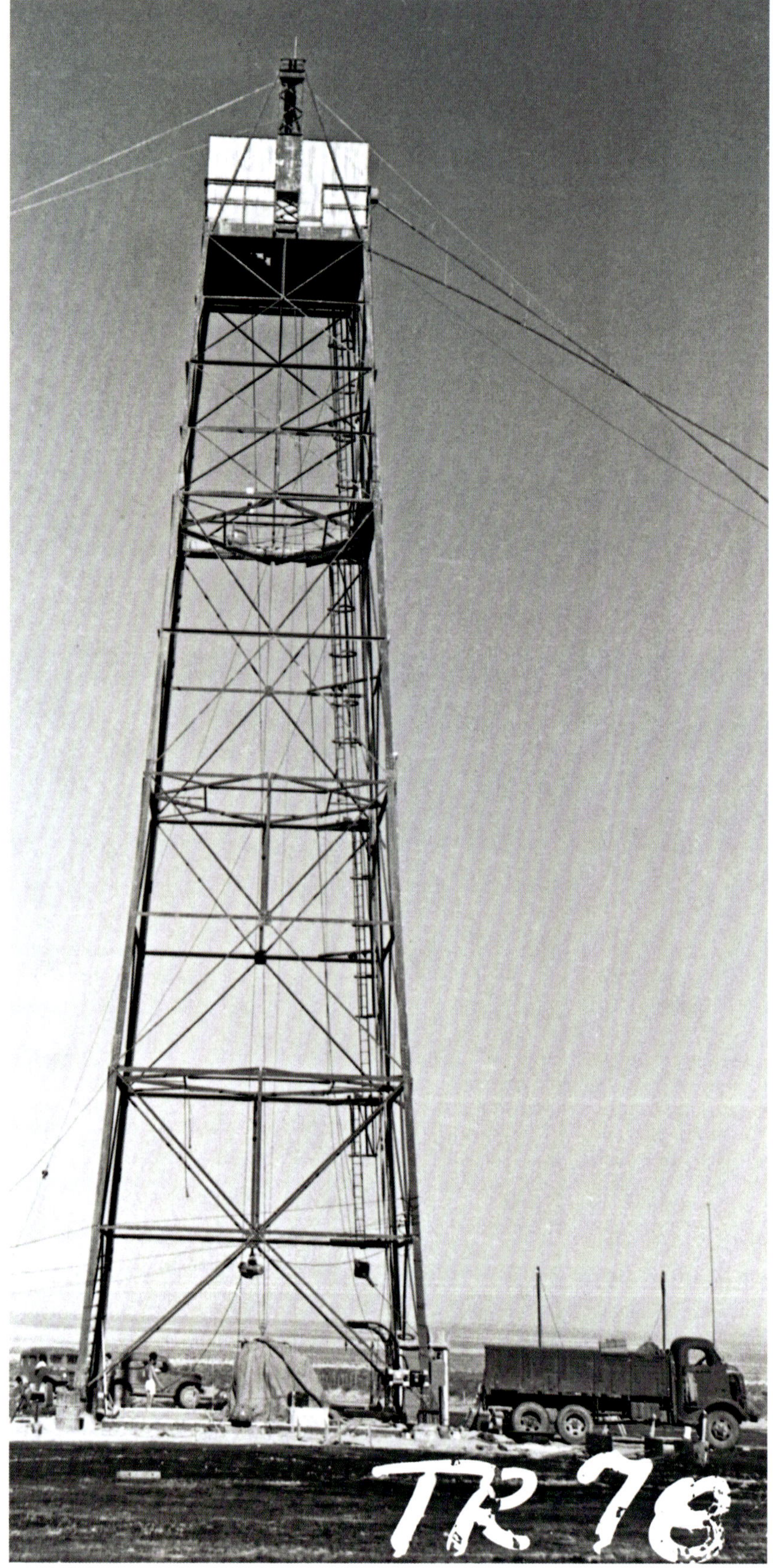

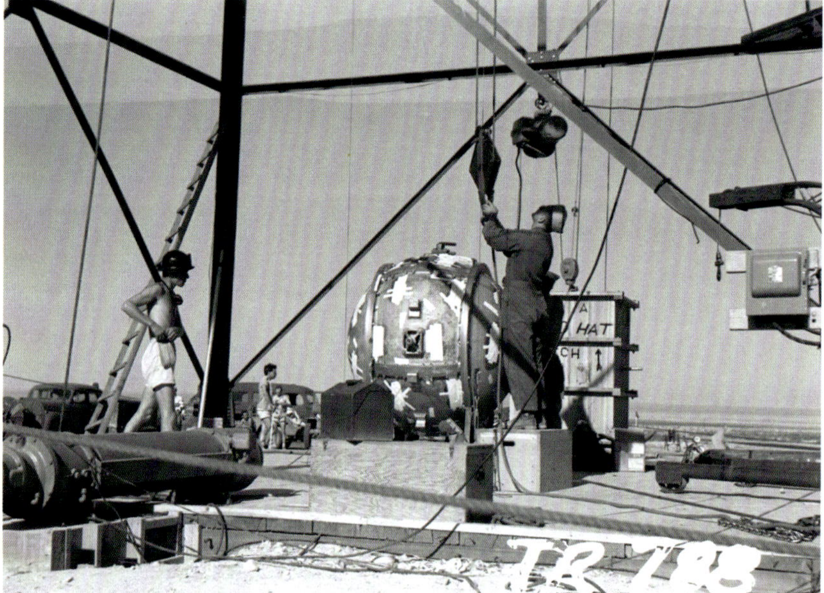

With assembly complete, workers rotated the Gadget onto its side using tongs and then attached a heavy-duty winch to lift it 100 feet overhead. As a precautionary measure—one that probably did more to ease anxiety than it would have to save the day—mattresses from base camp, shown below, were stacked under the Gadget in case the multi-ton device came crashing back to the ground.

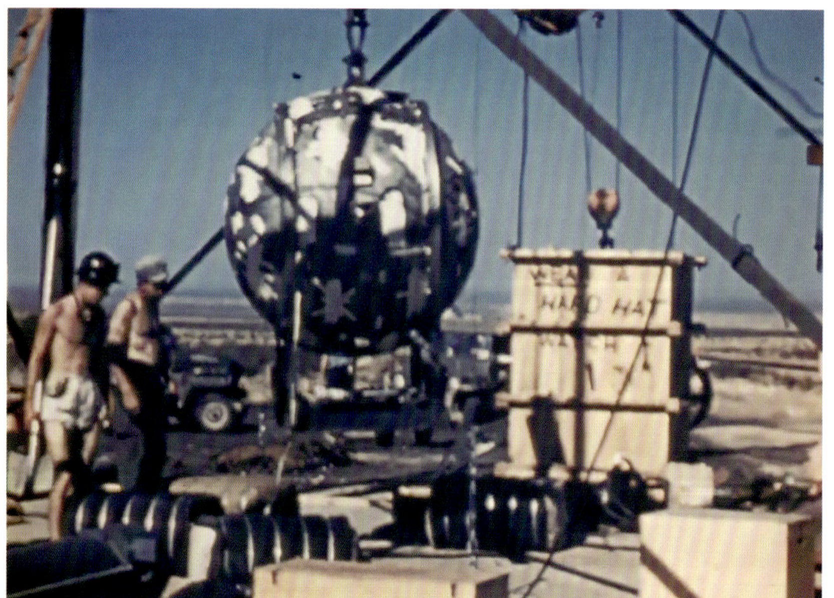

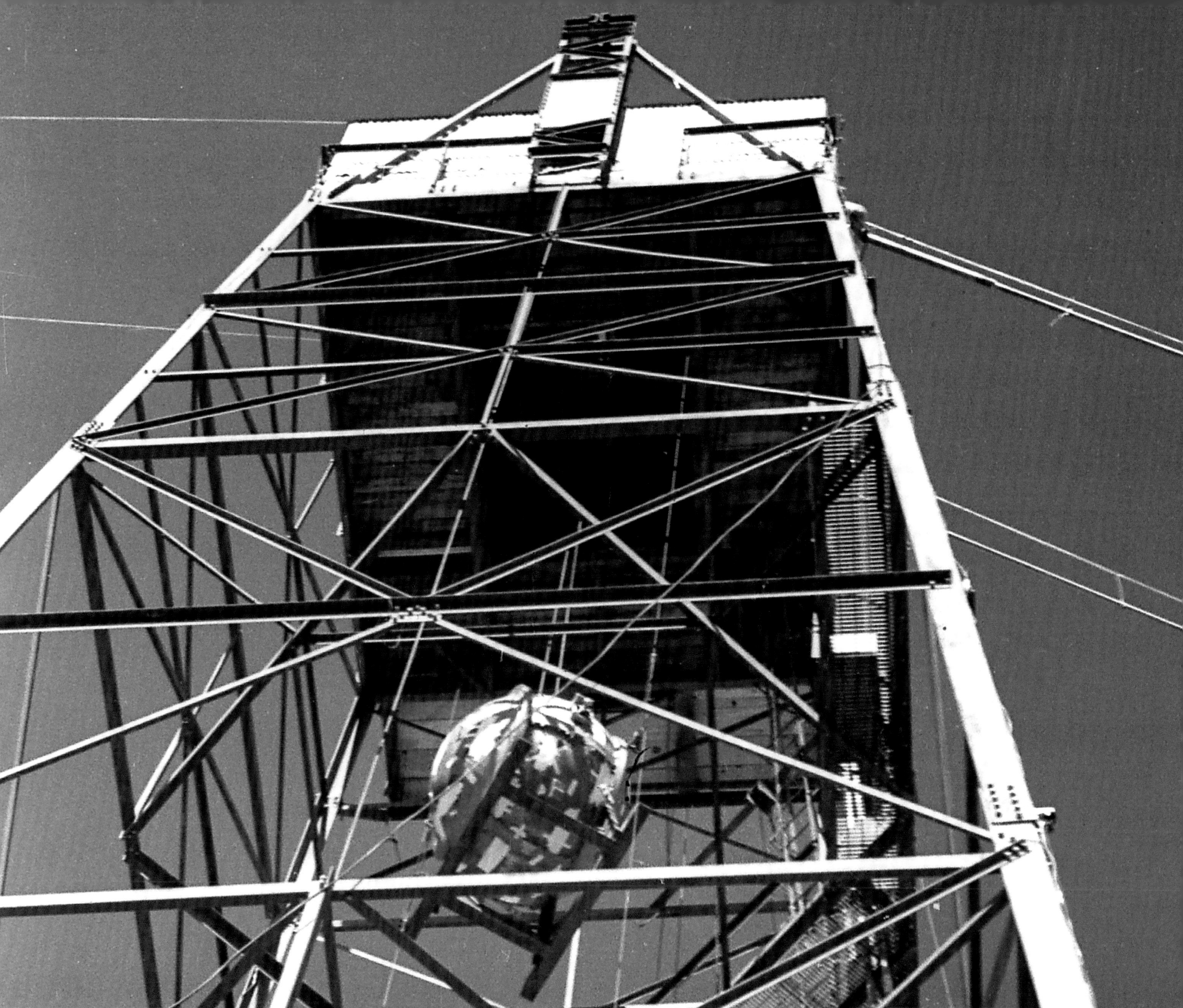

The destination was a 225-square-foot platform, outfitted with a Gadget-sized trapdoor for initial entry that would then become a floor underneath the test device. The platform, called the shot cab, was walled and covered to provide protection from the elements for the Gadget and workers.

As the Gadget was making its one-way journey to the top of the tower, Edward Creutz was completing preparations for detonation of its non-nuclear twin (*right*) back at Los Alamos. The test used the same type of detonator and firing switch that would be used to set off the Gadget two days later.

Each of the Gadget's thirty-two high-explosive lenses had to detonate at the exact same time to ensure symmetrical implosion of the core. Inconsistent timing had plagued the early implosion program, and achieving the needed synchrony required invention of a reliable, instantaneous detonator that would also not accidentally fire if exposed to static electricity or a physical shock—and could withstand the severe cold and vibration of falling from a bomber at high altitude.

A group at the laboratory led by future Nobel laureate Luis Alvarez developed a mechanism called the exploding bridge wire detonator, patented by Lawrence Johnston. The detonator housing is shown below. Per Johnston's design, a high-voltage current of electricity would travel through large, heavily insulated cables until it reached a thin "bridge wire." When the current reached the bridge wire, the load would vaporize the wire, generating an instantaneous shockwave that set off the high-explosive lens underneath.

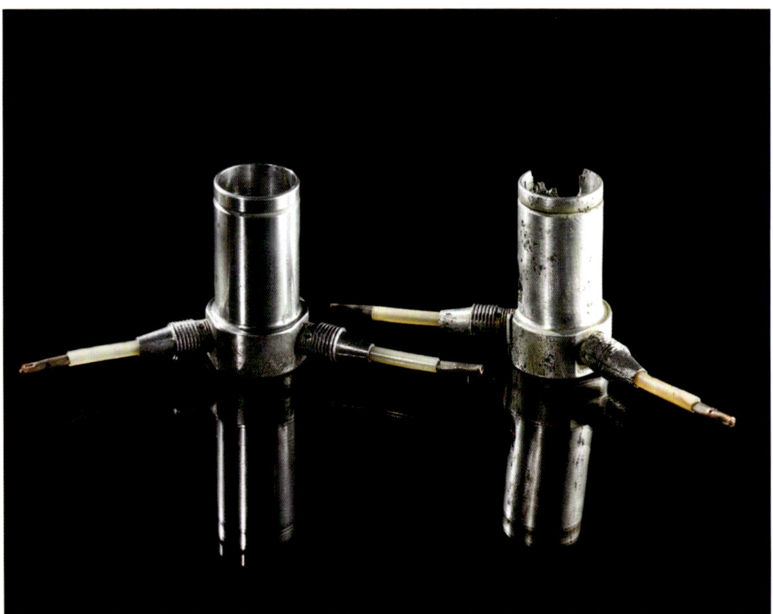

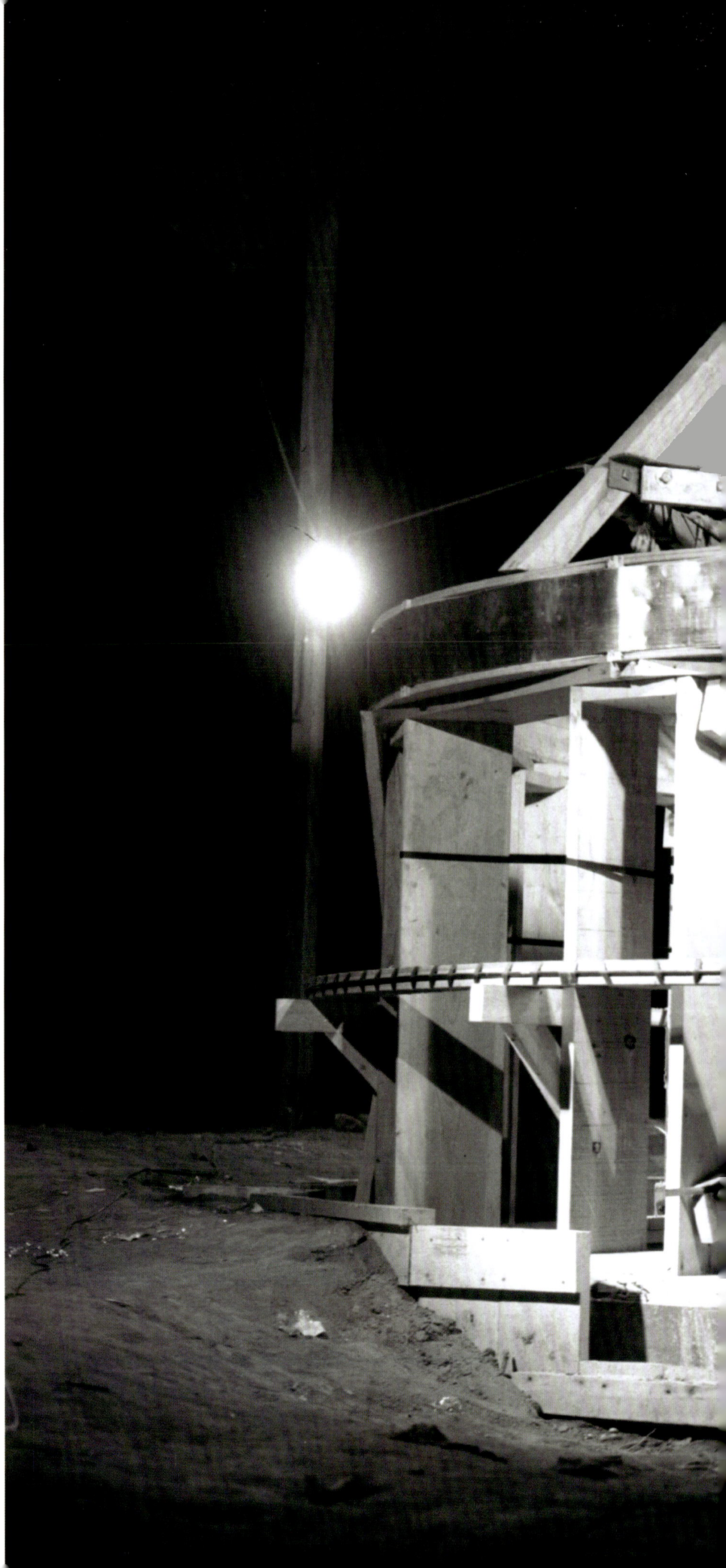

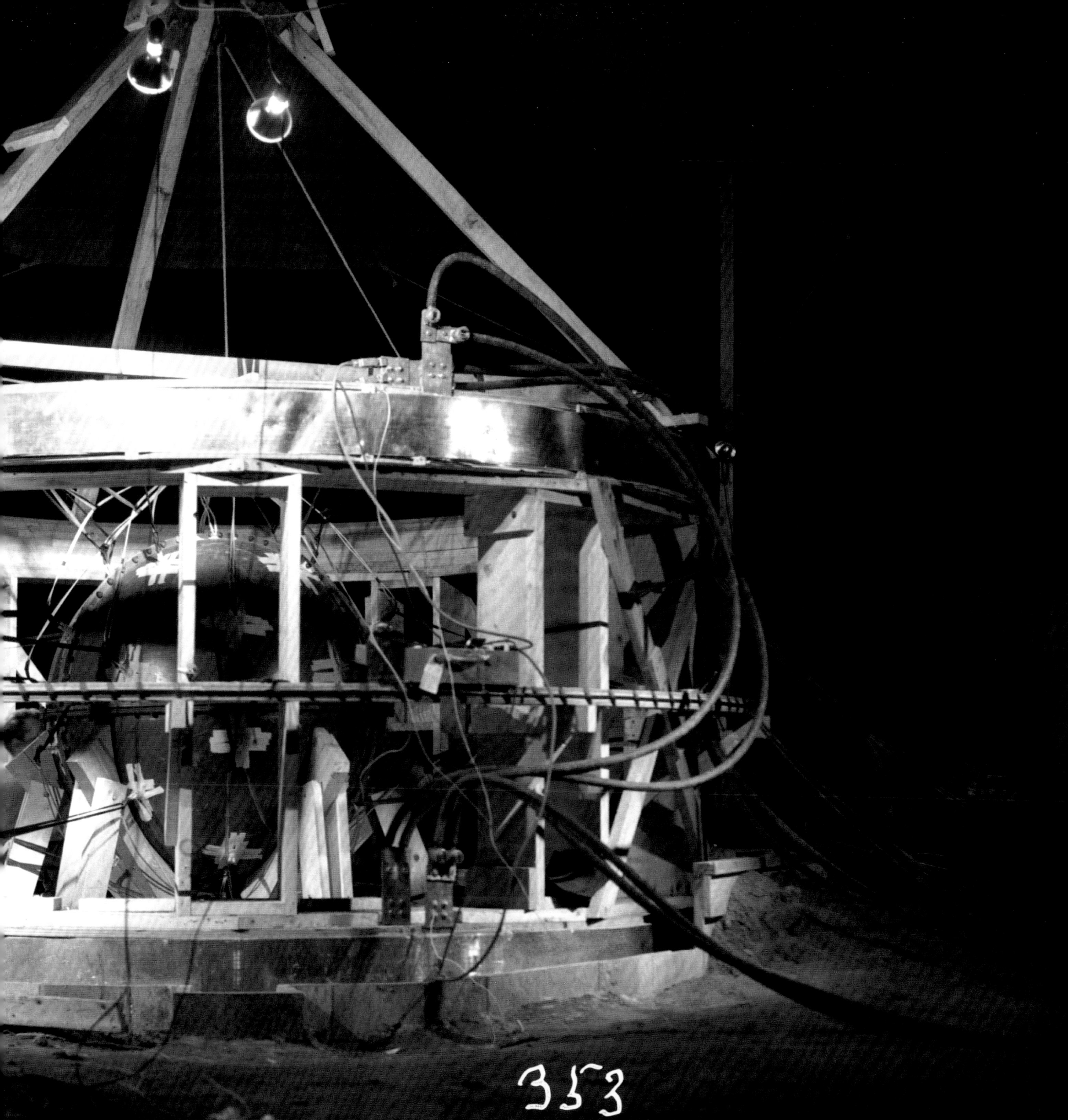
353

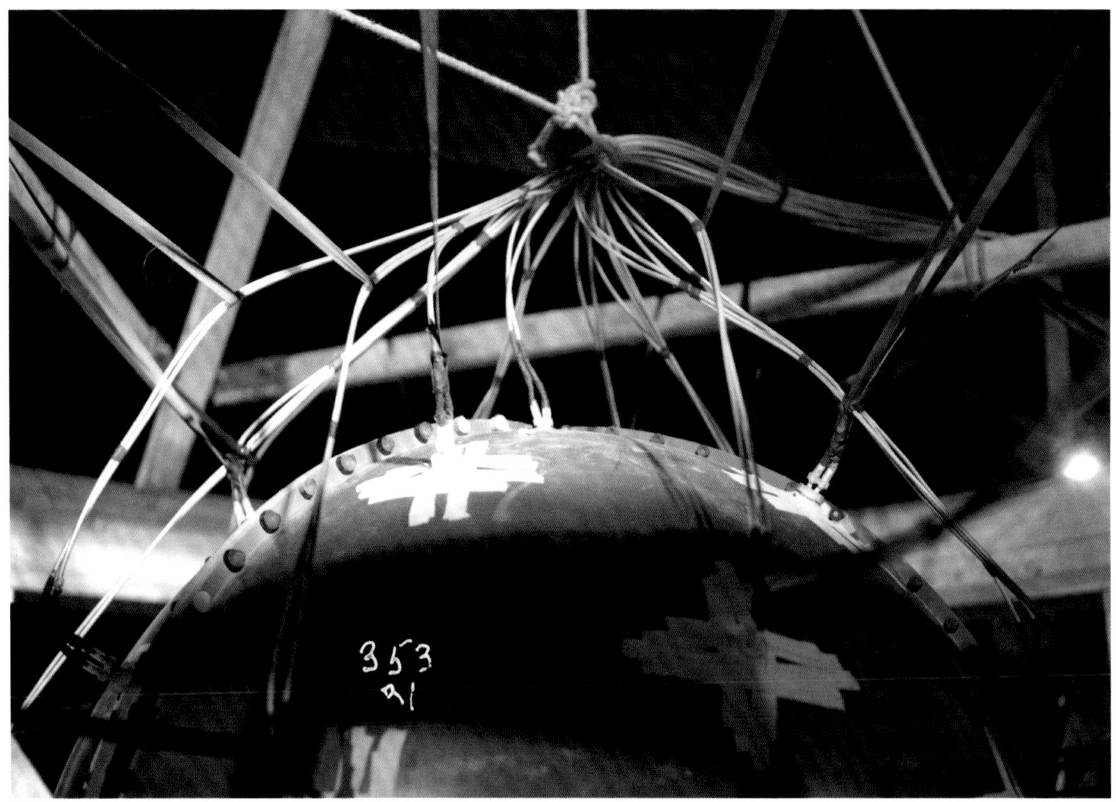

In the Creutz test, the performance of the lens system would be observed indirectly through use of a magnetic pickup coil, which would track compression of an inert metal sphere representing the plutonium core. Although the Creutz device was mostly identical to the Gadget, its casing was made of plastic rather than metal so as not to interfere with the magnetic field.

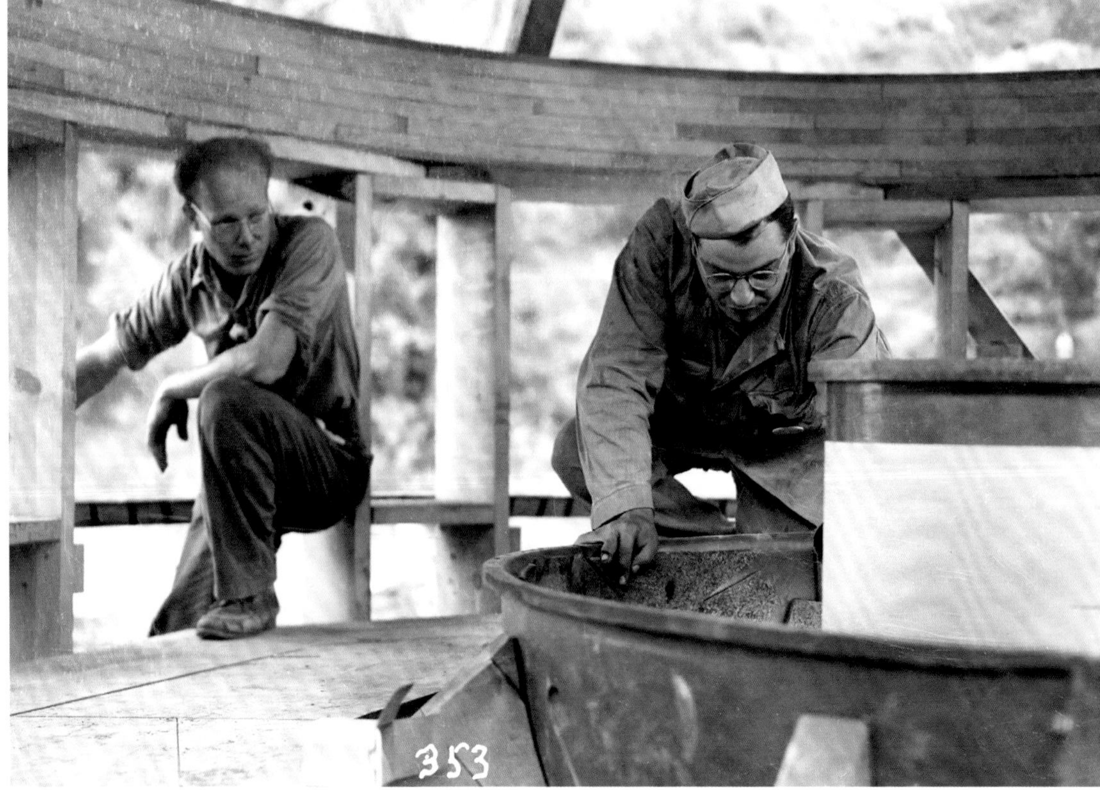

At left, George Kistiakowsky looks on as a technician leans over a piece of the plastic casing to calibrate the coil.

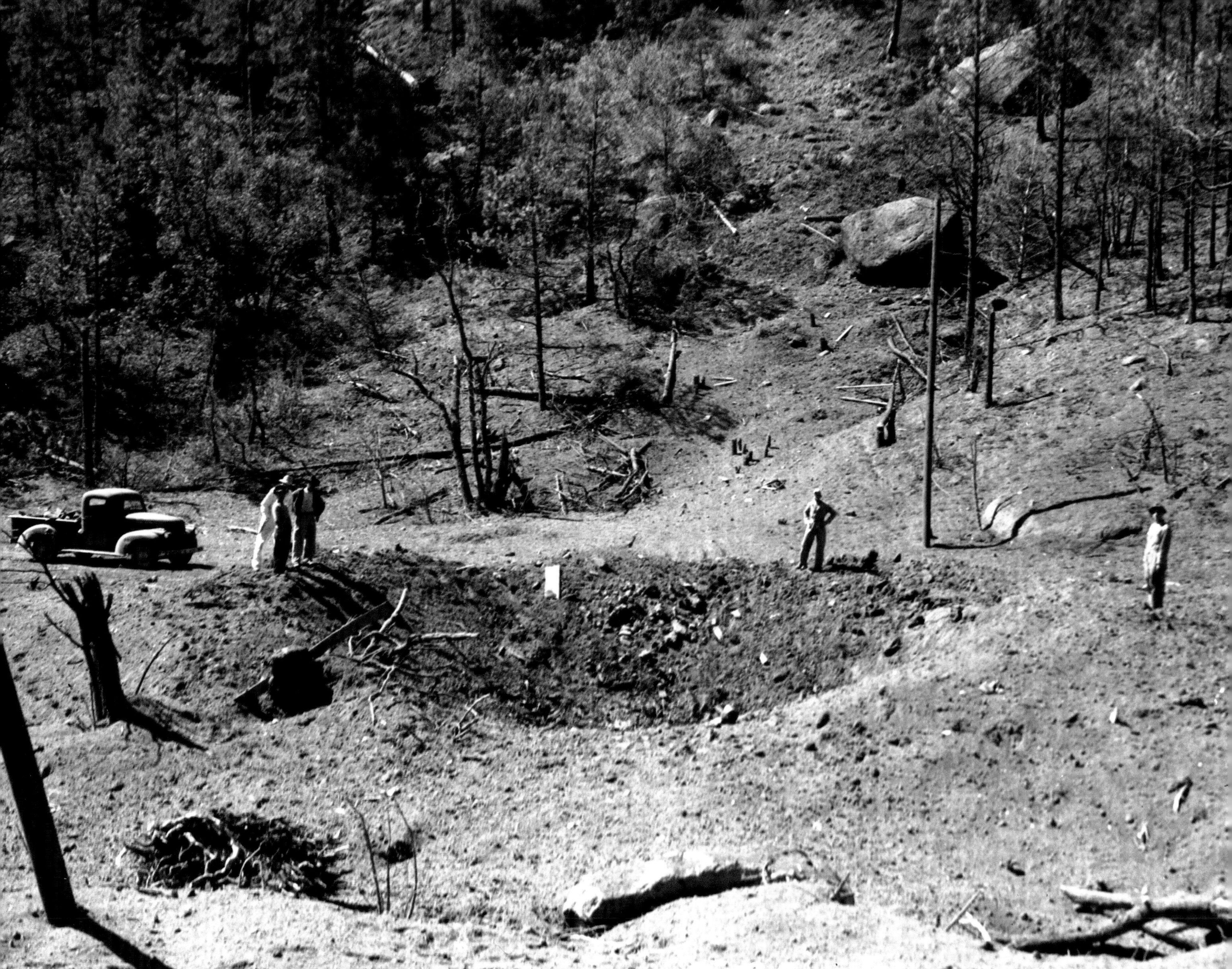

The July 14 Creutz detonation created a large crater in Pajarito Canyon and a mountain of doubt about the lens system. The results seemed to indicate that the implosion waves would not compress the core fast enough to produce a sustained fission chain reaction.

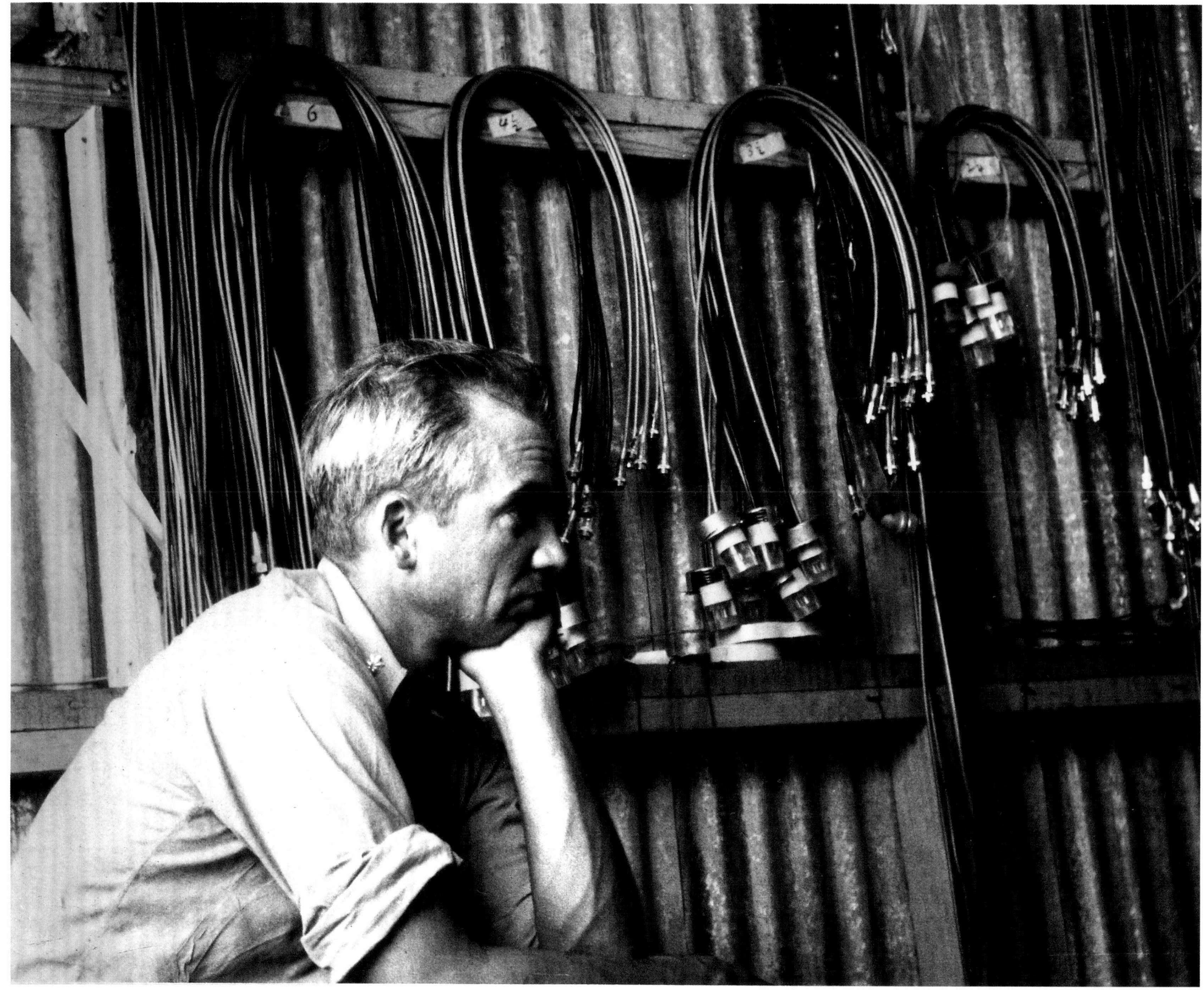

News of the Creutz test failure arrived at the Trinity site that same morning. With more uncertain times perhaps returning to mind—and nervous glances in the direction of Jumbo—scientists were suddenly back to fearing a fizzle. All they could do, however, was continue with "Operations aloft," as Bradbury's Hot Run checklist referred to preparations at the top of the tower. Bradbury is pictured above inside the shot cab, with detonator cables hanging behind him, ready to be plugged into the Gadget. (Los Alamos Historical Society Photo Archives.)

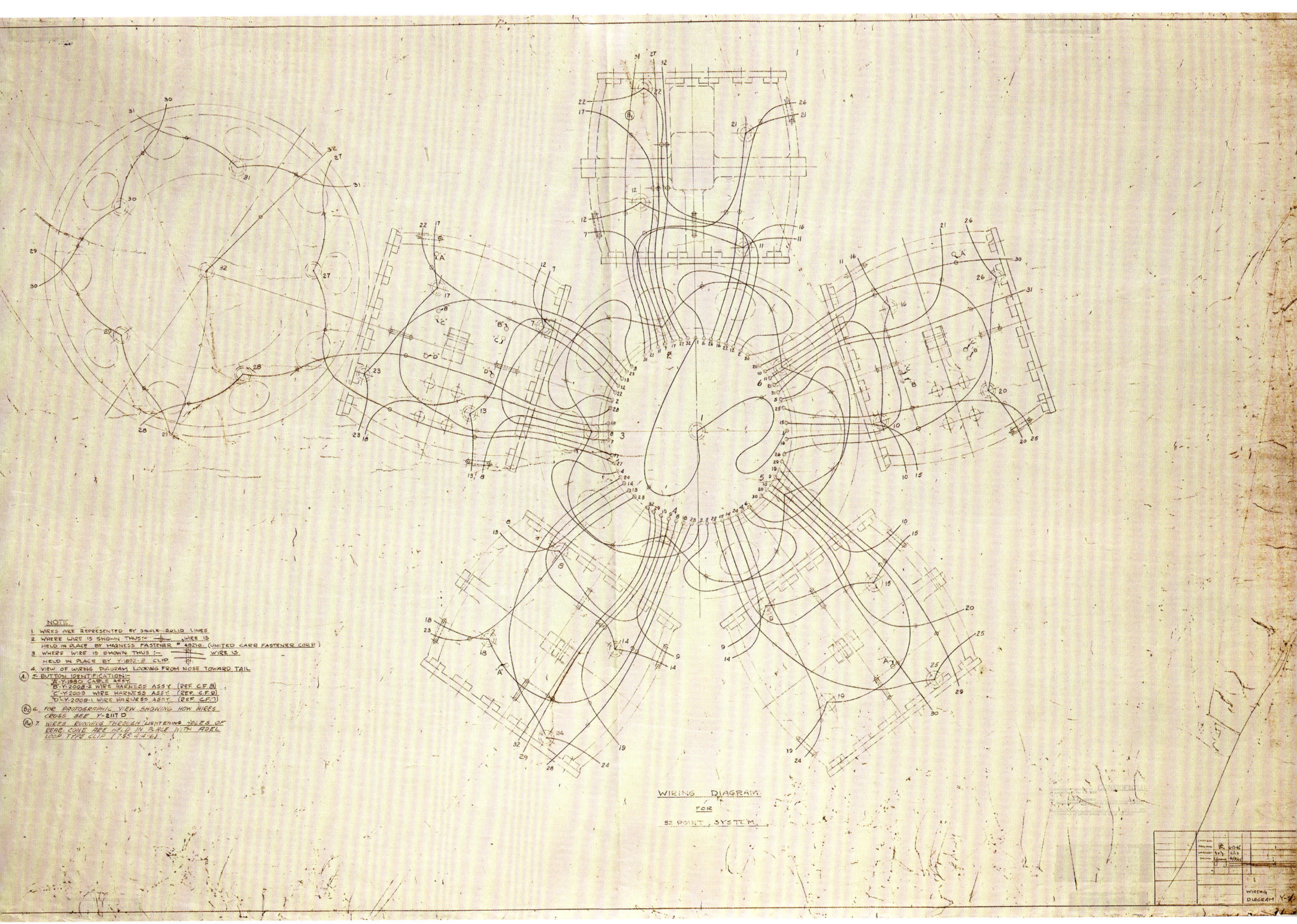

A diagram of the device's circuitry is shown above. Upstream of the detonators, an on-off mechanism, like a heavy-duty light switch, was required to activate the 5,000 volts of electrical current that would simultaneously reach and vaporize each detonator's bridge wire. This component, called the "spark gap switch" or, officially, the X unit, was invented by Donald Hornig, a young chemist in the Explosives Division. Hornig modified the circuitry of a high-voltage flash photography device designed for military surveillance planes so that instead of triggering a bulb to flash, the current surging through the X unit would trigger the Gadget's detonators to fire.

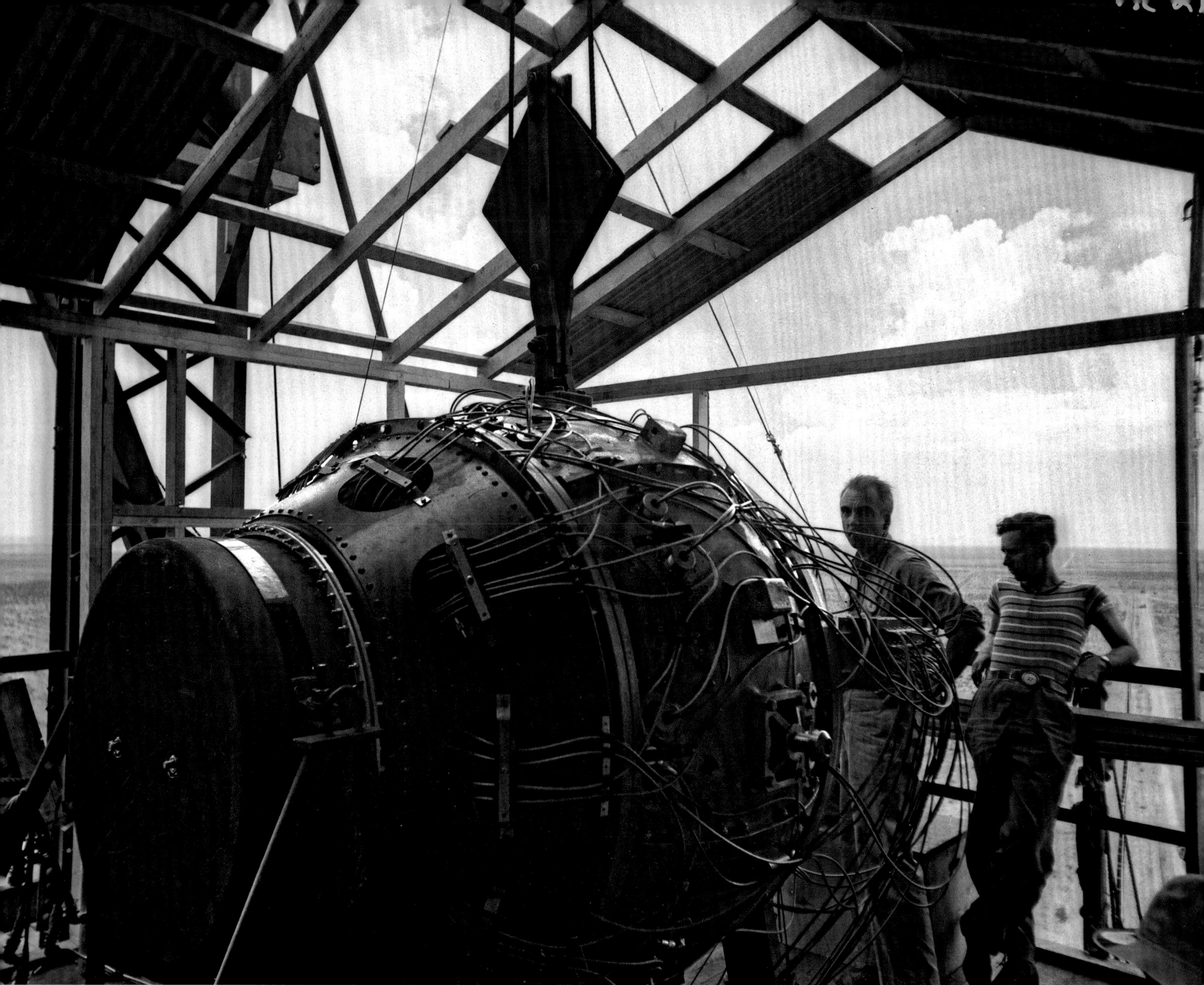

Installing the Gadget's wiring was a tedious and dangerous operation, completed throughout an entire day as members of the Explosives Division stood by in the shot cab to "criticize potential rough handling," according to the Hot Run. Bradbury and an unidentified man are pictured above next to the fully wired test device. The thick cables provide pathways for electrical current from Hornig's wheel-shaped X unit (on the left side of the Gadget) to reach the detonators.

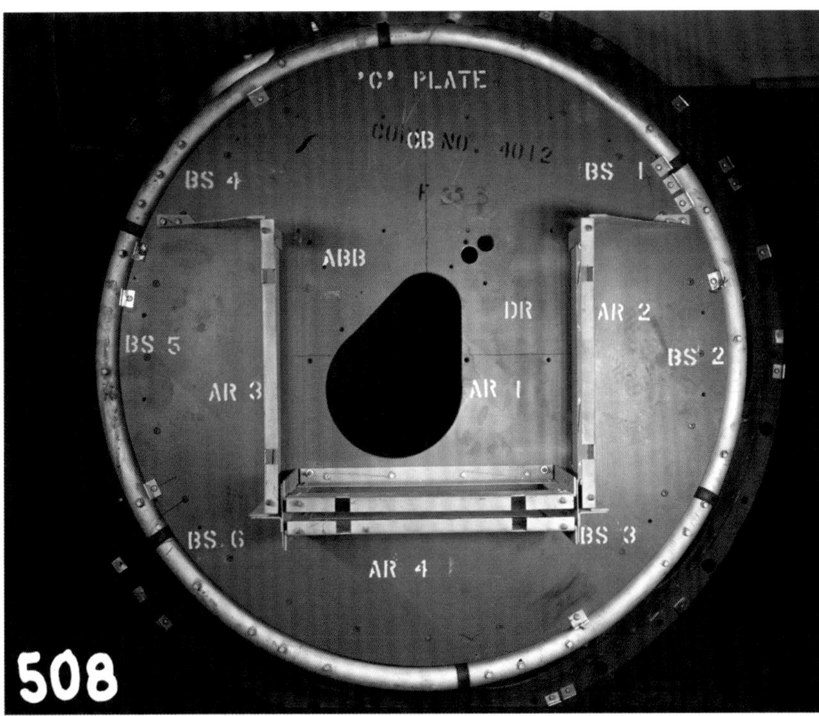

A cone-shaped support assembly (*right*) was placed over one of the Gadget's polar caps to house the X unit, shown above and below before and after attaching the components.

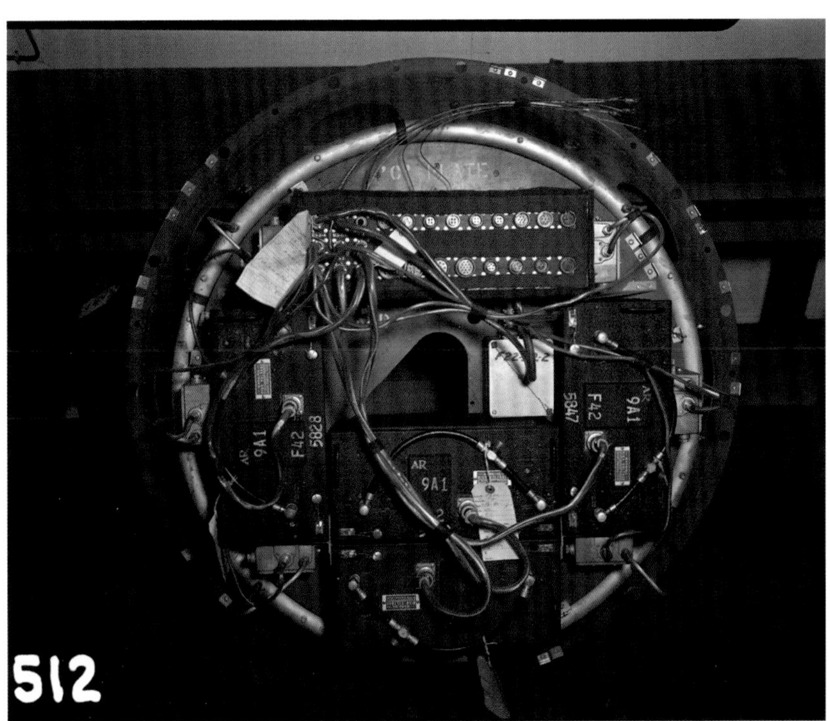

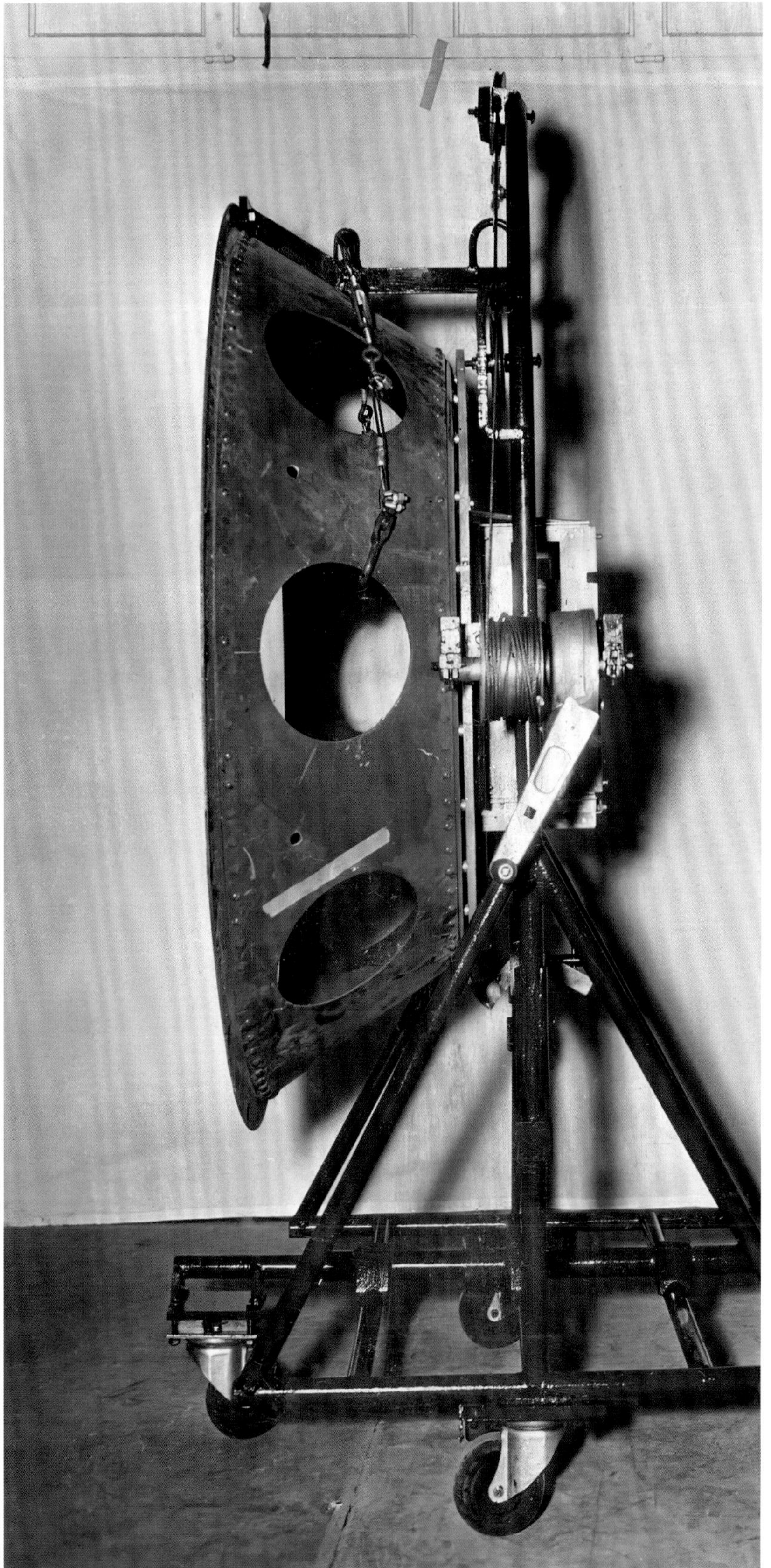

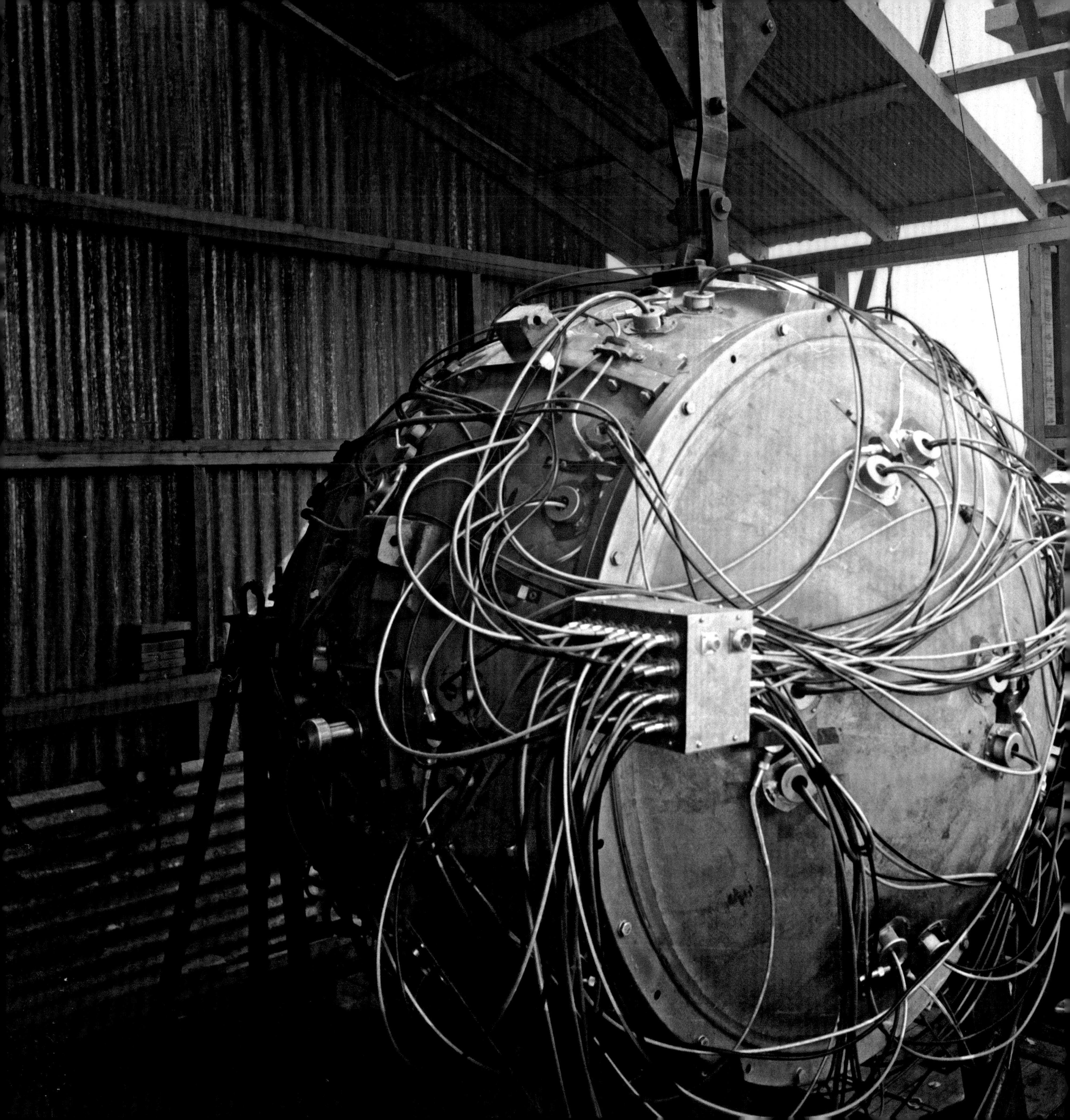

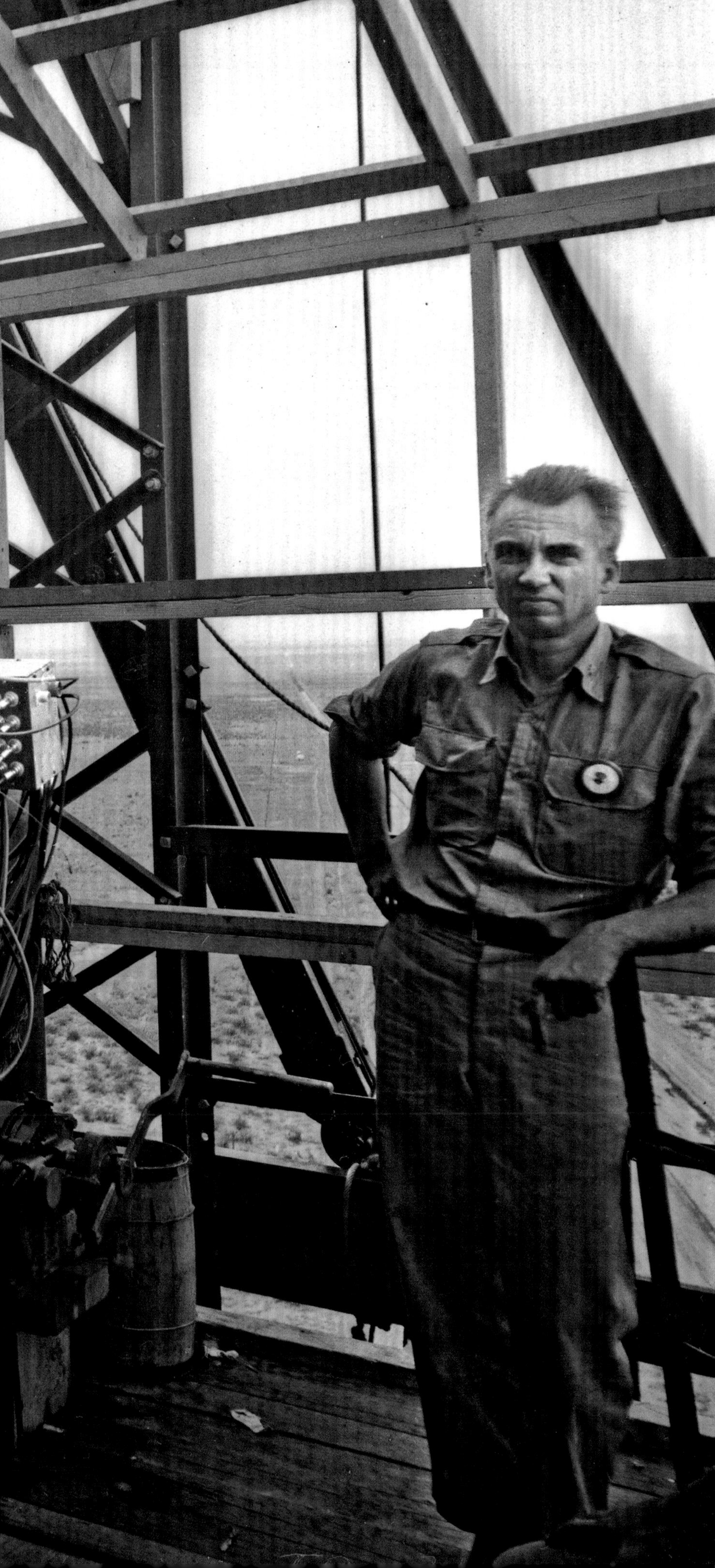

By 5:00 p.m. on July 14, the Gadget was ready. Besides the absence of a ballistic case, the only giveaway that this device was for a science experiment was the presence of informer switches—the reason for the two rectangular boxes and much of the rat's nest of wires. They were attached to the Gadget to answer the open question of whether the detonators could fire in true unison to begin the implosion. If something went wrong, the sensors' feedback would be critical to revising the system for a second test device.

If Trinity went the way of the Creutz test, it seemed like a second trial of the Gadget might be inevitable. The next morning, however, an update arrived from Los Alamos. Hans Bethe, a future Nobel laureate and leader of the Theoretical Division, had been running calculations through the night hoping to arrive at a different conclusion.

Much to the relief of Kistiakowsky, under fire since the apparent failure of his lens system, Bethe concluded the implosion design was such that the results obtained from the Creutz test might occur even if compression was achieved to perfection. After a major scare, it turned out the lens test signaled neither success nor disaster for the Trinity Gadget.

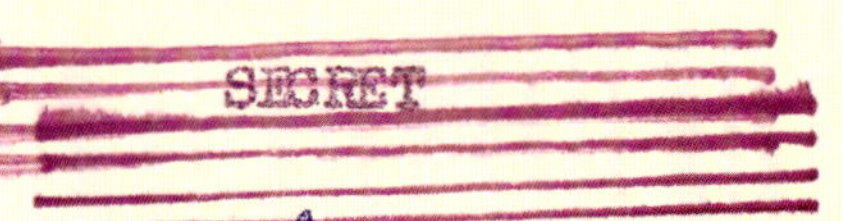

CLASSIFICATION CANCELLED
PER DOC REVIEW JAN. 1973

- 4 -

Saturday, 14 July, 0900 — Operations aloft

1. Wiring of X unit proceeds direction of and by He-Me people. Note that X unit should have cables attached to cone prior to this time.
2. Detonators are staked to coax by Caleca of detonator group.
3. Detonators are placed by Caleca to conform with requirements of informer switches. HE people stand by to criticise potential rough handling.
4. Detonators and informers in place, verified by Greisen.
5. X unit and informer unit safed--verified by Bradbury or Kistiakowsky.
6. X-7 will provide all detonators, informer switches, informer cables (adequate length), informer apparatus (where they get it is their business). X-5 will supply prepared detonator cables. X-6 will obtain detonator springs and other necessary gear from appropriate sources. X-6 will supply all fittings for wiring. Schaffer to check that all mechanical parts (nuts, bolts), etc. are supplied.
7. Note that once detonators are on sphere, no live electrical connection can be brought to X unit, informer unit, or anywhere else on sphere. Hence all testing must be done before sphere is lifted to tower. After that it is too late.

Saturday, 14 July, 1700 — Gadget complete

Sunday, 15 July, all day — Look for rabbit's feet and four leafed clovers. Should we have the Chaplain down there? Period for inspection available from 0900-1000

Monday, 16 July, 0400 — BANG!

Tensions remained high on the final day before the test shot. Although now regarded as inconclusive, the Creutz results had shaken confidence; meanwhile, two X units had experienced major malfunctions during rehearsals—both, it turned out, from human error—and as July 16 approached, thunderstorms threatened to derail the schedule or damage the numerous electrical components on which firing switches and diagnostic apparatuses depended.

Drawn up during lighter times, the instructions for the day of the fifteenth were apt: "Look for rabbit's feet and four leafed clovers. Should we have the Chaplain down there?"

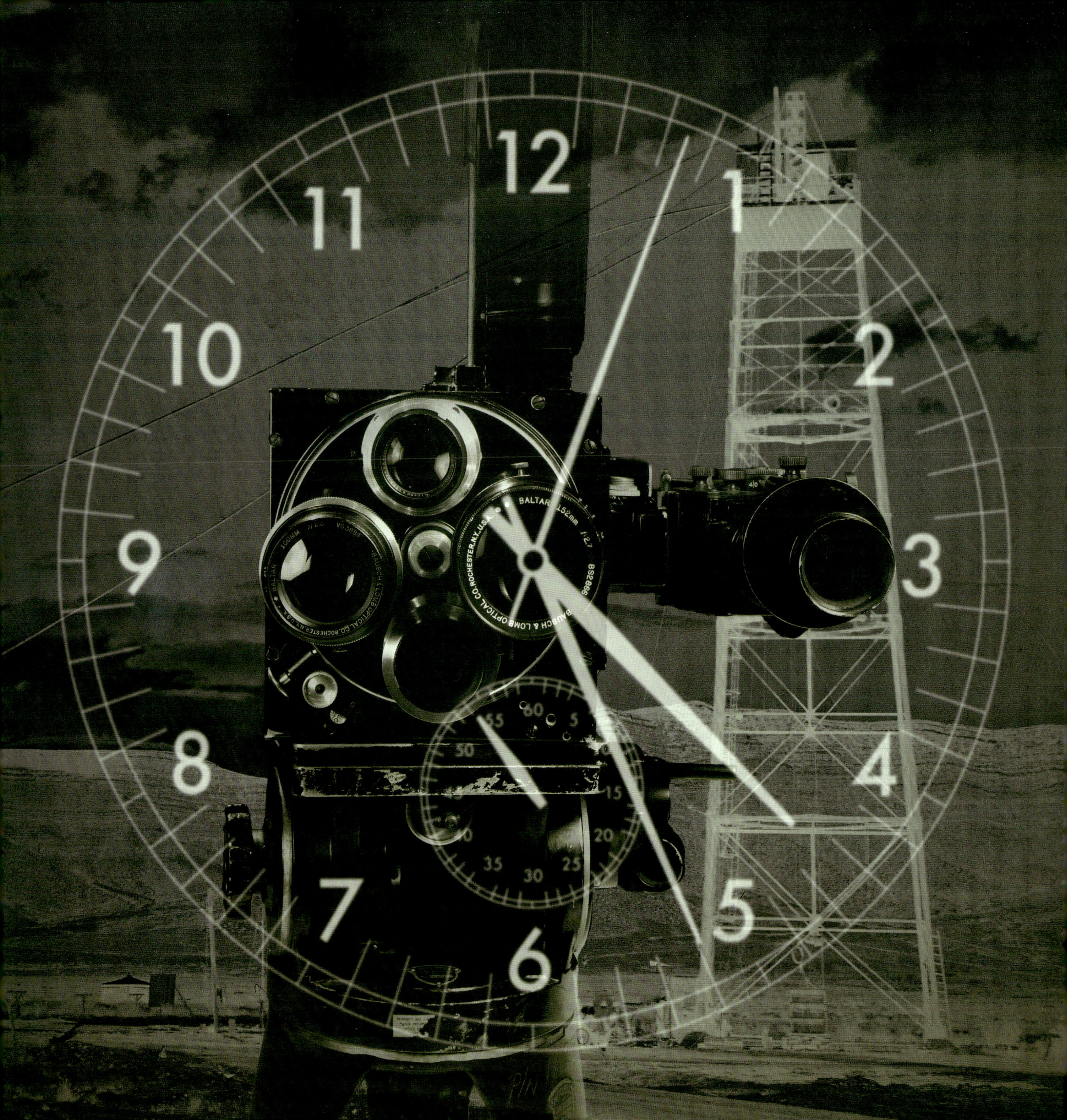
12
11
1
10
2
9
3
8
4
7
5
6
BALTAR 152mm f:2.7
BS2866
BAUSCH & LOMB OPTICAL CO. ROCHESTER, N.Y. U.S.A.
60
55
5
50
15
20
25
30
35

05 The Countdown

While the exact day of the Trinity test was a moving target, with July 16 identified as the "soonest possible" date, the timing of the shot had been carefully and intentionally scheduled for four o'clock in the morning. Cameras were prepped for dark conditions, with illumination systems set up across the basin. The hope was that most people in the area not associated with the top-secret project would be sound asleep; they would read in the next day's news about an "ammunitions dump" having exploded on the bombing range. But on the eve and morning of the test, the weather refused to cooperate. Violent, scattered thunderstorms moved through portions of the Jornada del Muerto.

The forecast weighed heavily in light of recent reports from the Health Group. Since the 100-ton test, there had been growing concern that nuclear fallout—radioactive debris and fission products ejected into the sky that would eventually make their way back to the ground—might be more severe than originally anticipated. The danger would be especially high if volatile winds sent the cloud over populated areas or if stormy weather intervened to push down the radioactive particles before they could decay harmlessly in the upper atmosphere.

Given these concerns, no serious consideration was given to firing in the rain, but when would it stop? Would the test require a delay of minutes, hours, or even days? With the nation's leader awaiting word from the desert and a new day only hours away for New Mexicans throughout the region, the project's lead meteorologist took center stage as advisor to Groves and Oppenheimer. It was an unenviable role for Jack Hubbard. If the Trinity device detonated on schedule, which now coincided with the middle of a storm, precipitation would quickly pull concentrated, highly radioactive material back to earth, posing a significant risk to those below. If the test was postponed, the president would enter the first day of negotiations at Potsdam without knowing whether the United States possessed a nuclear capability.

Fortunately, there was a third way forward. If the skies cleared during the night and the test was delayed until just before dawn, this would greatly improve the safety outlook and still leave time to pass along news of a successful test (or a fizzle) to the president. Shifting the test into waking hours would dramatically increase the number of unintended observers, but the competing priority of protecting those observers from a radioactive downpour held far greater standing. Thus, security was sacrificed for safety and the demands of international politics: Trinity would—hopefully—unfold as dawn approached. Hubbard had the unfortunate duty throughout the night of briefing the weather situation to Oppenheimer, who remained on edge, and Groves, who was even more agitated than usual.

As the hours passed, senior scientists, many of whom still had important roles to play during and after the test, waited in personnel shelters for the culmination of their tireless efforts, while at base camp, Groves and his scientific advisors, Vannevar Bush and James Conant, who had been the driving forces behind the nuclear enterprise, lay prone on the ground hoping for one of history's greatest technical gambles to pay off. Enrico Fermi attempted to ease the tension, jokingly taking bets on the possibility of the test igniting the atmosphere and ending all human life. Initially perturbed, Groves later came to appreciate this source of levity as Trinity's sleep-deprived audience plodded toward an uncertain finish. For members of the radiation monitoring teams and evacuation detachments who were scattered throughout the region, the detonation would not provide a moment of emotional release but, rather, the beginning of their critical mission to protect the surrounding population.

To the north and west of ground zero, dozens of cameras awaited the firing signal from the control bunker. With the storm clouds having finally scattered, Bainbridge and physicist Joseph McKibben initiated a twenty-minute timer at 5:09 a.m. This was the last action required; unless someone intervened, the detonators would fire. Donald Hornig, having just performed the nerve-wracking task of connecting the firing switch to the Gadget on top of the tower as lightning cracked the sky, was assigned another important role. If the test needed to be stopped for any reason, Hornig would throw the "knife switch" to shut down the operation.

With about thirty seconds to go and the countdown on an automated timer, Bainbridge left the control bunker, got on the ground, put on his welder's goggles, and looked away from ground zero. Meanwhile, the transmission carrying the countdown became entangled with public radio. Various accounts report different soundtracks for Trinity, but it's likely "The Star-Spangled Banner" rang across the test site. Brig. Gen. Thomas Farrell wrote later, "The scene inside the shelter was dramatic beyond words. . . . It can be safely said that most everyone present was praying. Oppenheimer grew tenser as the seconds ticked off. He scarcely breathed. He held on to a post to steady himself."

Just after 5:29 a.m., the clock ran down to zero. No order to activate the knife switch ever came. Free from his duty inside the bunker, Hornig rushed out into the desert to witness the dawn of a new era.

* * * * * * *

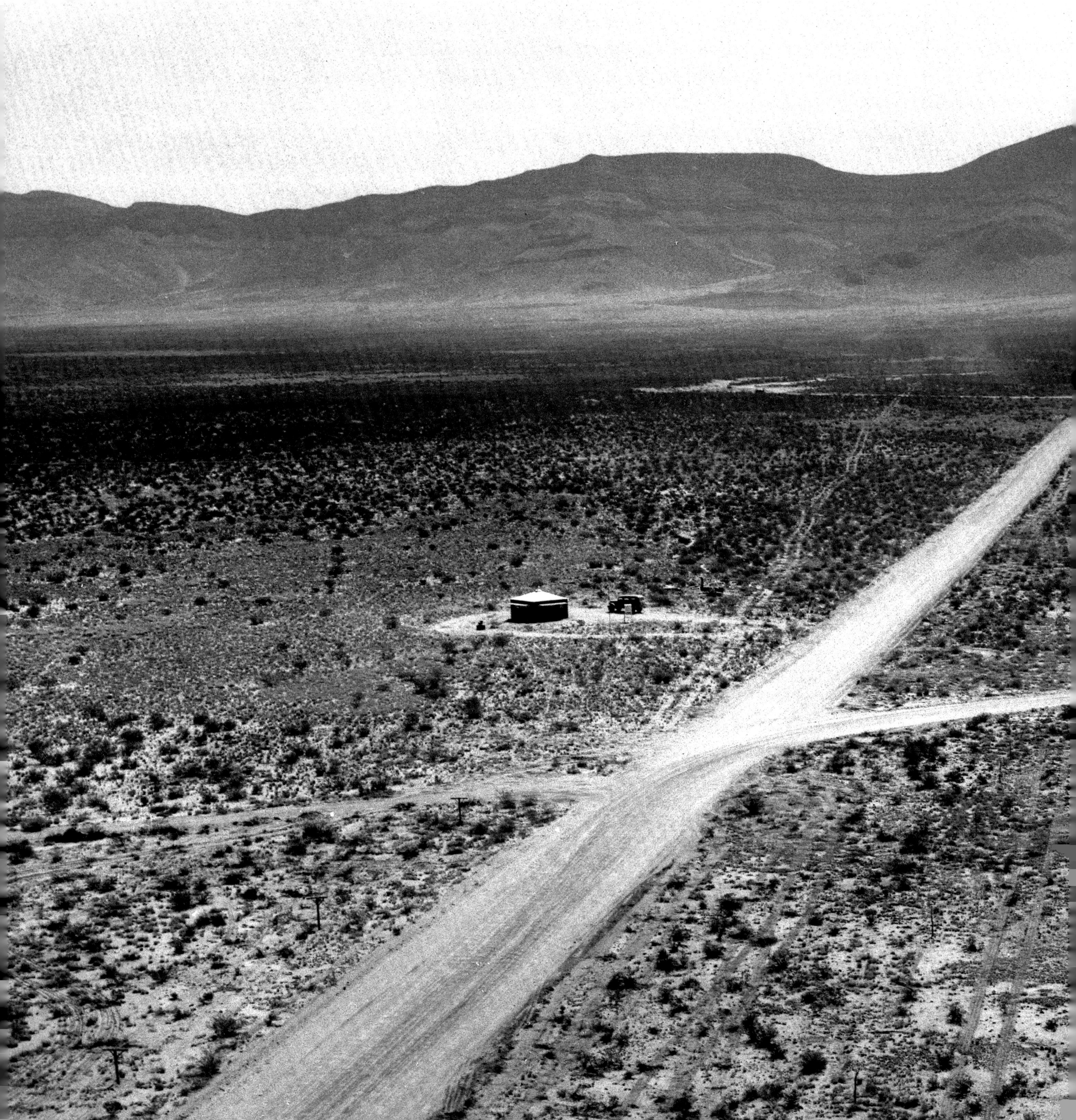

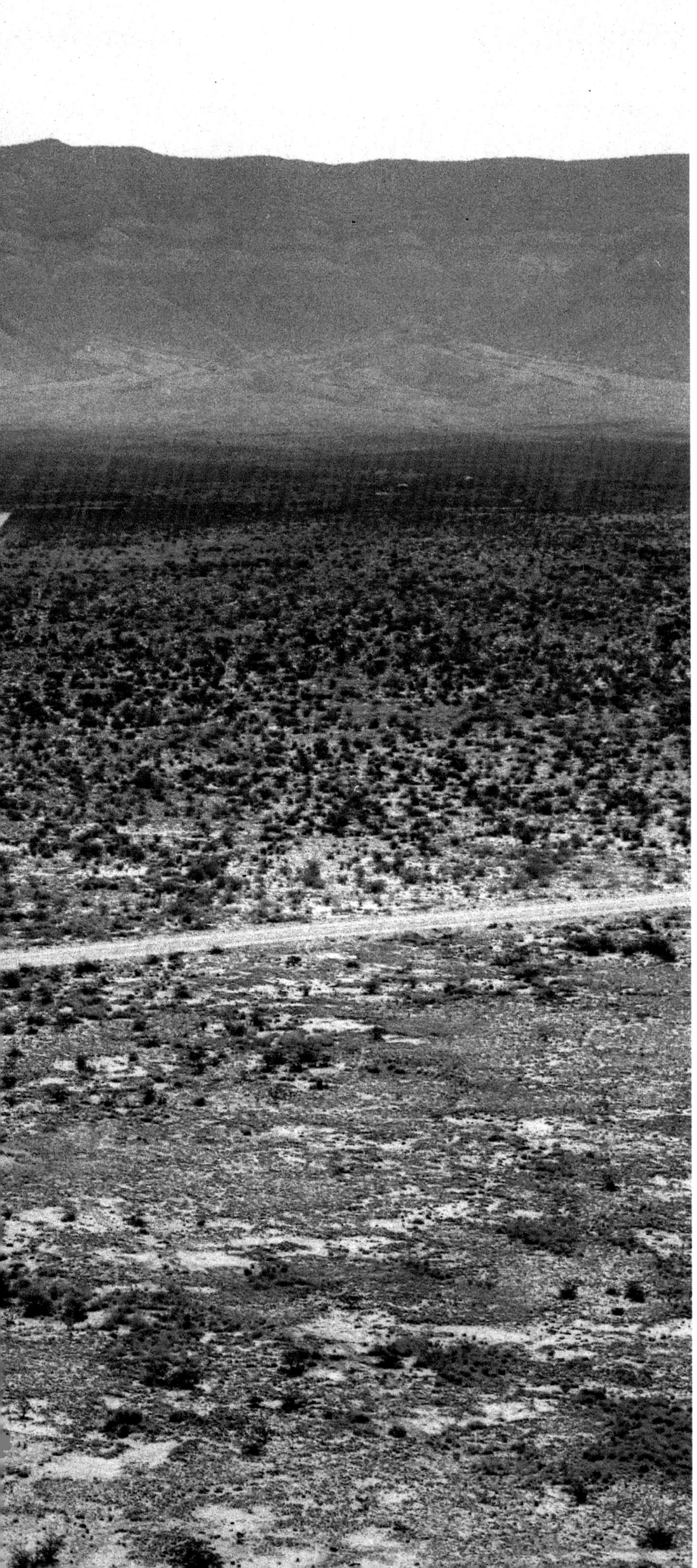

On the afternoon of July 15, General Groves arrived at the Trinity site to find gusty winds and misty skies. By 8:00 p.m., with only eight hours to zero time, scattered, violent storms had descended upon the dusty basin—bringing local showers that could last all night and into the morning.

The conditions sparked extreme concern among scientists and weather experts alike, and talk of postponement had begun circulating at base camp even before the rain set in. Hoping for a last-minute break in the storms, Groves and Oppenheimer chose to defer an official delay of the test shot until after a meeting with meteorologist Jack Hubbard that was scheduled for 2:00 a.m. Hubbard, stationed in the hutment at left, 100 yards east of the tower, would remain near the Gadget late into the night, launching weather balloons into the atmosphere.

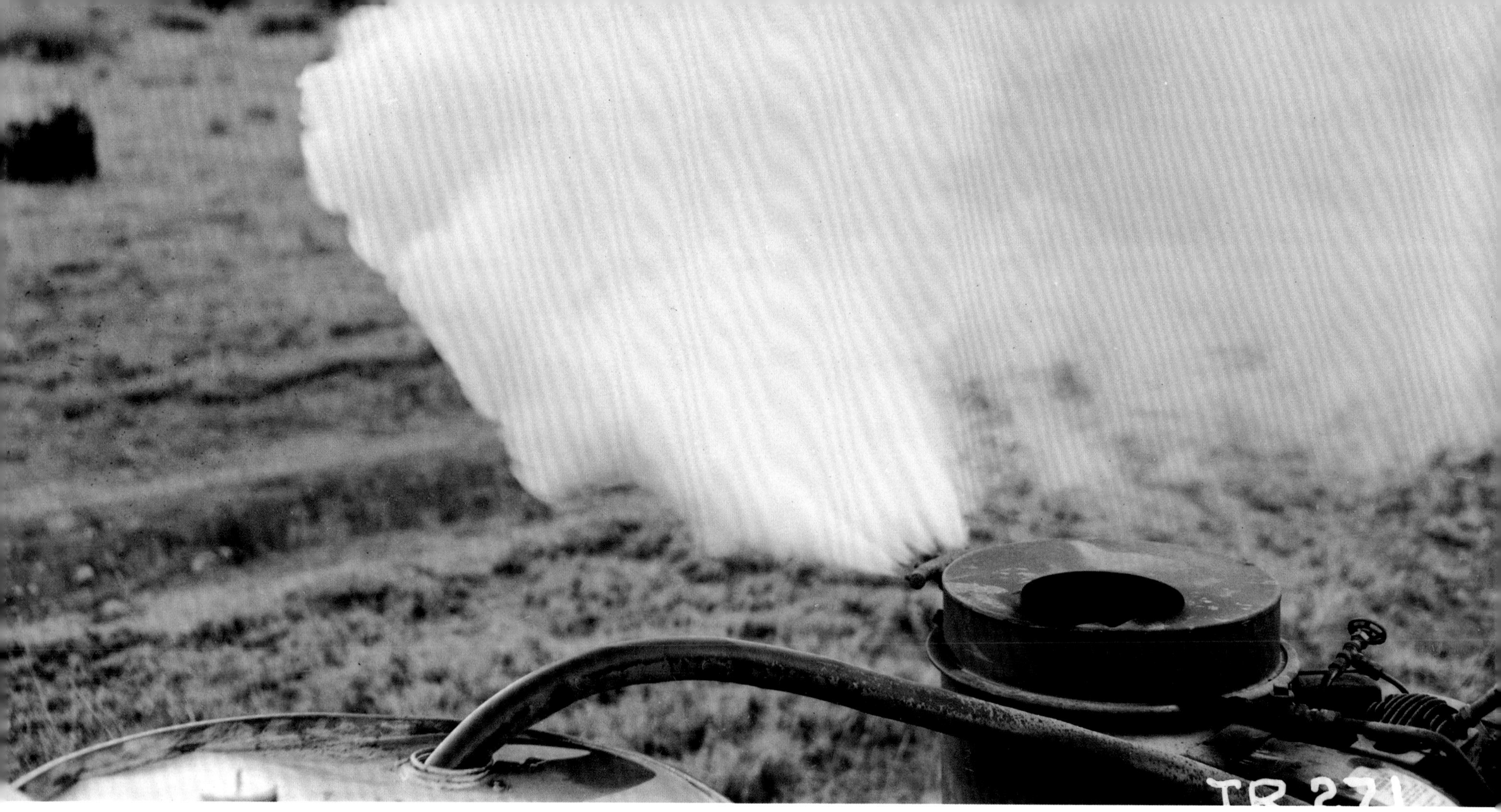

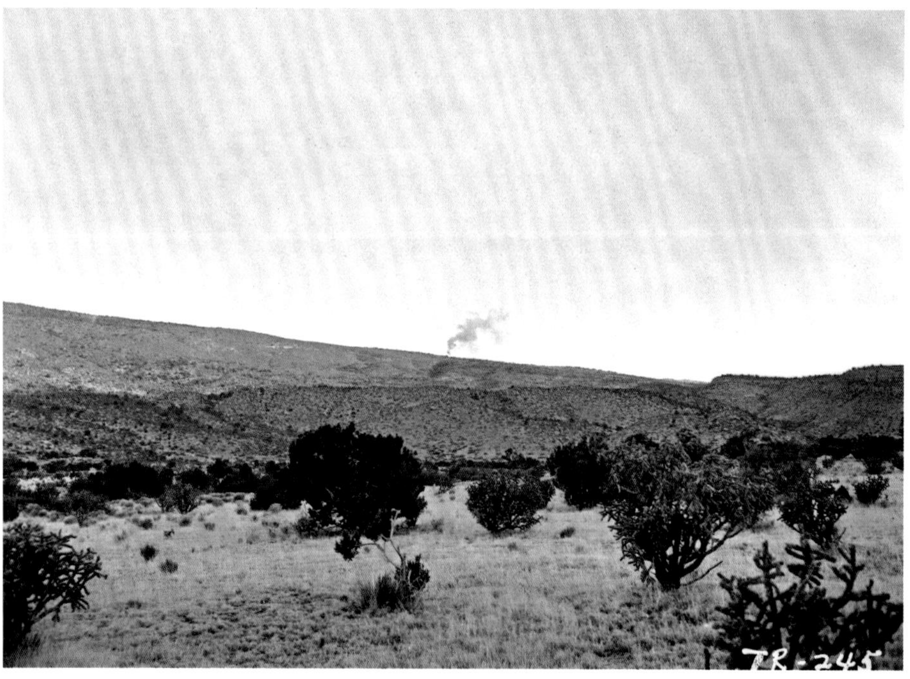

While rain and wind would certainly interfere with shot diagnostics, the main concern regarding the weather was for public safety. Since the 100-ton test in early May had accelerated safety planning for radioactive fallout, scientists had been diligently gathering data on the behavior of human-made "clouds" at the Trinity site. Throughout July, Frank Oppenheimer, brother of laboratory director J. Robert Oppenheimer and himself a physicist (and chairman of the Trinity test personnel safety committee), had conducted dispersion tests using smoke pots and generators. The results of these studies kept Hubbard apprised of wind patterns as well as helping to build a collection of data for the Health Group as to when and where the airborne radioactive particles might settle.

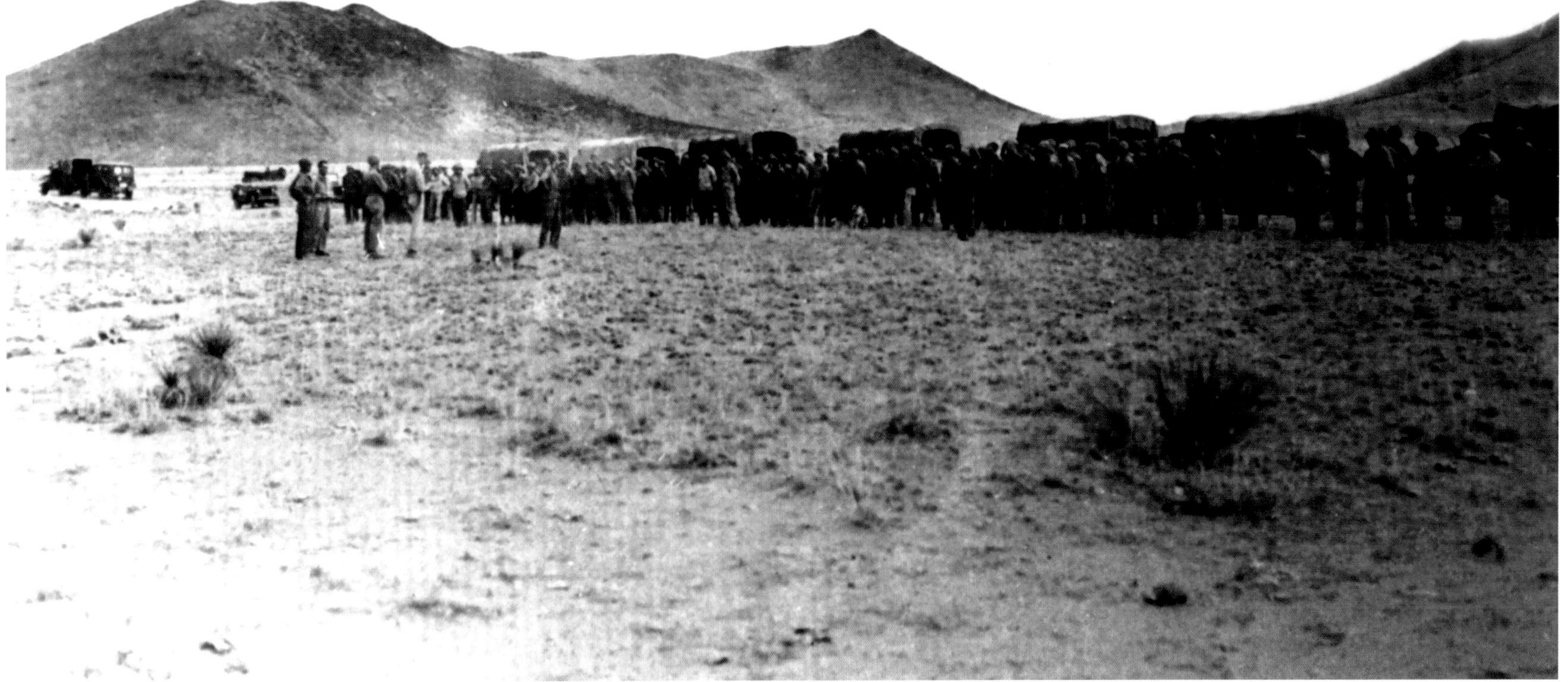

Growing unease about the potential for local contamination—even under good weather conditions—had prompted the Health Group to formulate a comprehensive civilian evacuation plan. If warranted, the plan would be enacted by Maj. T. O. Palmer and a detachment of 144 soldiers (*above*) who had access to 140 vehicles, 500 gallons of drinking water, rations, and other emergency supplies. The detachment set up a campsite about nine miles northwest of ground zero, ready to house 450 civilians.

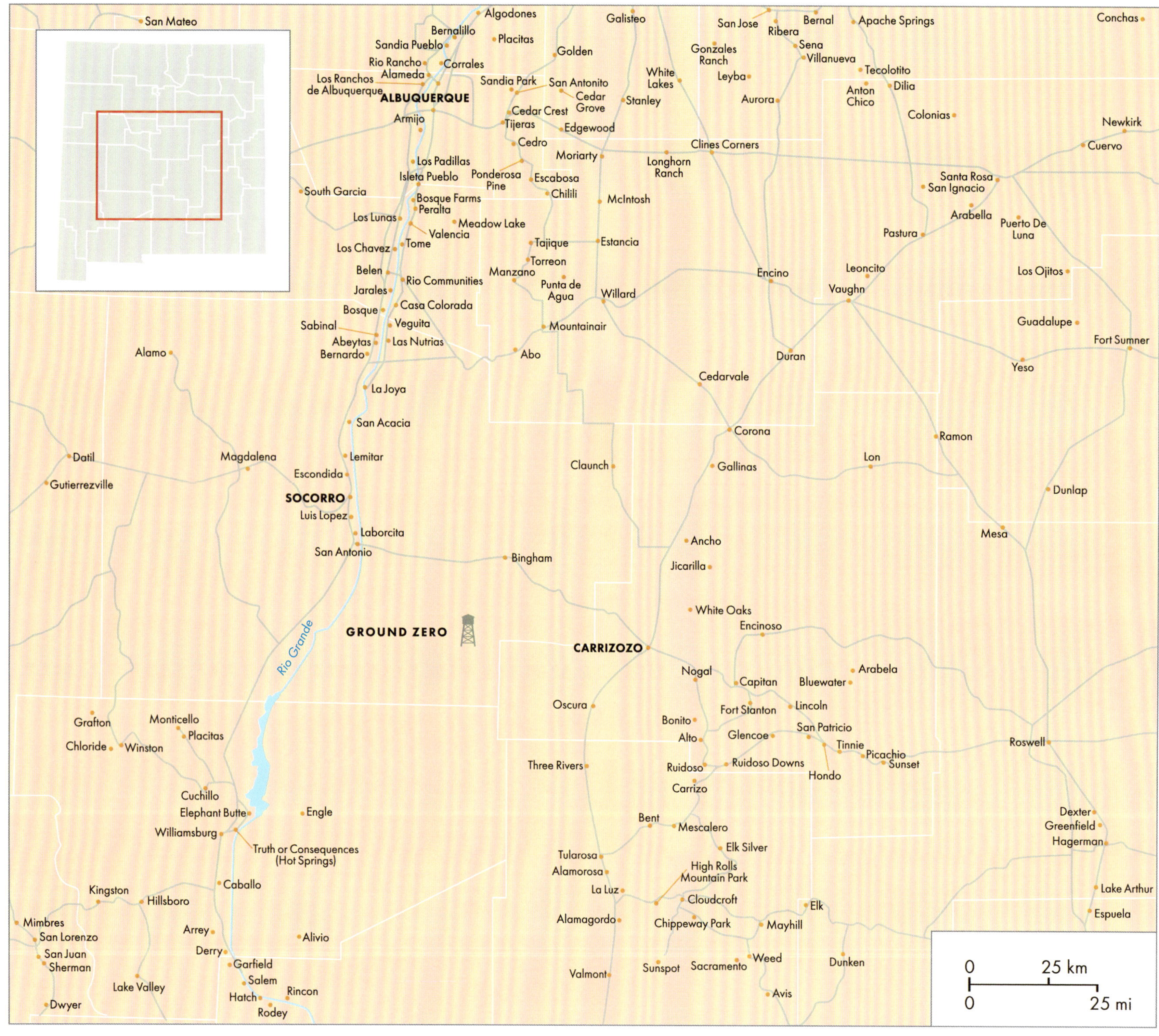

An extensive radiation monitoring program was set up throughout the Trinity site and surrounding region. Some technicians were assigned posts on-site while others would travel in pairs, chasing the drifting cloud with gamma ray counters and air sampling equipment to ensure readings stayed below an established threshold in populated areas. These data would inform any decision to evacuate civilians. Socorro and Carrizozo, to the north and east of ground zero, were considered the most at-risk population centers.

4

SECRET

Attached hereto is a table of data compiled by S/SGT Leonard reporting in greater detail the record of radiation levels taken while in company with the reporting Agent.

N. J. McElwreath
Special Agent C.I.C.
S/Sgt. Robert R Leonard

Signed by

In addition to gathering data for real-time assessment of evacuation needs, monitoring teams were instructed to keep detailed, handwritten, and signed records in case medical lawsuits were later filed. Film cards that could passively detect and record radiation exposure were also mailed to nearby post offices.

As the one who would have to answer for all costs incurred by the Manhattan Project, Groves was also concerned about property damage that might result from the test shot, depending on the strength of the ground shock wave. Geophones were in place to measure earth vibrations near ground zero, but Groves was concerned about vibrations that might spread much further than the test site and potentially lead to damage claims.

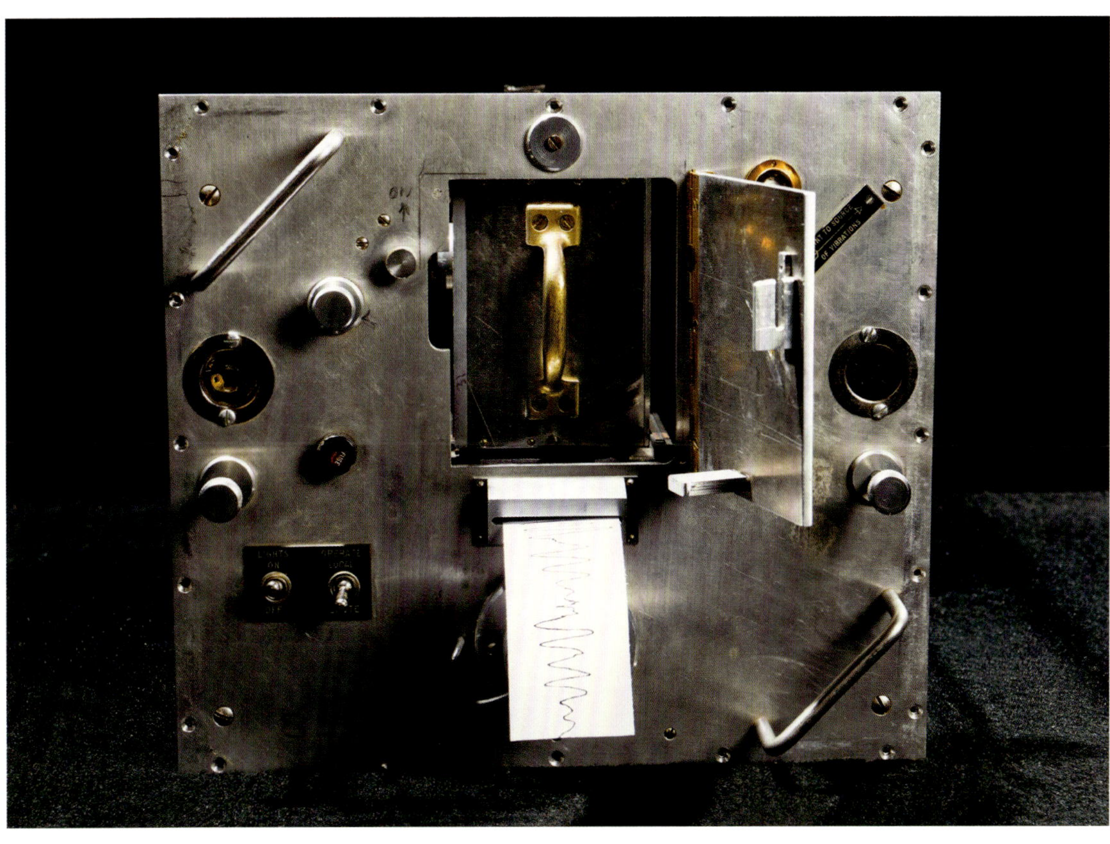

To create a record of distant earth vibrations, or lack thereof, Groves enlisted the services of seismologist L. Don Leet. Leet was highly skeptical that these observations were necessary; the scaled 100-ton test had validated his position that ground shock would be negligible, if not imperceptible, at the distances Groves was worried about and in fact would not cause any damage beyond about 1,000 yards.

Nevertheless, Leet dutifully arranged for the measurements. He fabricated five strong-motion seismographs with built-in cameras (*left*). These were specially designed to pick up high seismic activity that, for more sensitive instruments listening for weaker motion, would be off the charts. The seismographs were stationed 9,000 yards north of ground zero; in the towns of Tularosa, Carrizozo, and San Antonio, New Mexico; and at Elephant Butte Dam. A fleet of standard seismographs was spread across the western United States, as far away as Tucson, Arizona; El Paso, Texas; and Denver, Colorado.

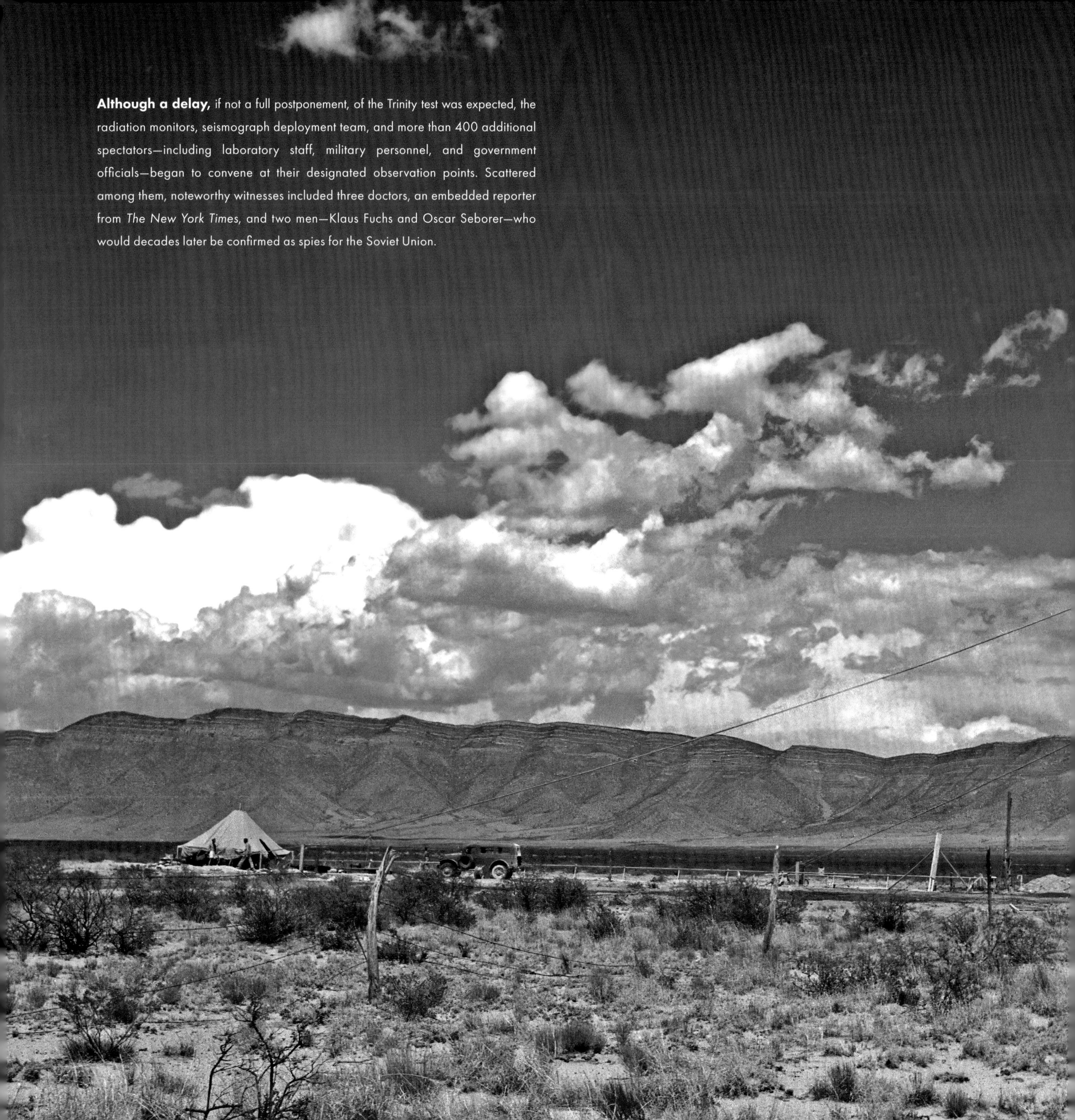

Although a delay, if not a full postponement, of the Trinity test was expected, the radiation monitors, seismograph deployment team, and more than 400 additional spectators—including laboratory staff, military personnel, and government officials—began to convene at their designated observation points. Scattered among them, noteworthy witnesses included three doctors, an embedded reporter from *The New York Times*, and two men—Klaus Fuchs and Oscar Seborer—who would decades later be confirmed as spies for the Soviet Union.

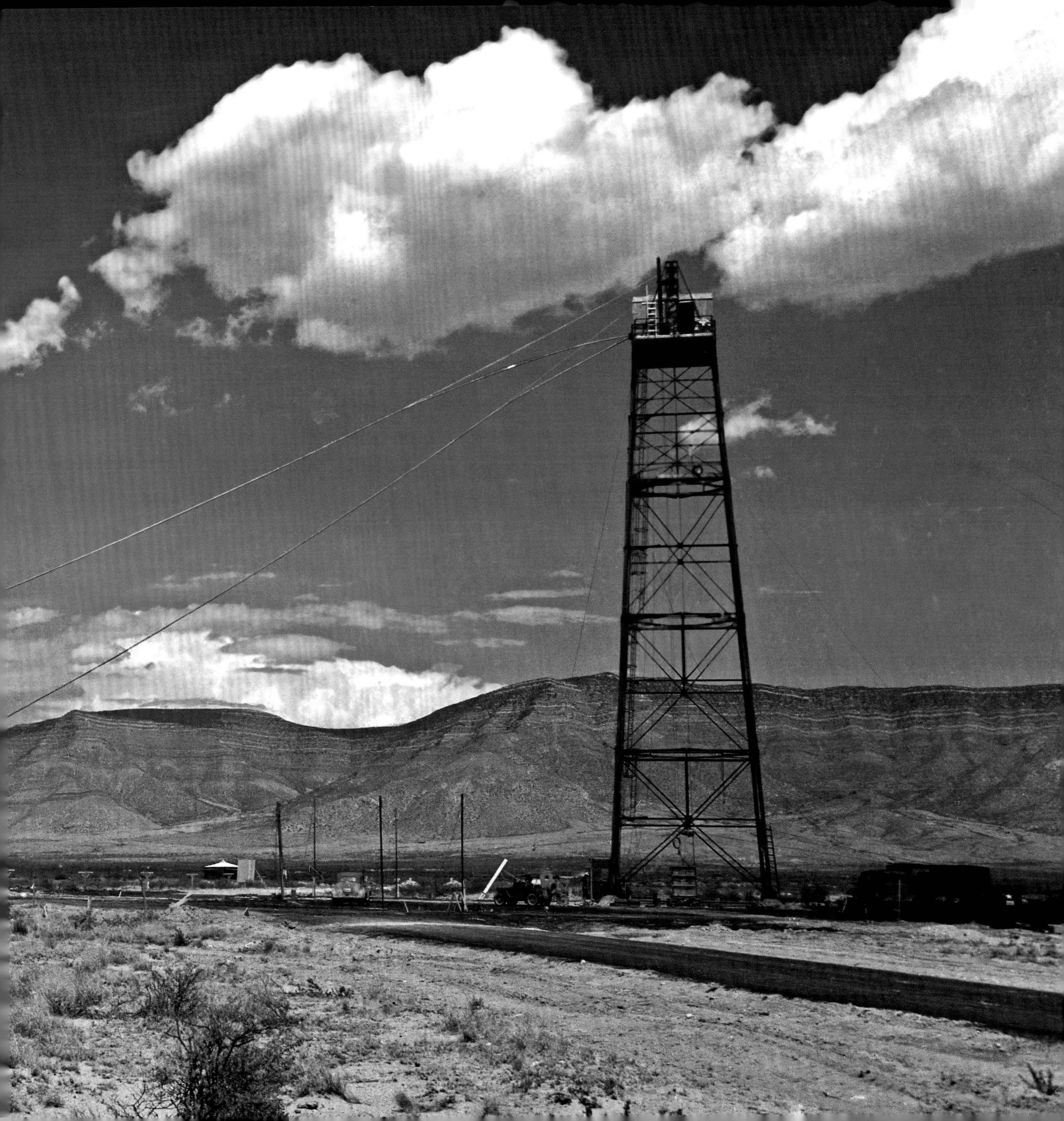

Directions for Personnel at Campana Hills Camp at Time of Shot 62
(Coordinating Council Camp) ~~at Campana Hills~~)
July 15, 1945 41

1. Do not leave main group at camp where there will be monitoring and evacuation facilities. There will also be contact with planes, shelters, and area monitors over the radioreceivers.

2. All personnel at Hill Camp will conform with the following Safety Regulations:

a) At a short signal of the siren at minus 5 minutes all personnel whose duties do not specifically require otherwise will prepare a suitable place to lie down on.

b) At a long signal of the siren at minus 2 minutes all personnel whose duties do not specifically require otherwise, will immediately lie prone on the ground, the face and eyes directed toward the ground and with the head away from "zero". Do not watch for the flash directly, but turn over after it has occurred and watch the cloud. Stay on the ground until the blast wave has passed (2 minutes).

c) At two short blasts of the siren, indicating the passing of all hazard from light and blast, all personnel will prepare to leave as soon as possible.

3. The hazard from blast is reduced by lying on the ground in such a manner that flying rocks, glass, and other objects do not intervene between the source of blast and the individual. Open all car windows.

4. The hazard from light injury to the eyes is reduced by shielding the closed eyes with the bended arms and lying face down on the ground. If the first flash is viewed a "blind spot" may prevent your seeing the rest of the show. A Welder's filter glass should be used upon first looking at the ball of fire after the initial flash has passed.

5. The hazard from ultraviolet light injuries to the skin is best overcome by wearing long trousers and shirts with long sleeves.

Base camp commander Lt. Howard Bush asked personnel who intended to leave immediately after the test to watch the shot from Campana Hill (also referred to as Campana Hills or Campania Hill), about twenty miles to the northwest.

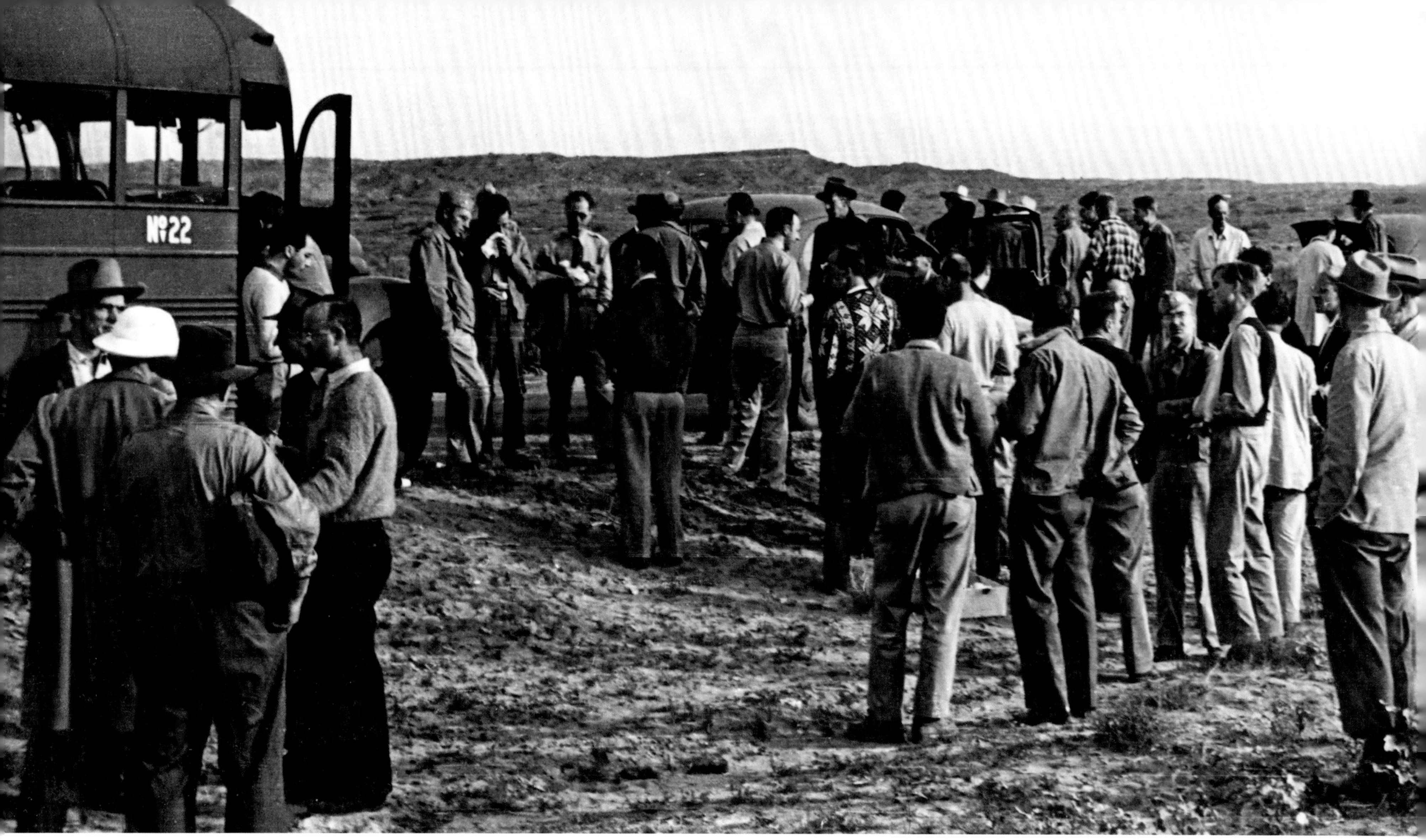

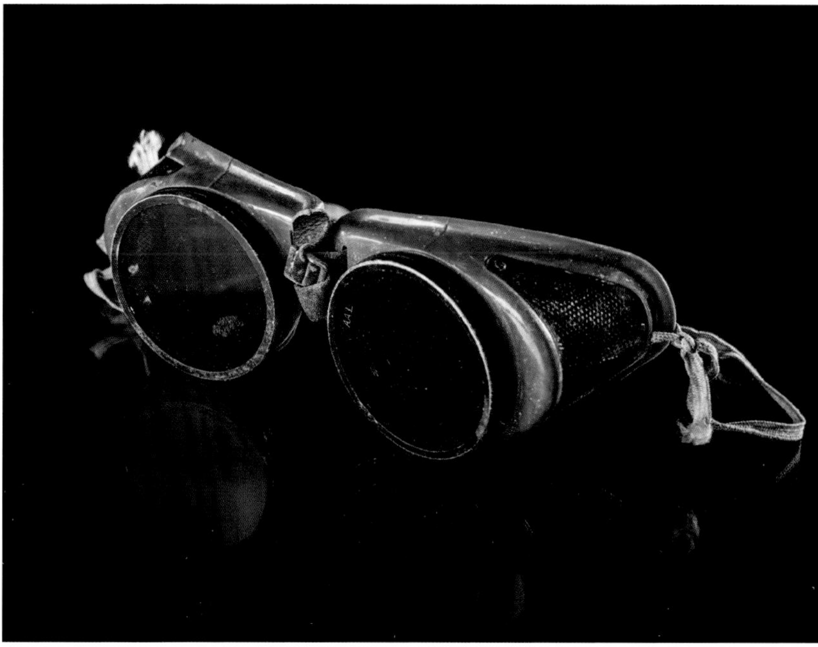

The Campana contingent would be joined by busloads of observers arriving late that evening from Los Alamos (*above*), including many of the theoreticians who had contributed to the Trinity test from afar. (Photo courtesy of Richard Wheeler, Winters, California.)

All were supplied with welder's glass as eye protection; some received goggles, like those at left, while others held pieces that were loose or taped to cardboard. The initial burst of ultraviolet light was predicted to be 300 times as bright as the sun. Because it wasn't known how the human eye would respond to such an event, the recommendation issued was on the side of caution: only after a few seconds could viewers safely turn toward the blast, and even then they must look through the welder's glass. To avoid "sunburn" from the UV radiation, observers were also encouraged to cover their skin with pants and long sleeves.

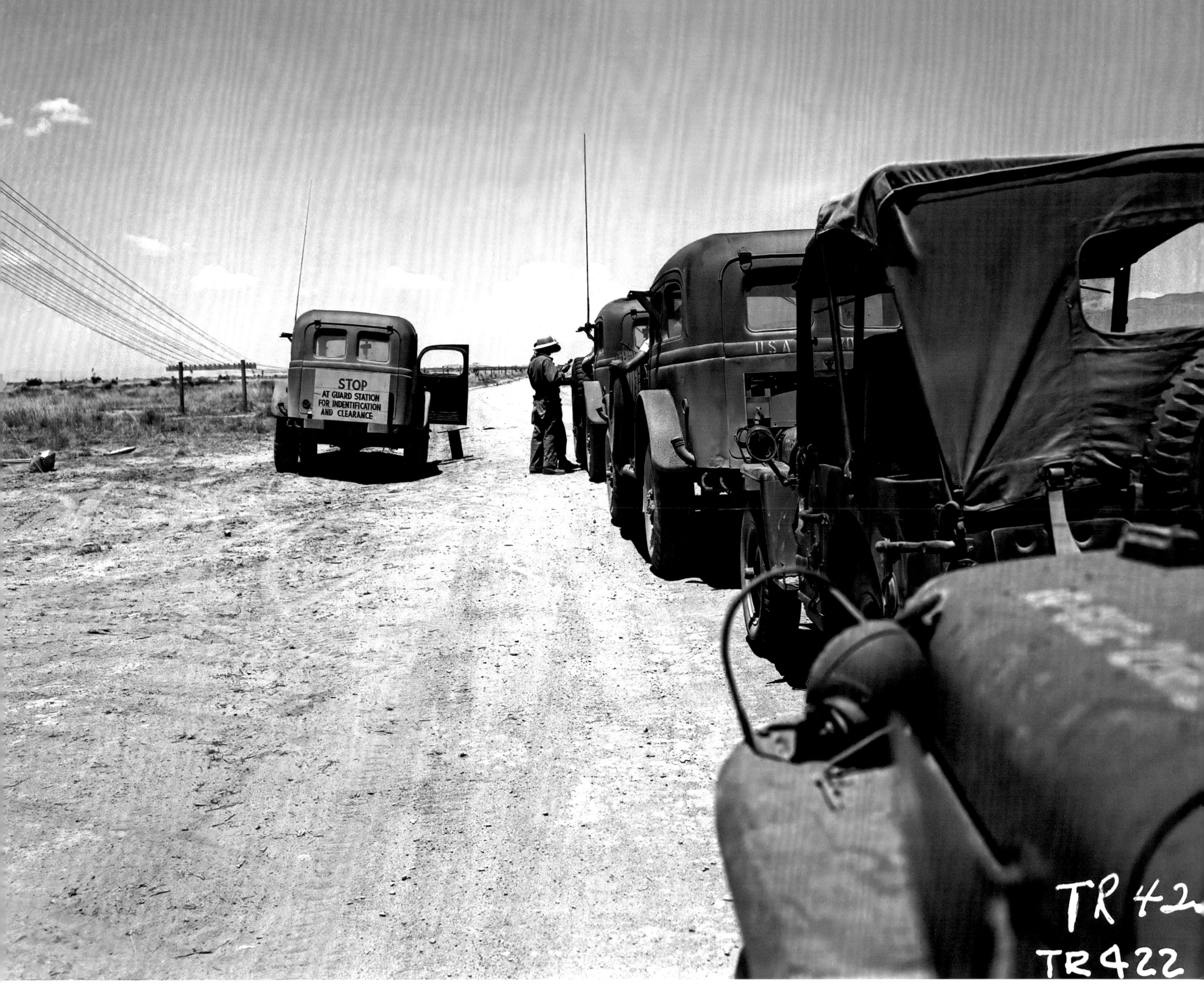

At noon, Military Police had closed the road from base camp to ground zero, setting up a checkpoint at Pennsylvania and Broadway (*above*). Between two and three hours before zero time, a representative from each post met with Lieutenant Bush to compare records of authorized workers having entered and exited the restricted area. They then conducted a final sweep of the access roads to clear out any remaining non-essential personnel. Per instructions from Bush, the guards would migrate to foxholes and personnel shelters before countdown to wait for the blast, ready to return afterward to their usual posts or roadblocks to control access to ground zero.

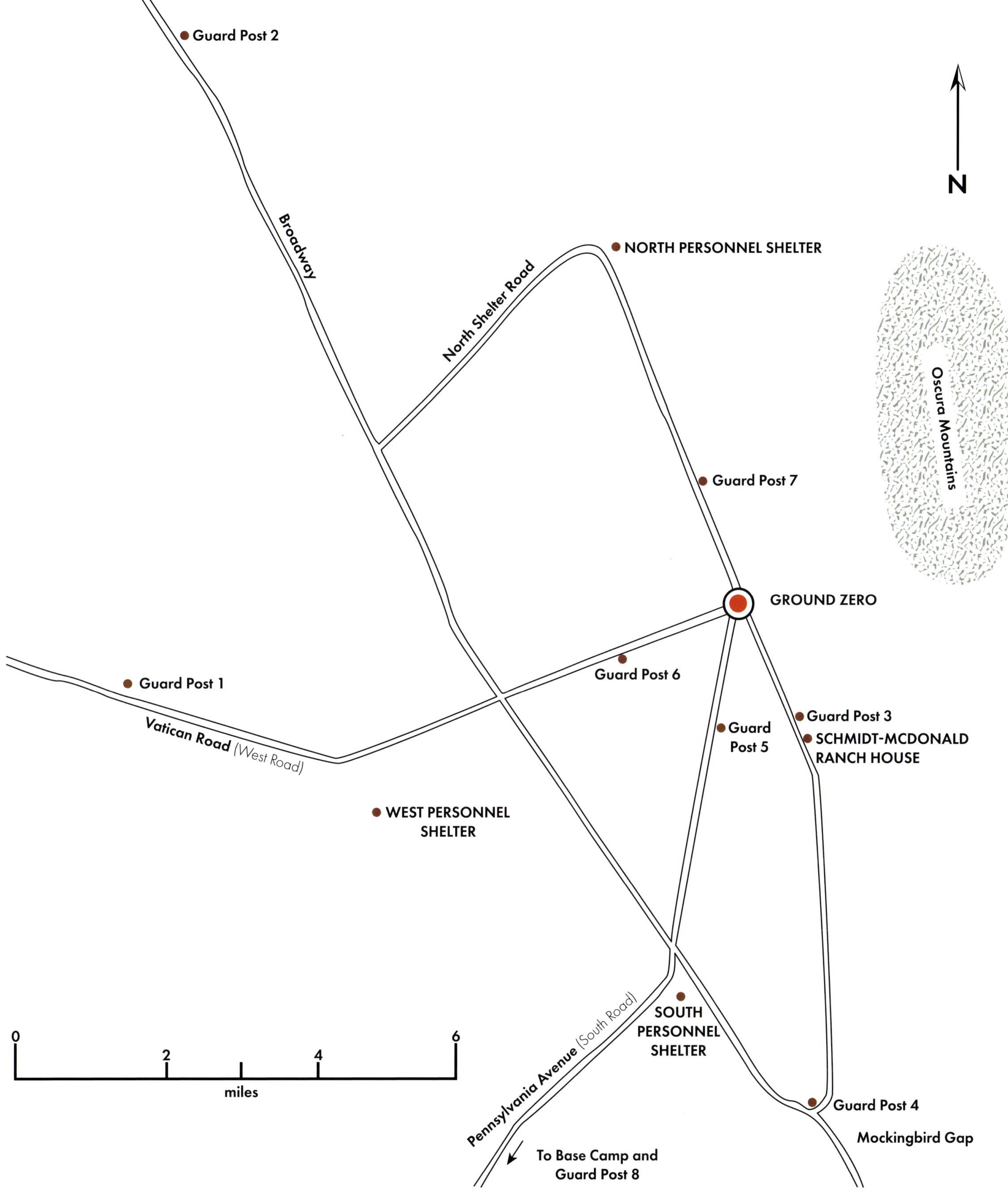
Guard Post 2
Broadway
N
NORTH PERSONNEL SHELTER
North Shelter Road
Oscura Mountains
Guard Post 7
GROUND ZERO
Guard Post 6
Guard Post 1
Vatican Road (West Road)
Guard Post 5
Guard Post 3
SCHMIDT-MCDONALD RANCH HOUSE
WEST PERSONNEL SHELTER
SOUTH PERSONNEL SHELTER
Pennsylvania Avenue (South Road)
0
2
4
6
miles
Guard Post 4
Mockingbird Gap
To Base Camp and Guard Post 8

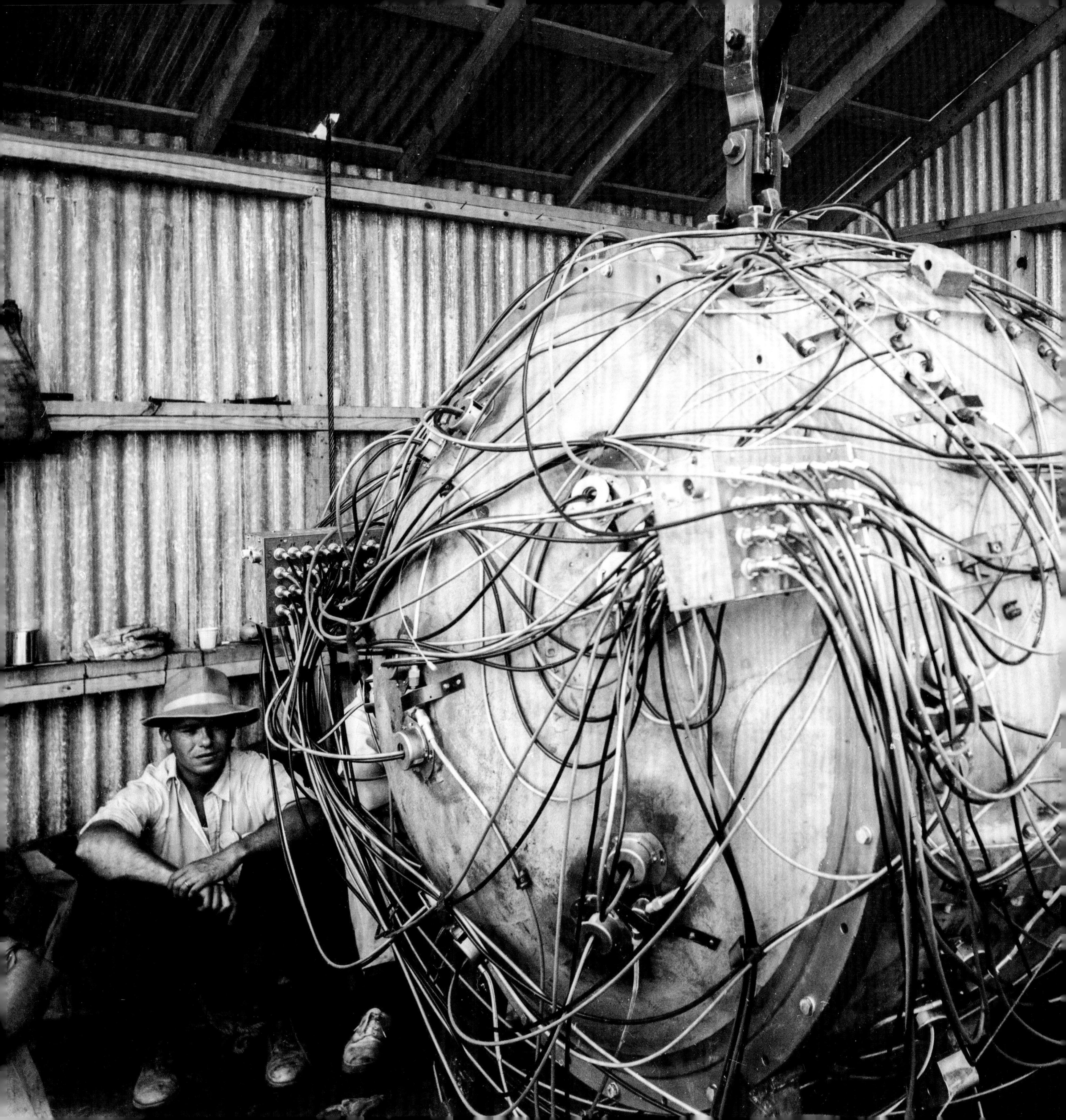

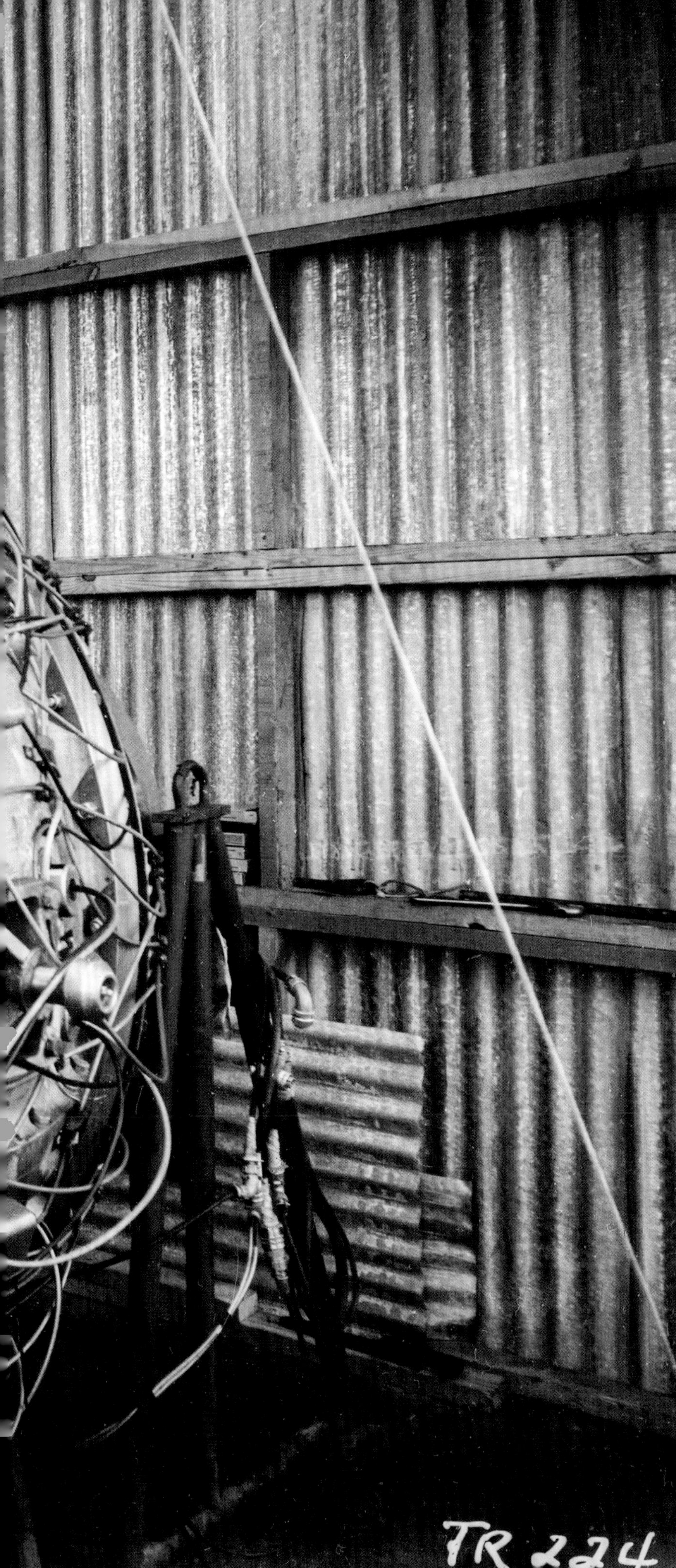

Even with strict security measures in place, Groves viewed the final waiting period—and every minute of delay—as an opportunity for some nefarious individual to tamper with the Gadget. Thus, a group including Donald Hornig (*left*) took shifts guarding the all-important device atop the tower. Although the 100-foot structure was properly grounded, Hornig, taking the last shift, remained anxious as lightning bolts illuminated the desert sky.

Troubles with the firing mechanism he invented had plagued the young chemist in the week before the test shot. One of his X units, sitting atop the tower, discharged spontaneously during a thunderstorm. An uproar ensued, until Hornig discovered that a loose grounding wire had permitted a build-up of static electricity from the lightning.

The second incident occurred only days before the test, during one of several rehearsals. When the signal was given, the attached firing unit did not respond. Hornig and Kistiakowsky dismantled the device, finding that the problem was not a defect. The X unit had been designed for only a single use, seeing as it would be destroyed by a proper detonation. The malfunctioning unit had been fired hundreds of times during testing and preparation and worn out a switch.

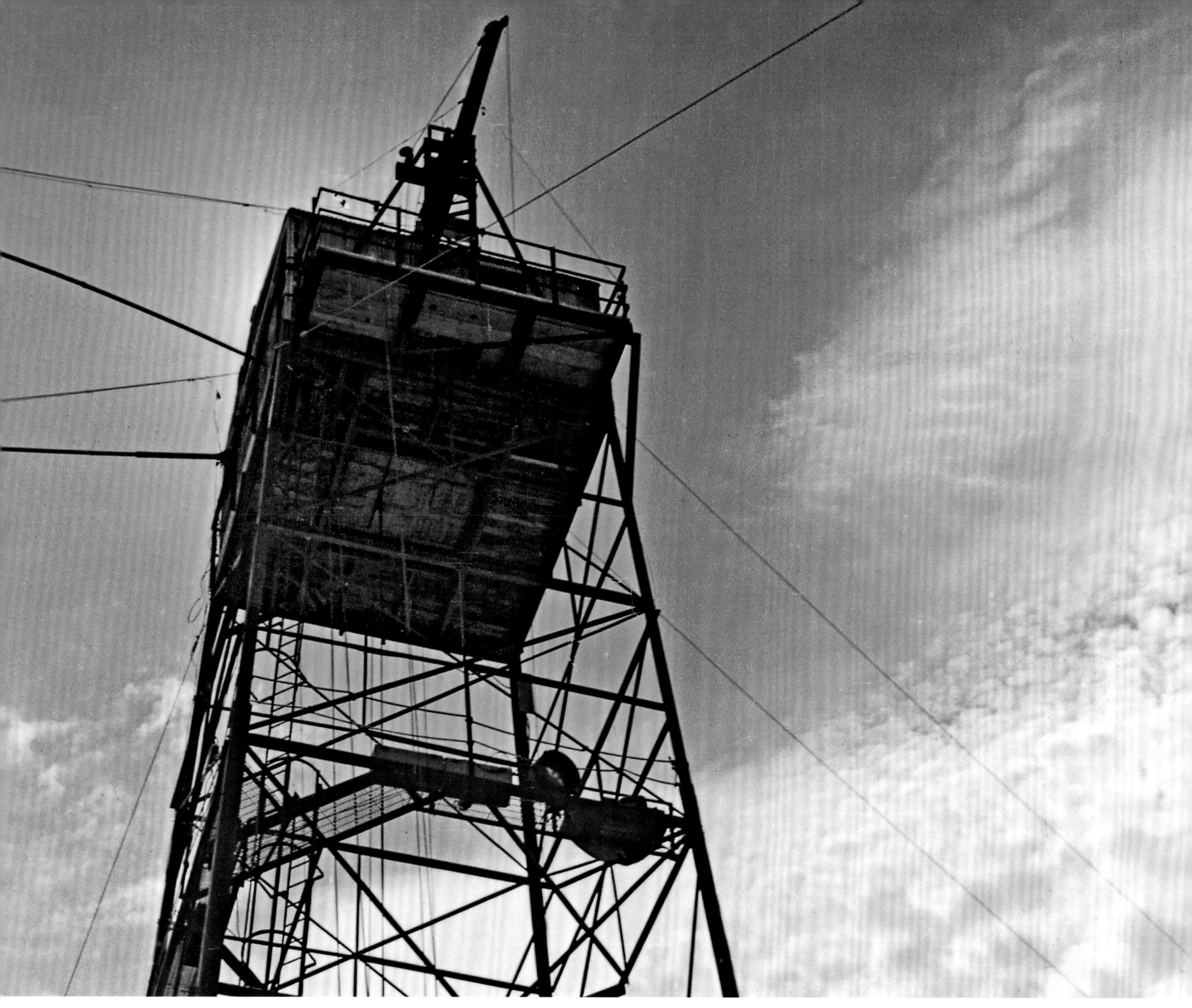

Despite uncertainty surrounding when, and if, the shot would go off, an arming party set off into the night to activate firing and timing switches between the controls at south 10,000 and the Gadget. Focused on the technical complexities of the test shot, this group was strongly opposed to any significant delay, anticipating that if they missed their window that morning, a postponement of several days would follow, creating opportunities for electrical shorts and all manner of water damage to set in.

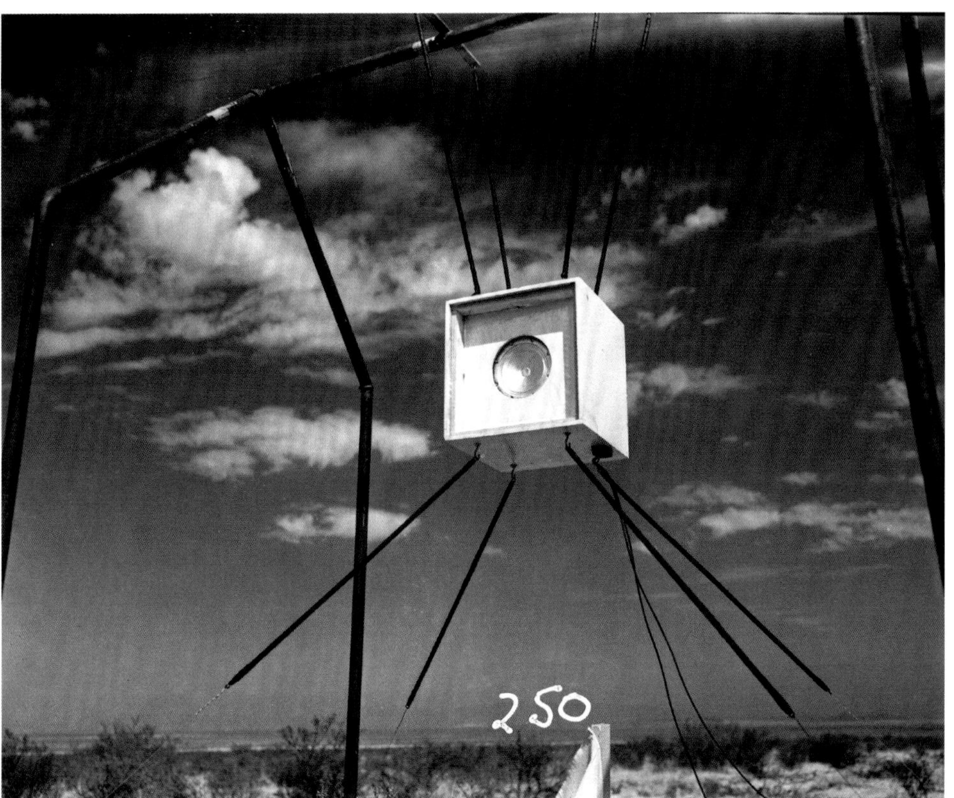

The group of seven, led by Bainbridge and including George Kistiakowsky, physicist Joseph McKibben, and a security detail, arrived at the tower around 11:00 p.m. On cue, Hornig connected the live X unit and climbed down to find his way to an observation point. He was the last to see the test device.

Sgt. W. Stewart set up a small reference charge for an experiment designed to track the speed of the shock wave and left to find his designated shelter. This very experiment (*left*), measuring excess velocity, would turn out to provide some of the most accurate yield data collected from the test shot. Another Military Police sergeant, J. T. Alderson, was responsible for gathering weather data at the tower over the next several hours.

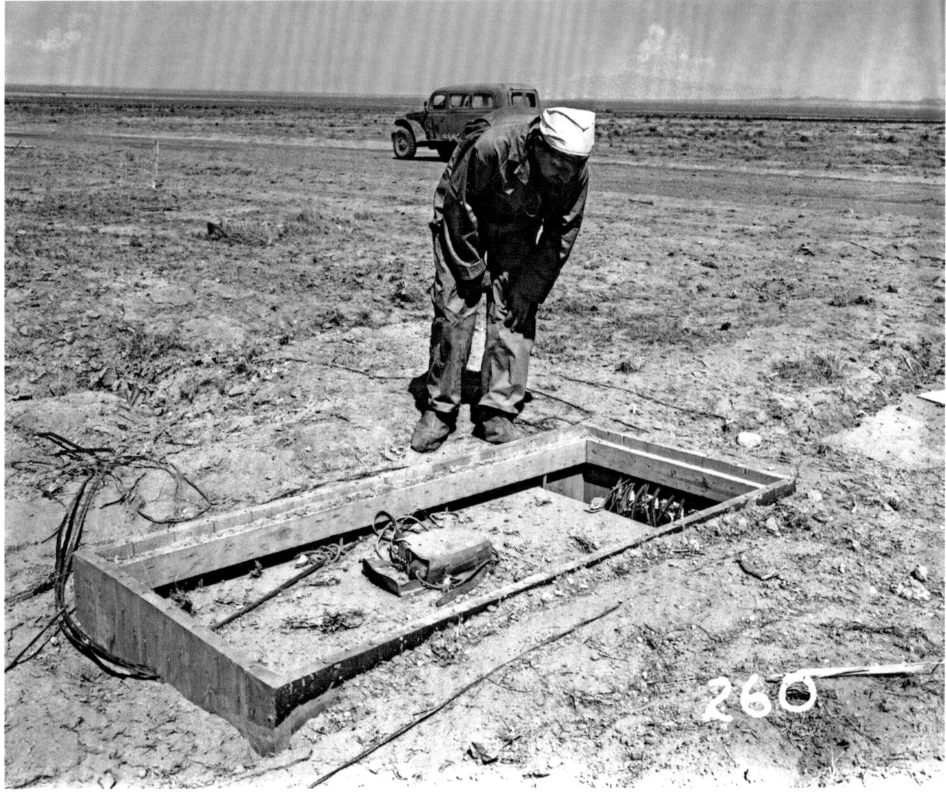

The arming party would remain at ground zero, on guard, until just before the countdown began. At that point, two army searchlights—providing illumination for photography and cloud tracking—would be activated. The bright lights were thought to be a sufficient deterrent to any would-be, last-minute saboteurs.

While they waited, Kistiakowsky climbed seventy-five feet up to an instrument platform (visible in the photo at far left, beneath the shot cab) to adjust a spotlight that served as an aiming point for Julian Mack at his post in the west 10,000 photography bunker. Bainbridge and McKibben broke away to activate timing and sequence switches at a small, wooden control station (*left*) 900 yards to the west.

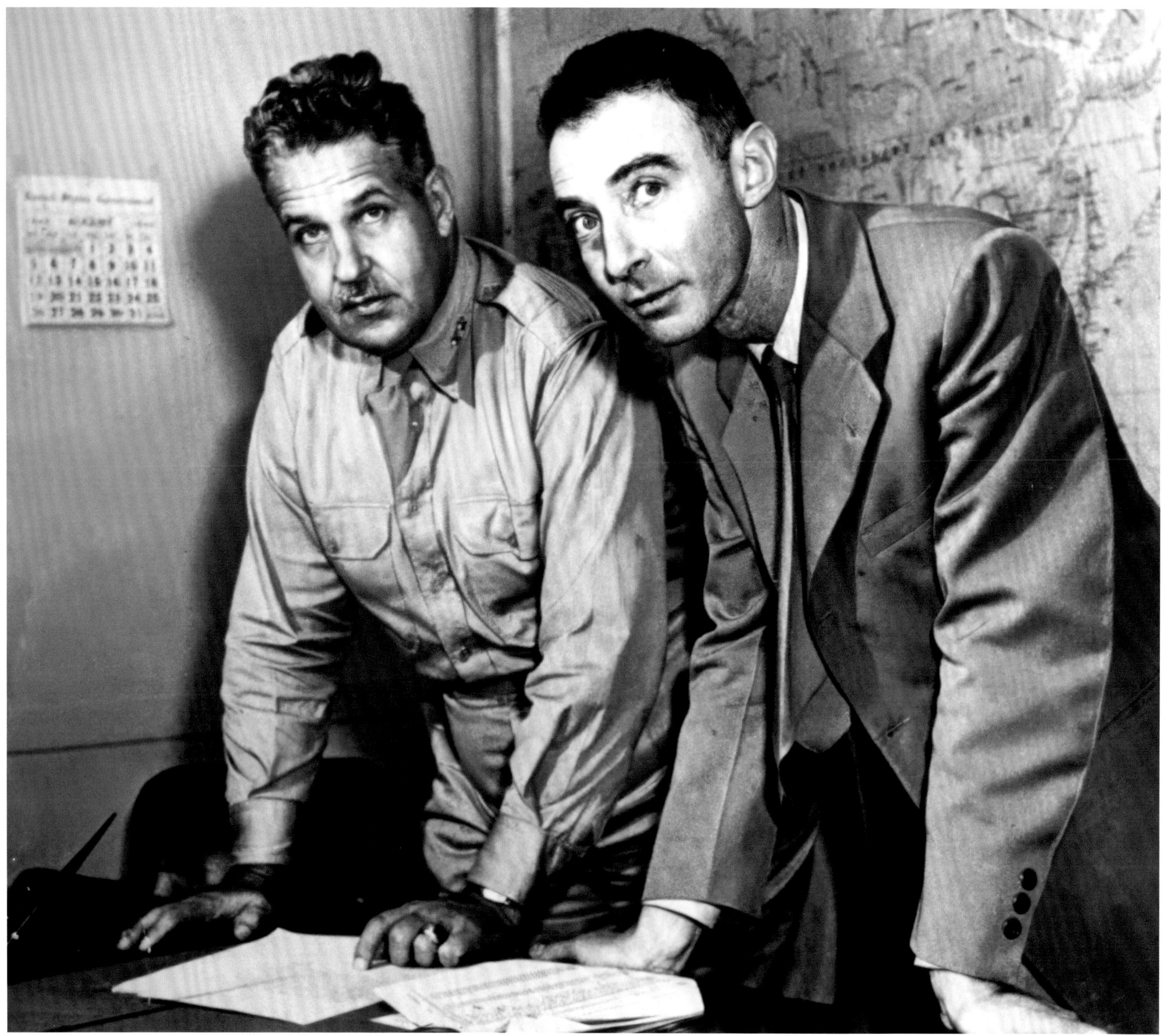

When the 2:00 a.m. weather meeting arrived, all was in place for firing the device. Only the final okay and a press of a button stood in the way of the first attempted detonation of a nuclear weapon. Groves, Oppenheimer, and Hubbard convened at base camp—which, about ten miles from ground zero, had escaped most of the downpour—to discuss the options. In addition to fearing technical breakdowns and a "letdown" among workers after so much build-up to this moment, Groves faced the pressure of Potsdam—both Truman's negotiation with Soviet leader Joseph Stalin and the ultimatum to Japan that would be delivered shortly after. (Marie Hansen/The LIFE Picture Collection/Shutterstock.)

The general agreed to an initial delay of one hour. As the cover of darkness slowly passed them by, the group moved deliberations to the control bunker, where Hubbard and Oppenheimer would witness the test. Groves was to watch from the safety of base camp, hosting a small group of VIPs and intentionally separating himself from Oppenheimer. For the sake of the project, at least one of the two leaders needed to survive if something went terribly wrong.

At 3:30 a.m., the rain finally began to clear. The storm clouds were forecast to scatter shortly before the dawn. At Hubbard's recommendation, zero time was reset again, now to 5:30 a.m.—when the first light of day would be peeking over the horizon.

At T minus thirty minutes to detonation, a final poll was taken. Farrell (voting in Groves's stead), Oppenheimer, Hubbard, and Bainbridge—any of whom had the power to veto the shot—all agreed to start the countdown.

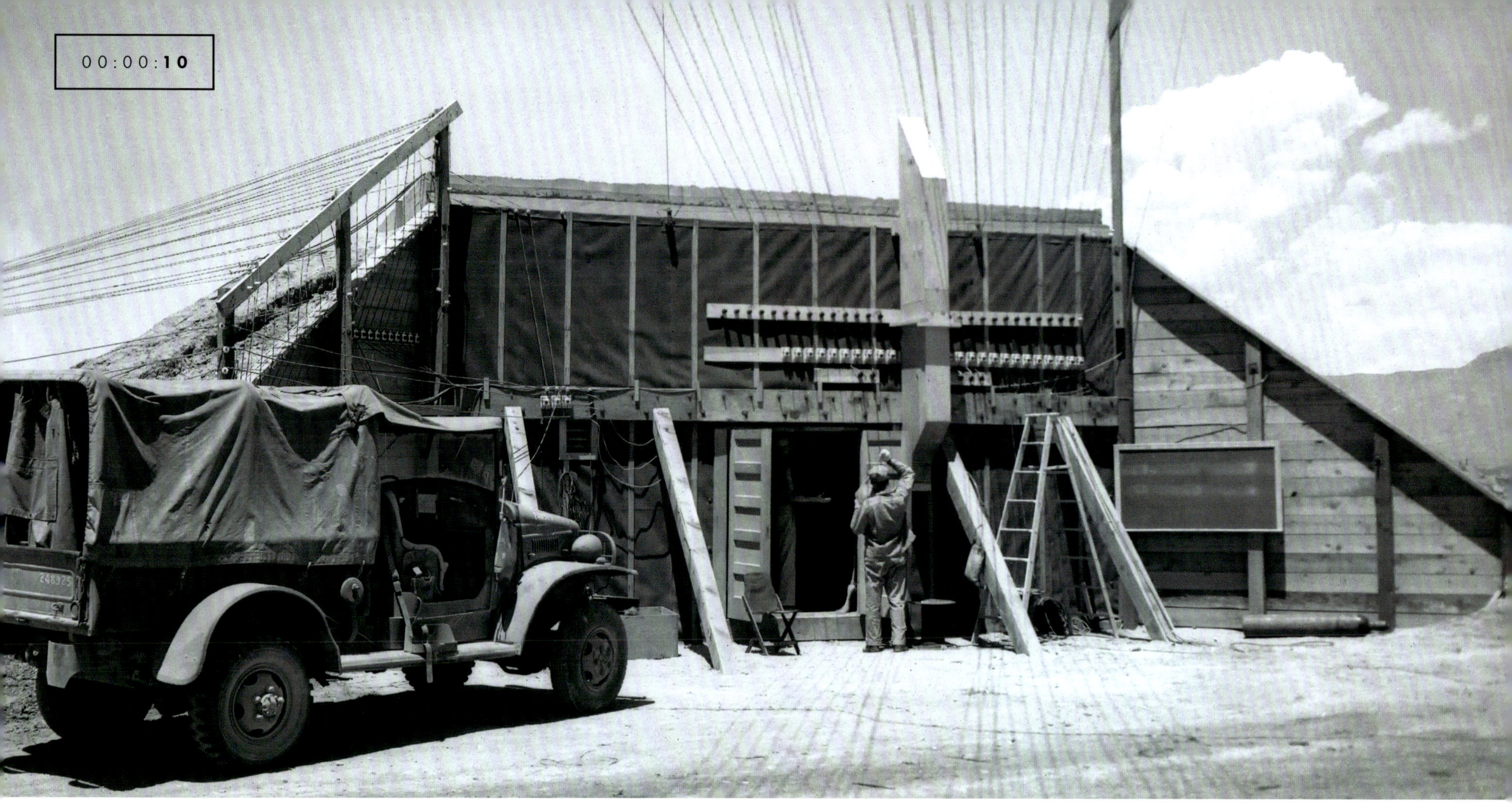

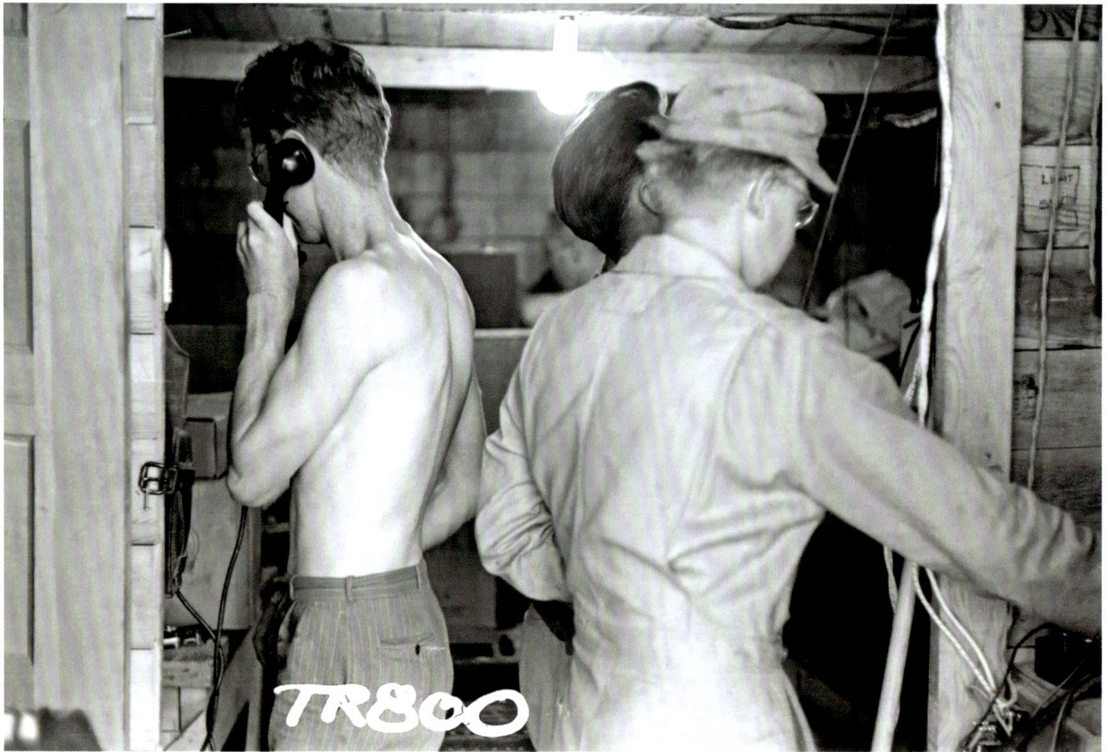

At 5:09 a.m., a voice from the control bunker spoke through the public announcement system, setting in motion the final steps of preparation. The twenty-minute countdown had begun, courtesy of physicist Samuel K. Allison—a witness to the first self-sustaining nuclear chain reaction, when the Chicago Pile went critical in 1942.

Still with jobs to do, many of the scientists in the crowded personnel shelters—which doubled as instrument safehouses—sprang into action. After readying the oscilloscopes and other devices, they would be able to gather round a viewing periscope mounted on the backside of each bunker, near the rear door. The doors were left open for ventilation but ultimately served as points of quick entry and exit for those who changed their mind at the last second about where to watch the explosion.

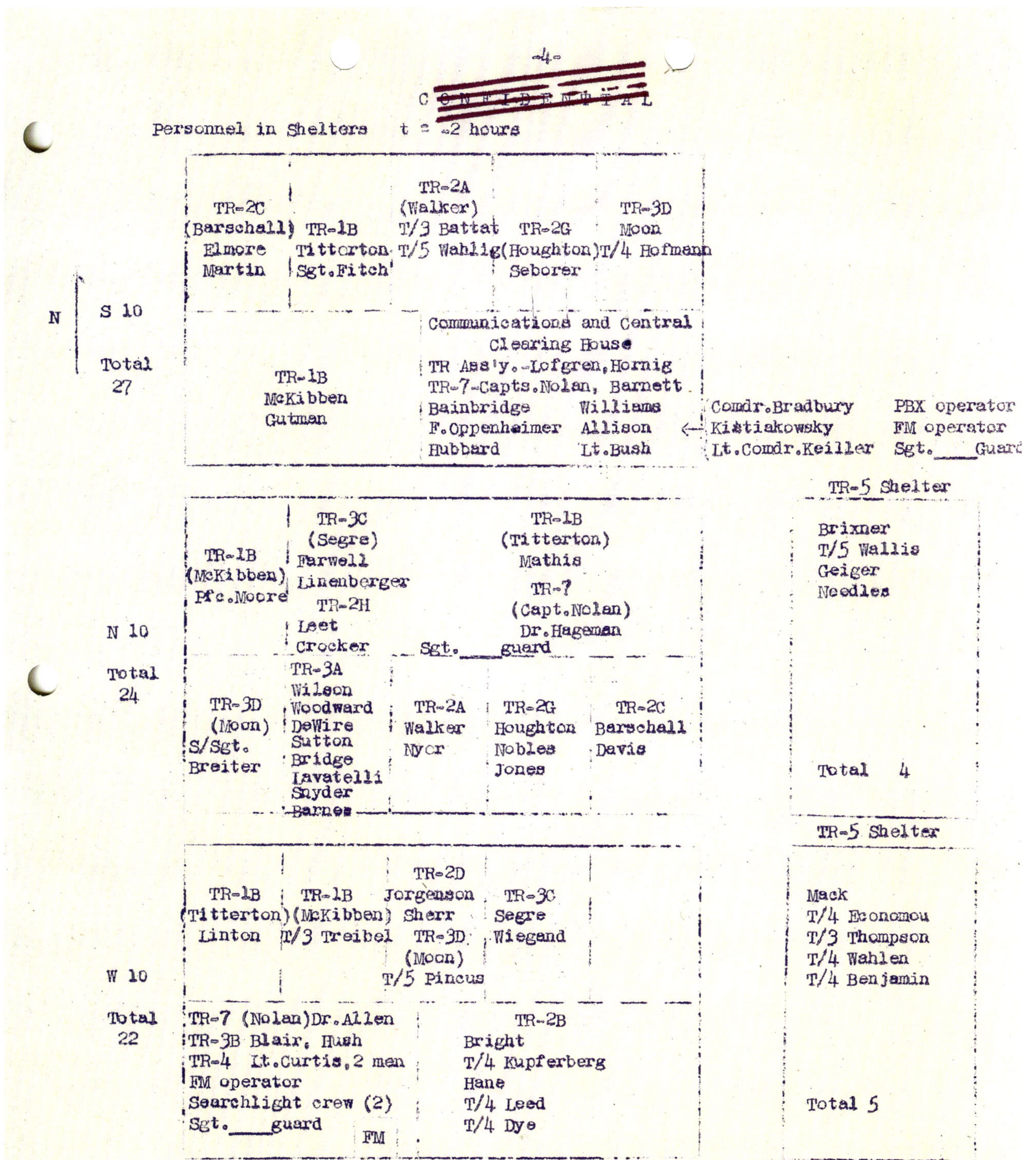

-4-

CONFIDENTIAL

Personnel in Shelters t = -2 hours

N

S 10

Total 27

TR-2C (Barschall) Elmore Martin

TR-1B Titterton Sgt. Fitch

TR-2A (Walker) T/3 Battat T/5 Wahlig

TR-2G (Houghton) Seborer

TR-3D Moon T/4 Hofmann

TR-1B McKibben Gutman

Communications and Central Clearing House
TR Ass'y.-Lofgren, Hornig
TR-7-Capts. Nolan, Barnett
Bainbridge, F. Oppenheimer, Hubbard, Williams, Allison, Lt. Bush

← Comdr. Bradbury, Kistiakowsky, Lt. Comdr. Keiller, PBX operator, FM operator, Sgt. ___ Guard

N 10

Total 24

TR-1B (McKibben) Pfc. Moore

TR-3C (Segre) Farwell Linenberger

TR-2H Leet Crocker

TR-1B (Titterton) Mathis

TR-7 (Capt. Nolan) Dr. Hageman

Sgt. ___ guard

TR-3D (Moon) S/Sgt. Breiter

TR-3A Wilson Woodward DeWire Sutton Bridge Lavatelli Snyder Barnes

TR-2A Walker Nyer

TR-2G Houghton Nobles Jones

TR-2C Barschall Davis

TR-5 Shelter
Brixner
T/5 Wallis
Geiger
Needles
Total 4

W 10

Total 22

TR-1B (Titterton) Linton

TR-1B (McKibben) T/3 Treibel

TR-2D Jorgenson Sherr

TR-3D (Moon) T/5 Pincus

TR-3C Segre Wiegand

TR-7 (Nolan) Dr. Allen
TR-3B Blair, Hush
TR-4 Lt. Curtis, 2 men
FM operator
Searchlight crew (2)
Sgt. ___ guard

FM

TR-2B Bright T/4 Kupferberg Hane T/4 Leed T/4 Dye

TR-5 Shelter
Mack
T/4 Economou
T/3 Thompson
T/4 Wahlen
T/4 Benjamin
Total 5

Altogether, the three shelters housed roughly 100 people. The number of occupants evidently increased after the above list was issued, and some of the shelter assignments changed (for instance, Seborer was not inside the control bunker). Each shelter had a chief scientist: at north 10,000, Robert Wilson; at west 10,000, John Manley; and at south 10,000, Frank Oppenheimer. Once the shot had gone off, a medical chief would take command, ready to give the order to evacuate if radiation monitoring devices began to alarm.

With six minutes to go, a signal rocket brought the countdown to the Military Police in their foxholes; the observers near Campana Hill, whose vantage point is shown at right; and Major Palmer's evacuation detachment. A second flare, meant to give these distant observers a three-minute warning, failed to detonate.

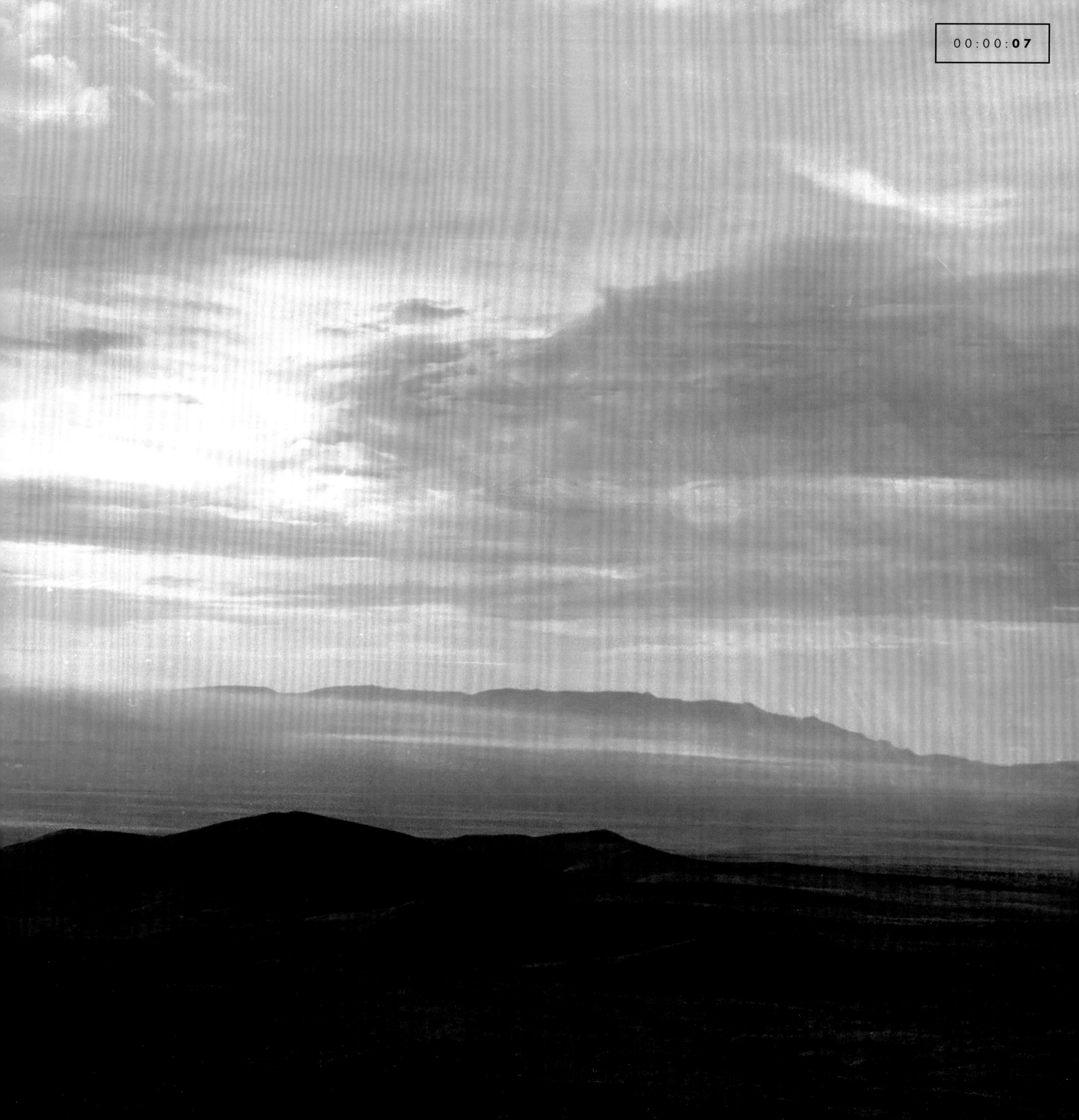
00:00:07

Across the way at base camp, a long siren dispatched another three-minute alert—this one successful—as observers lay facing away from the tower and clutching their welder's glass.

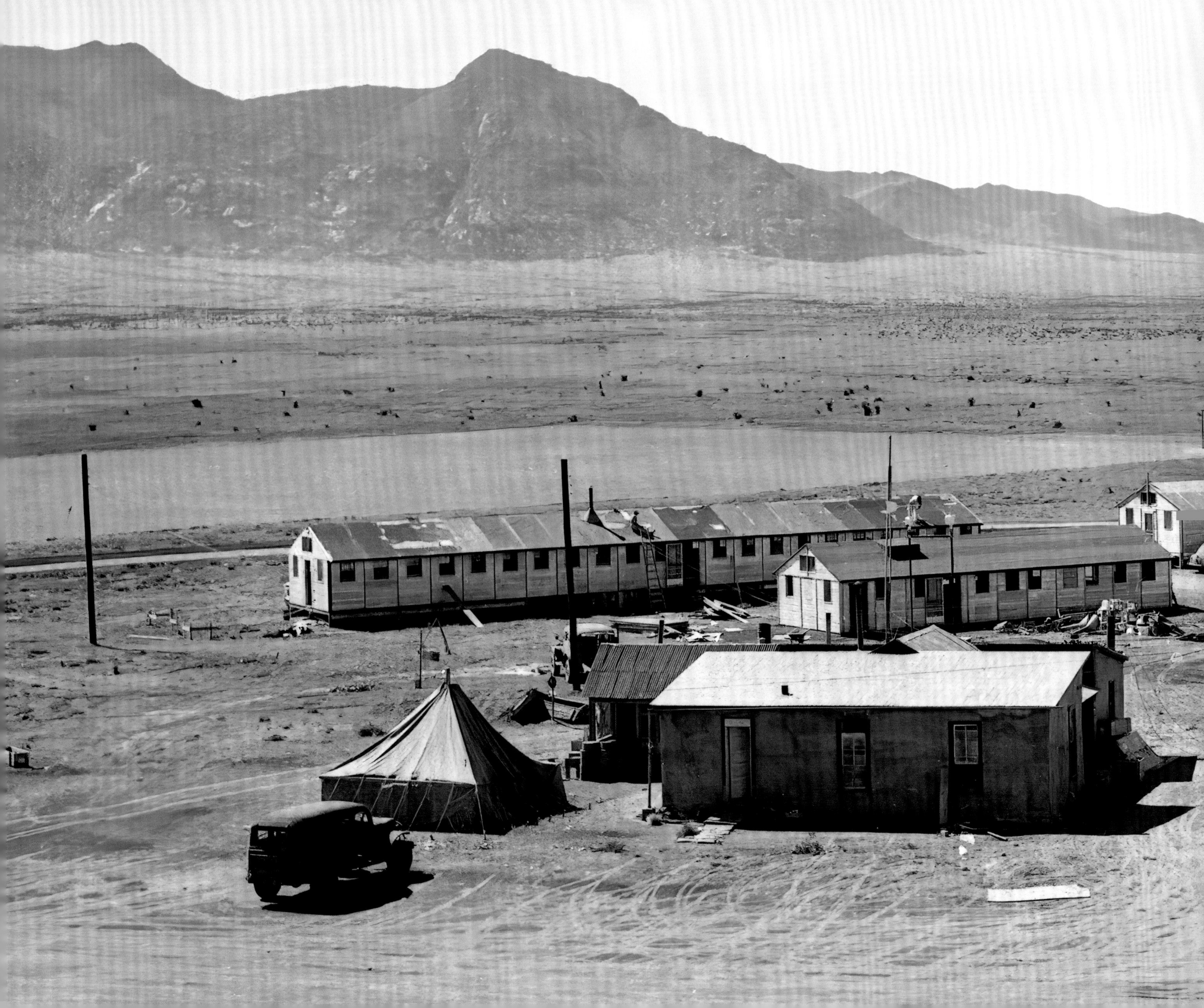

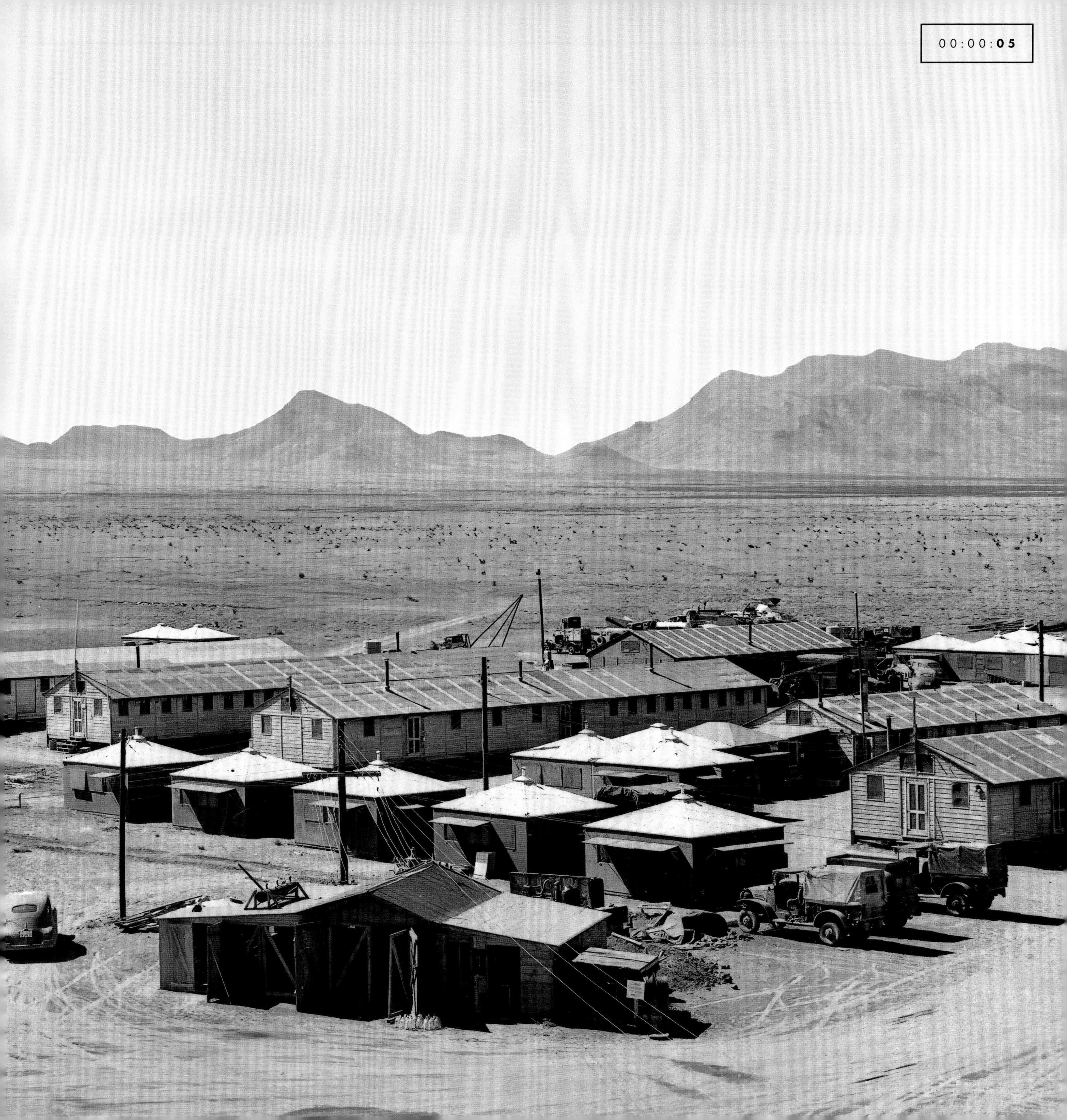

00:00:**04**

In the control bunker, as the minutes melted away, Bainbridge nervously contemplated the prospect of a misfire. If the device failed to detonate, it would be his duty as test director to climb the tower in search of faults. Nearby, Oppenheimer supposedly stated, "Lord these affairs are hard on the heart."

At forty-five seconds, an automatic timer kicked in, triggering the X unit to draw in charge from storage batteries, readying to fire. Cameras across the test site whirred to life as a gong began to toll, marking time across the public radio waves at forty seconds, thirty, twenty.

Despite instructions to remain under cover, Kistiakowsky ran outside and climbed atop the bunker. Allison's voice rang out a ten-second warning.

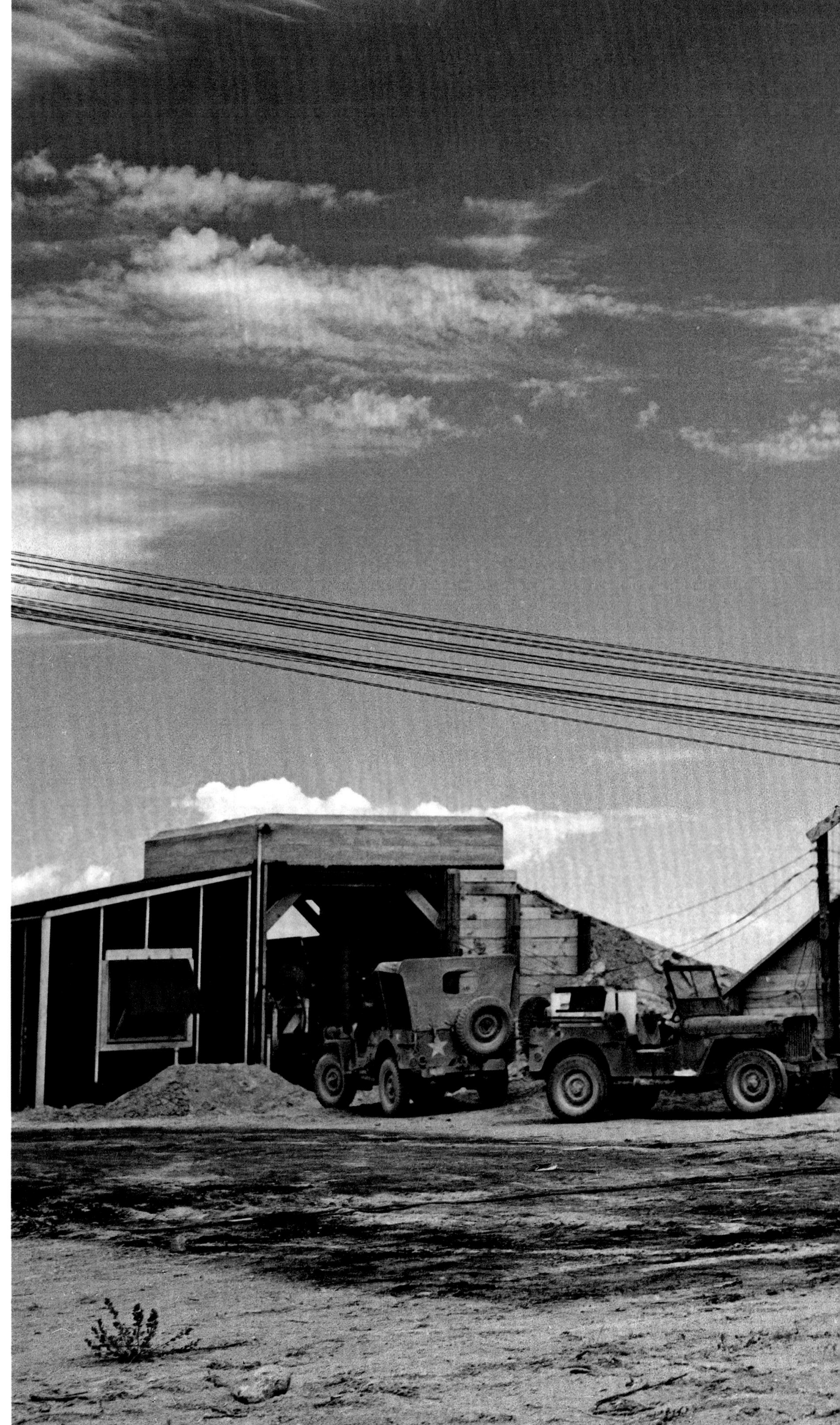

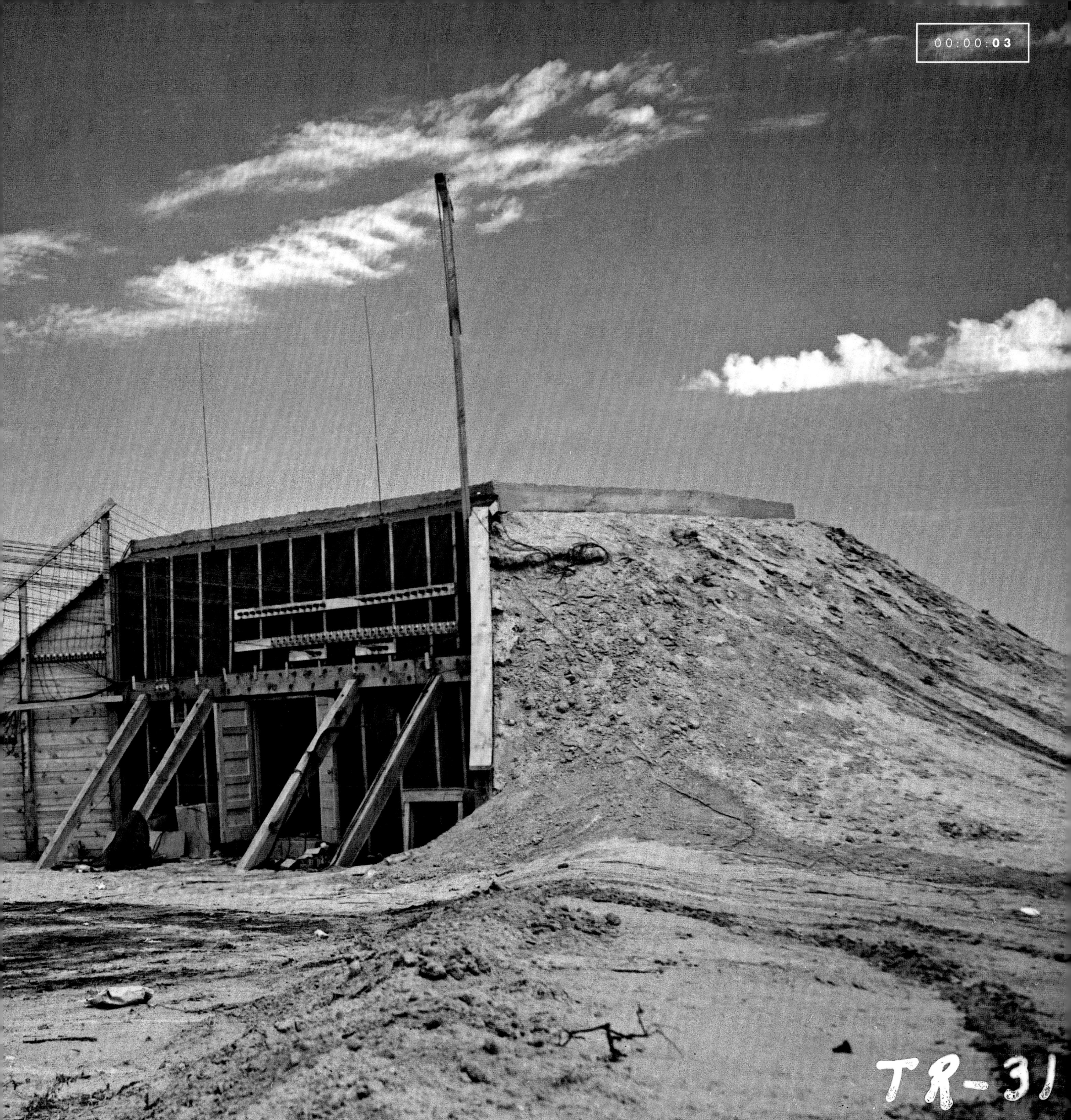
00:00:03
TR-31

00:00:**02**

As the clock ticked down to zero, Allison spoke again . . .

“Now.”

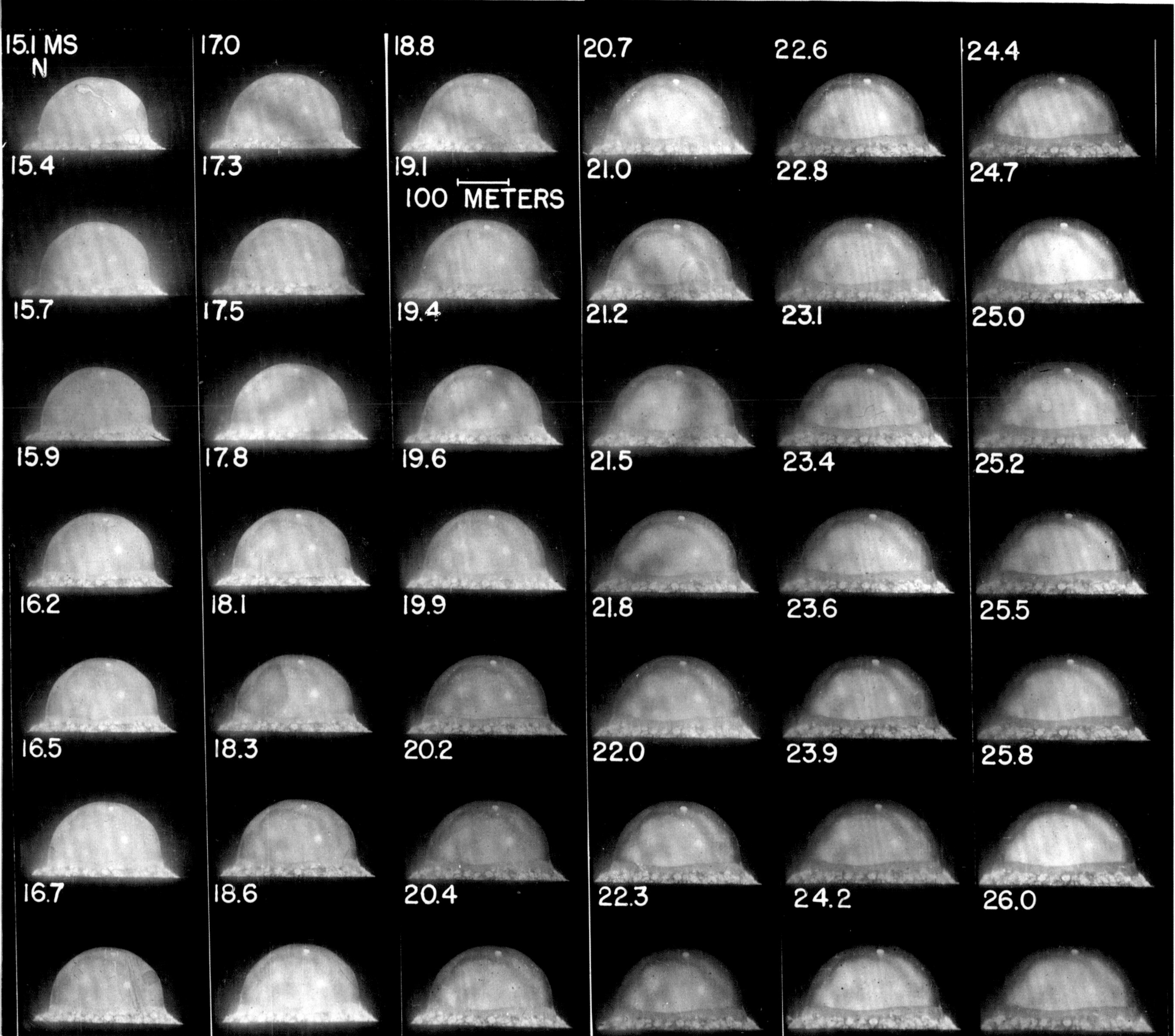
15.1 MS
N
15.4
15.7
15.9
16.2
16.5
16.7
17.0
17.3
17.5
17.8
18.1
18.3
18.6
18.8
19.1
100 METERS
19.4
19.6
19.9
20.2
20.4
20.7
21.0
21.2
21.5
21.8
22.0
22.3
22.6
22.8
23.1
23.4
23.6
23.9
24.2
24.4
24.7
25.0
25.2
25.5
25.8
26.0

06 5:29 A.M.

In the north 10,000 photography bunker, Berlyn Brixner was listening to the countdown on a loudspeaker, his head inside a turret loaded with cameras and film. He was one of the only people instructed to look toward the blast—through his welder's glasses—ready to follow the path of the fireball as it launched into the sky. The two Mitchell movie cameras at his station would deliver the best footage to come of the Trinity test, used by Los Alamos scientists to make some of the first measurements of the effects of a nuclear explosion.

When the detonators fired, the cameras captured what Brixner could not have seen—the very first light of a violent, silent sea of energy unfurling into the basin. As thirty-two blocks of high explosives erupted all together, their incredible force surged inward toward the sleeping plutonium core, compressing the dense sphere of metal instantaneously from all sides and bringing its atoms impossibly close together. A carefully timed burst of neutrons sowed momentary, uncontrolled chaos and then, as quickly as it began, the fission chain reaction ended. Footage from a high-speed Fastax camera at Brixner's bunker (*left*), shot through a thick, glass porthole, shows a translucent orb bursting through the darkness less than a hundredth of a second after detonation, as a rush of heat, light, and matter blew apart the Gadget.

When the brightness faded enough for witnesses to make out ground zero, they saw a wall of dust rise up around a brilliant, shape-shifting, multicolored ball of flames—forming a fiery cloud that shot into the sky atop a twisting stream of debris. The camera footage tells a story no less dramatic but hundreds of times more intricate, preserving the moment for scientists to return to again and again to measure and describe the behavior of the fireball and other visible effects with exacting detail. On balance, the photography effort was a huge success, despite only eleven of the fifty-two cameras producing satisfactory images. By arranging those cameras at intentionally staggered distances, complementary angles, and with a broad spectrum of frame rates and focal lengths, the Spectrographic and Photographic Measurements Group was able to piece together a remarkably complete picture of their subject.

According to the group's leader, Julian Mack, the more than 100,000 frames that were captured still "give no idea of the brightness, or of time and space scales." Mack has attributed fortune, as much as foresight, to the photographic record that was made, especially during the earliest phase of the blast. Indeed, the explosion was several times more powerful than predicted, and the intensity of its effects overwhelmed many of the cameras and diagnostic instruments that had been carefully calibrated for a now-obsolete range of yields. The human observers were similarly overcome. "The shot was truly awe-inspiring," Norris Bradbury has said. "Most experiences in life can be comprehended by prior experiences, but the atom bomb did not fit into any preconception possessed by anybody. The most startling feature was the intense light."

It is a common sentiment that words and even pictures pale in comparison to the experience of the explosion. Even so, soldiers, scientists, and many other witnesses have added their firsthand accounts—often absorbing and poetic—to complement the trove of hard data collected during the test shot. They describe an intense and blinding brightness that filled the basin with daytime; an ominous, darkening cloud rearing its head in eerie silence; the wait for the invisible wave rushing out from the heart of the Gadget; and the mighty roar that arrived at last, in a thunder, and seemed never to leave. Physicist Isidor Rabi, watching from twenty miles away near Campana Hill, remembers, "It blasted; it pounced; it bored its way right through you."

James Chadwick, head of the British contingent of scientists who joined the Manhattan Project, later said, "Although I had lived through this moment in my imagination many times during the past few years and everything happened almost as I had pictured it, the reality was shattering." George Kistiakowsky found himself certain that "at the end of the world—in the last millisecond of the Earth's existence—the last human will see what we saw."

* * * * * * *

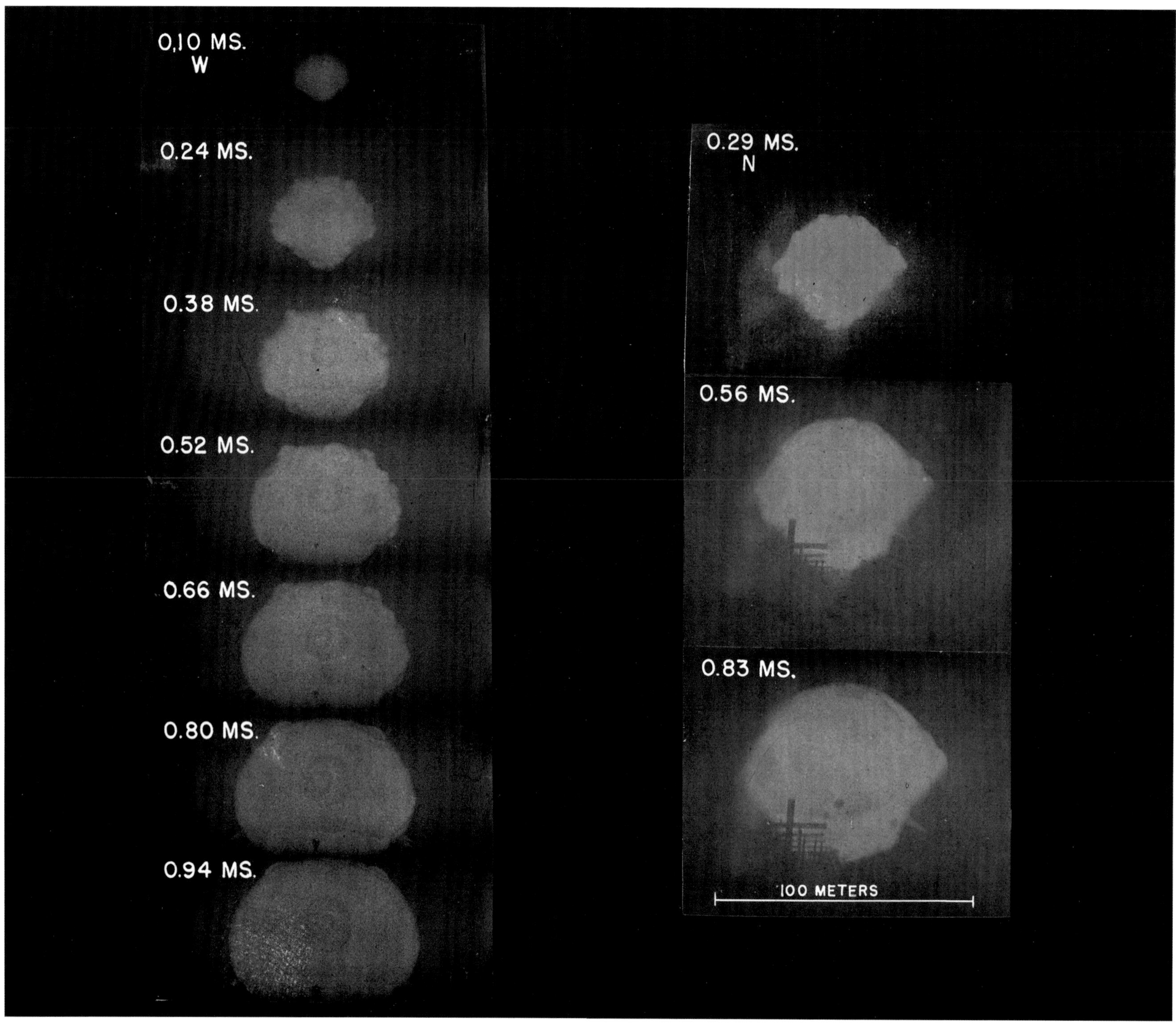

From inside small, mobile bunkers 800 yards to the north and west of ground zero, two Fastax cameras captured the very first millisecond of the explosion. Both the eight-millimeter and larger sixteen-millimeter film strips above, exposed at two different speeds, show the birth of the fireball atop the tower, seeming to pour upward from the Gadget. The initially cone-like shape, caused by interference from the shot cab platform, quickly rounded into a sphere. The ball of fire made contact with the ground, 100 feet below, after just six ten-thousandths of a second (in between the frames labeled 0.52 and 0.66 milliseconds).

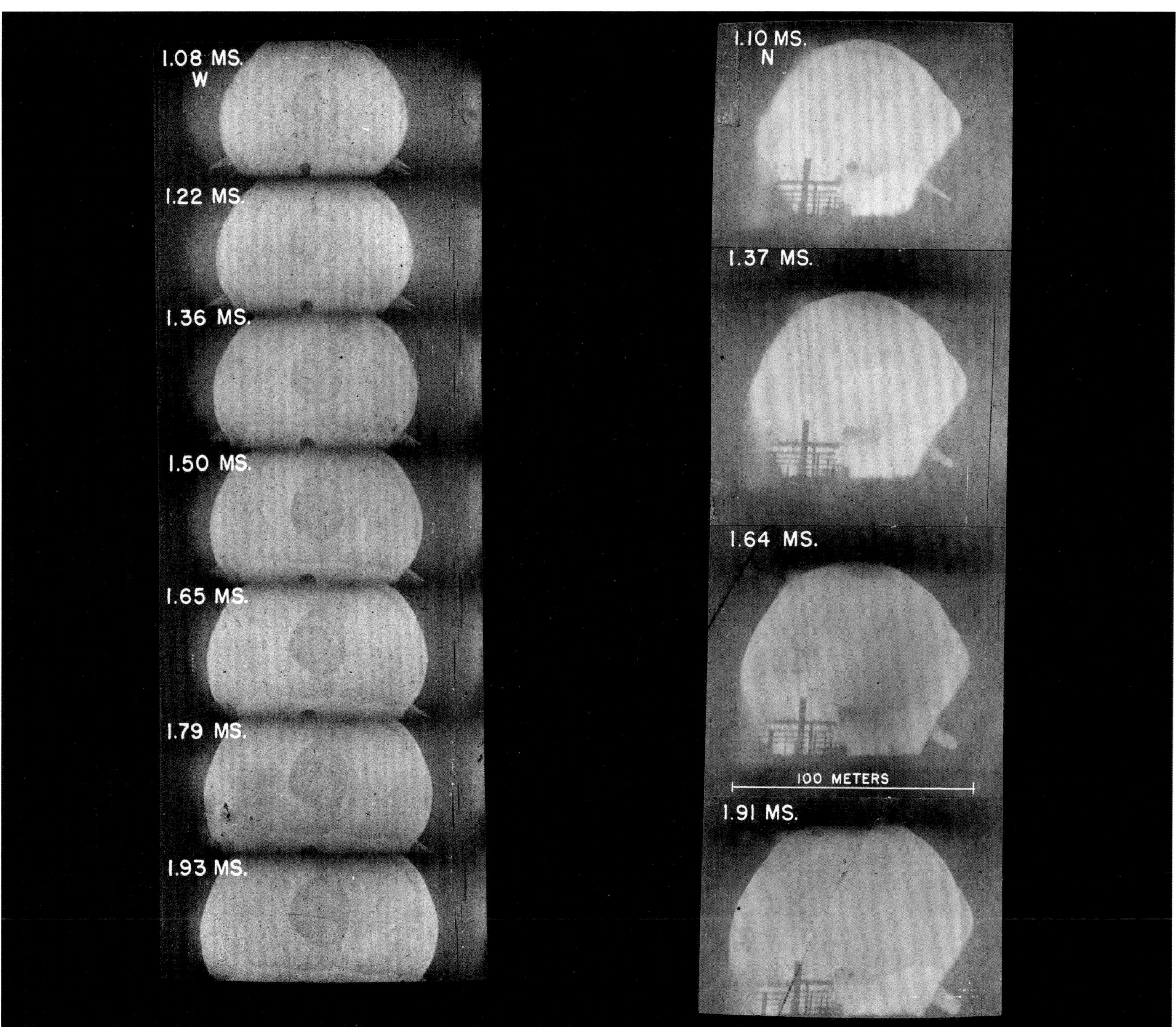

As one millisecond turned into two, the fiery sphere continued expanding with near-perfect symmetry, appearing misshapen only because of the frame lines on the Fastax film. The spike protruding to the right has been attributed to the guy wires leading down from the tower. Although the fireball appears washed out in the images above, Julian Mack writes in his official report on the photographic record of the Trinity test that "the object you see is, for the first few frames, many times brighter than the sun, and, for a considerable fraction of a second after that, brighter than any light ever produced on Earth." Physicist Joseph McKibben later recalled that the incredible brightness flooded in the open back door of the control bunker, drowning out the floodlights used for taking movies of the control panel.

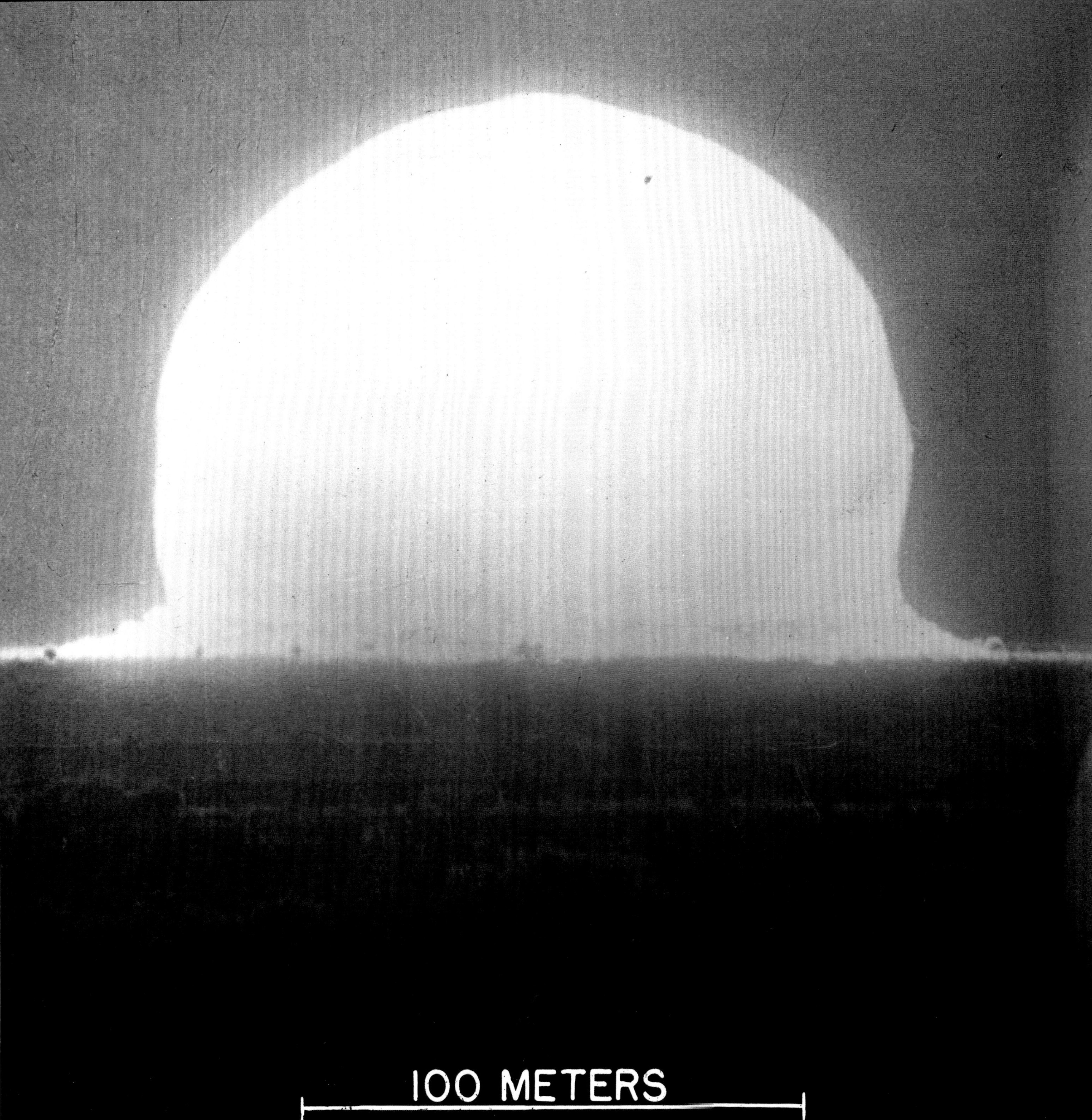
100 METERS

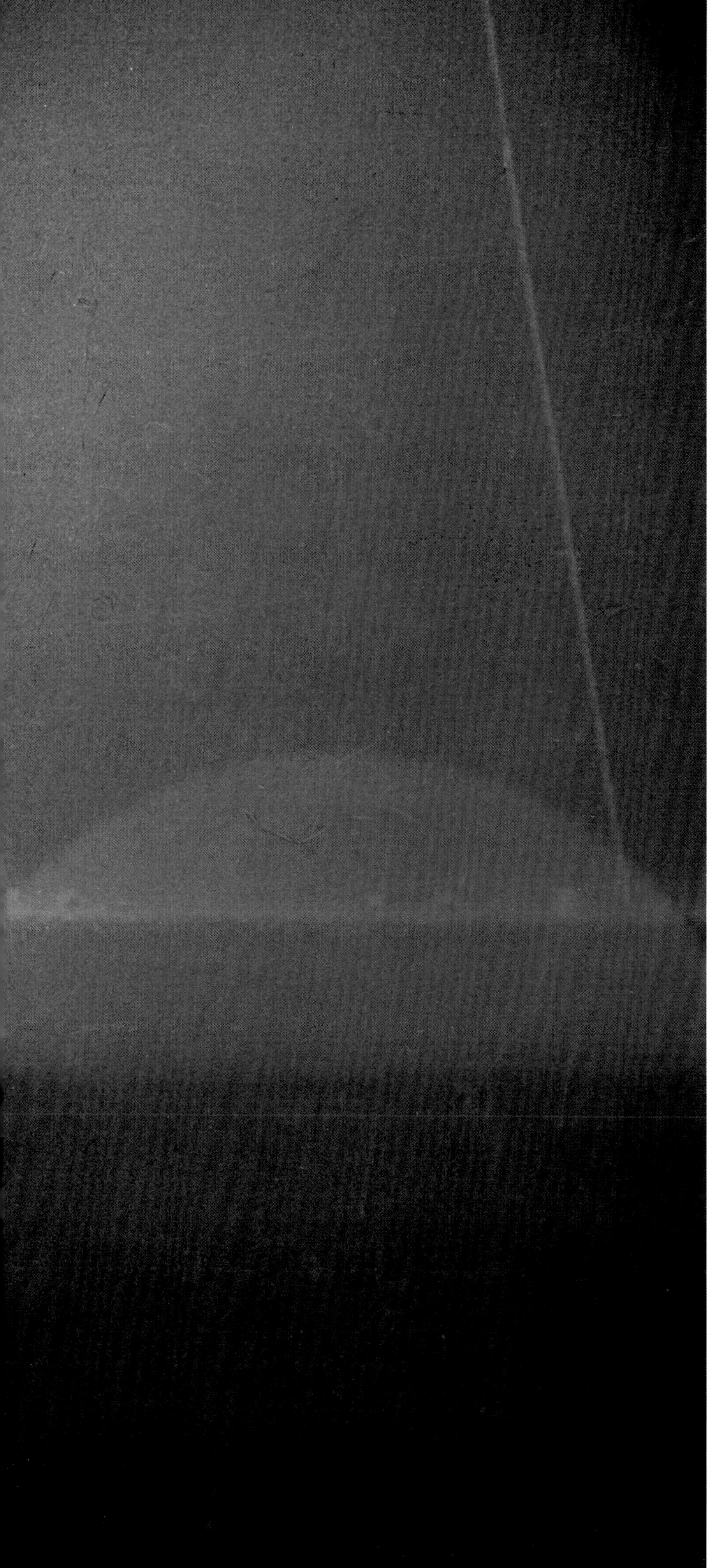

Footage from the photography station 10,000 yards north of ground zero is shown at left. The Mitchell movie camera behind these images, filming from on top of the bunker, was equipped with an eighteen-inch telephoto lens that delivered a close-up view of the fireball and its surroundings. This high-speed camera picked up at the sixth millisecond and allowed Mack to freeze time every 0.1 seconds thereafter for analysis.

The circle on this exposure is an artifact of a hole cut in the film, where a clock would have been visible through the side of the camera. No such apparatus was used in this case, so the hole was covered—well enough to shield against ordinary daylight but not against the "super daylight" that engulfed the camera.

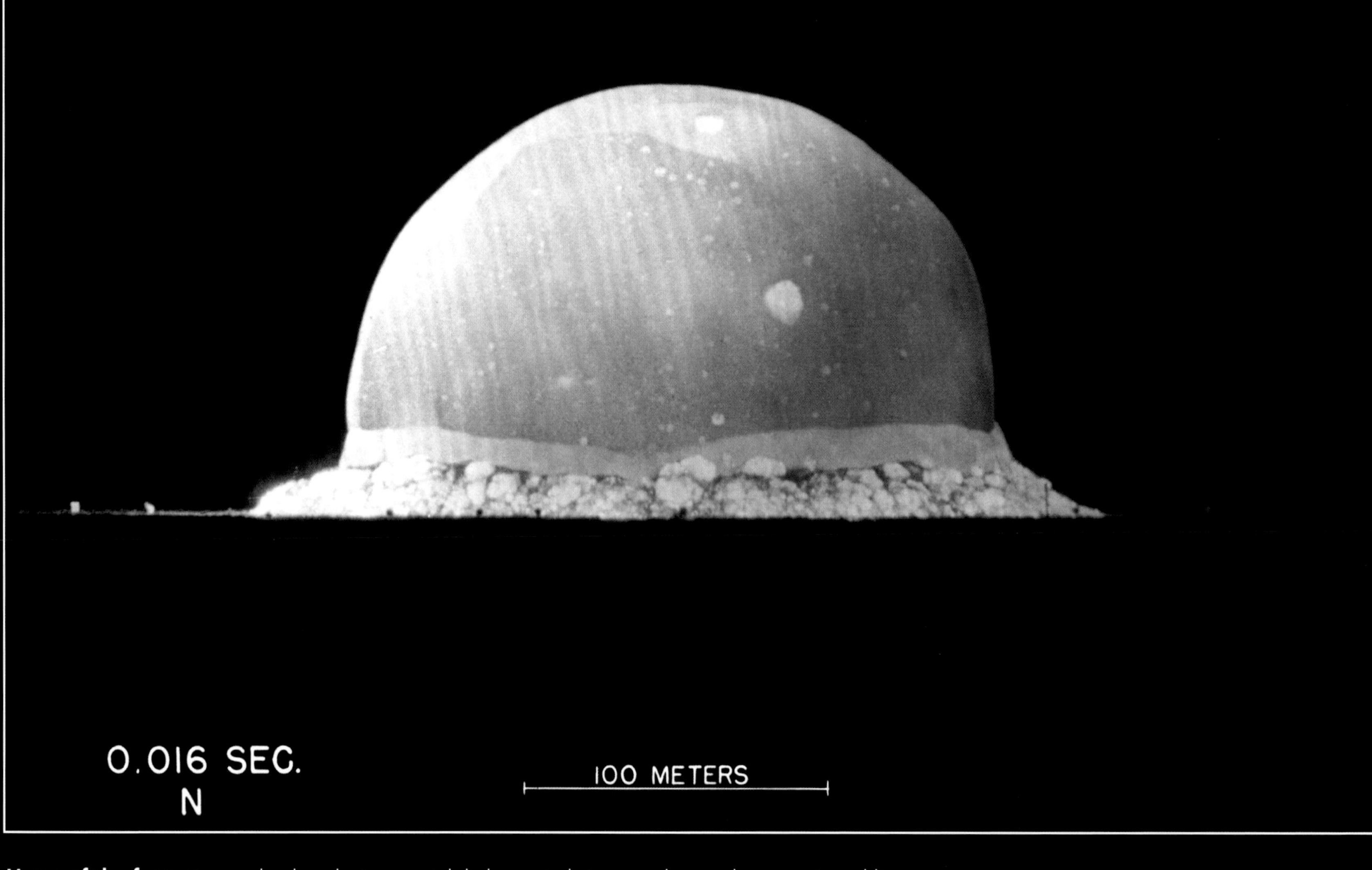

Many of the frames were developed using special darkroom techniques to sharpen the contrast, enabling detailed study of the fireball, shock wave, and other phenomena—both expected and unexpected.

In the image above, the two square objects nearest the edges of the frame, visible on either side of the explosion (though more faintly on the right), are billboards erected as reference points. The billboards are exactly 200 meters from the center of the explosion and 400 meters from each other, informing the 100-meter scale placed on the photographic emulsions here and elsewhere.

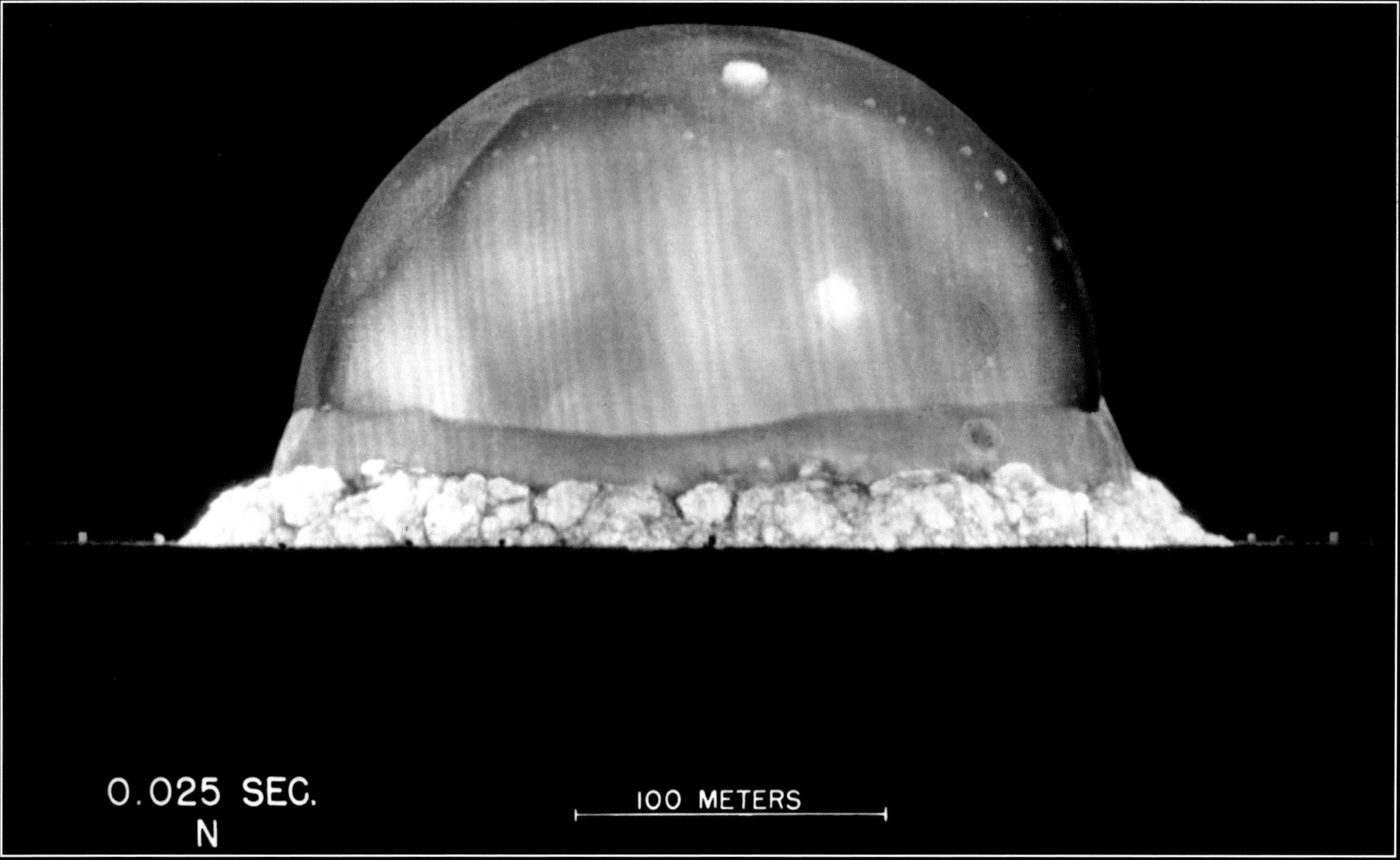

0.025 SEC.
N

100 METERS

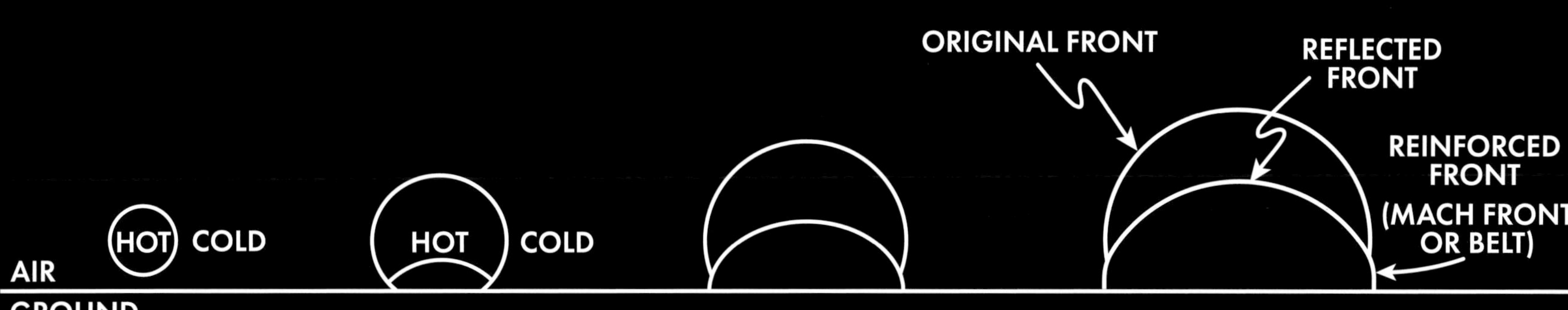

One of the unexpected effects revealed in the Mitchell footage was a roiling cloud of dust taking shape immediately along the ground, appearing to be even hotter than the growing ball of fire. Partly hidden by the dust skirt, a belt is also shown emerging around the base of the fireball. In his report, Mack explains that the belt, or Mach front, was formed by a portion of the shock front (the advancing edges of the spherical shock wave) bouncing off the ground and promptly catching up with the original shock front. Their points of overlap, as illustrated above, form a second front that is both stronger and faster than the original.

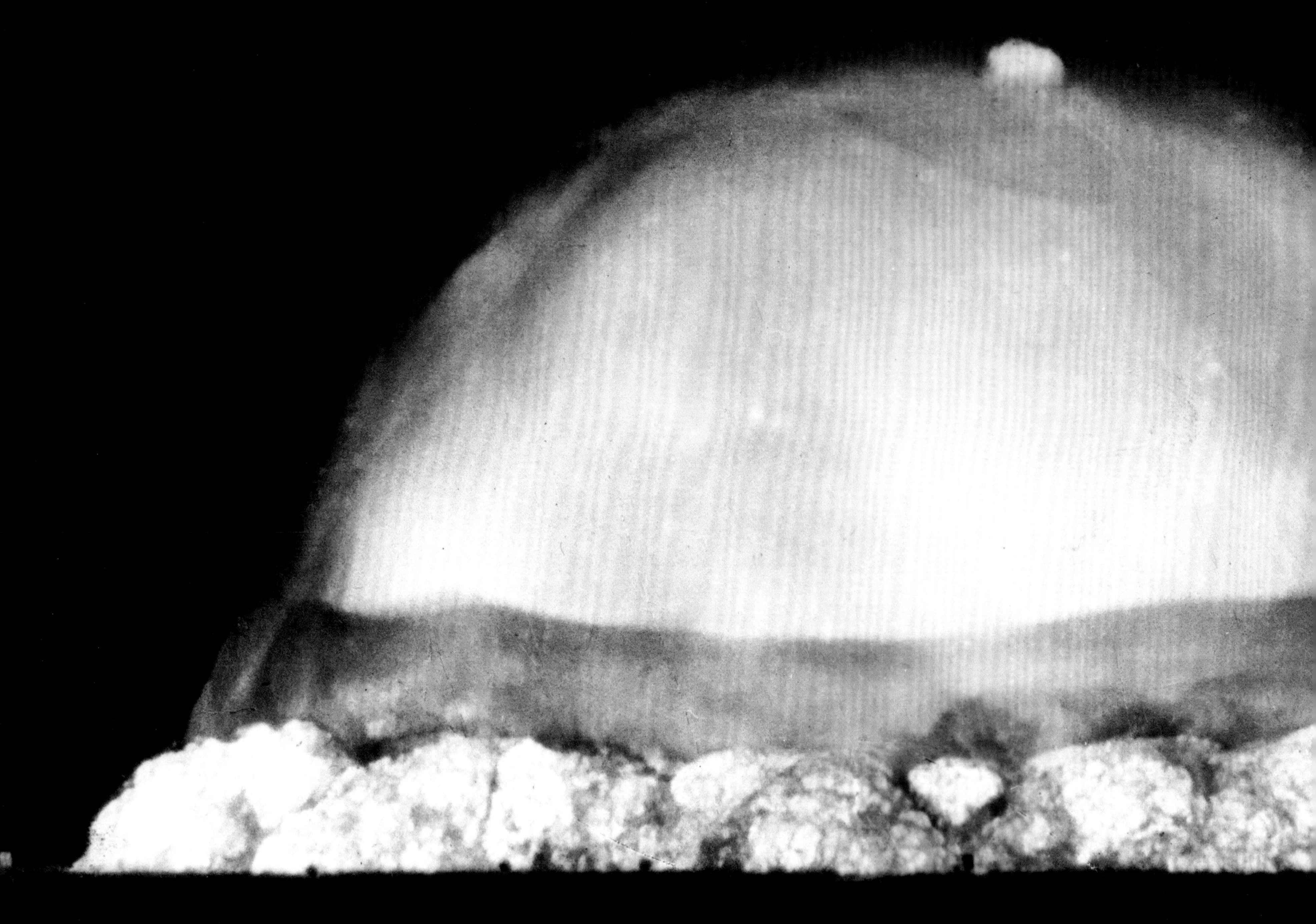

0.034 SEC.
N

100 METERS

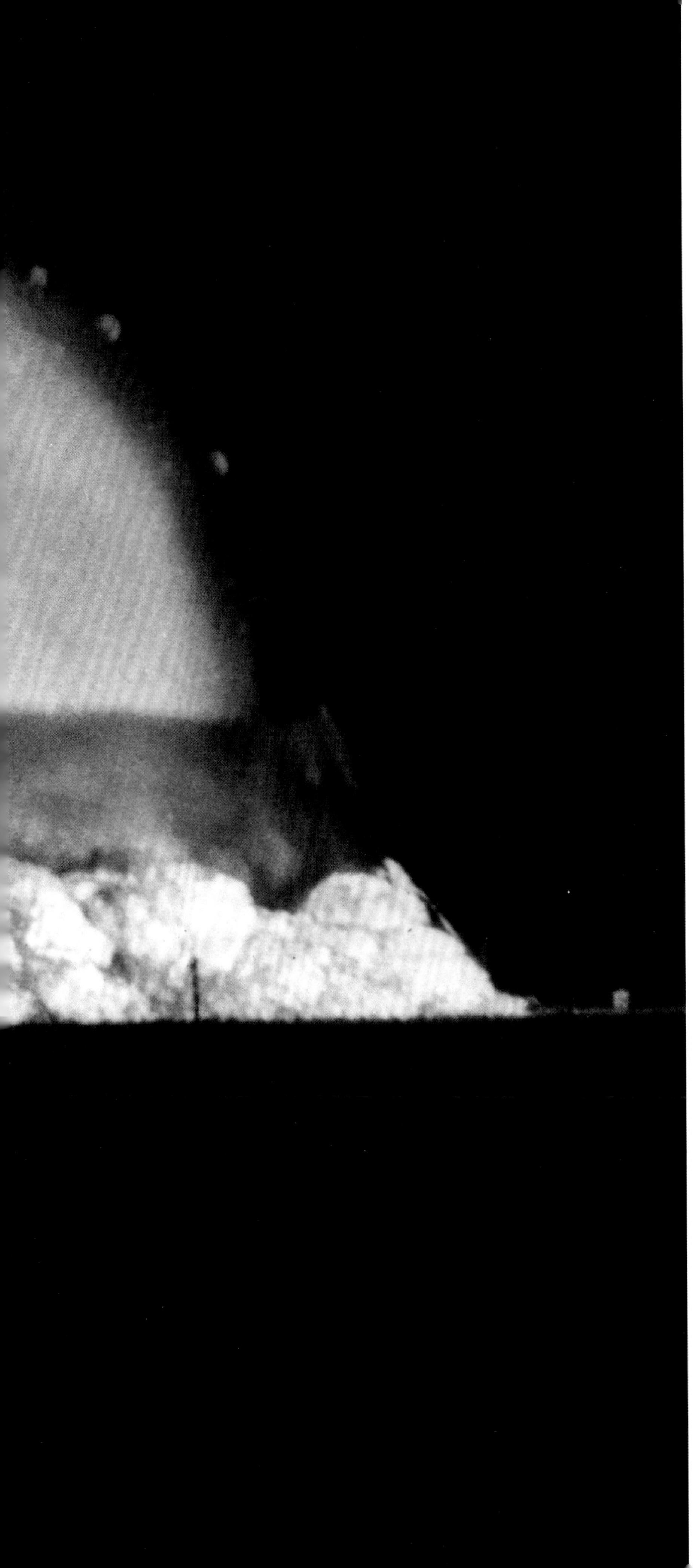

In this image from thirty-four milliseconds after the explosion, the dust skirt has expanded almost to the two billboards. The shock wave, which looks like the glass of a snow globe, has begun to separate from the slower flame front. The shock wave carries the vibrational force that would knock some viewers at the 10,000-yard shelters to the ground upon its arrival, about twenty-five seconds after detonation.

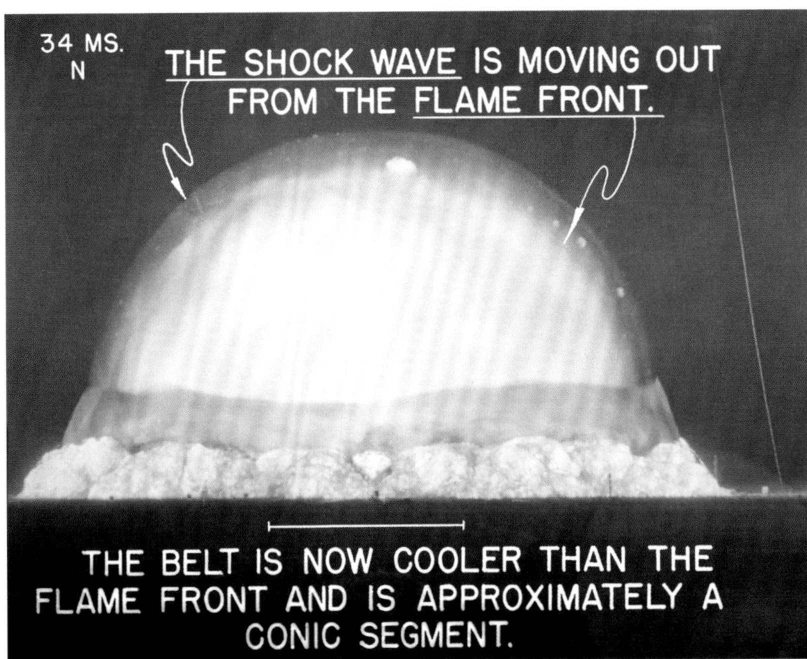

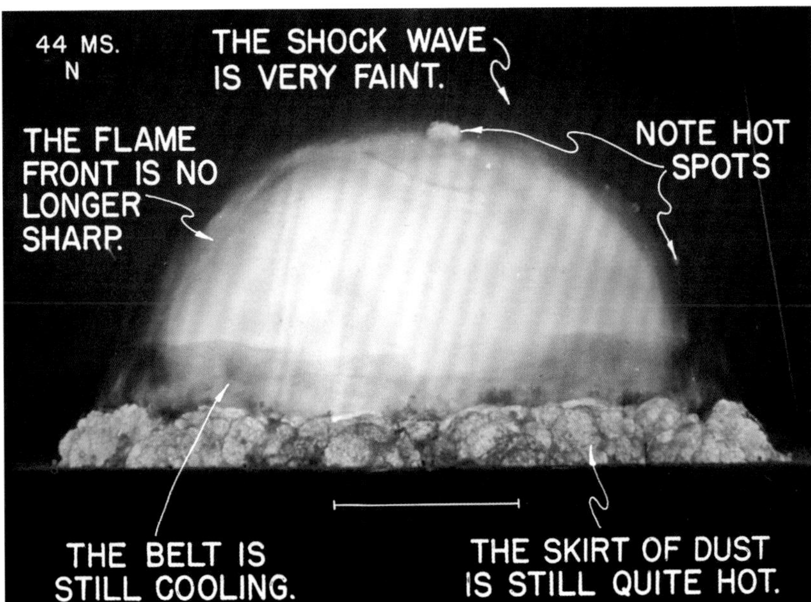

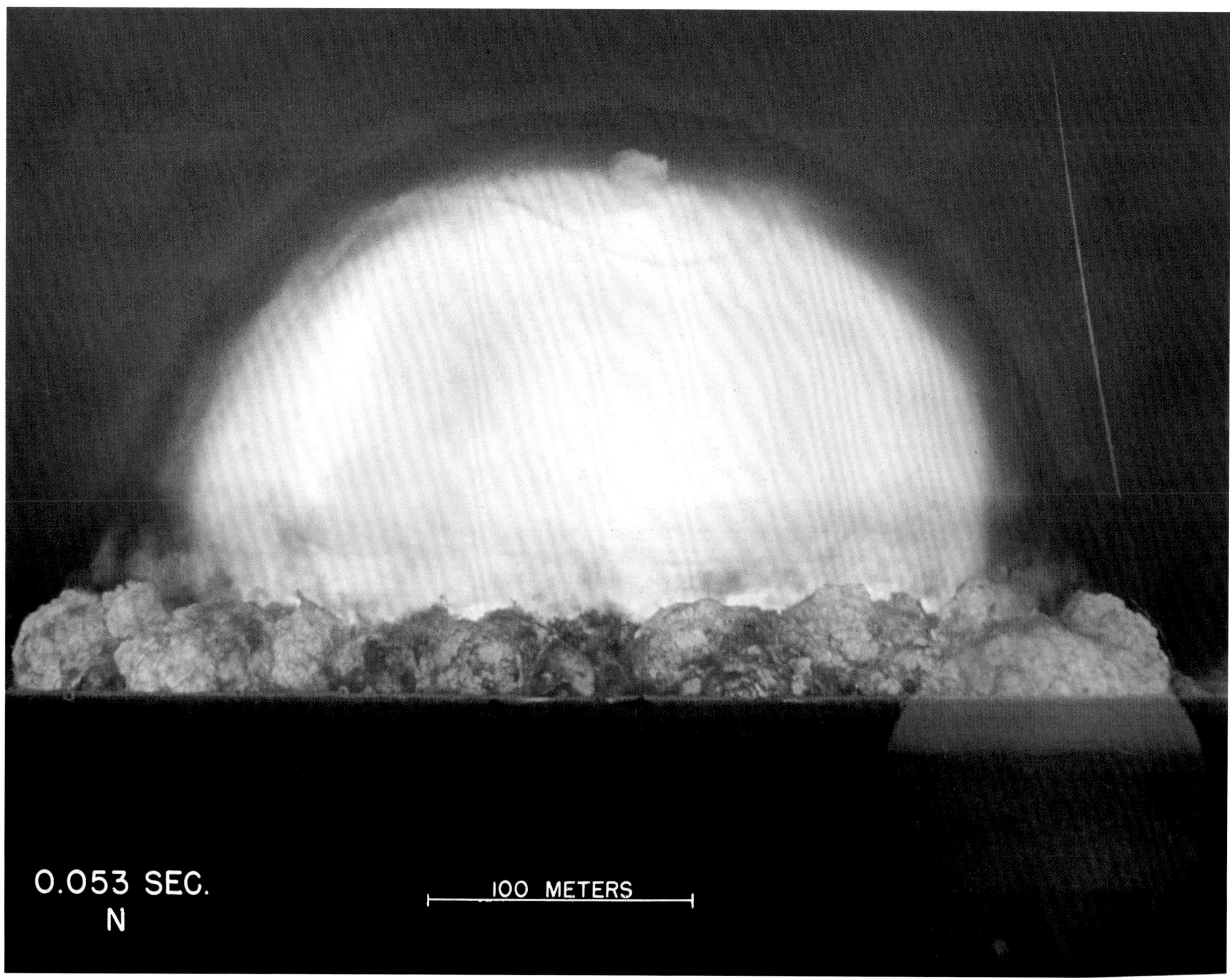

By fifty-three milliseconds, the dust skirt has darkened—signaling that is has cooled—and the ball of fire has begun to lose its initially well-defined shape. The shock wave becomes increasingly faint as it expands away from the center of the explosion, and a shadowy dark wave is seen splitting off from it.

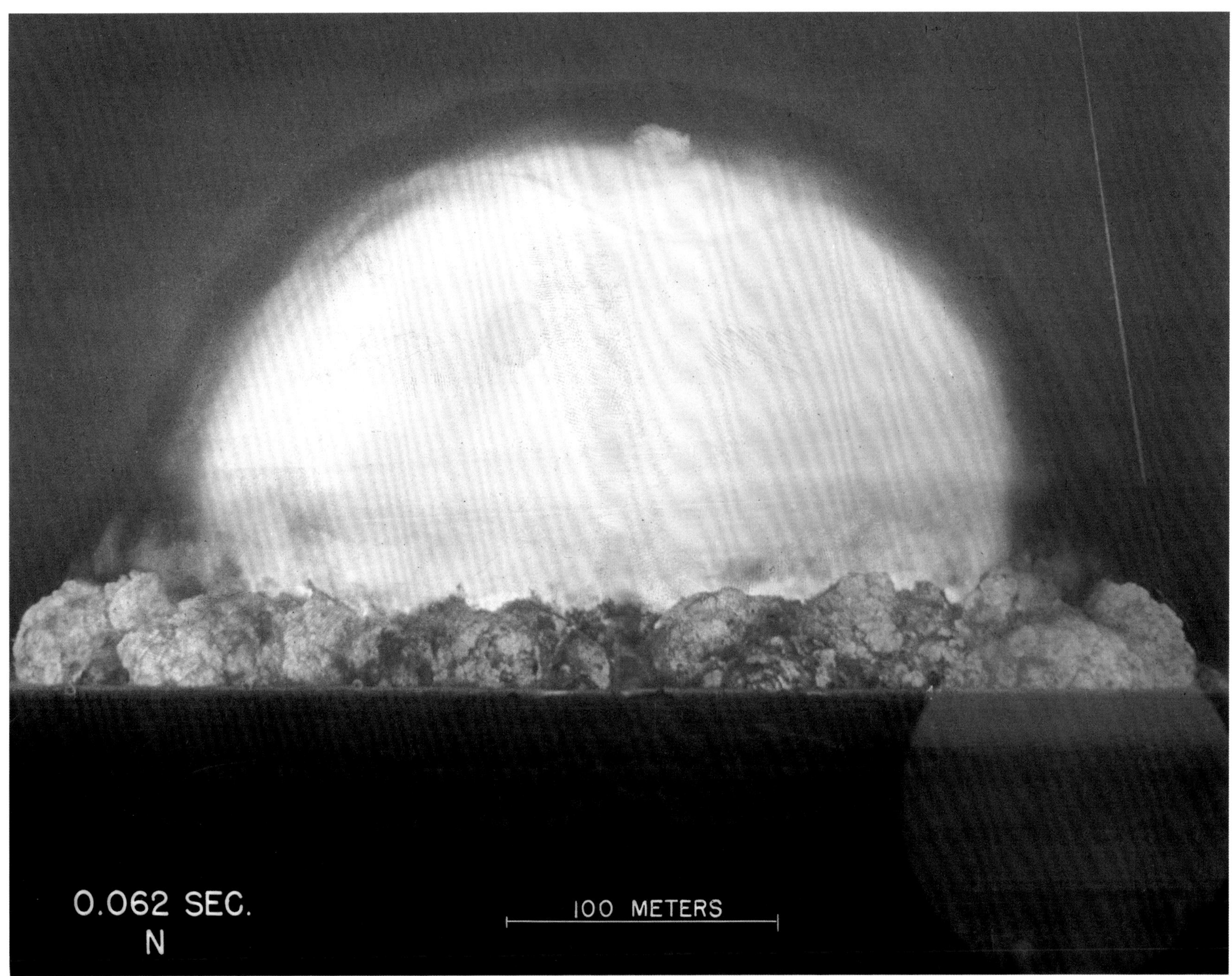

The dark wave, like the dust skirt, came as a surprise to scientists. They hypothesized at the time that it consisted of some kind of light-absorbing matter, most likely nitrogen compounds formed by the intense pressure and heat of the shock front.

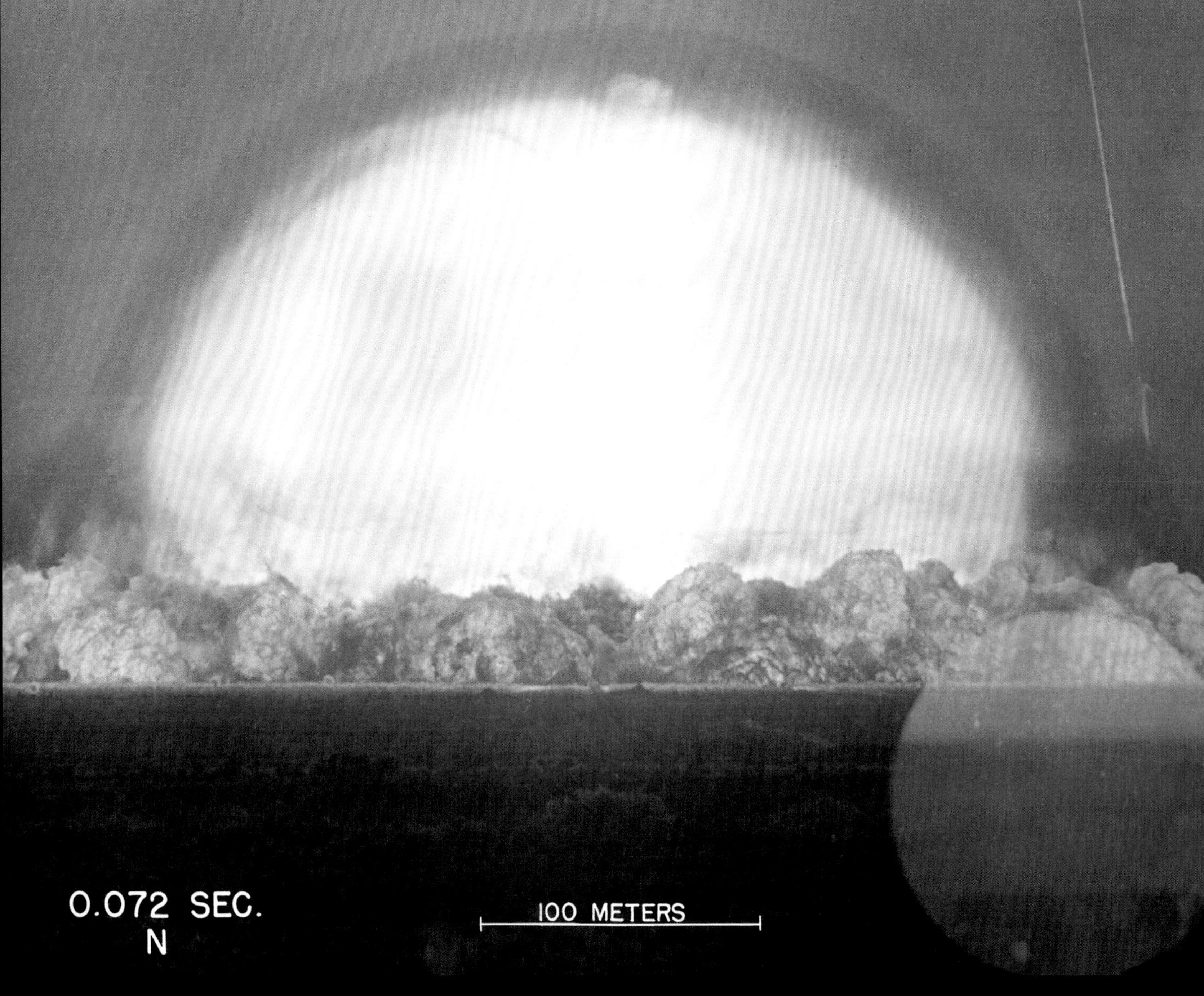

Once the shock wave has traveled far enough away from the fireball, it becomes invisible. An optical illusion, however, enabled scientists to map its location for another several milliseconds. In the image above, notice the appearance of a break in the cable at right. The cable is anchored 213 meters from the center of the blast, just past the now-engulfed reference billboard; an airborne balloon tethered to it is out of view.

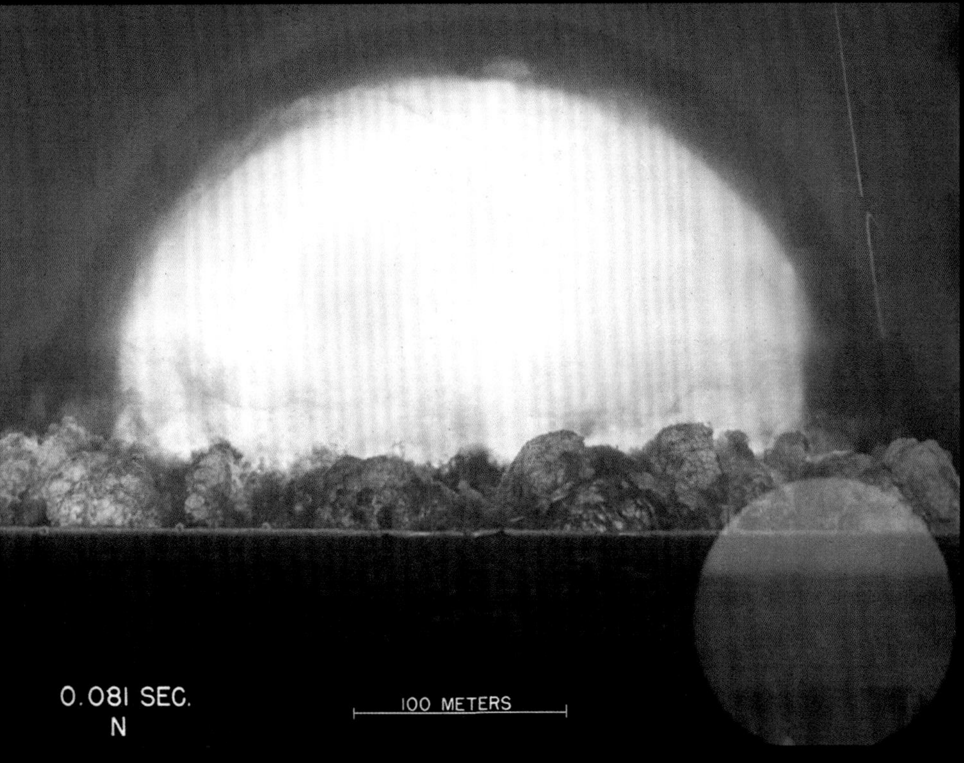

The apparent break in the cable is caused by refraction, as light traveling north to the camera lens passes through the invisible shock wave. Mack reported that the section of cable seen through the shock wave appears displaced to the right, while parts within two meters of the wave front are missing. The section between two and four meters from the edge of the shock wave shows up twice—giving the appearance that the lower segment of cable is hooked.

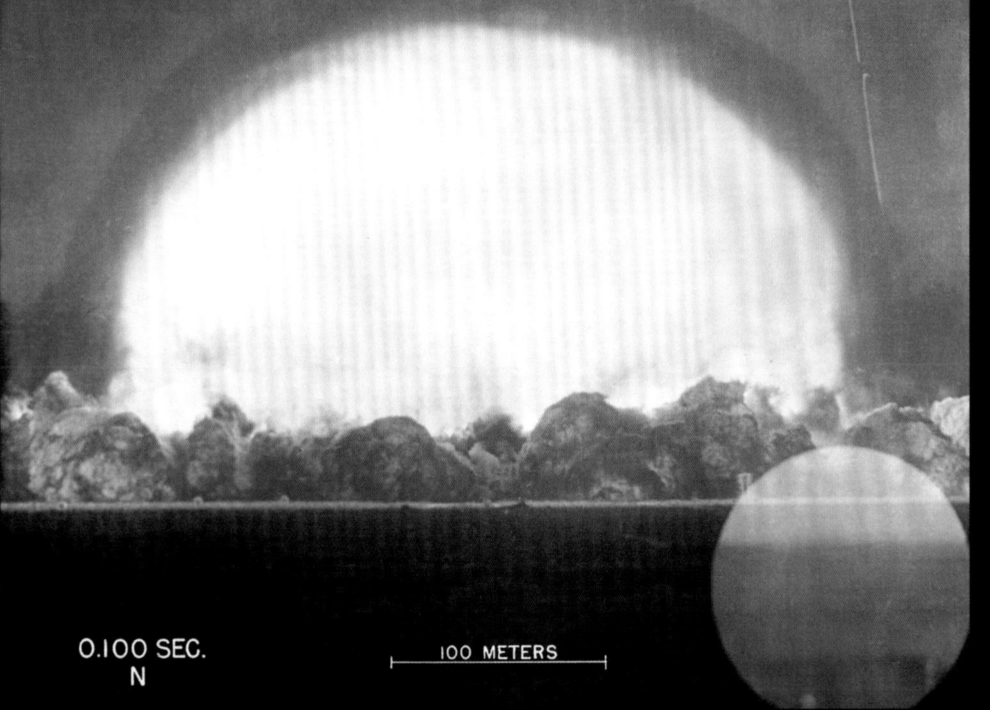

As the shock front continues expanding, it intersects with the cable at progressively higher points. By one-hundred milliseconds, or 0.1 seconds, the illusory break has climbed almost out of frame.

This method of following the shock wave proved to be diagnostically valuable, as the experiments designed to trace its location were disrupted by the intensity of the explosion and did not provide any data. Footage from a twenty-four-inch Mitchell camera at west 10,000, which provided an even closer view than the eighteen-inch Mitchell, was used to make shock wave velocity measurements based on refraction.

One full second after detonation—900 milliseconds after the previous photograph—the fireball, with a radius of about 300 meters, is largely obscured by the dust skirt and has risen almost out of frame.

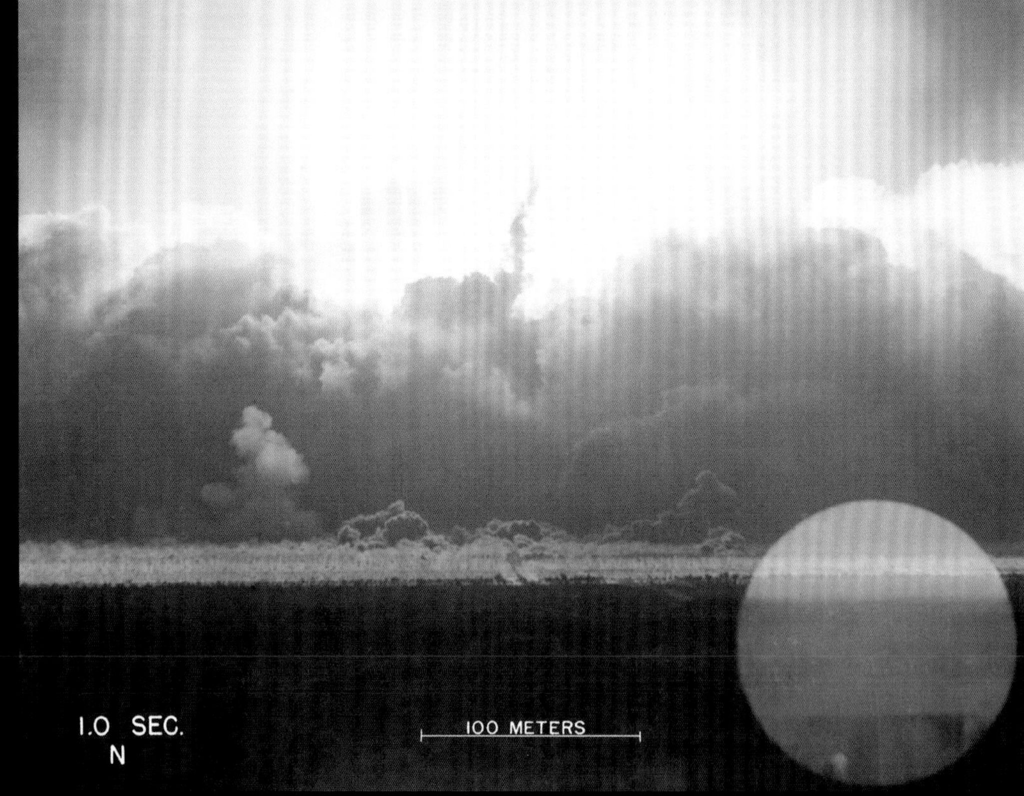

Between one and two seconds, the rising action of the fireball pauses while the skirt continues to grow taller. After this point, the eighteen-inch Mitchell, with a fixed focal length and aim, loses sight of the explosion.

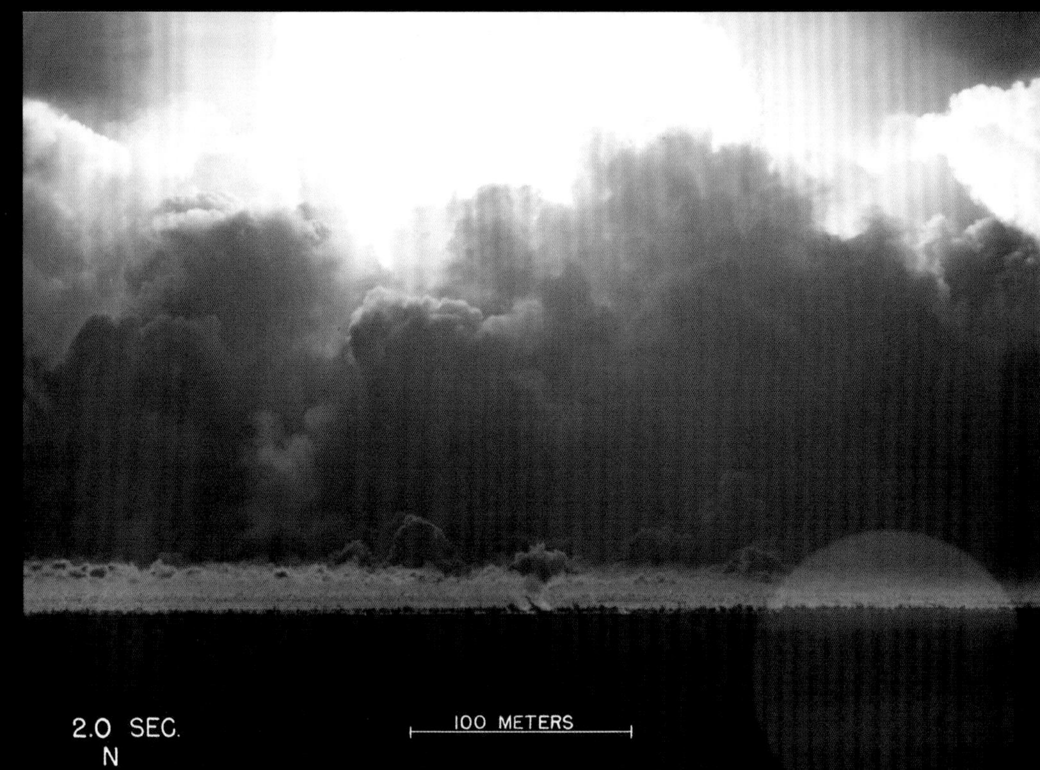

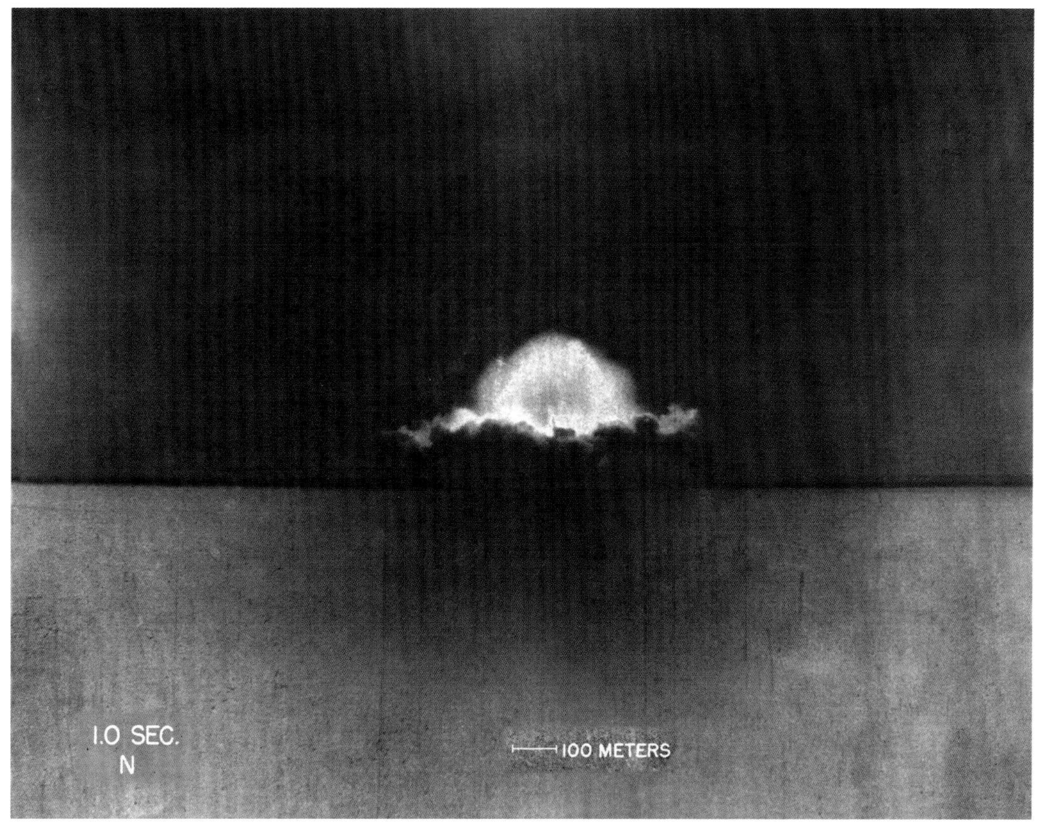

Footage from another Mitchell camera at the same bunker provides a basically identical picture but from a step backward. With a focal length of three inches, this camera was set up to capture a broader, zoomed-out view, showing the dust skirt from afar and the suspended fireball.

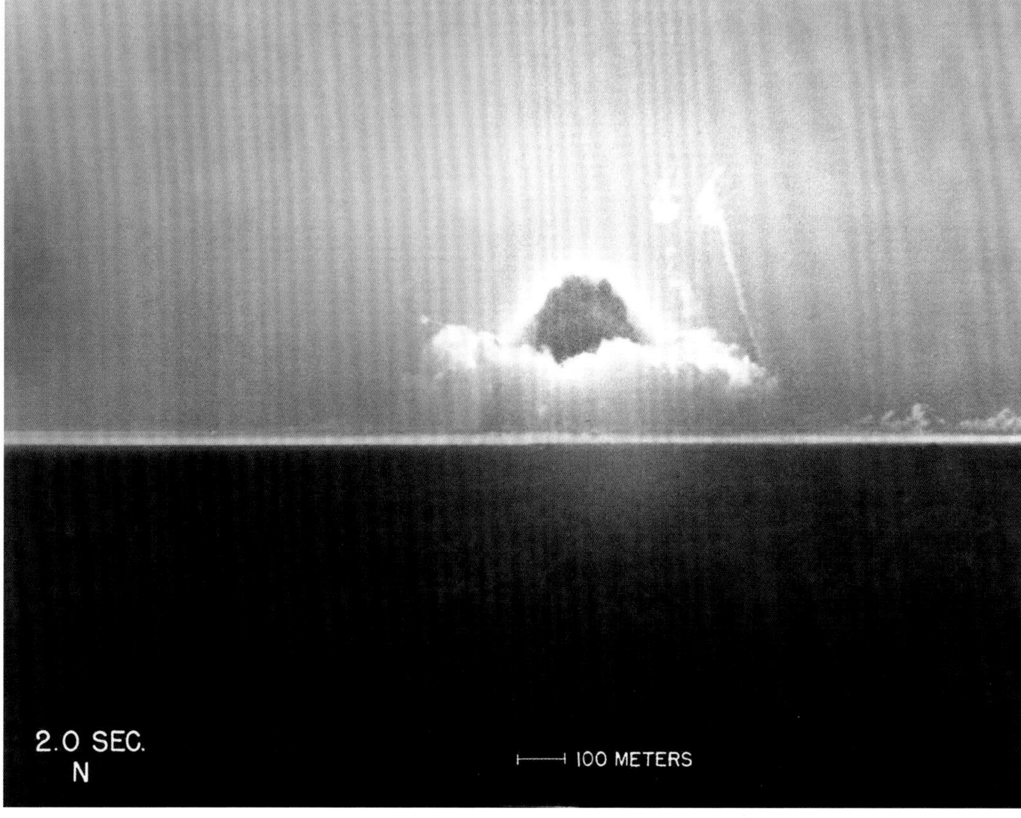

At two seconds after the explosion, the fireball is solarized, or overexposed—giving it a deceptively dark appearance. The barrage balloons and cables have vaporized from the heat of the blast.

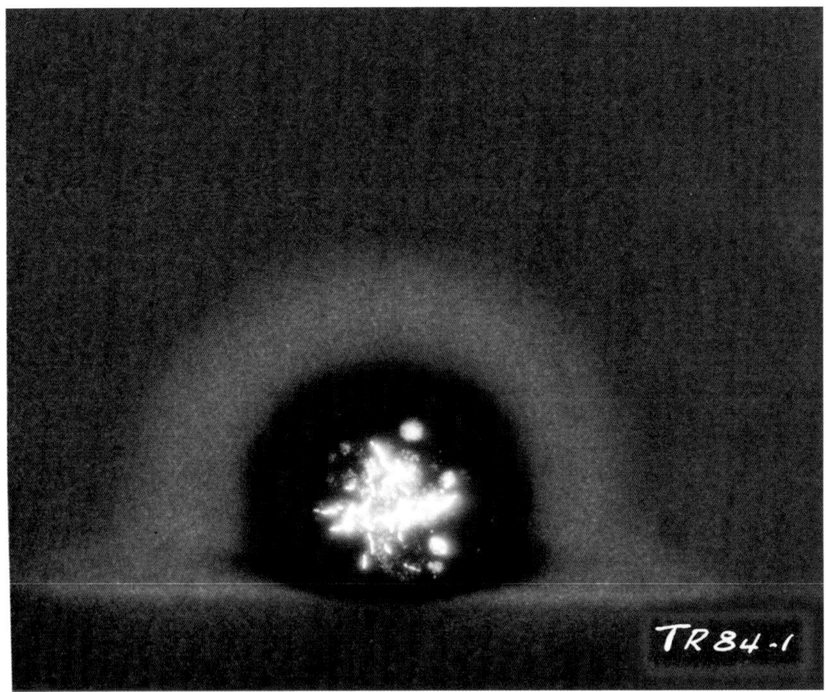

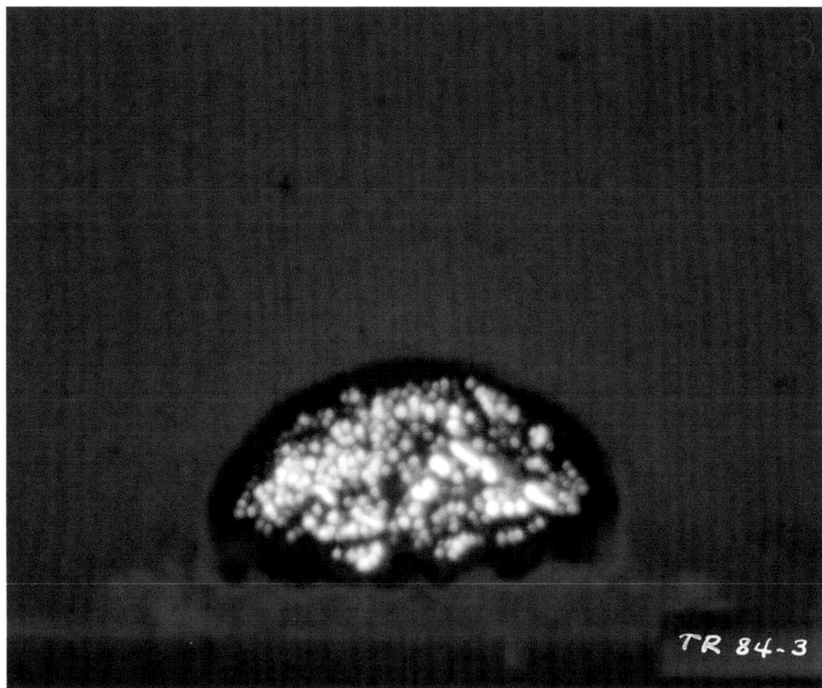

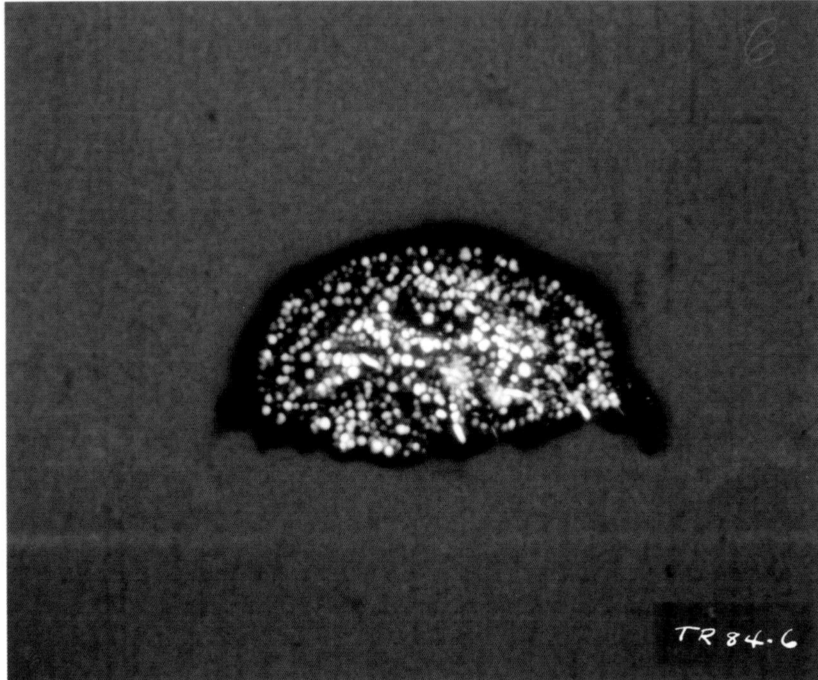

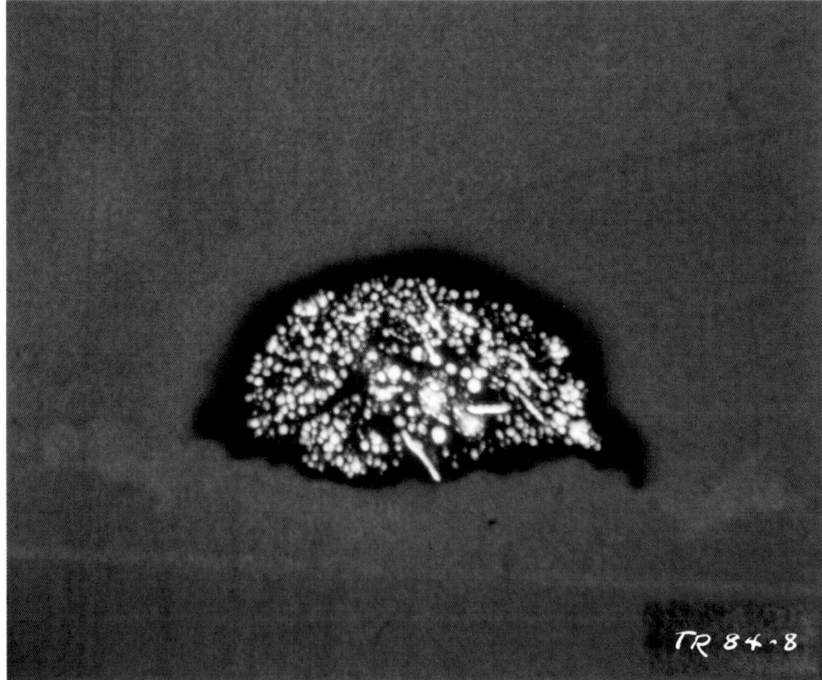

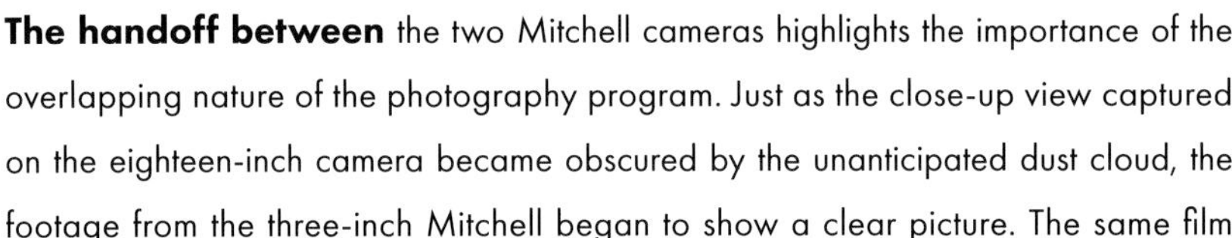
The handoff between the two Mitchell cameras highlights the importance of the overlapping nature of the photography program. Just as the close-up view captured on the eighteen-inch camera became obscured by the unanticipated dust cloud, the footage from the three-inch Mitchell began to show a clear picture. The same film only moments earlier was badly overexposed, with the underestimated, searing brightness of the fireball having burned the emulsion. A selection of these images, from frames one through sixteen, is shown above.

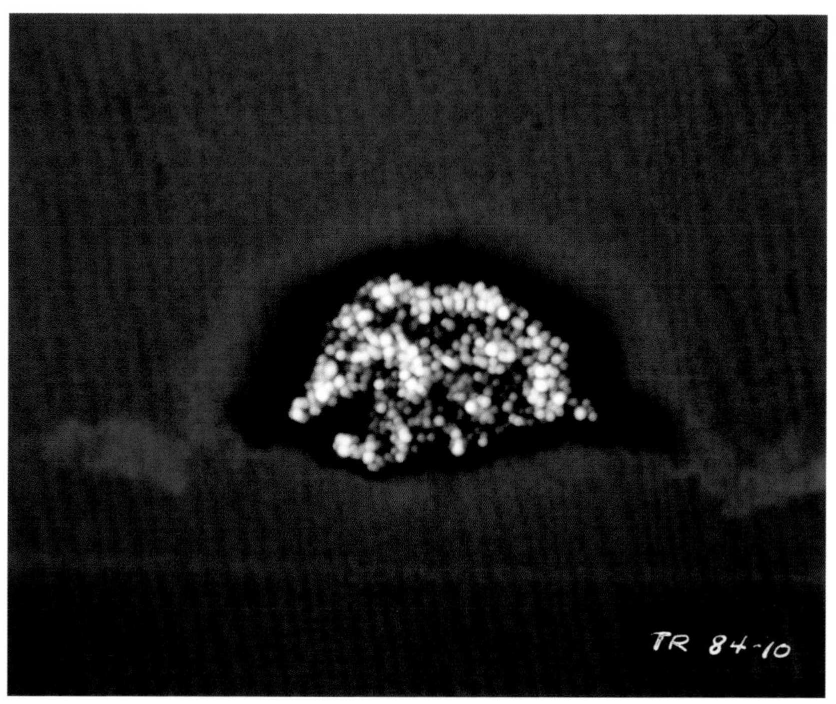
TR 84-10

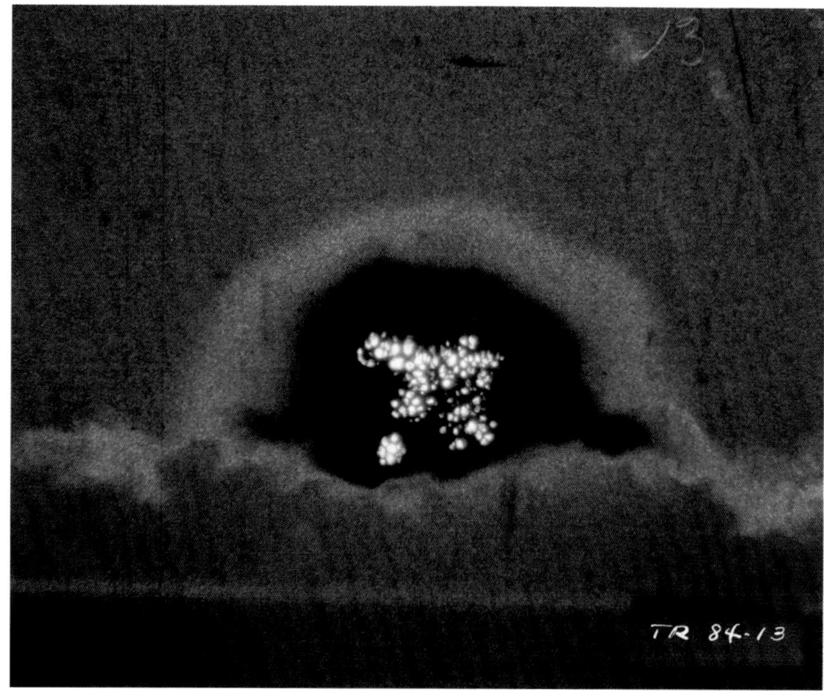
13
TR 84-13

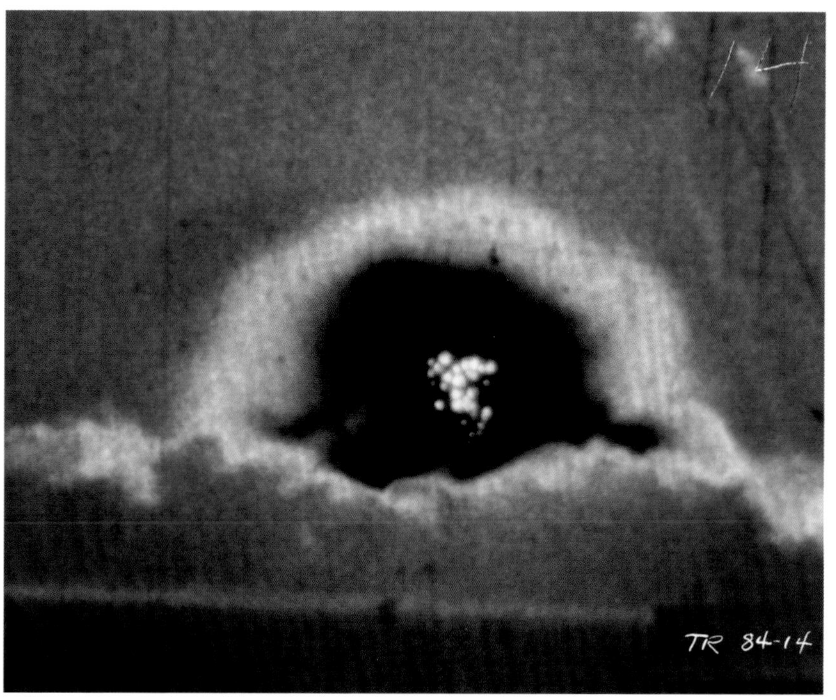
14
TR 84-14

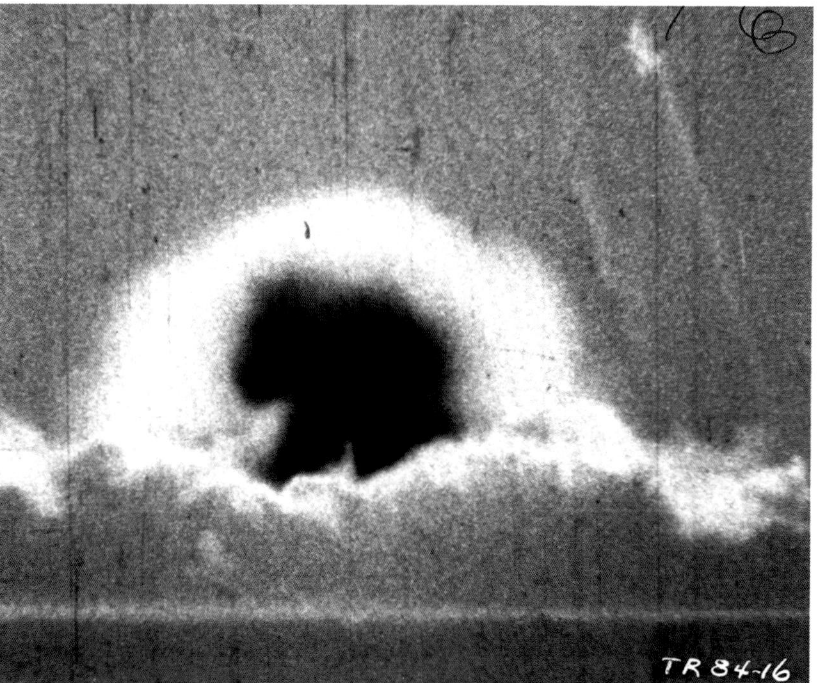
16
TR 84-16

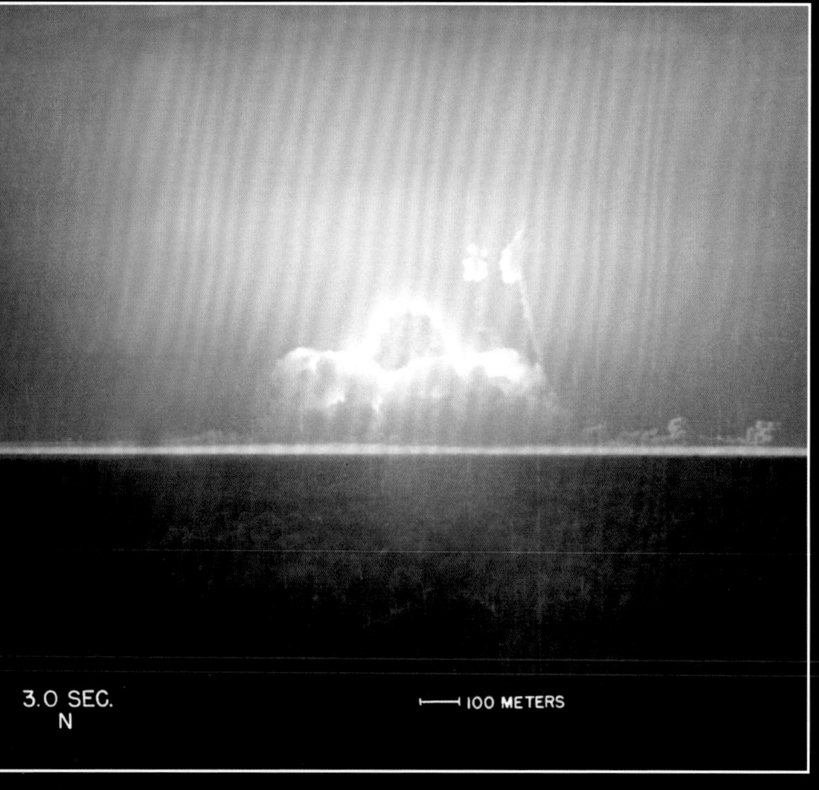

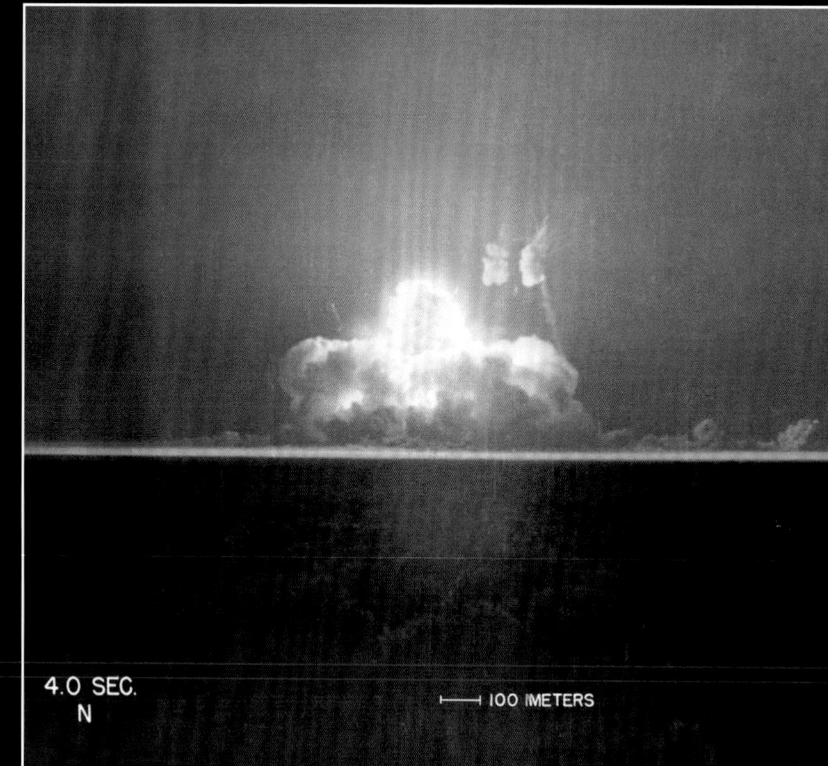

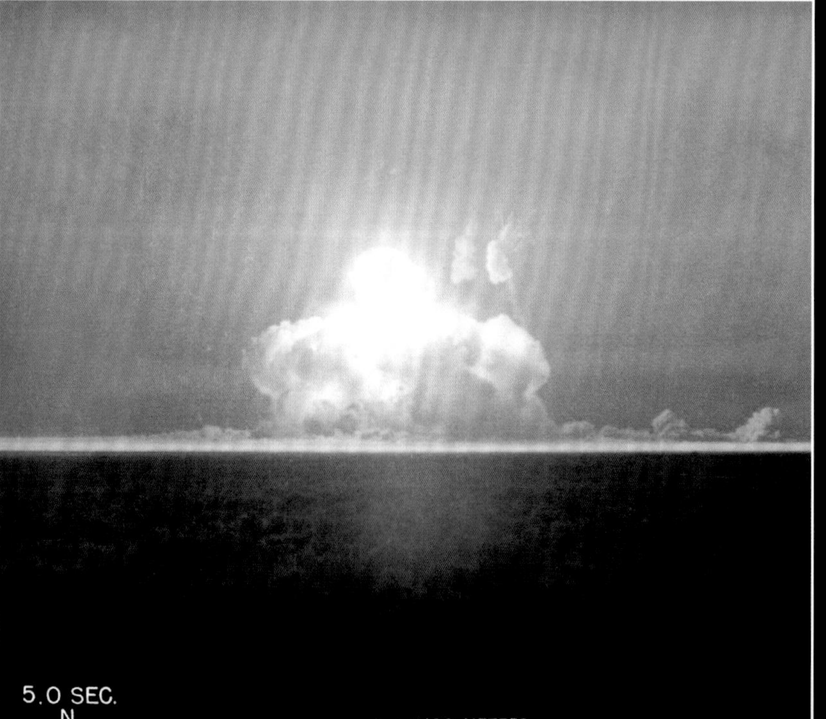

Around three-and-a-half seconds after detonation, the dust skirt and fireball can be seen coalescing to produce a new cloud of smoke. The lower portion of the skirt begins to form a neck, steepening at the sides, while the top third becomes a donut-shaped ring, or torus—the first signs of the strong convection current that has already begun carrying hot gases and dust up through the center of the skirt.

This rising matter begins to spill out over the torus and arrange itself into a huge, bulbous cloud, melding with the ball of fire—now itself caught in the current—as both rise quickly into the sky.

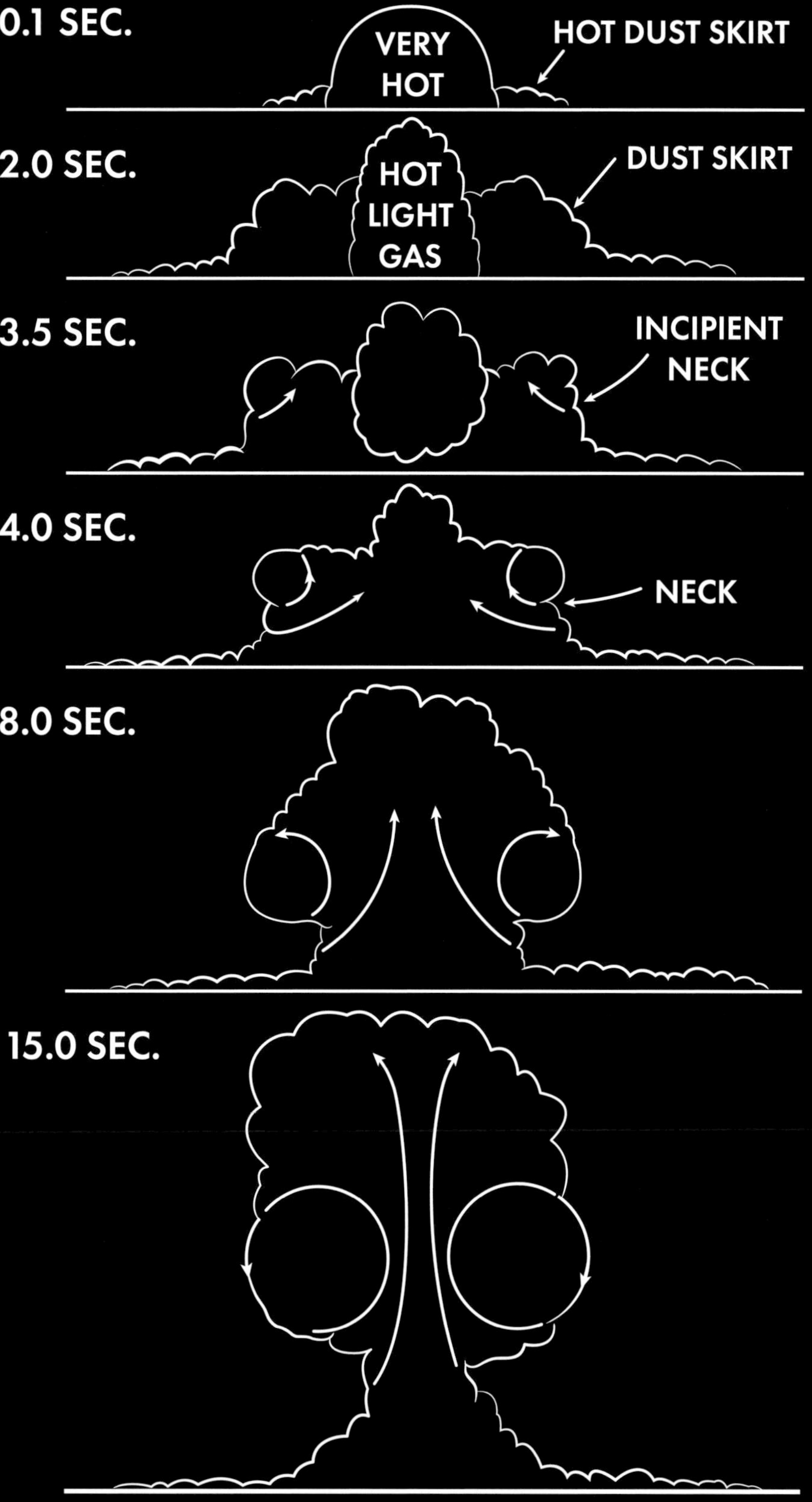
0.1 SEC.
VERY HOT
HOT DUST SKIRT
2.0 SEC.
HOT LIGHT GAS
DUST SKIRT
3.5 SEC.
INCIPIENT NECK
4.0 SEC.
NECK
8.0 SEC.
15.0 SEC.

The neck begins to rise as well, as it continues to constrict, decreasing gradually in radius from 360 meters at the time of its formation to about 100 meters by twenty-three seconds.

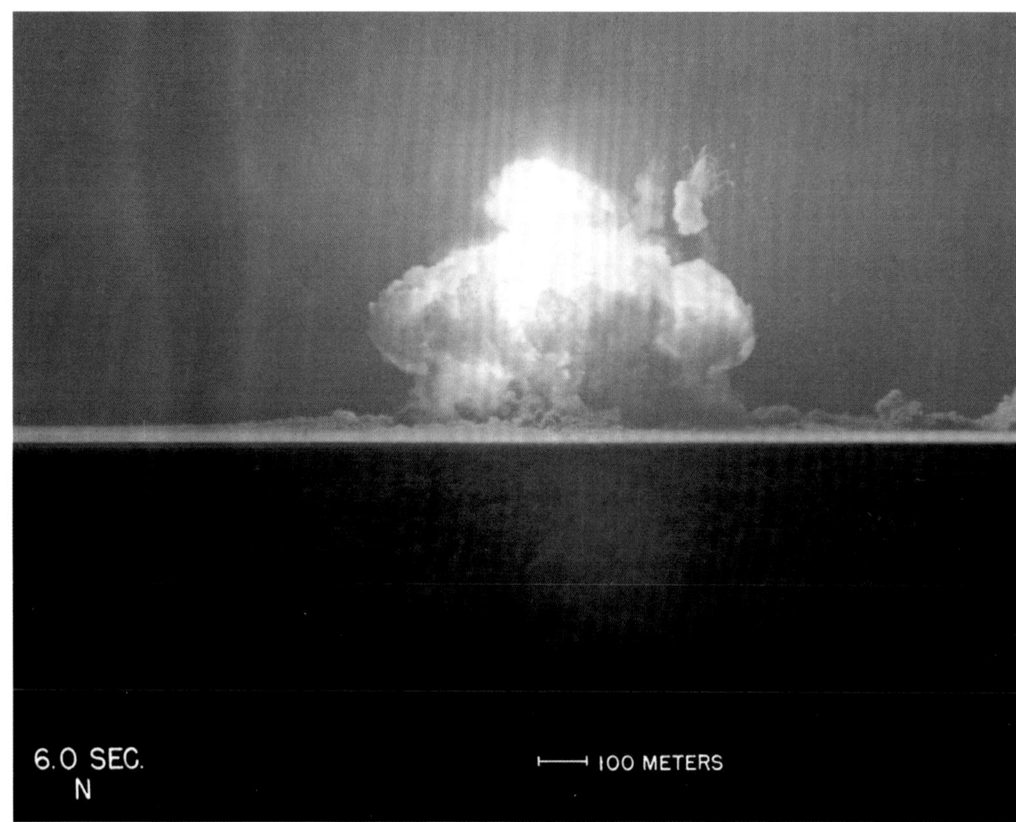

Having heeded instructions to shield their eyes from the initial burst of light, observers—many of whom were at the same distance as the camera recording these images—were now removing their welder's goggles and looking on with what they later described as terror, awe, and disbelief.

Physicist Isidor Rabi recalled looking "toward the place where the bomb had been; there was an enormous ball of fire which grew and grew and it rolled as it grew; it went up into the air, in yellow flashes and into scarlet and green."

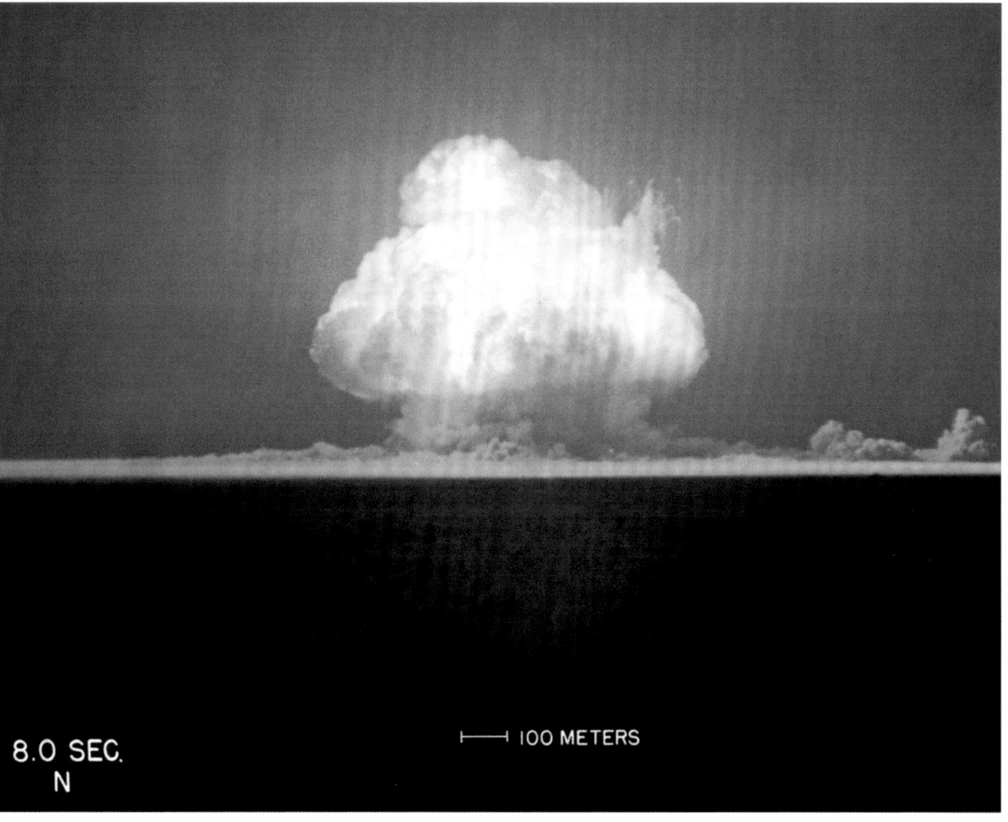

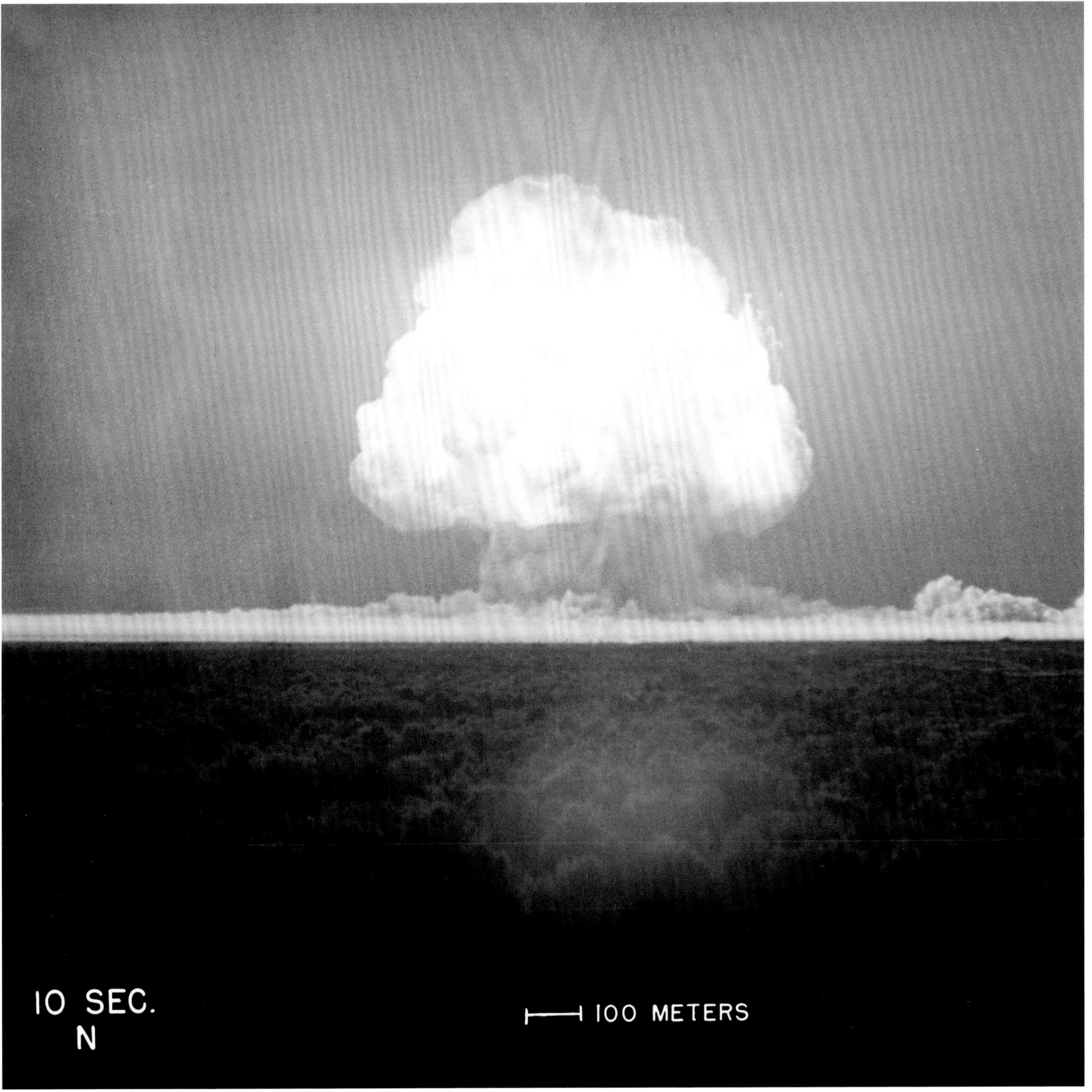
10 SEC.
N
100 METERS

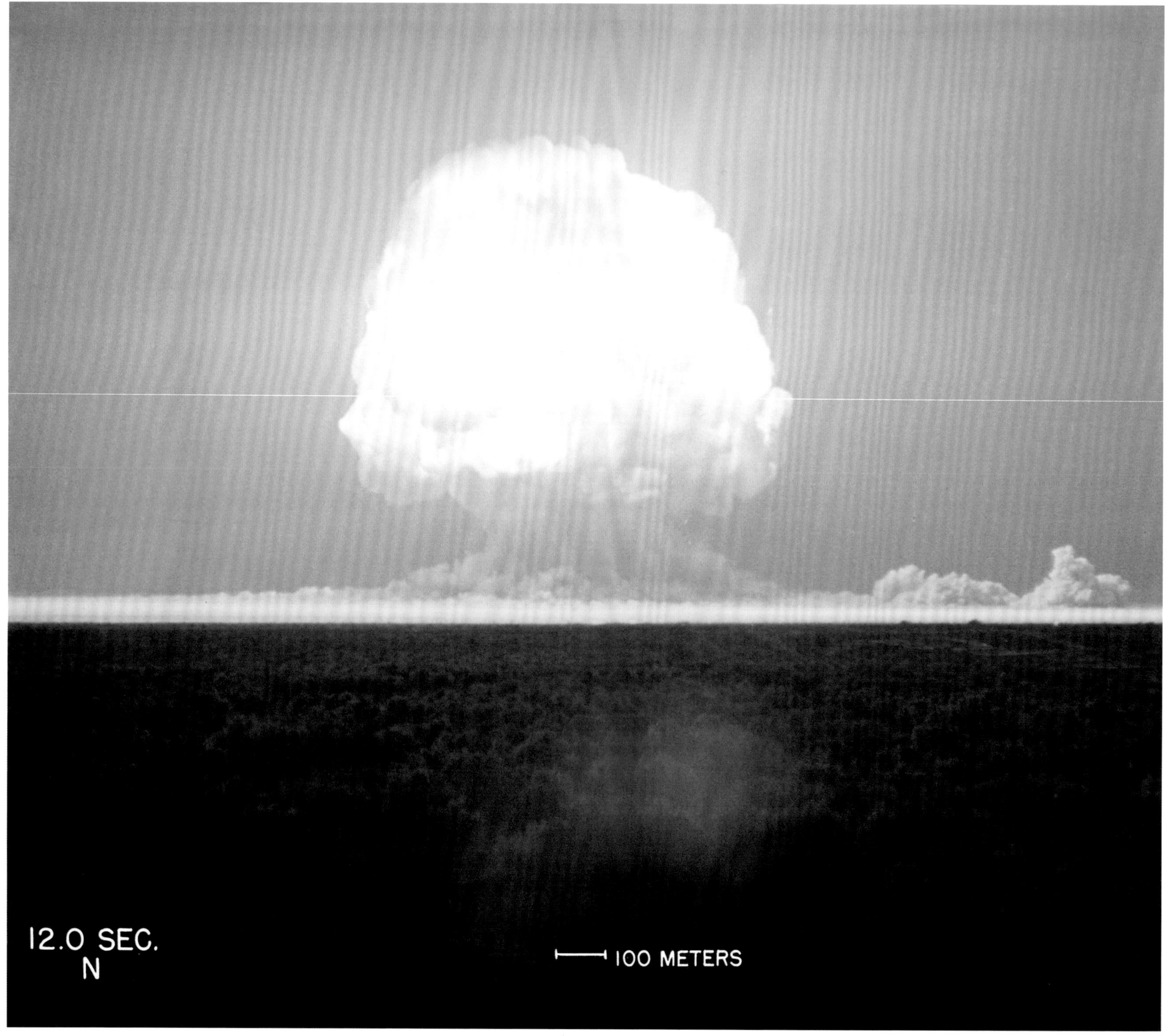

Between ten and fifteen seconds, the cloud of smoke rounds into a sphere, a shape it holds as it continues ascending.

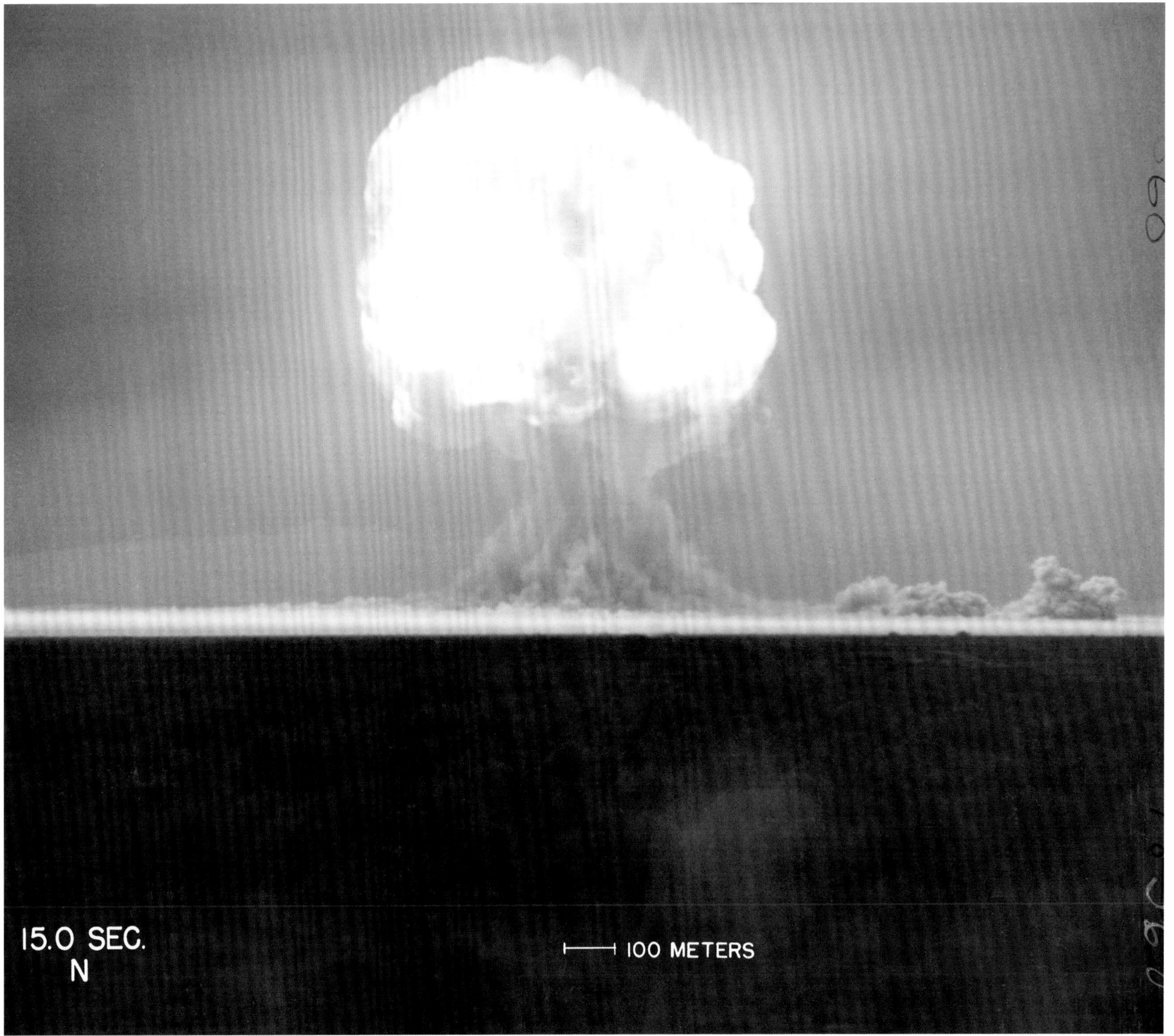
15.0 SEC.
N
100 METERS

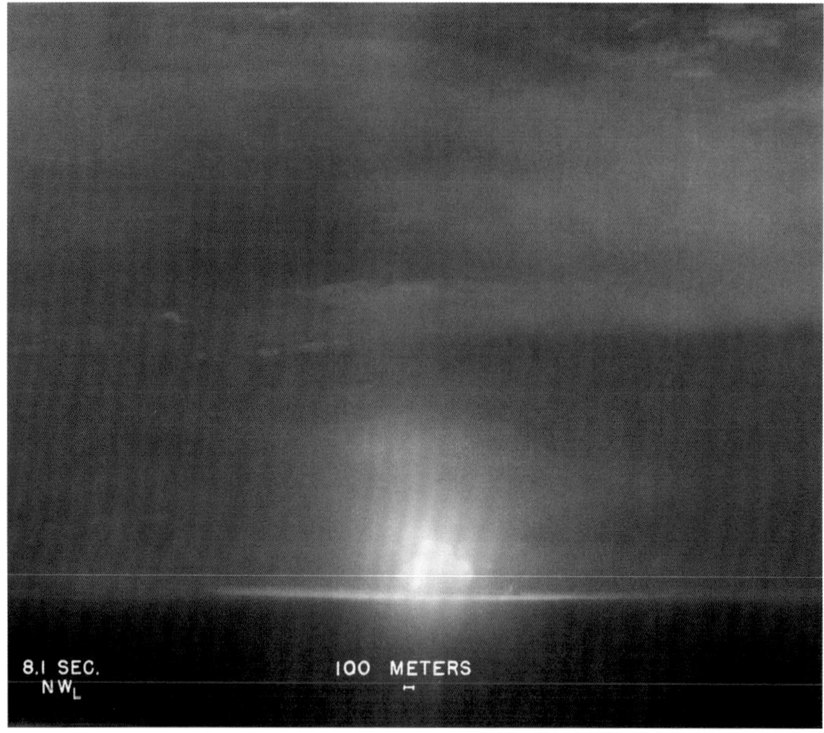

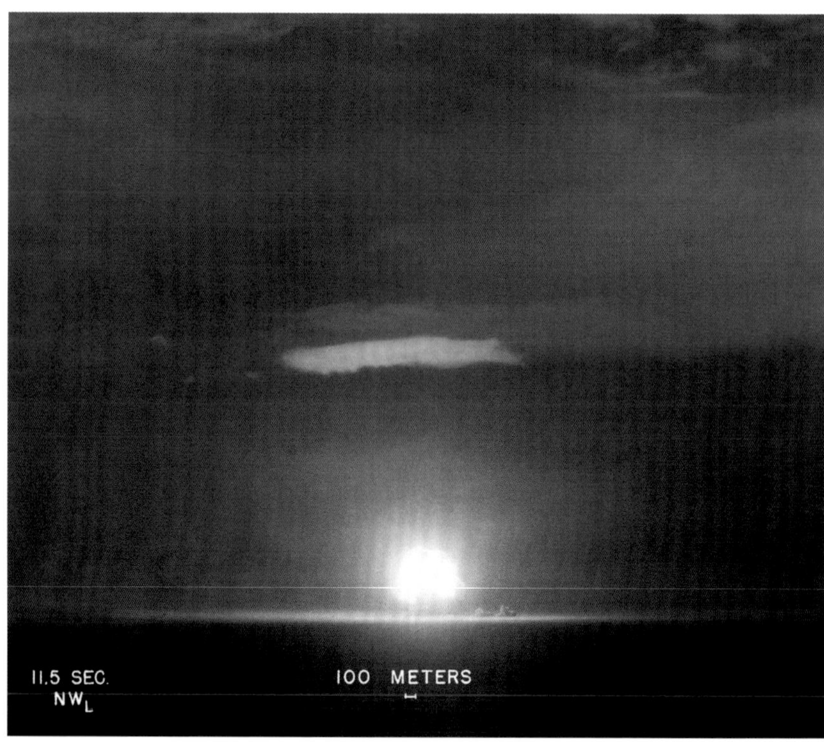

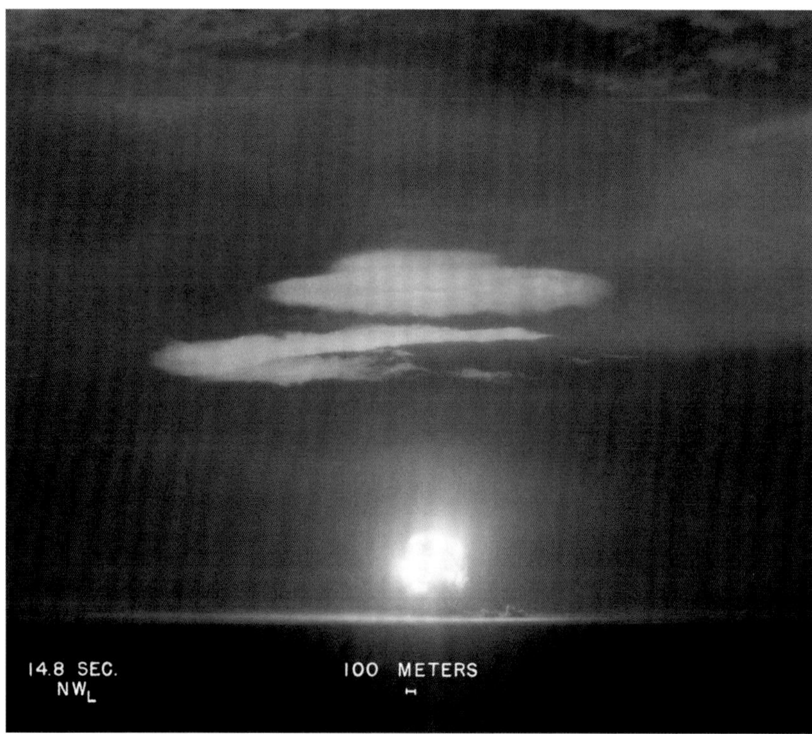

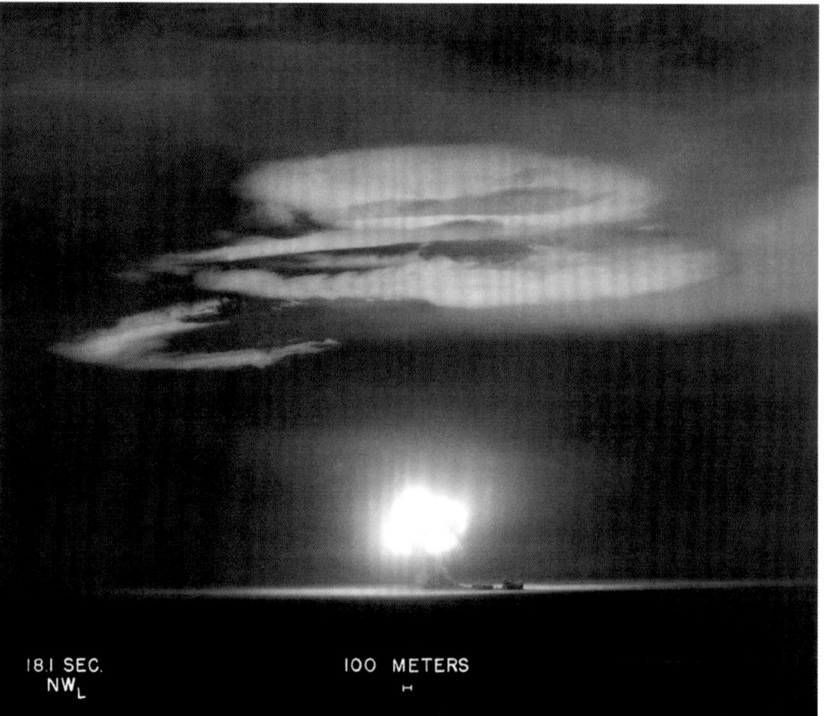

Photographs taken twenty miles away at Campana Hill show the explosion fifty times smaller than in footage from the three-inch Mitchell at north 10,000. Ring-shaped condensation clouds are visible over the blast beginning at eleven seconds—created as the shock front passed through moist layers of the atmosphere high above the ground. These exposures were made by hand in one of the few cases where automatic timers were not used.

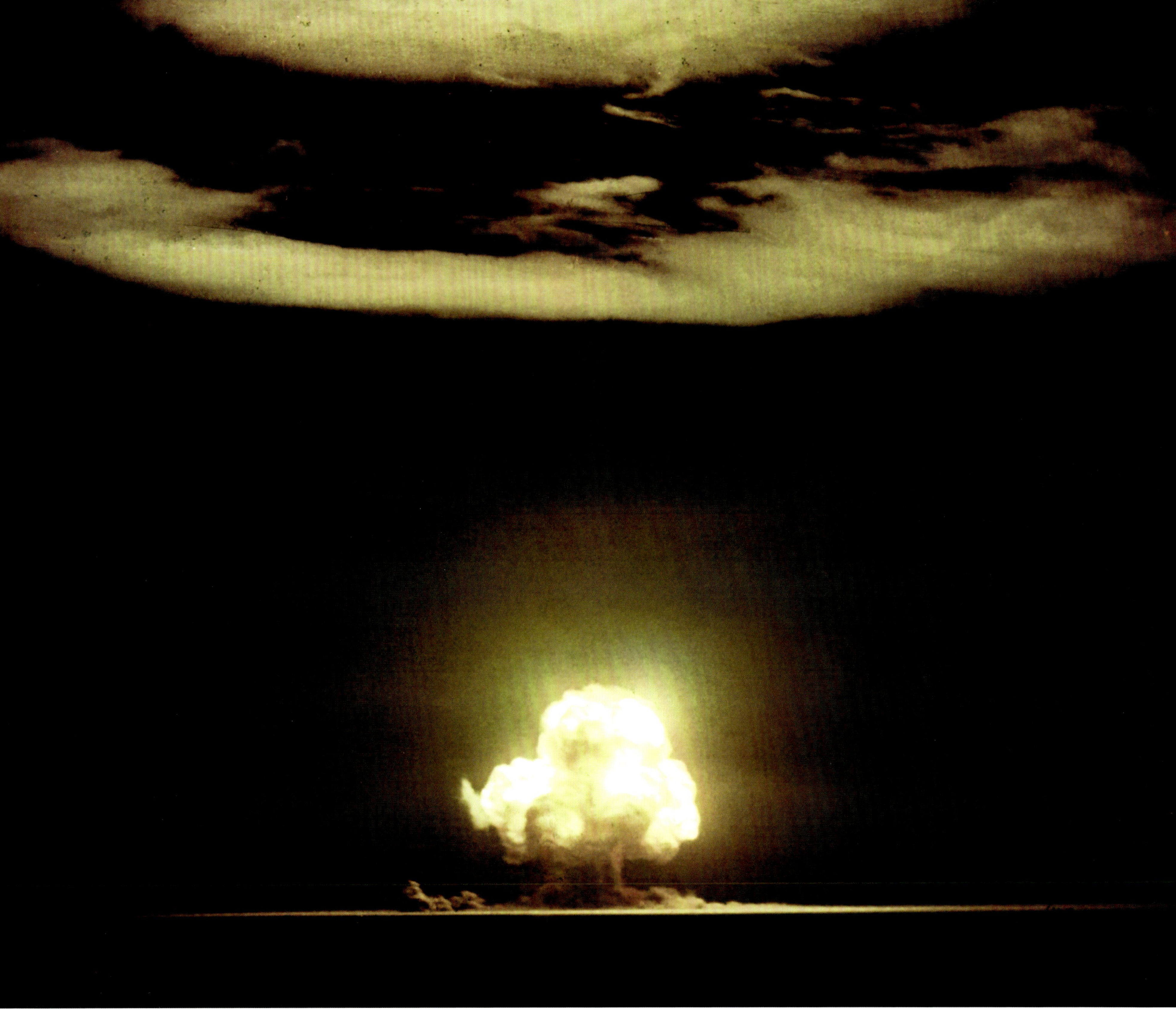

From his position near base camp, southwest of the explosion, technician Jack Aeby also took a series of photos by hand. Aeby, a member of the Special Engineer Detachment and an amateur photographer, received permission from his group leader Emilio Segré to document the test with his personal camera, a Perfex 44. He captured what is now the only quality, color image of the Trinity test. The numerous Kodak cameras shot on colored film as well but with poor results.

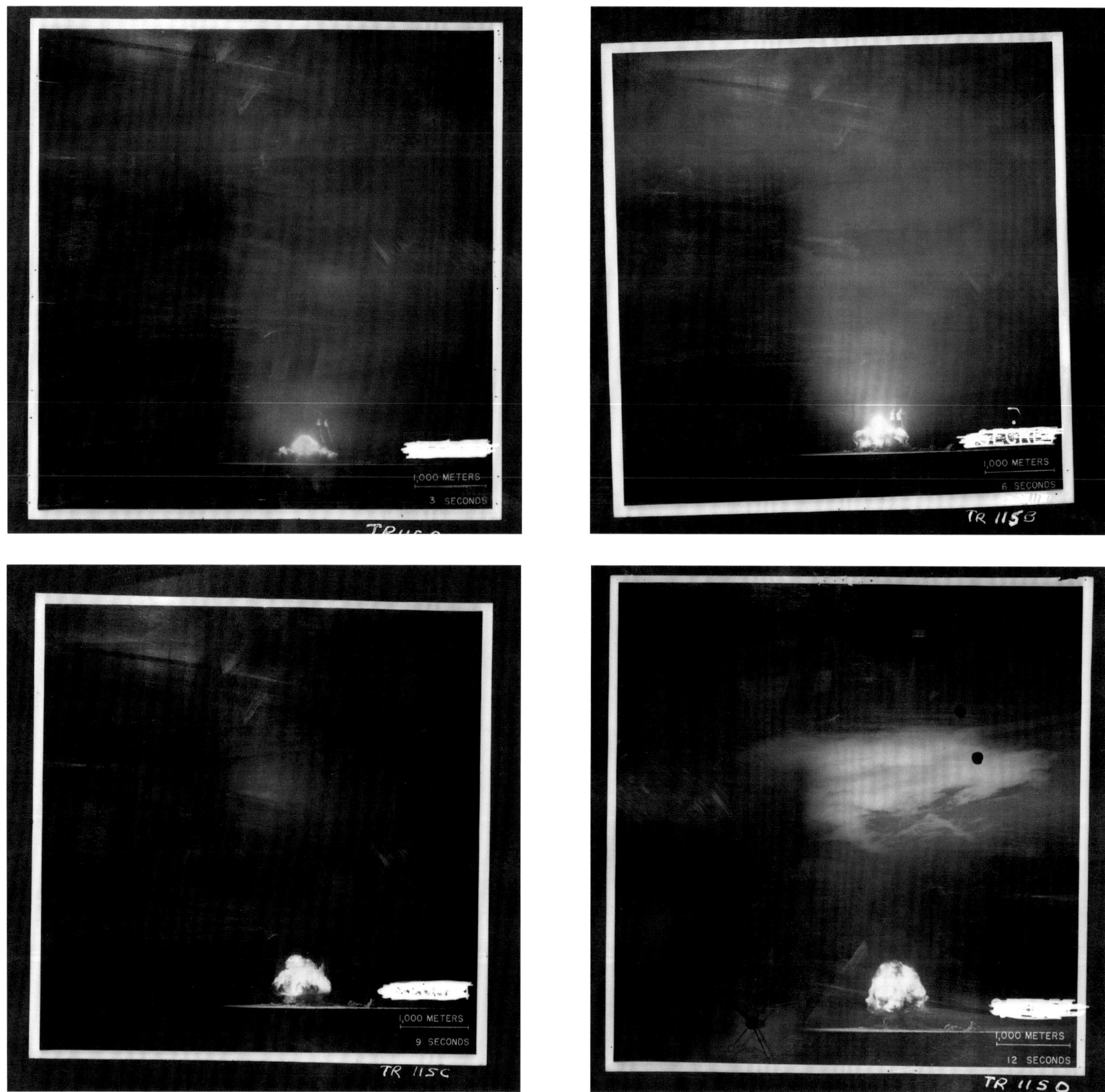

The above set of exposures, taken by a timed camera inside the photography bunker at north 10,000, show the fireball's expansion in three-second increments. The clock face absent from the eighteen-inch Mitchell footage is visible here.

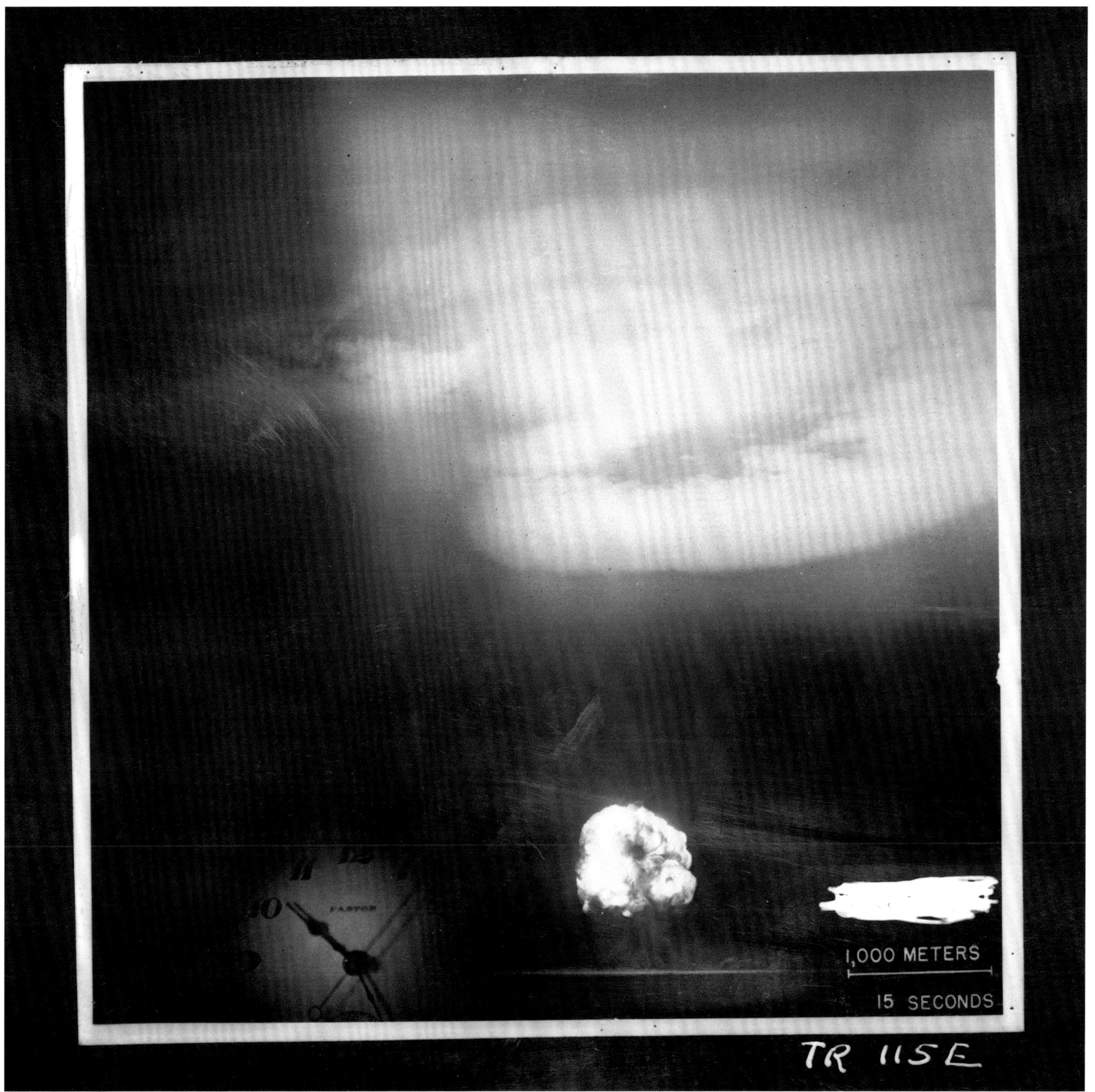
1,000 METERS
15 SECONDS
TR 115E

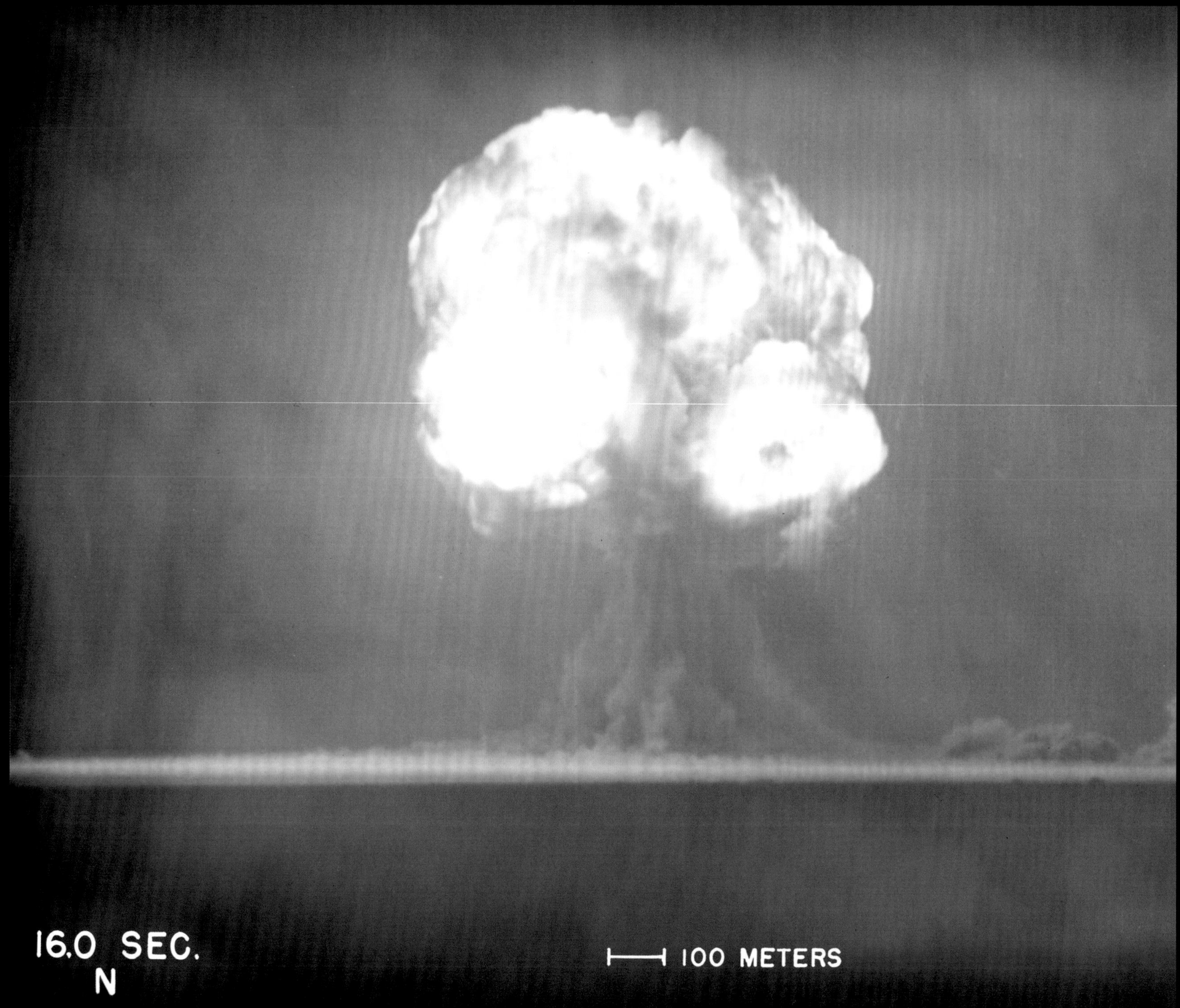

At sixteen seconds, the surrounding landscape and plume of smoke begin to fade from view in the camera footage, with localized flames accounting for most of the waning light.

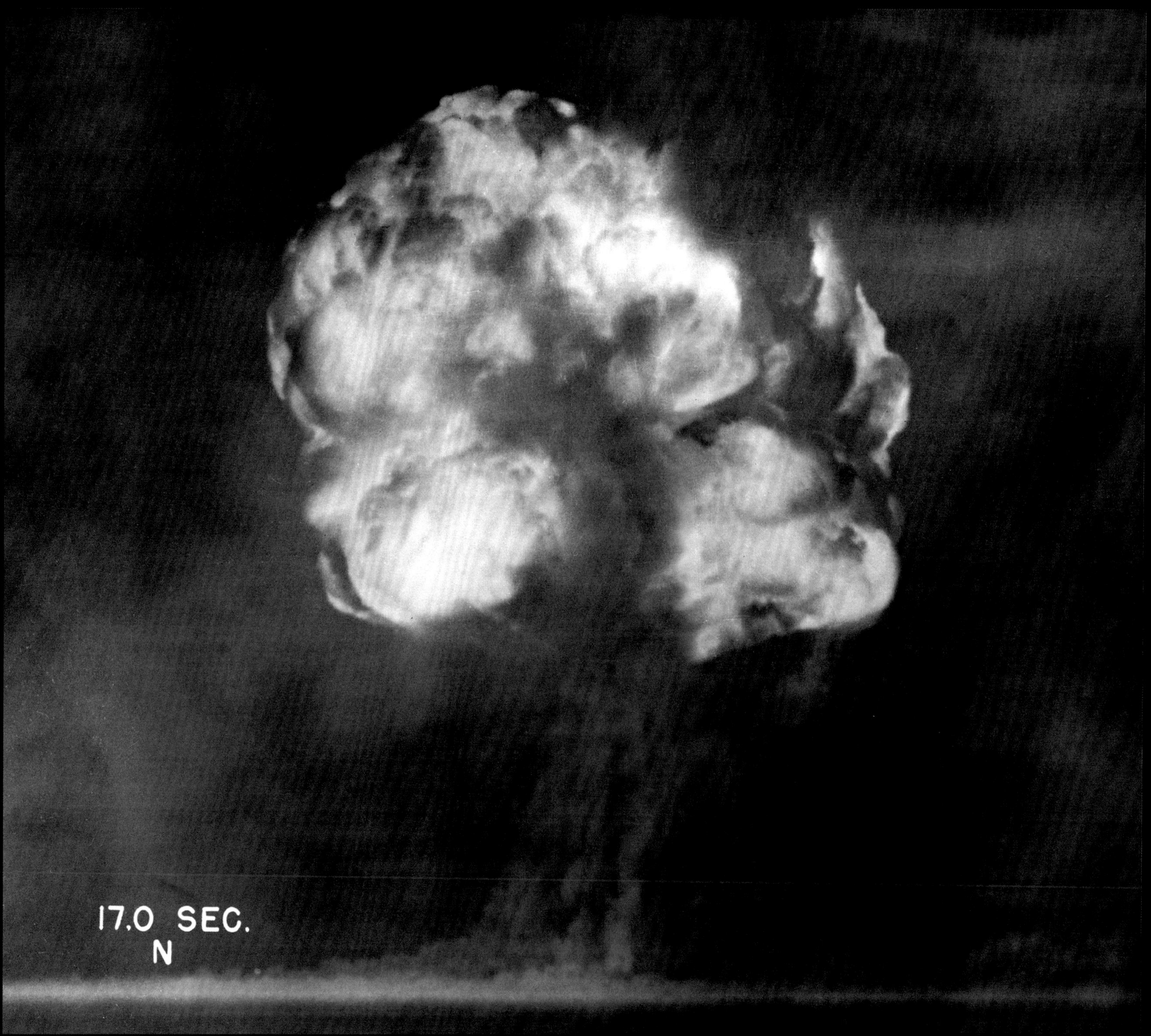

With combustion in the fireball nearing completion, the display has so far proceeded in silence. The shock wave and, close behind it, the roar of the blast, are still on their way outward from ground zero.

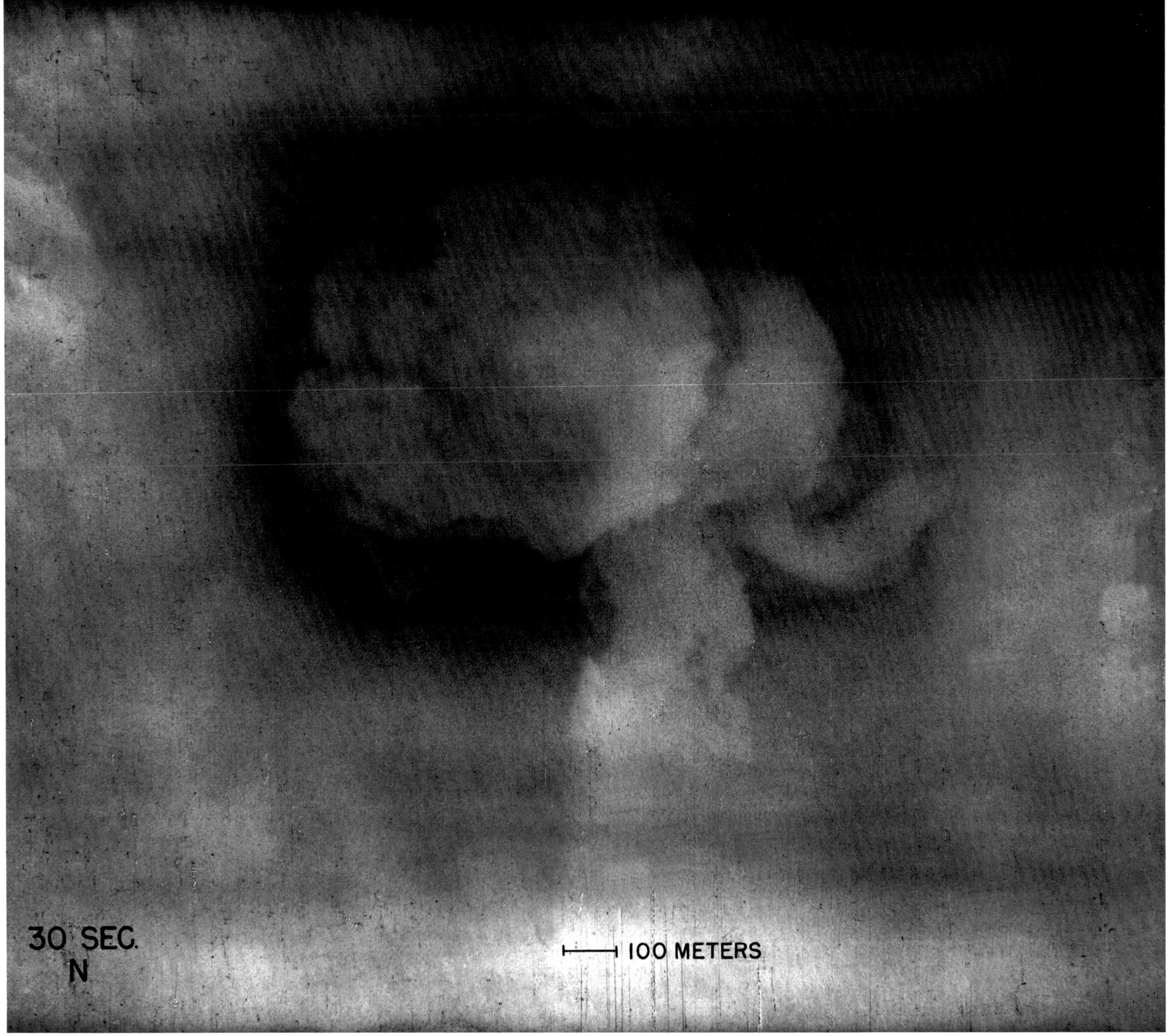

Thirty seconds after detonation, in the absence of firelight, the cloud remained illuminated for about 100 meters in all directions by the bluish glow of ionizing radiation—high-energy, massless particles like gamma rays and x-rays that interact with surrounding matter. The "huge cloud of transparent, purplish air" hung over the Trinity site as the shock wave rushed in, according to Bainbridge, who later described the scene as unforgettable: "a foul and awesome display."

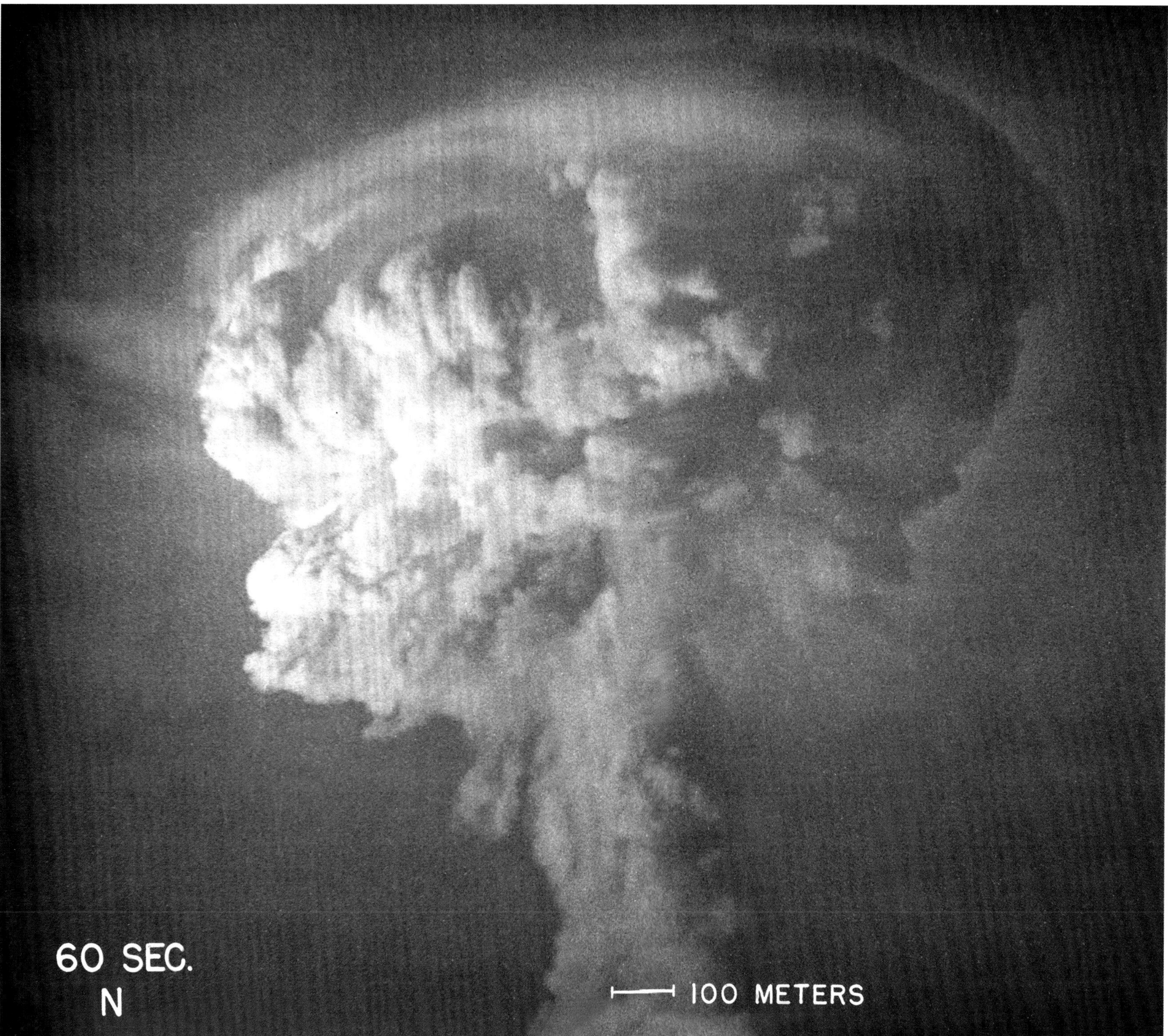

The cloud, now over two miles high, was illuminated again by a timed flash bomb at sixty seconds. In Frank Oppenheimer's memory, "It just seemed to hang there forever. Of course it didn't. It must have been just a very short time until it went up. It was very terrifying. And the thunder from the blast. It bounced on the rocks, and then it went—I don't know where else it bounced. But it never seemed to stop."

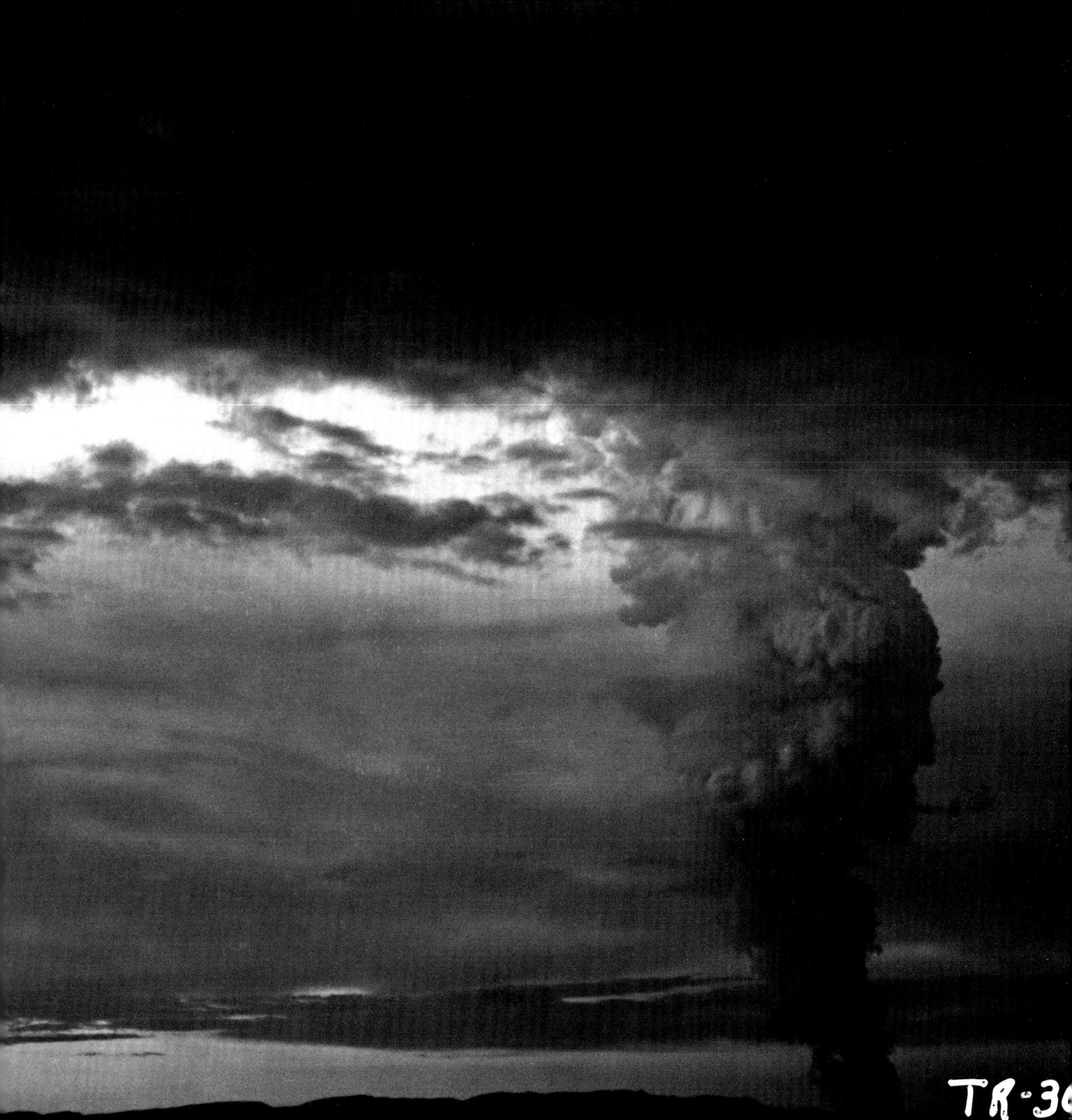
TR-3

At a towering height of over 50,000 feet, the cloud was finally pulled apart by winds in the stratosphere.

As the remnants of the first nuclear explosion drifted away, several hundred people were left to process the complicated meaning of their success, having entered abruptly a reality very different from the one into which the Manhattan Project was born. J. Robert Oppenheimer recalled decades later, "We knew the world would not be the same. A few people laughed, a few people cried. Most people were silent."

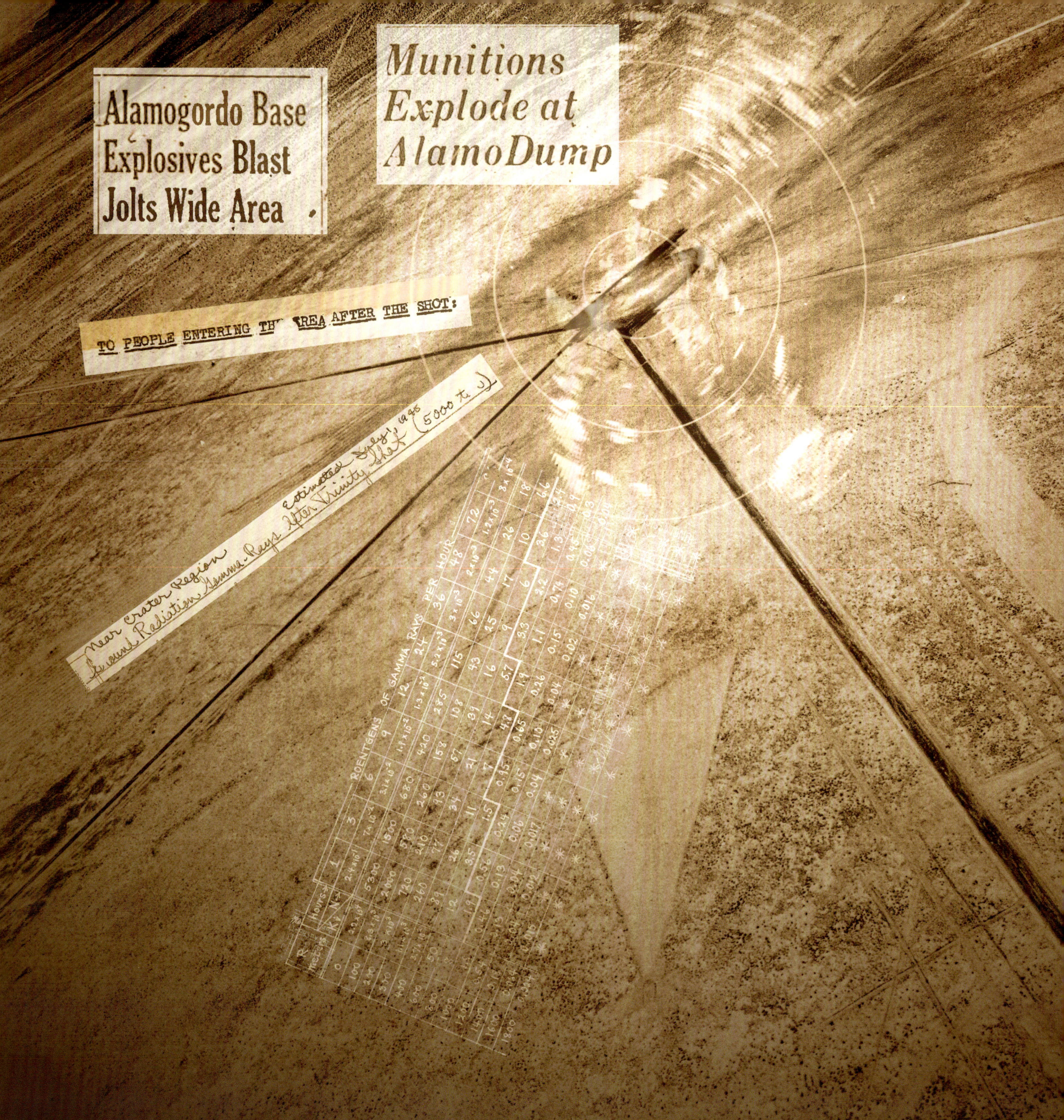

Alamogordo Base Explosives Blast Jolts Wide Area
Munitions Explode at Alamo Dump
TO PEOPLE ENTERING TH AREA AFTER THE SHOT:
Near Crater Region Estimated July 1, 1945
Ground Radiation Gamma-Rays After Trinity Shot (5000 t. u.)
ROENTGENS OF GAMMA RAYS PER HOUR

07 Ground Zero

Most scientists present for the Trinity test were attending strictly as observers, and they returned immediately to Los Alamos by the busload. Victor Weisskopf, a theoretical physicist involved in safety planning and experiment design, described the range of emotions that reverberated among his colleagues: "Our first feeling was one of elation, then we realized we were tired, and then we were worried." Mathematician Stan Ulam, who chose not to attend the test, noted of the returning witnesses, "You could see it on their faces . . . something very grave and strong had happened to their whole outlook on the future."

While the long ride to the laboratory afforded time for contemplation, personnel remaining on-site to attend to post-test responsibilities had to set aside the memory of the shocking spectacle—and mounting trepidation about the future—to focus on the task at hand: quantifying the success of the Trinity shot. There were soil samples to gather, field instruments to retrieve, films to send for development, and analyses to be made of all recoverable data.

The results were already in for one experiment—perhaps Trinity's most famous, and least sophisticated. As the blast wave rumbled through Enrico Fermi's location at base camp, the physicist dropped pieces of paper from about six feet in the air and measured how far the passing wave displaced them. Observing a shift of about eight feet, he calculated the Gadget had unleashed the equivalent of 10,000 tons of TNT. Radiochemical analysis of plutonium remnants and fission products in the crater soon indicated that his rough estimate, already double the expected energy release, was off by a factor of two: the actual yield was closer to 21,000 tons.

Due in some part to the magnitude of the explosion and in large part to the rain delay, New Mexicans throughout the region witnessed the brilliance of Trinity's light. Rumors abounded the morning of the sixteenth, as residents suggested an incoming meteor or exploding bomber might have illuminated the predawn sky. Although Manhattan Project leaders had hoped the test would go unnoticed, the number of witnesses necessitated issuing one of two carefully prepared press releases; which one would depend on the data gathered by radiation monitoring technicians, springing into action as the mushroom cloud began its long process of dispersion.

In the hours after the detonation, the monitors found that all areas known to be inhabited were testing below the established safety threshold for exposure. The official statement thus read: "Following a blast felt over hundreds of miles Monday morning, explosion of 'a considerable amount of high explosive and pyrotechnics' in a remote area of the Alamogordo air base reservation was reported by Col. William O. Eareckson, commandant." The second press release, which explained an evacuation was necessary due to the presence of artillery shells containing poisonous gas, was not needed.

Nearly 5,500 miles away in Potsdam, Germany, a coded message added good news from Trinity to President Truman's arsenal as he prepared to enter deliberations with Allied leaders: "Operated this morning. Diagnosis not yet complete but results seem satisfactory and already exceed expectations. Local press release necessary as interest extends great distance. Dr. Groves pleased." As the conference unfolded, Truman partially revealed to Stalin the closely held secret: that the United States possessed "a new weapon of unusual destructive force." Courtesy of a thriving spy network, however, the Soviet leader was well aware of the Manhattan Project; whatever reaction Truman was expecting, Stalin gave none, suggesting casually that the president "make good use of [the weapon] against the Japanese."

The diplomatic implications of a nuclear capability would factor greatly in the coming peacetime, and in the decades to follow. For the time being, however, military applications commanded the foreground. George Kistiakowsky, who had readied the explosive lenses for the test device with a dentist's drill only days before, recalls of July 16, "The big shots at the ranch were appropriately congratulatory later on. But that was the anticlimax. Before the day was over, I was back in Los Alamos to get the X Division going on building the next Gadget." With the rehearsal complete, a trio of high-explosive lens assemblies and one plutonium core would soon be on their way to an island in the distant Pacific.

* * * * * * *

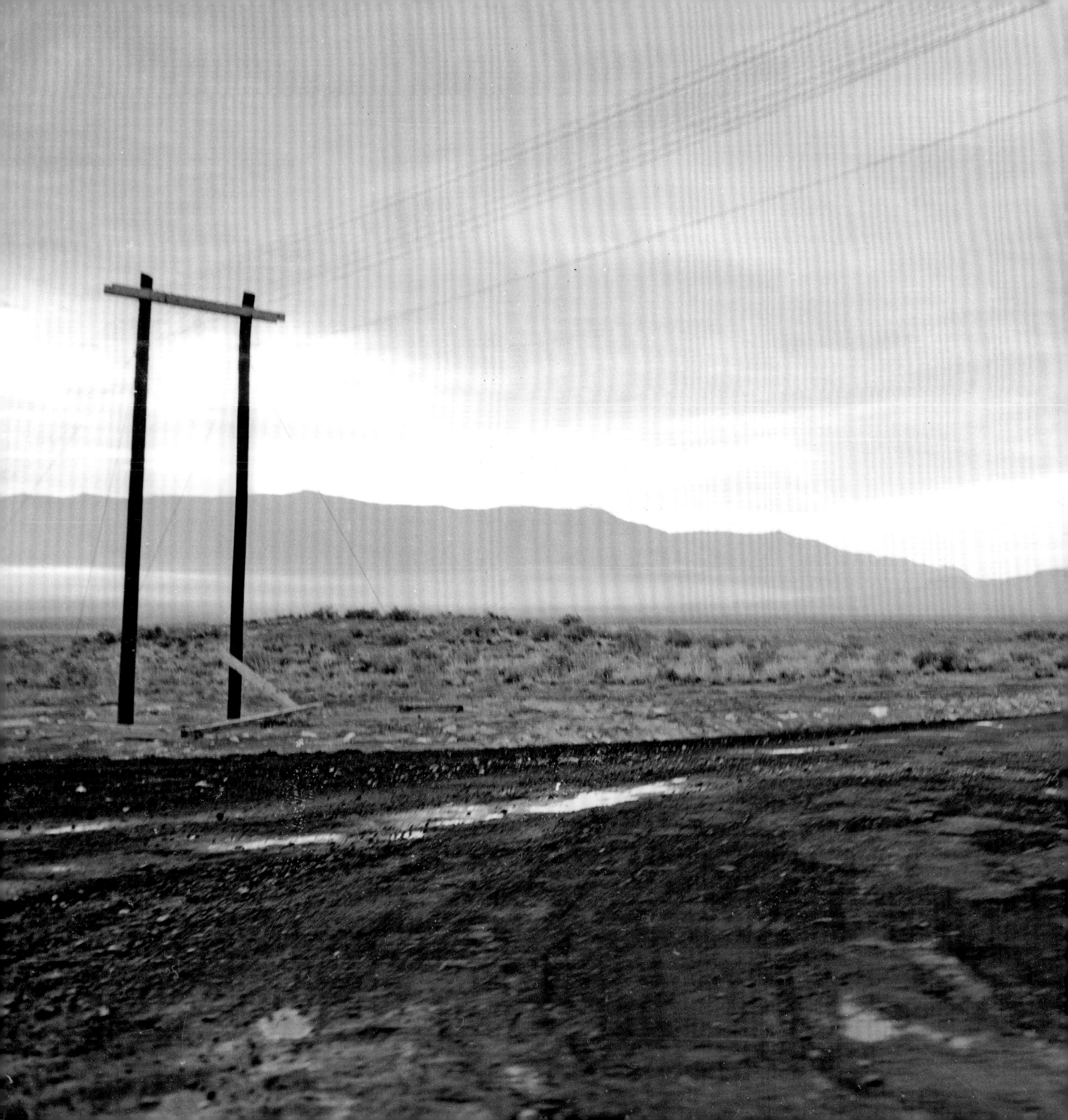

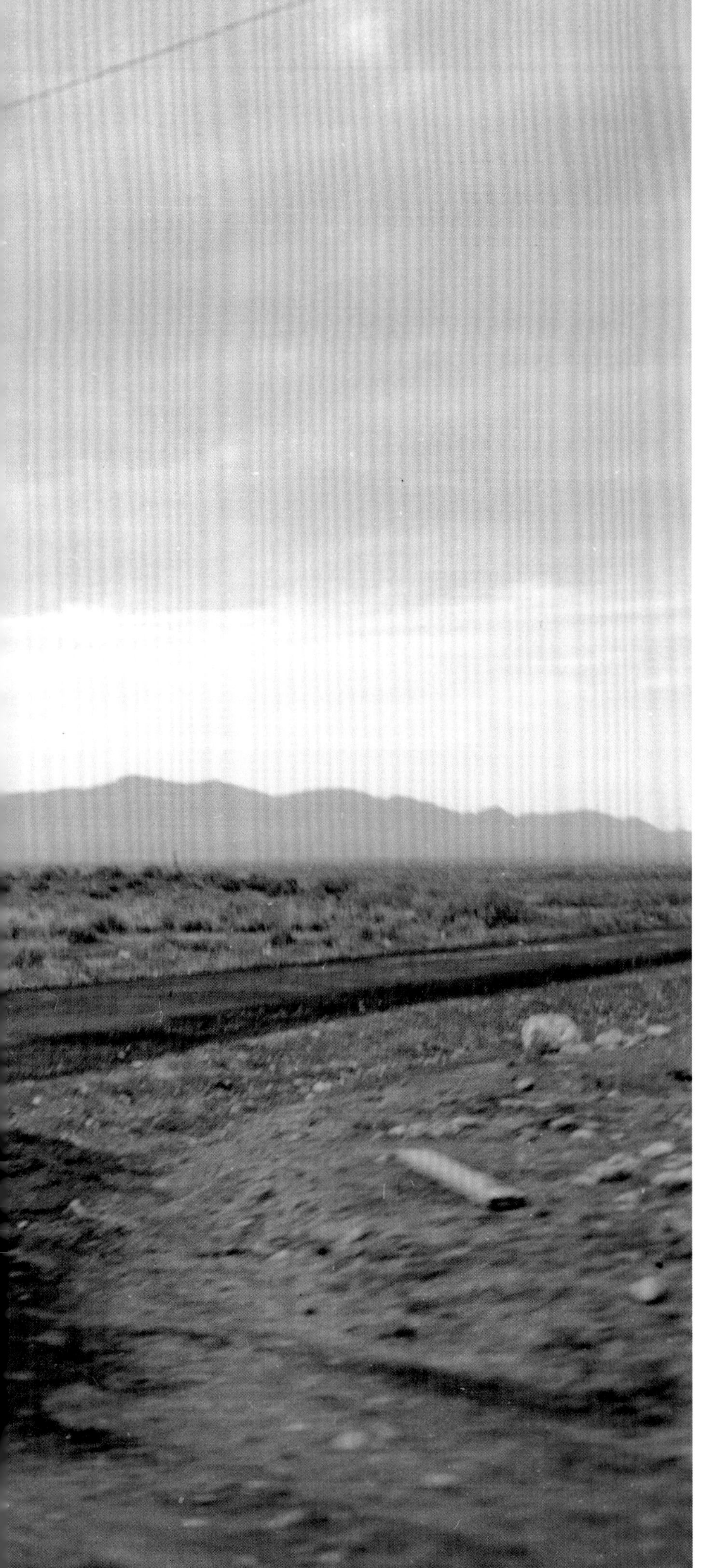

When the sun rose on July 16, a pall of smoke hung over the basin. Notwithstanding a false alarm that triggered evacuation of the north 10,000 personnel shelter, the observation points remained safe from radiation and the roads were gradually cleared so that personnel could return to base camp.

Scientists—both those with official duties, such as retrieving instruments and camera films, and those without—were eager to explore ground zero, lining up at the checkpoints where Military Police controlled access to the vicinity of the crater. However, anyone entering the contaminated area would have to exercise extreme caution. While many of the radioactive particles left behind by the blast would break down into harmless elements over time, these materials posed a very real threat in the immediate aftermath.

A network of monitoring devices called gamma sentinels were stationed across the test site, each providing a real-time measurement of gamma radiation. Designed by British physicist Philip Moon, the sentinels, together with ongoing monitoring by the Health Group, would help inform the safety of travel to areas with measurable radioactivity.

Gamma sentinels within 4,000 yards of ground zero were set up inside small, buried shelters with heavy lids that were opened after the blast by the action of a counterweight—the suspended rectangular box, at right.

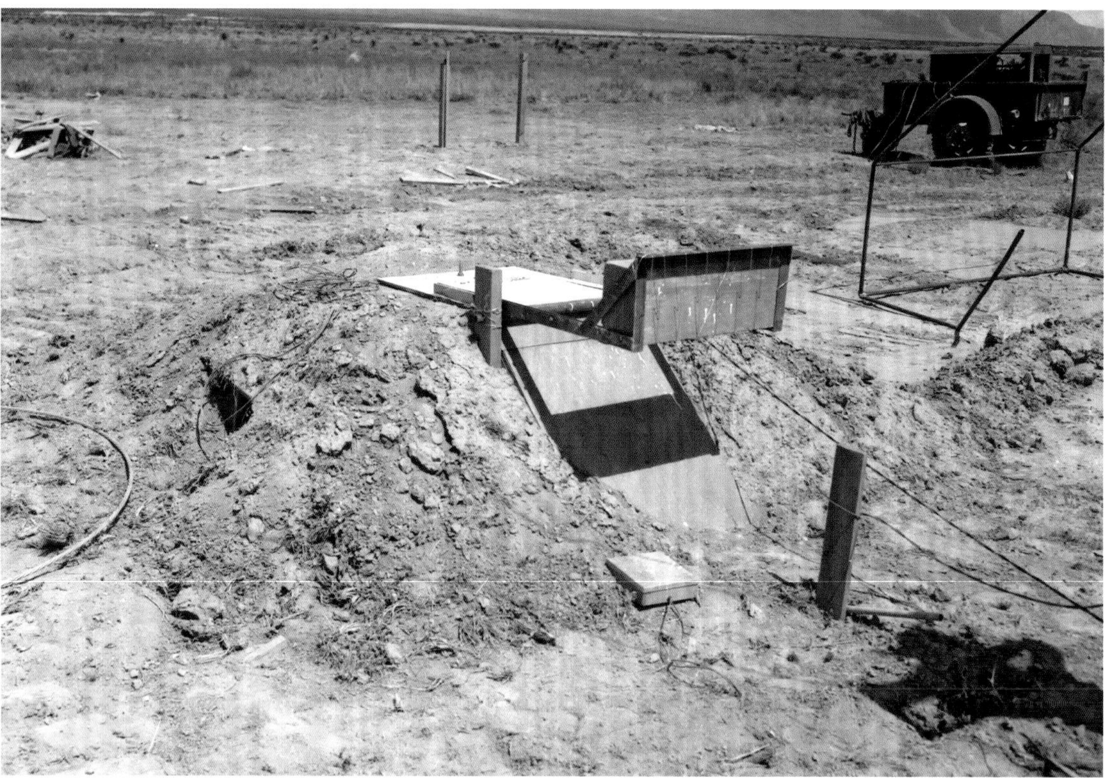

When the monitoring unit was lifted off its shelf, a set of switch buttons were released that activated the device. A sentinel having been exposed by the counterweight system is shown at right.

Moon reported after the test that the stations at 400 and 800 yards were damaged, along with the device at south 1,500, but the eleven other stations—including those at north, south, and west 10,000—remained operable.

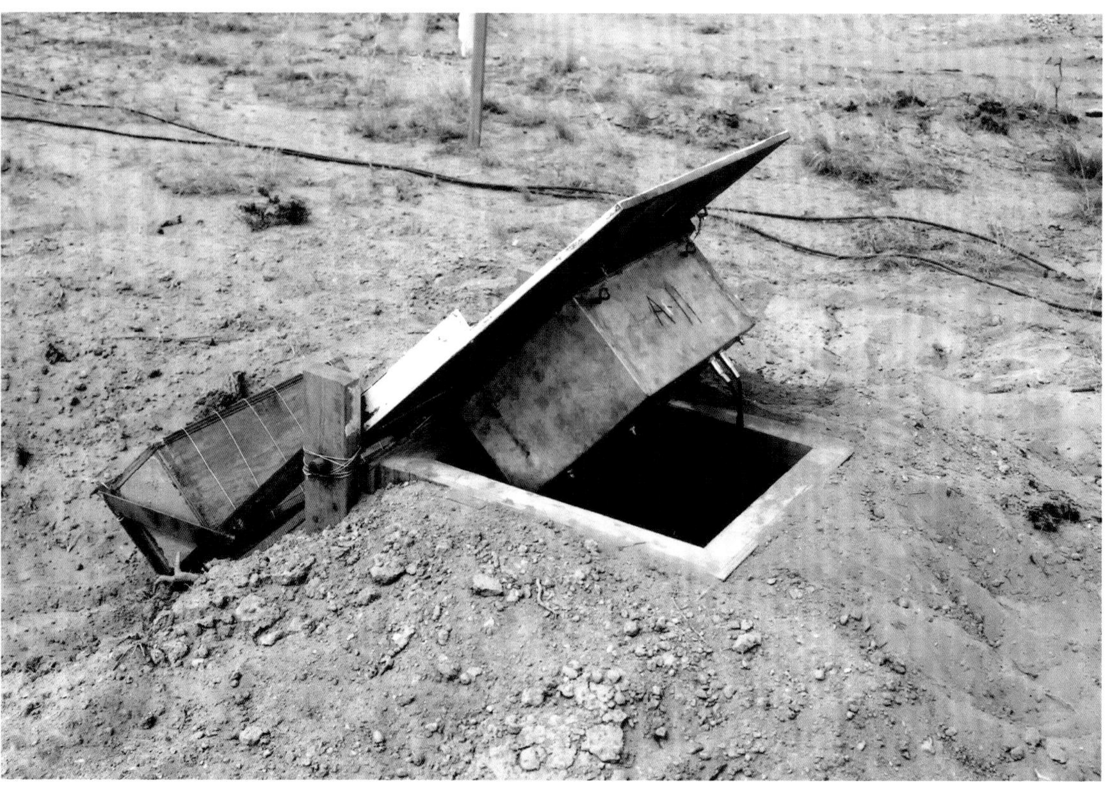

COUNTERWEIGHT
SWITCH BUTTONS
PIVOT
SAND
SAND
BOLT
CONCRETE
SPONGE RUBBER
TO TOWER

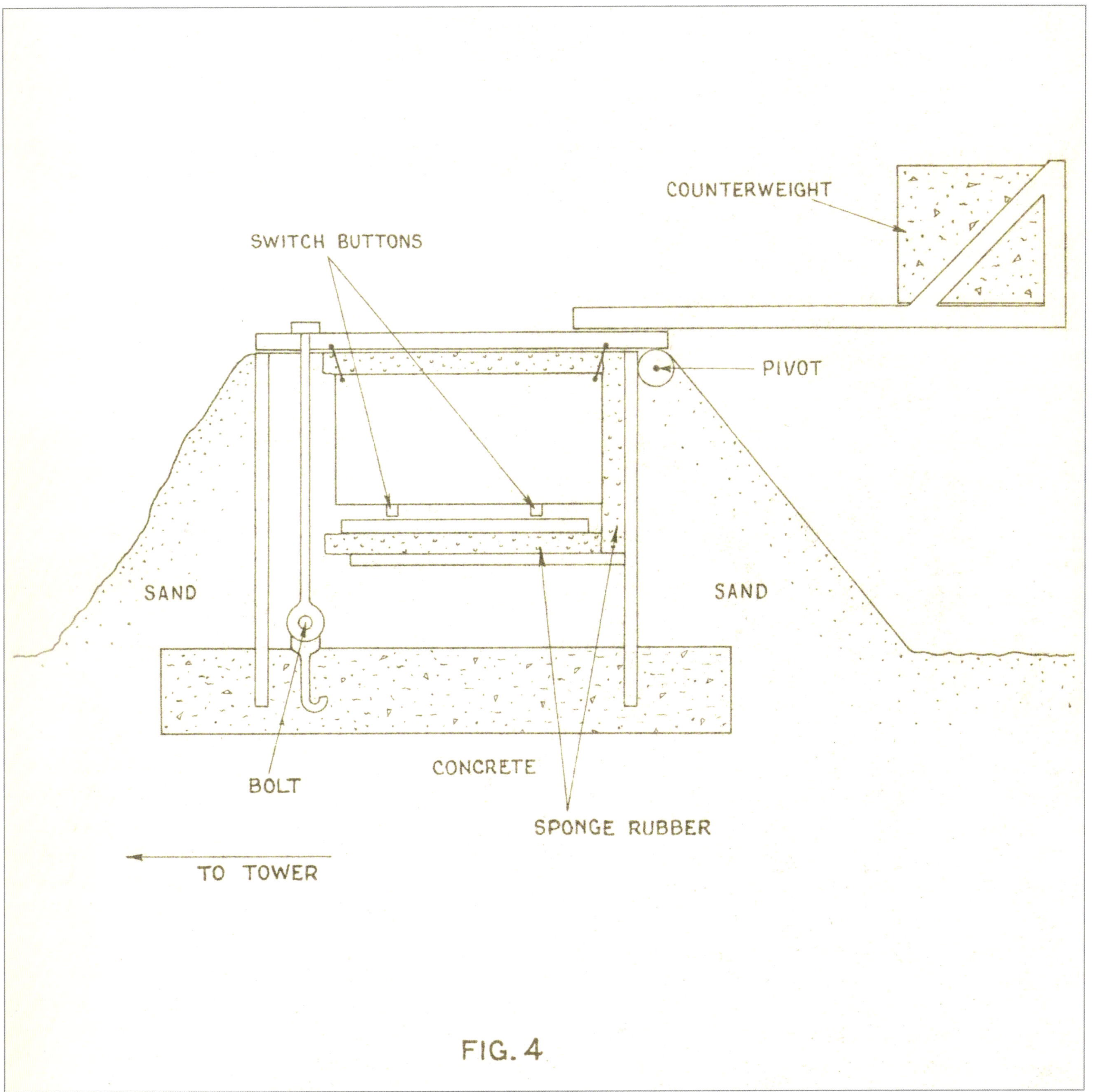

FIG. 4

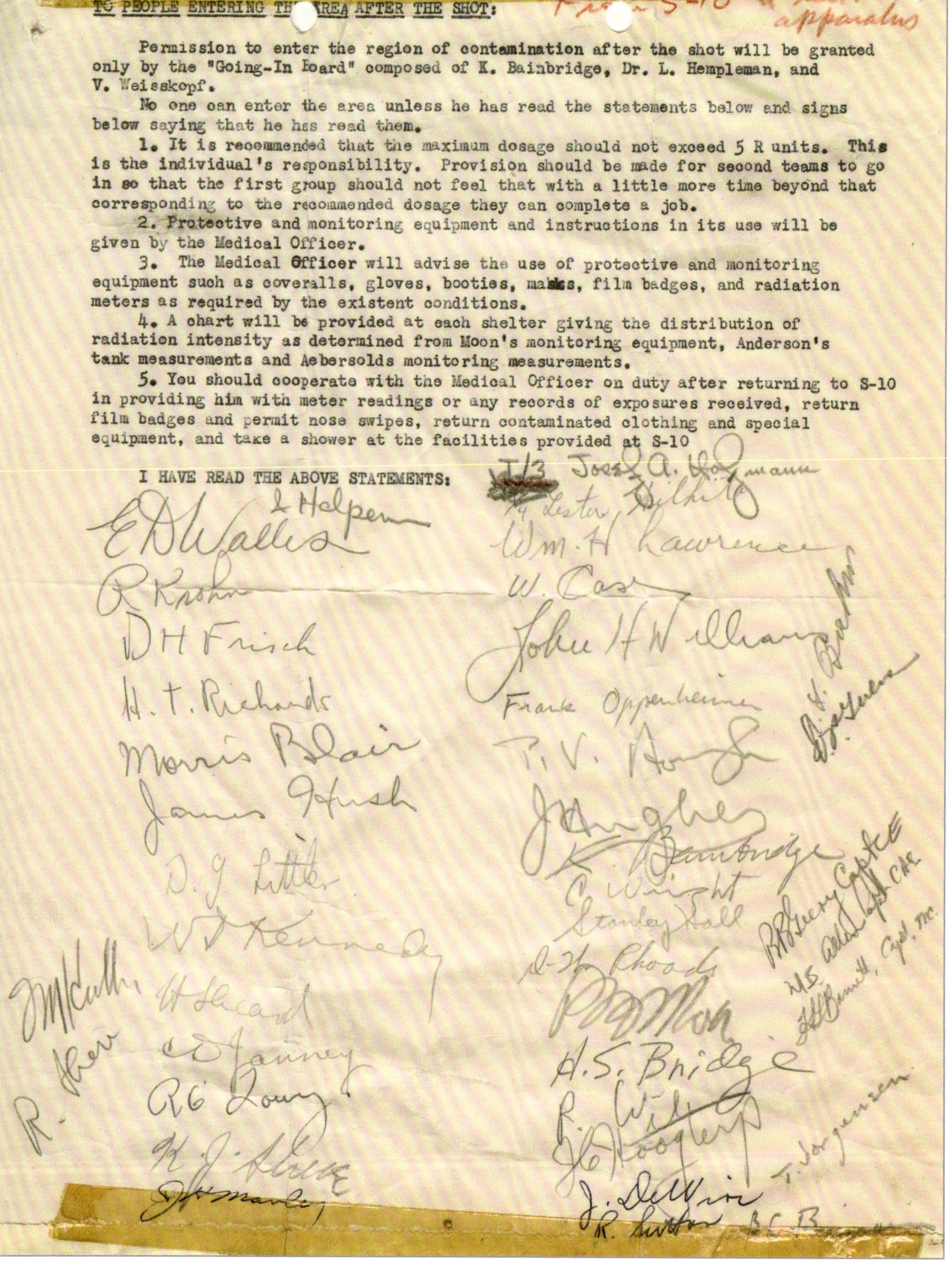

TO PEOPLE ENTERING THE AREA AFTER THE SHOT:

Permission to enter the region of contamination after the shot will be granted only by the "Going-In Board" composed of K. Bainbridge, Dr. L. Hempleman, and V. Weisskopf.

No one can enter the area unless he has read the statements below and signs below saying that he has read them.

1. It is recommended that the maximum dosage should not exceed 5 R units. This is the individual's responsibility. Provision should be made for second teams to go in so that the first group should not feel that with a little more time beyond that corresponding to the recommended dosage they can complete a job.

2. Protective and monitoring equipment and instructions in its use will be given by the Medical Officer.

3. The Medical Officer will advise the use of protective and monitoring equipment such as coveralls, gloves, booties, maks, film badges, and radiation meters as required by the existent conditions.

4. A chart will be provided at each shelter giving the distribution of radiation intensity as determined from Moon's monitoring equipment, Anderson's tank measurements and Aebersolds monitoring measurements.

5. You should cooperate with the Medical Officer on duty after returning to S-10 in providing him with meter readings or any records of exposures received, return film badges and permit nose swipes, return contaminated clothing and special equipment, and take a shower at the facilities provided at S-10

I HAVE READ THE ABOVE STATEMENTS:

A "Going-In Board" was established—consisting of Kenneth Bainbridge, Health Group leader Louis Hempelmann, and theoretician Victor Weisskopf—to determine whether a party had a valid reason for visiting ground zero in the very first days after the test. In addition to receiving approval, each person who went into the contaminated area had to sign an acknowledgment of responsibility (*left*).

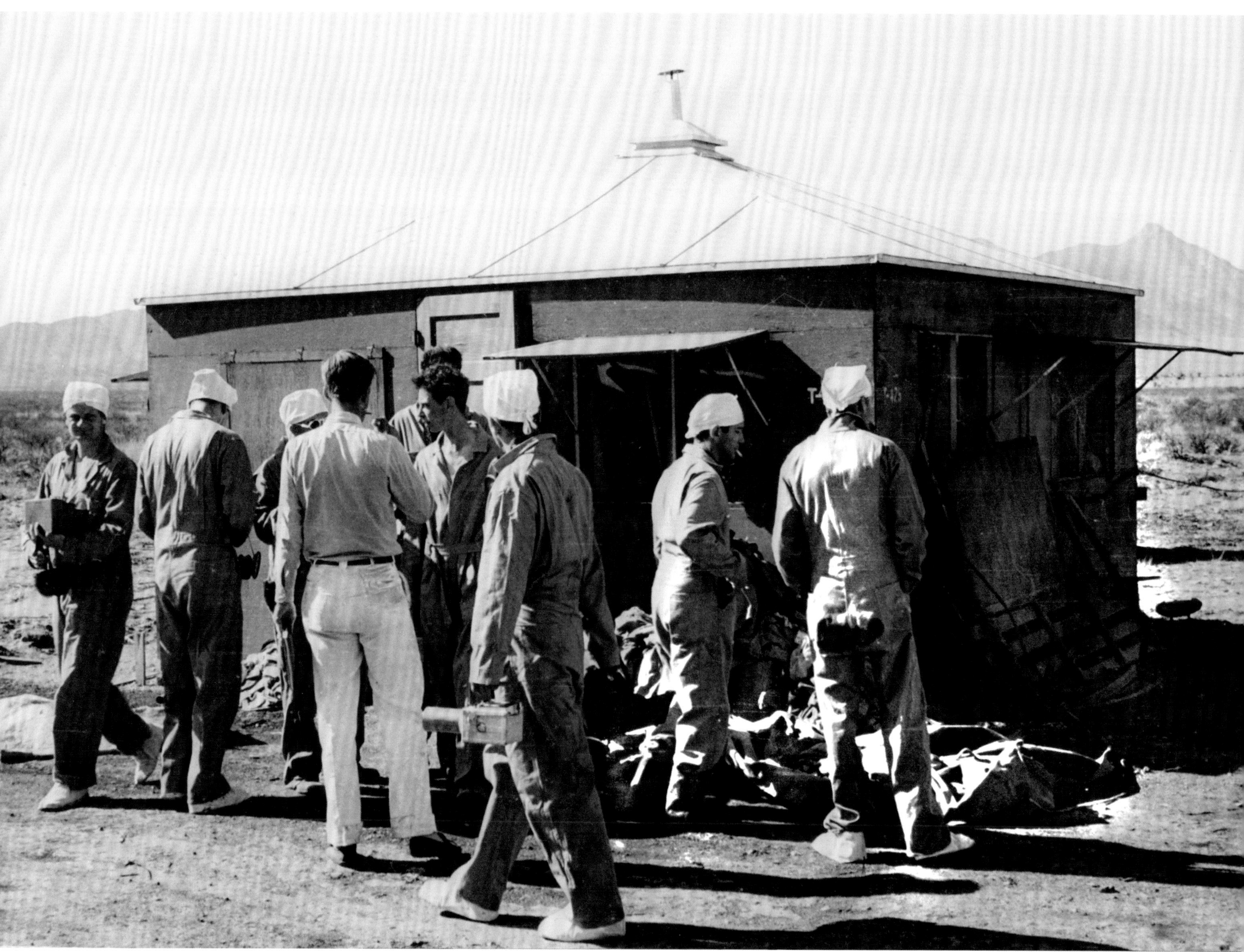

Members of going-in parties were instructed to wear protective clothing—body suits, gloves, head and feet coverings, and respirators—as well as film badges that would darken proportionately to the amount of radiation received and provide a record of their total exposure. The maximum allowable dose for a single instance of exposure, measured in a unit called roentgens, was set at five roentgens per hour.

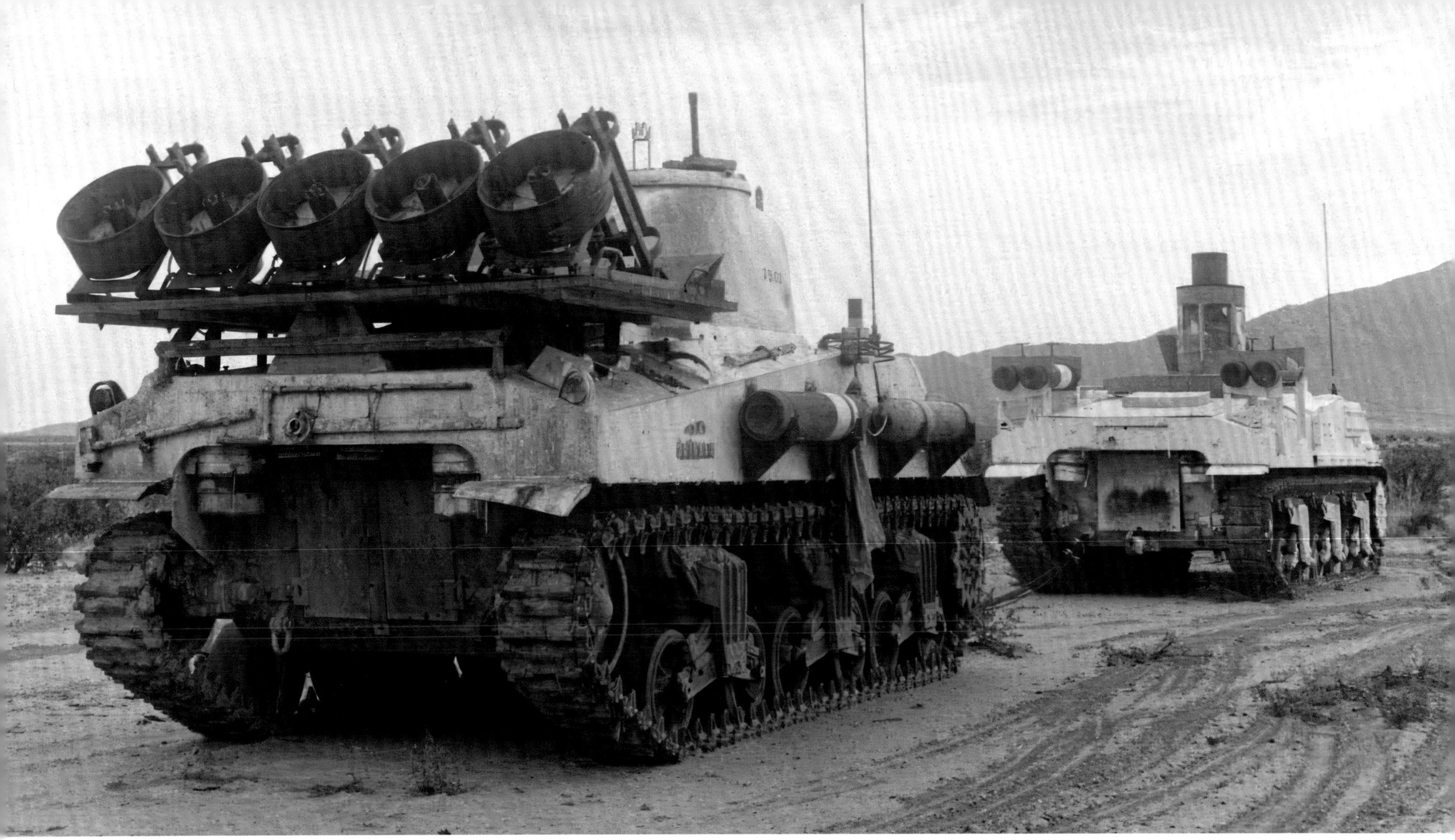

With the most time-sensitive mission, the drivers of the Sherman tanks were the first to enter the contaminated area. It was crucial to pick up soil samples as quickly as possible because the radioactive particles that scientists hoped to observe would decay rapidly into other elements.

Four hours after the shot, the lead-shielded tank (pictured above, to the rear) made its first run, lurching toward ground zero. About 100 yards from the center of the crater, the tank had to turn back due to high radiation levels. Herbert Anderson, who led the radiochemical sampling effort, described finding that "the ground was covered with a fused layer of green stuff" (*right*).

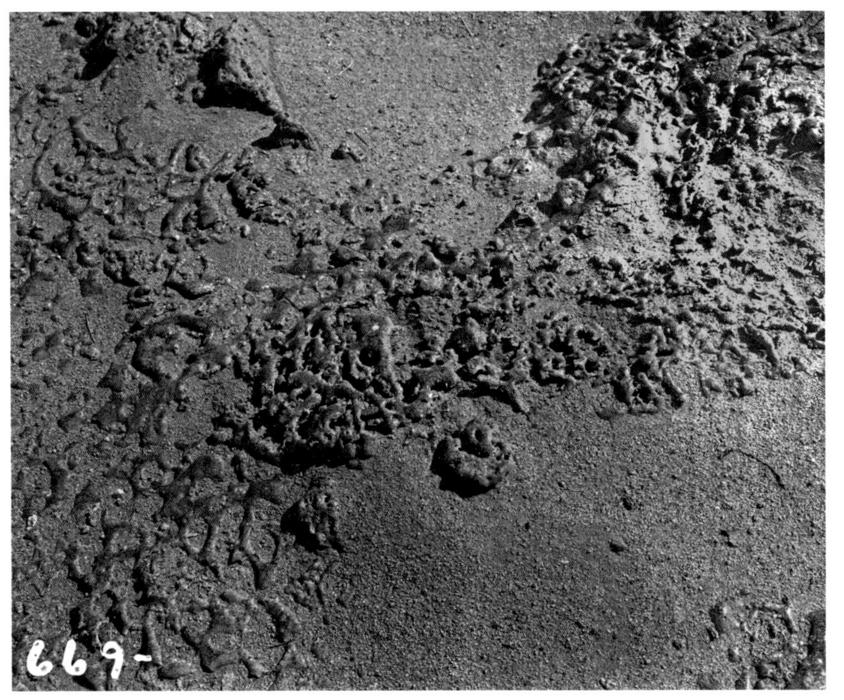

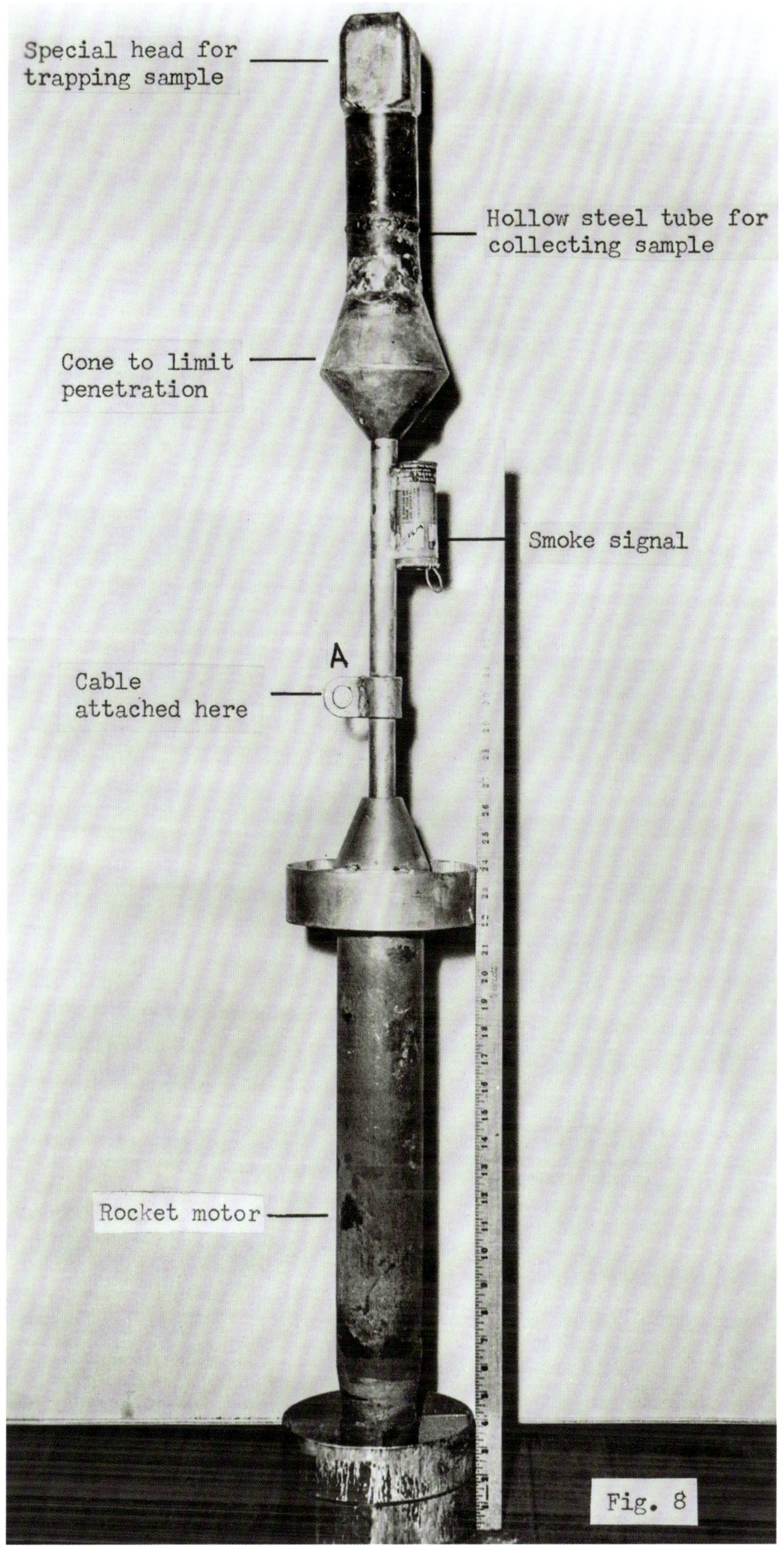

The crew went in again just over twelve hours after detonation, this time making it to the center of the crater. Radiation levels inside the tank, even with its 11,000 pounds of lead shielding, were prohibitively high: it was estimated that passengers received a full dose of five roentgens on the twelve-minute collection run.

The second tank, equipped with sampling rockets, stayed 500 yards outside the crater, where radiation levels were much lower. The tank broke down but was able to launch and reel in collection tubes, a number of which provided "excellent" samples according to the Radiochemistry Group.

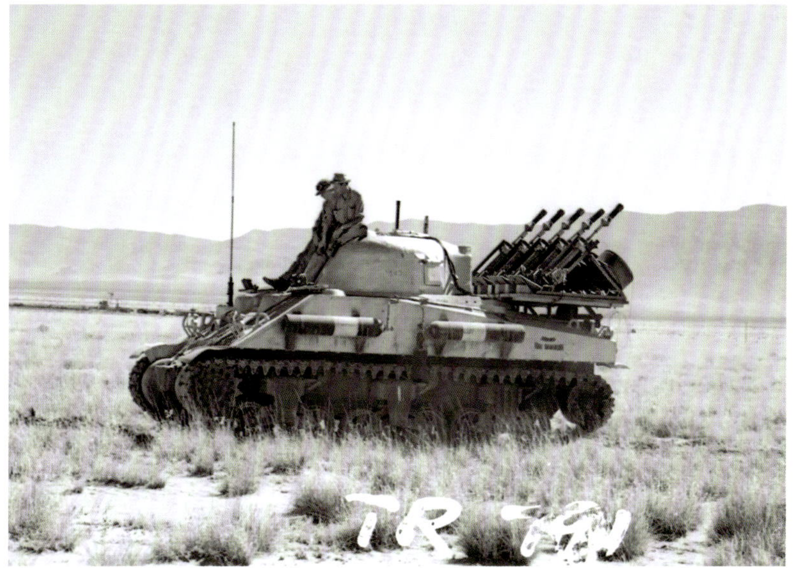

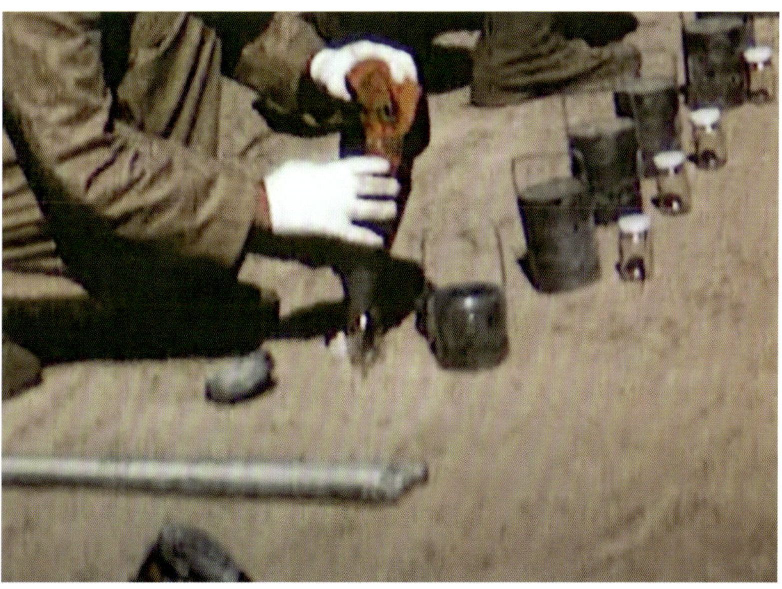

The "green stuff" observed by Anderson is now well known as trinitite, a radioactive glass formed from sand taken up into the superheated fireball. The sand crystals rained down in a molten slurry, covering the area surrounding the crater in puddles that resolidified into a thin sheet, with scattered droplets extending for several hundred yards. Groves reported that the green glass could be seen from as far away as five miles.

The scientist at right, standing in a patch of trinitite, is Boyce McDaniel, who monitored fission rates in the Gadget's plutonium core while the device sat atop the shot tower.

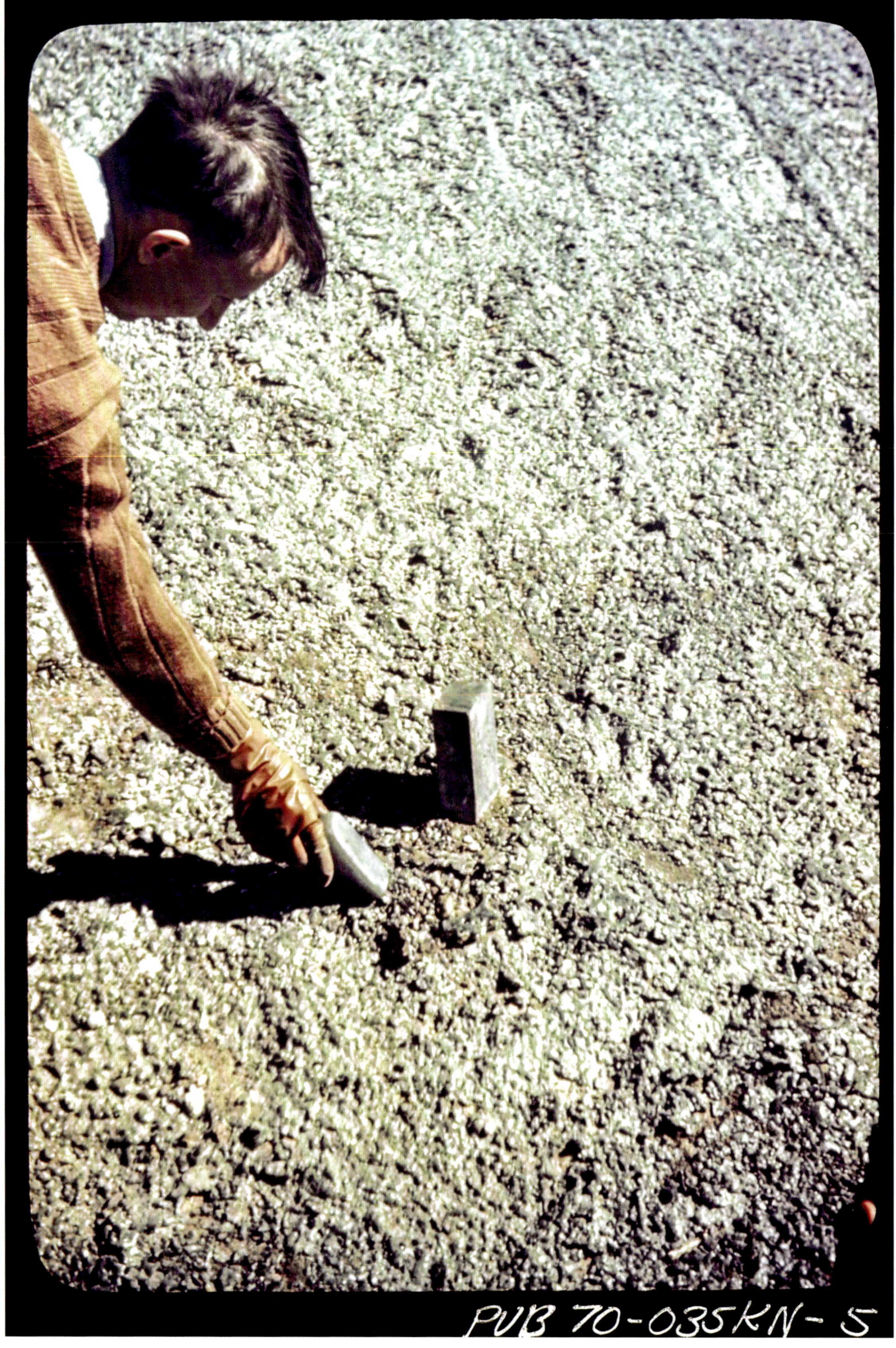

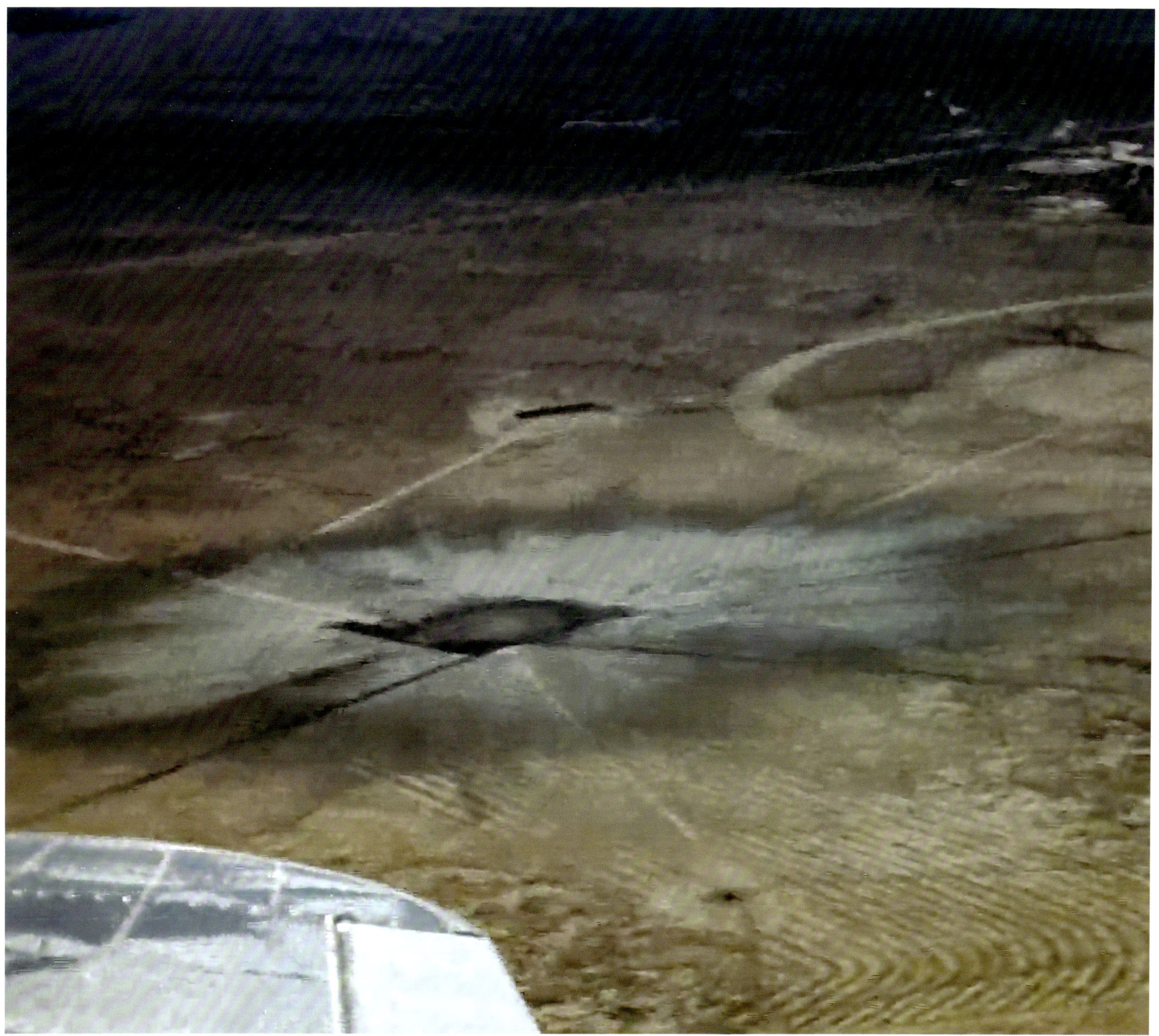

Aerial footage taken shortly after the detonation shows the distinctive starburst pattern in which the trinitite fell. Although most is green, some is grayish-black, likely because the sand took on tiny pieces of the shattered metal tower. Other trinitite is red, thought to be due to the presence of copper from the coaxial cables that ran down from the tower to the north.

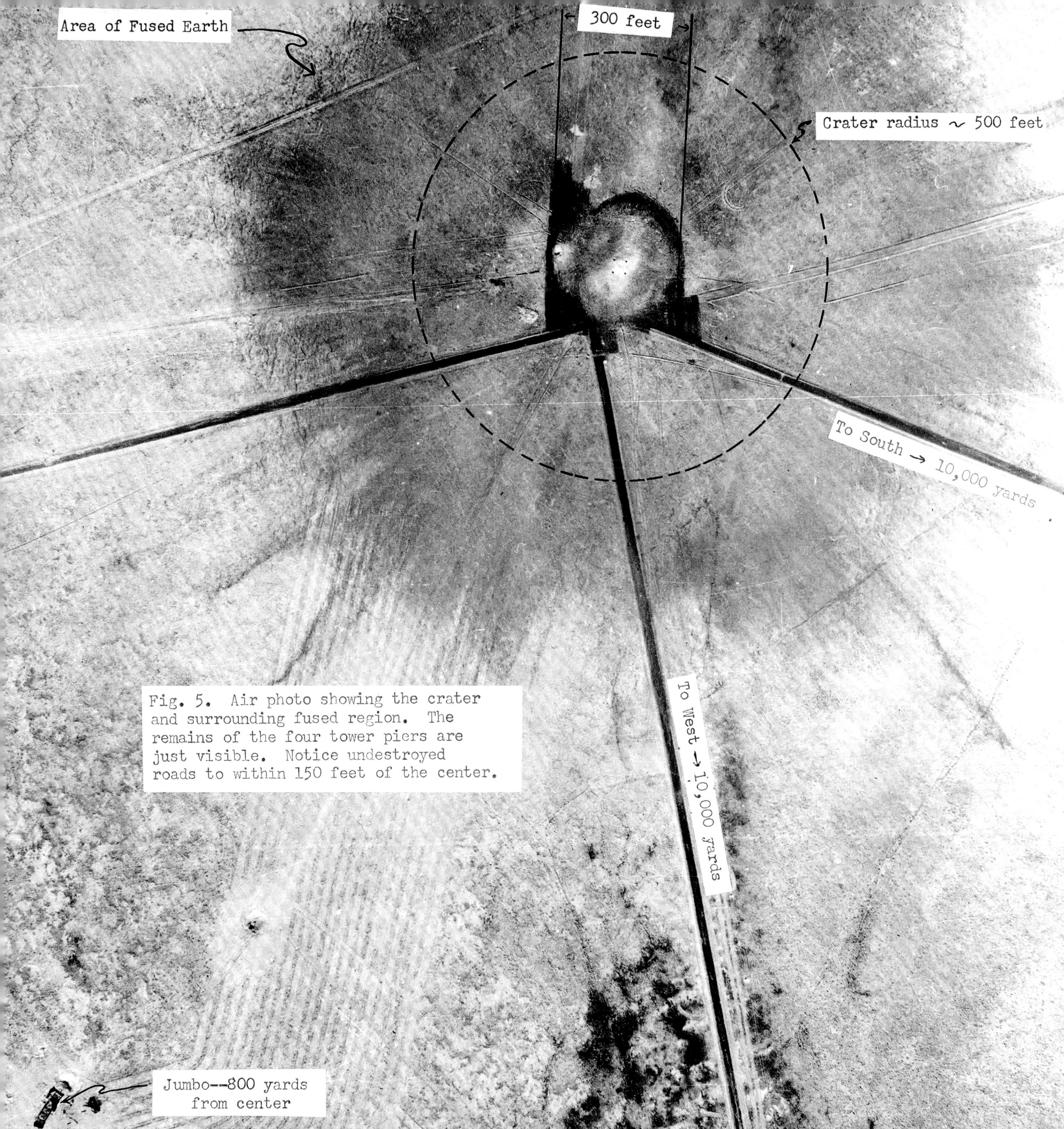

Fig. 5. Air photo showing the crater and surrounding fused region. The remains of the four tower piers are just visible. Notice undestroyed roads to within 150 feet of the center.

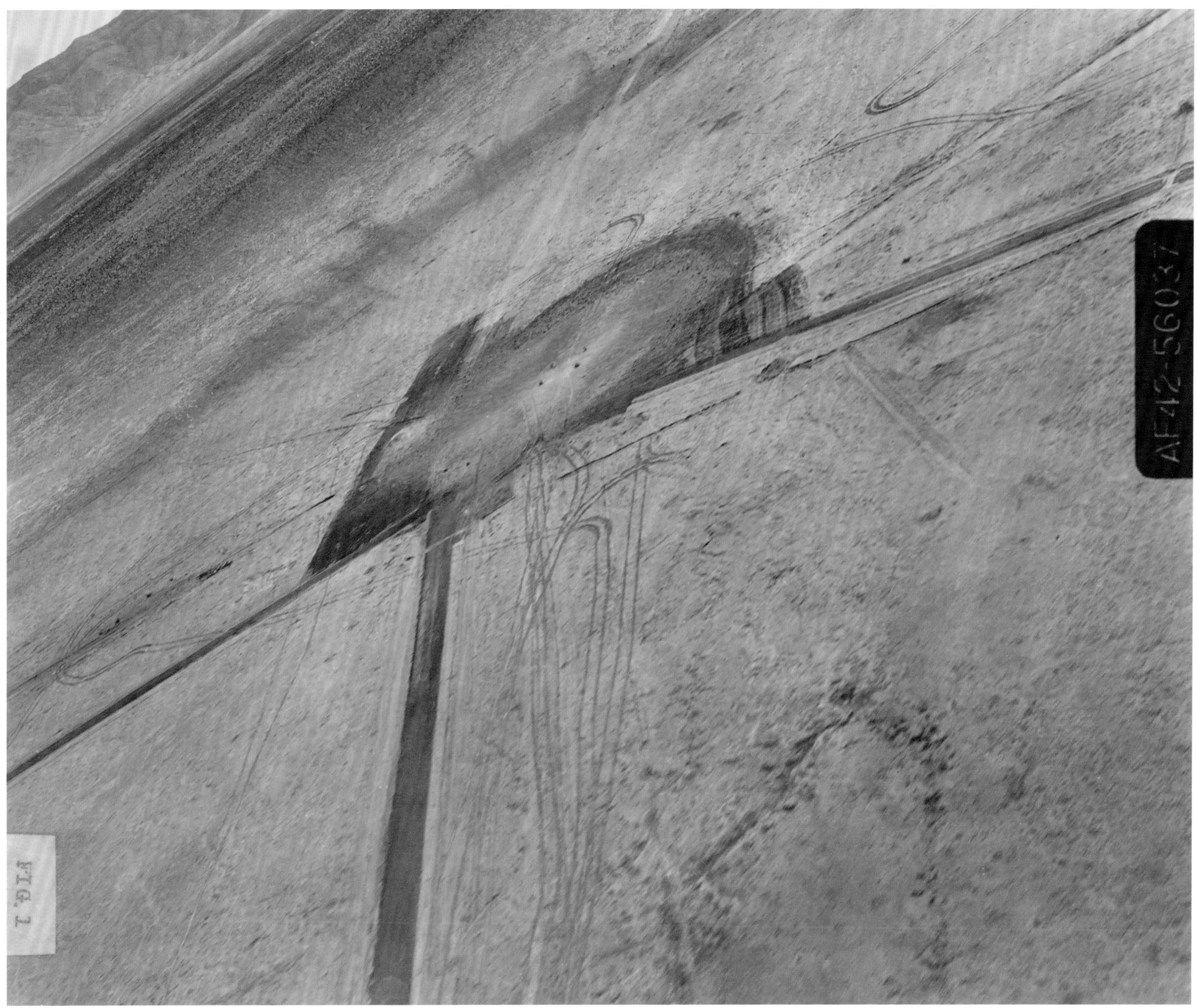

The Gadget left behind a crater around 1,100 feet across and 9.5 feet at its deepest. According to Groves, "all vegetation had vanished." The 100-foot steel tower had also vanished. Its upper portions were vaporized, and most of the material below the shot cab disintegrated into tiny pieces. The structure's remains—mangled pieces of the concrete footings—appear as small black dots in the aerial photograph at left, taken twenty-eight hours post-detonation. The same four dots are also shown above, centered in a sloping pit created by the force of the blast. Note the tank tracks going into and out of the crater.

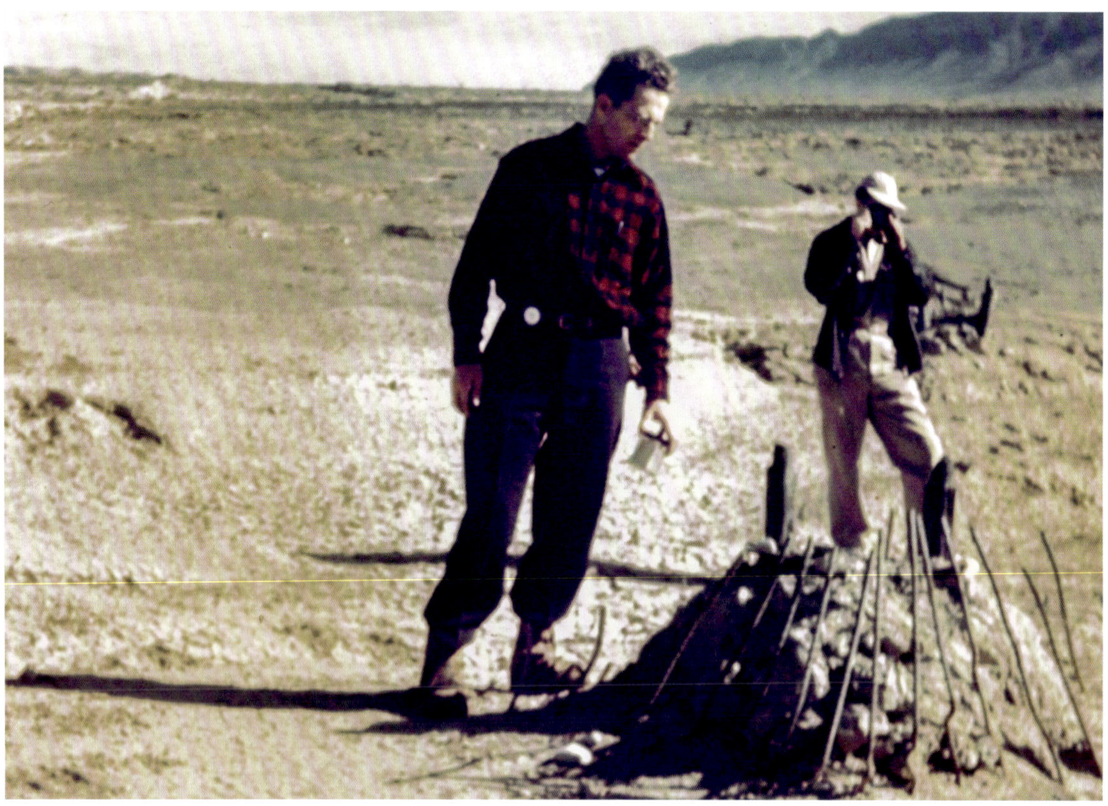

The tower footings were pushed several feet into the ground, which had itself been depressed and compacted by the blast wave. At left, physicist Raemer Schreiber and a photographer stand near one of the footings while visiting ground zero about a month after the test.

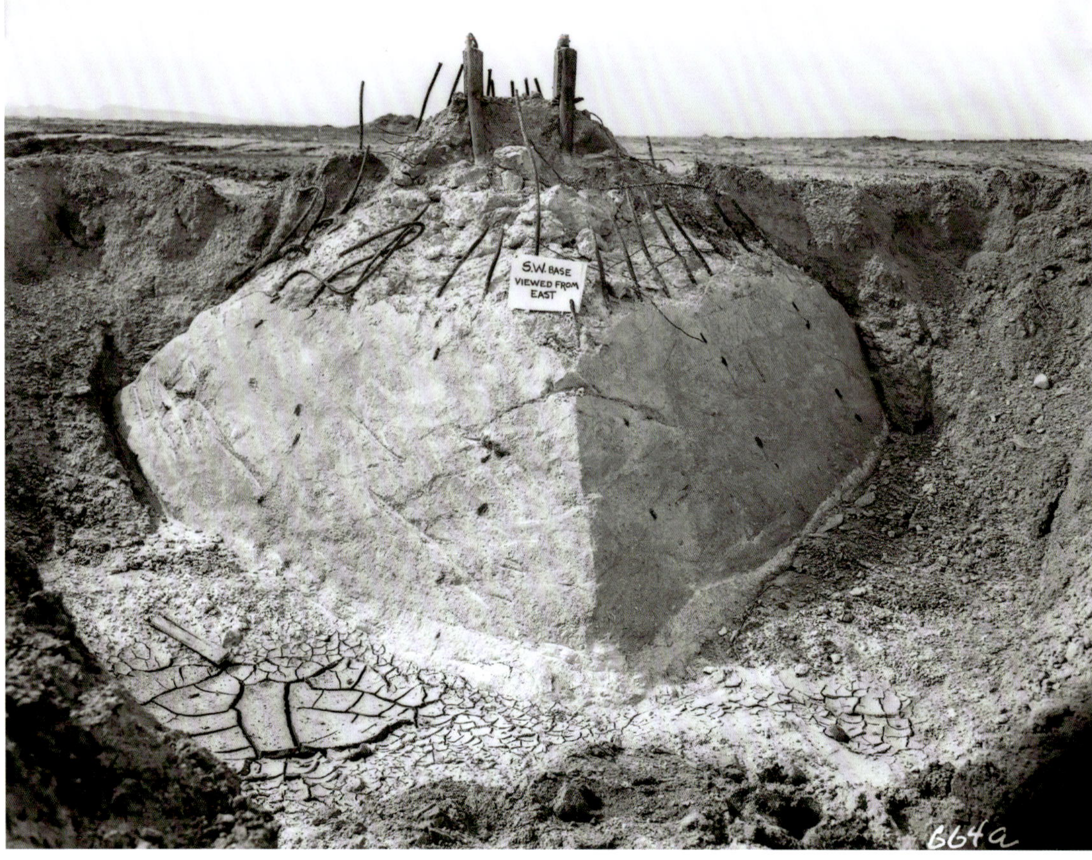

The area around the northwest and southwest footings was excavated in October of 1945, when radiation levels had dropped significantly. The dig revealed that beneath the pulverized concrete and twisted rebar exposed above ground, the buried sections of footing were cracked but intact.

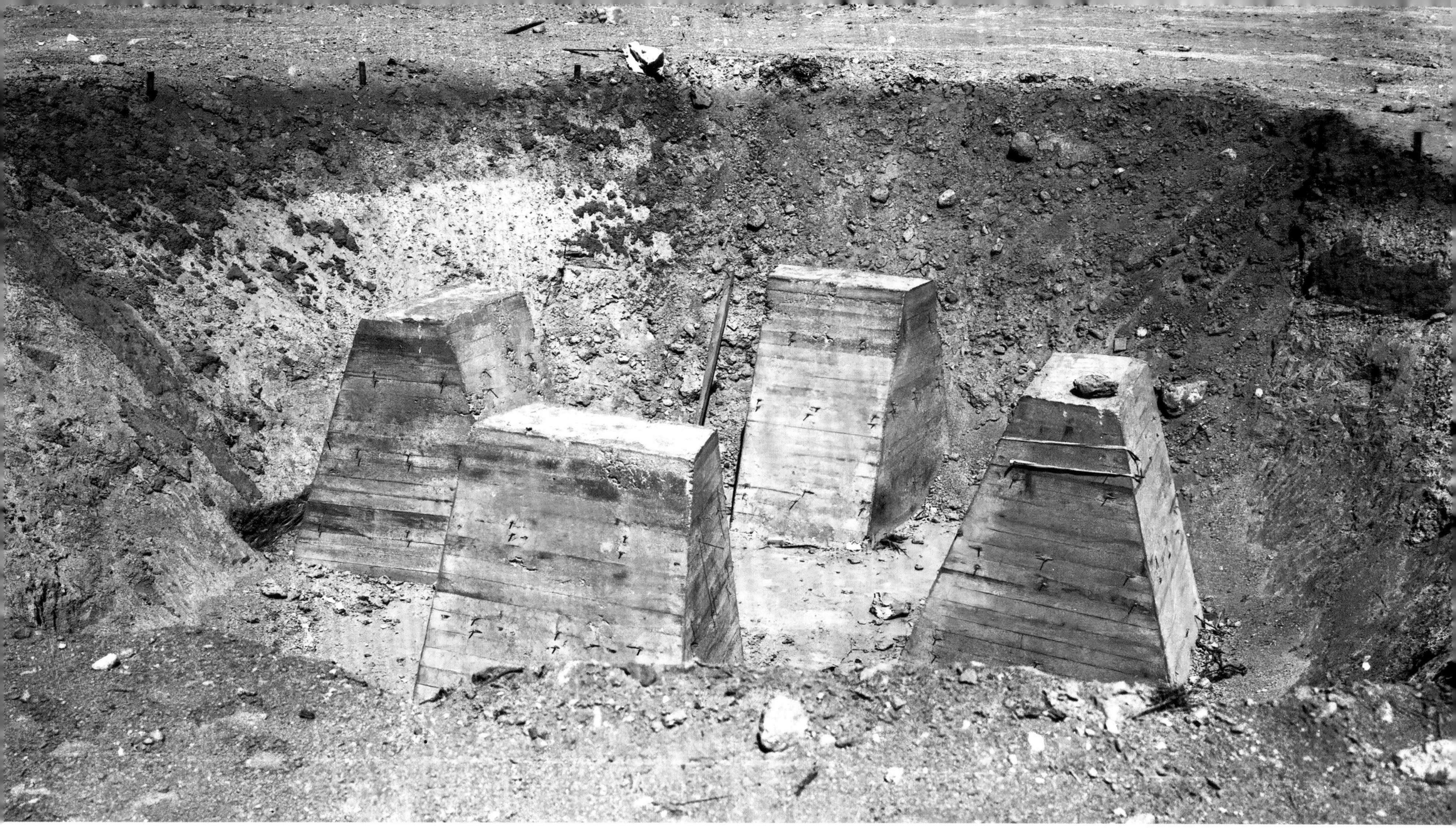

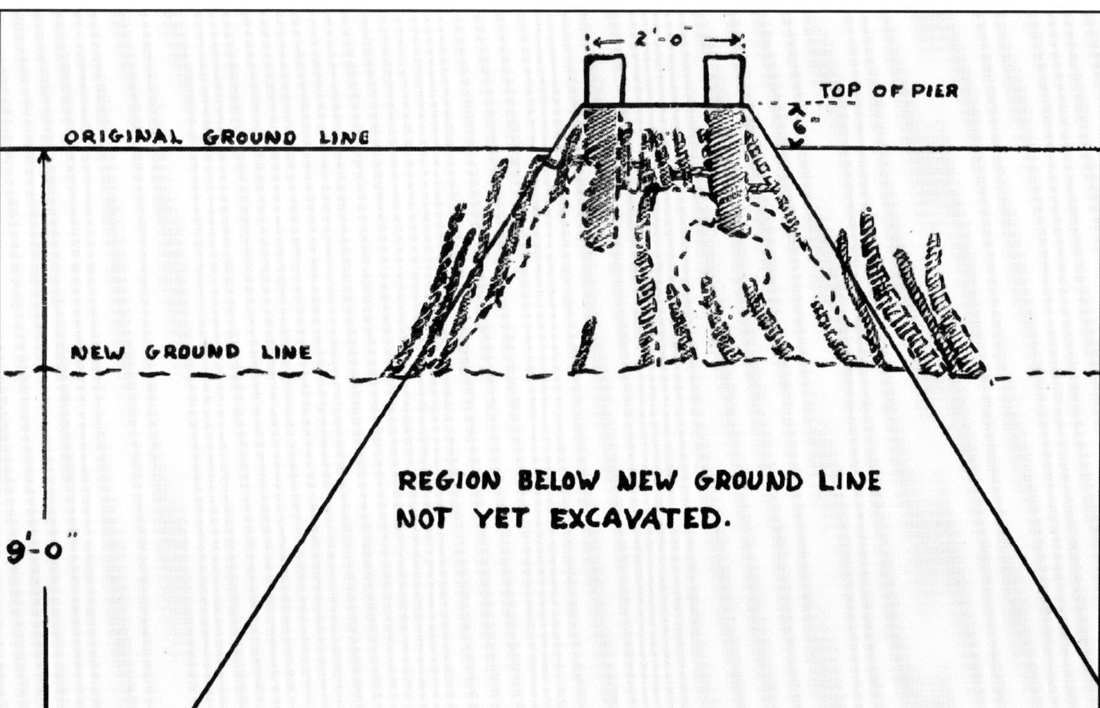

The tower footings as originally poured are shown above. The schematic illustrates the new ground line that was created by the blast.

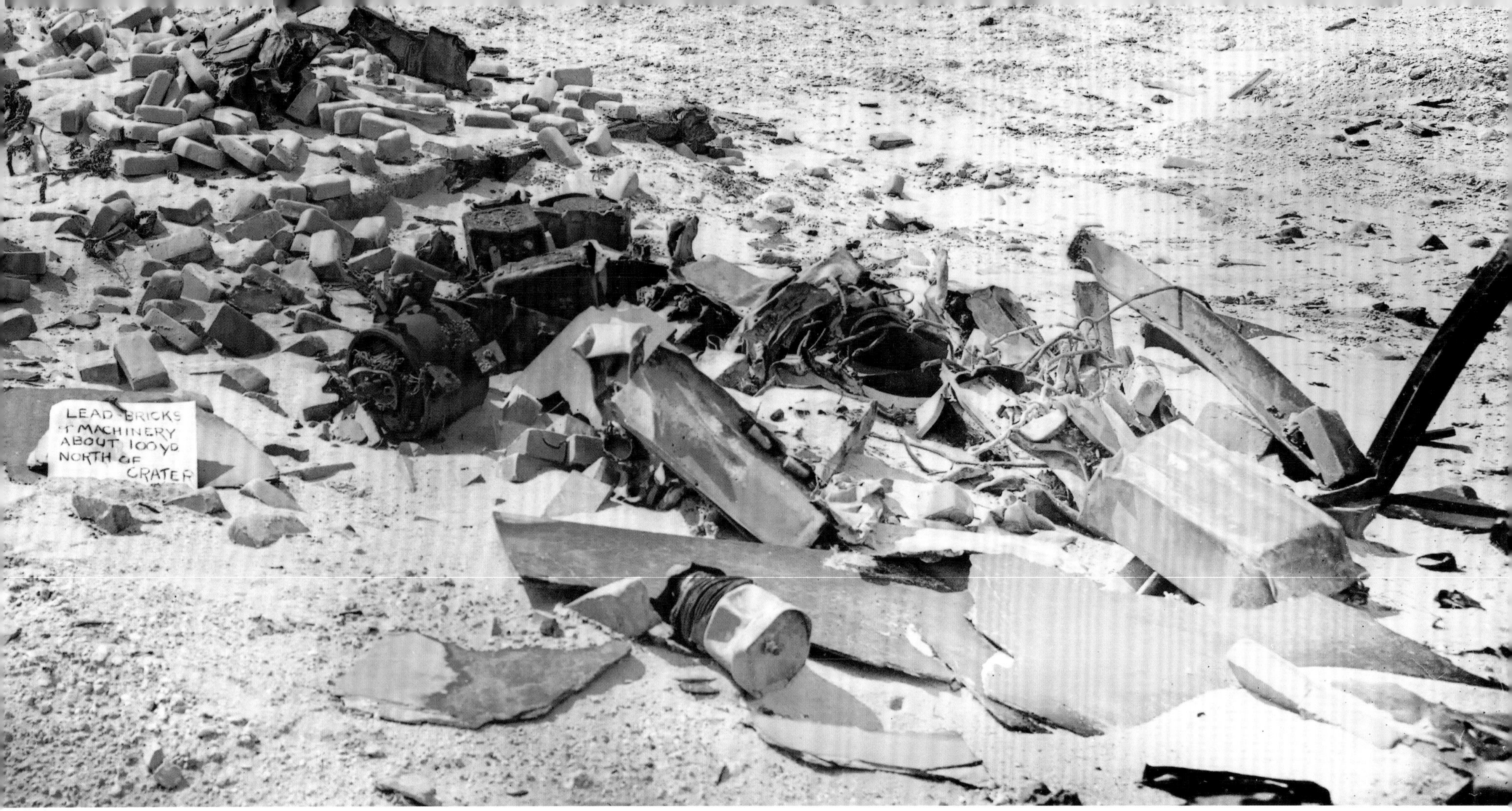

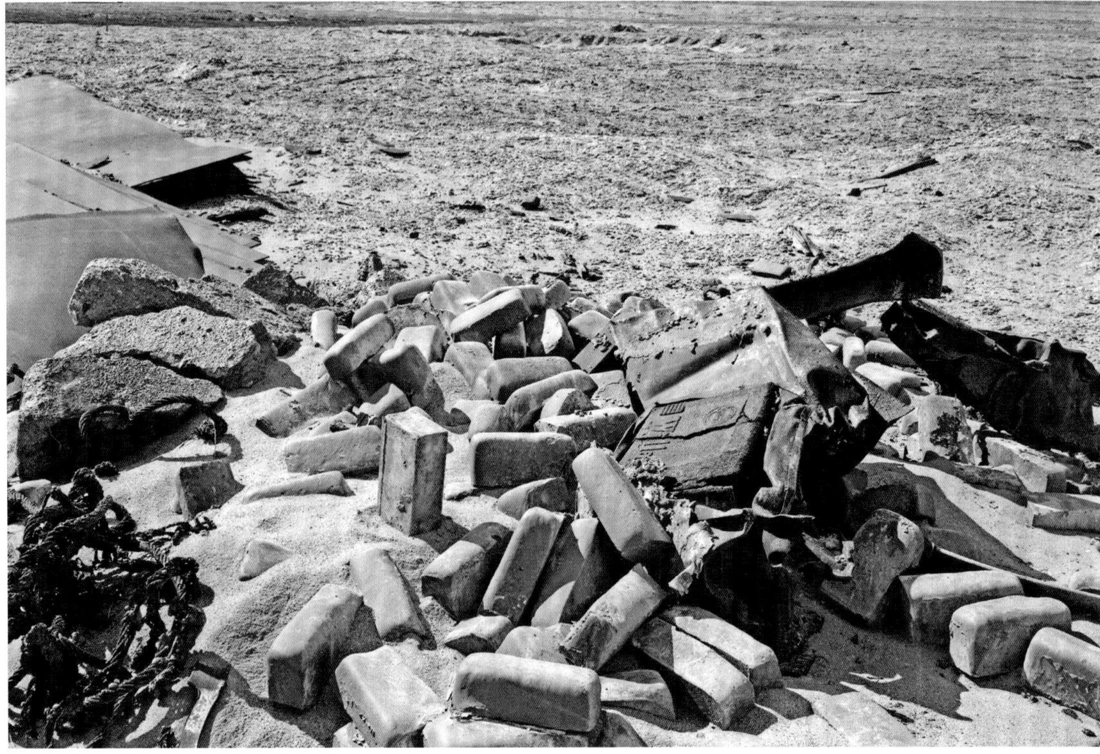

As the Gadget's shock wave and heat radiated downward and outward, they destroyed just about everything within 200 yards. At north 100, the lead bricks around a small shelter were melted. An undamaged brick has been stood up in the center of the pile for comparison, at left.

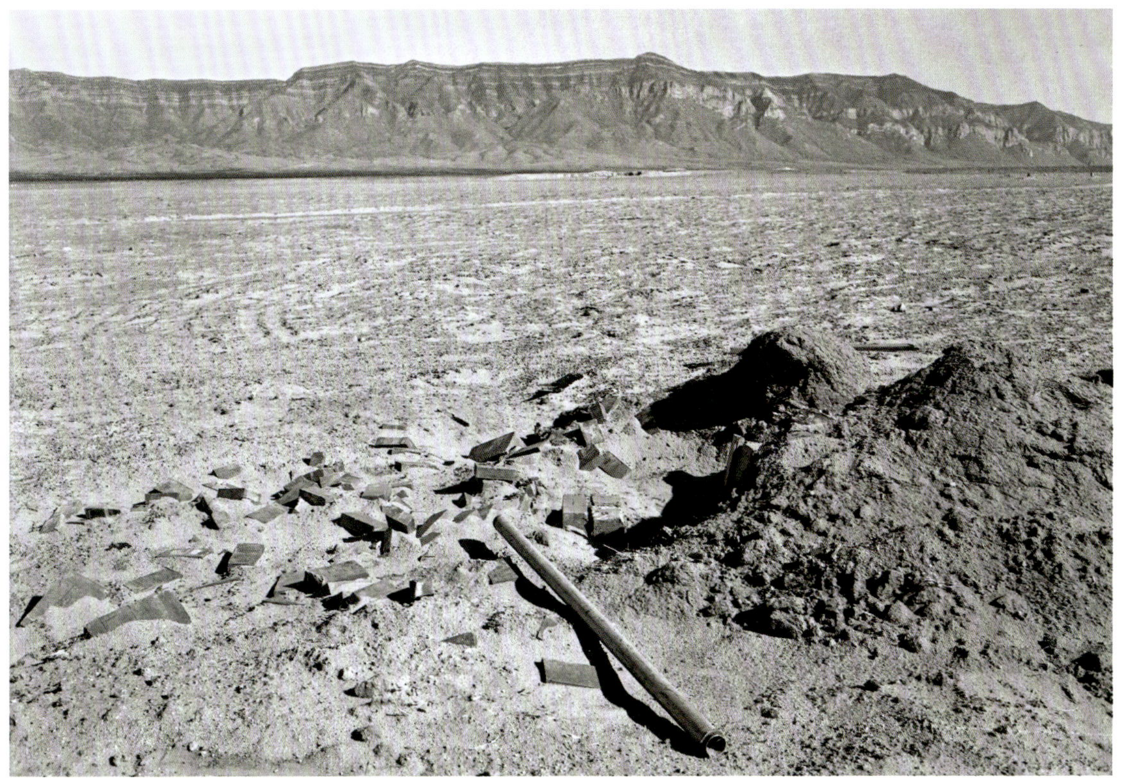

The rubble shown here was a chimney set up 150 yards south of the blast. Below is the damaged roof of the Schmidt-McDonald Ranch House barn, just over two miles from ground zero. (Photo courtesy of the Bretscher Family.)

Groves wrote to US Army Chief of Staff George C. Marshall that based on the destructive force seen and measured at Trinity, the same type of bomb, detonated 1,800 feet in the air, could be expected to destroy practically all structures within an area of one to two miles around the center of the explosion.

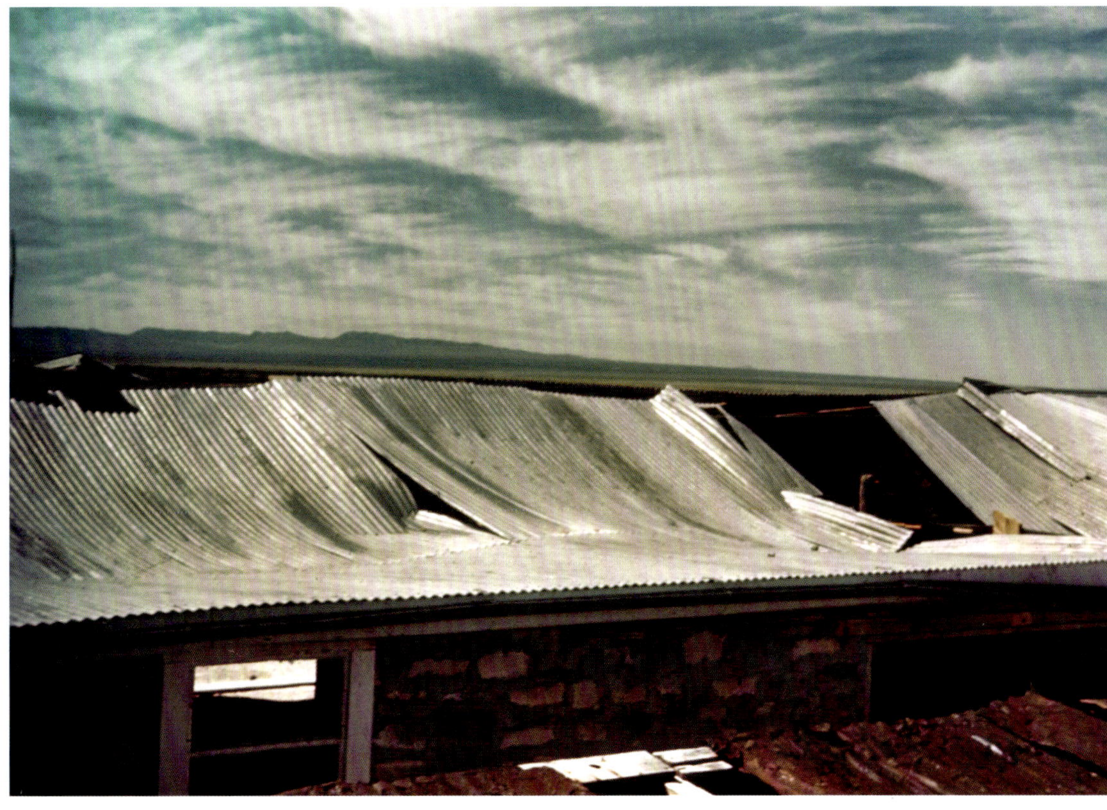

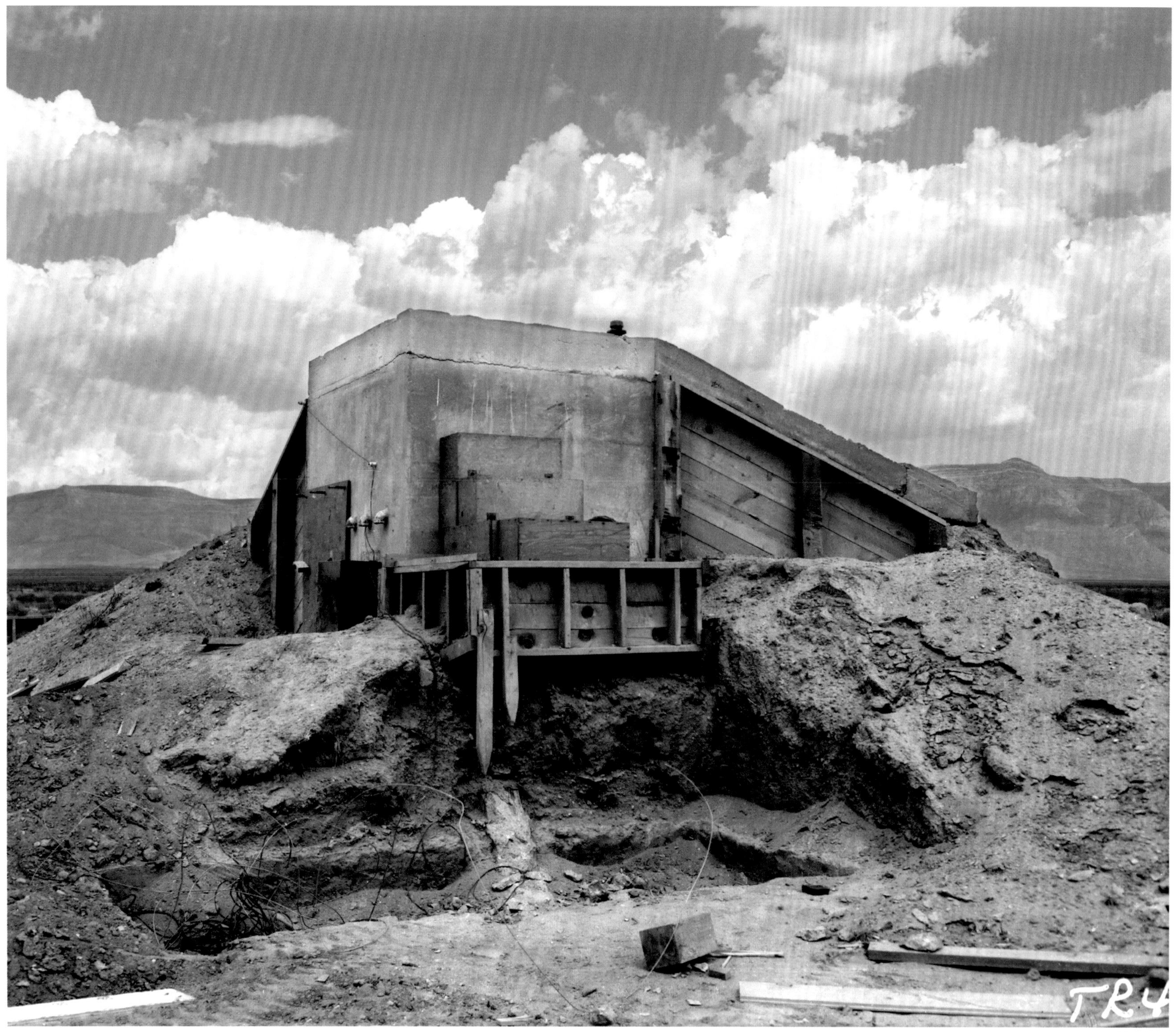

Even the more heavily fortified structures were visibly damaged. The photography bunker at north 800, just under half a mile from ground zero, is shown above. Its twin at west 800 (*right*) sustained a major crack in one of the walls.

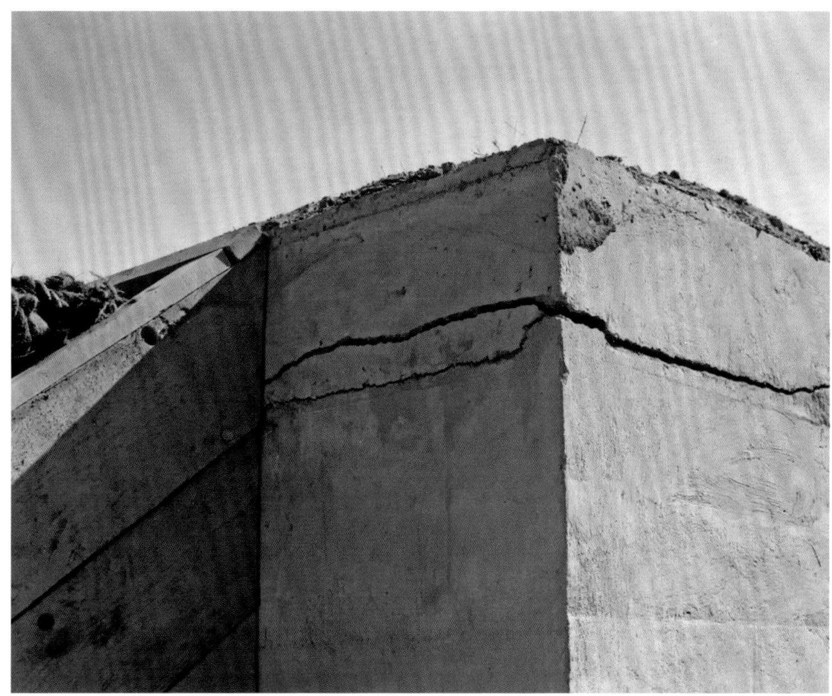

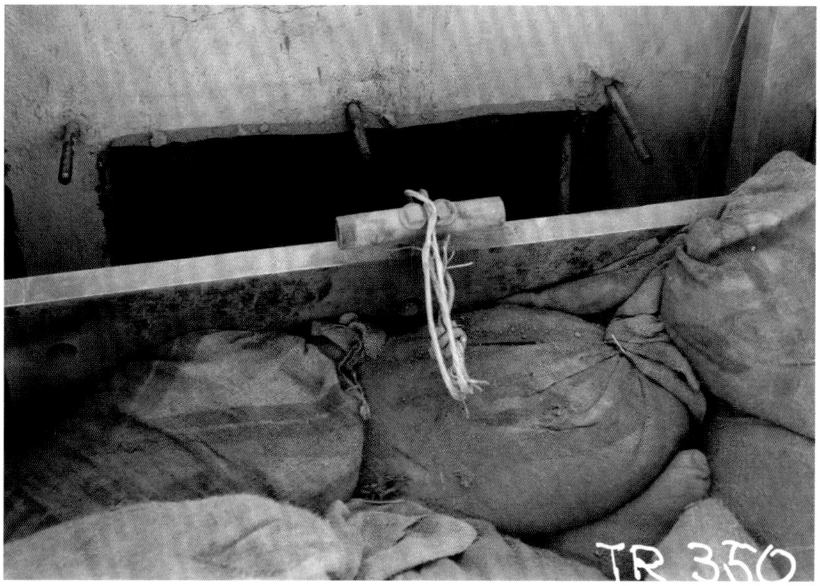

The crew below is moving sandbags away from a small door to access the interior.

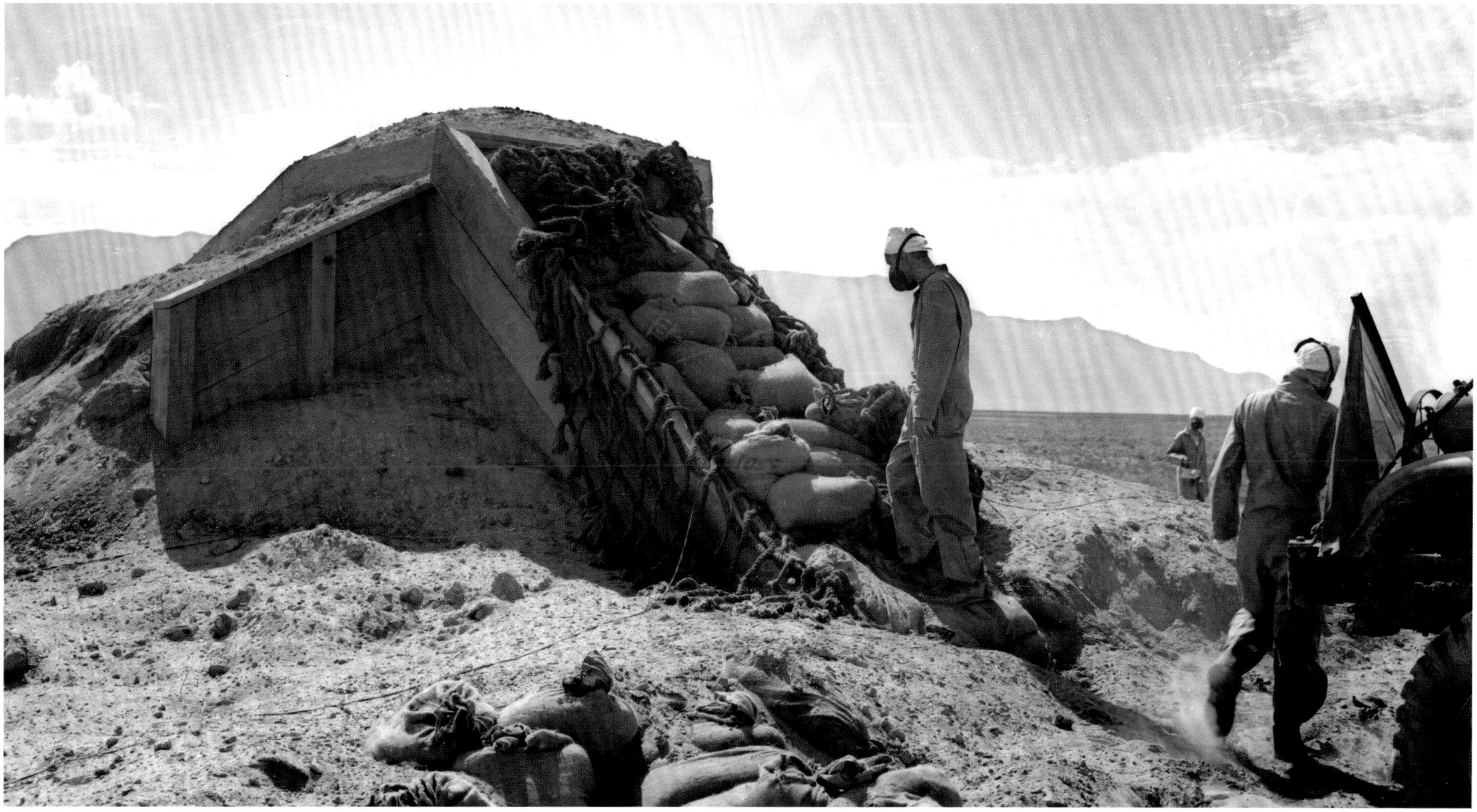

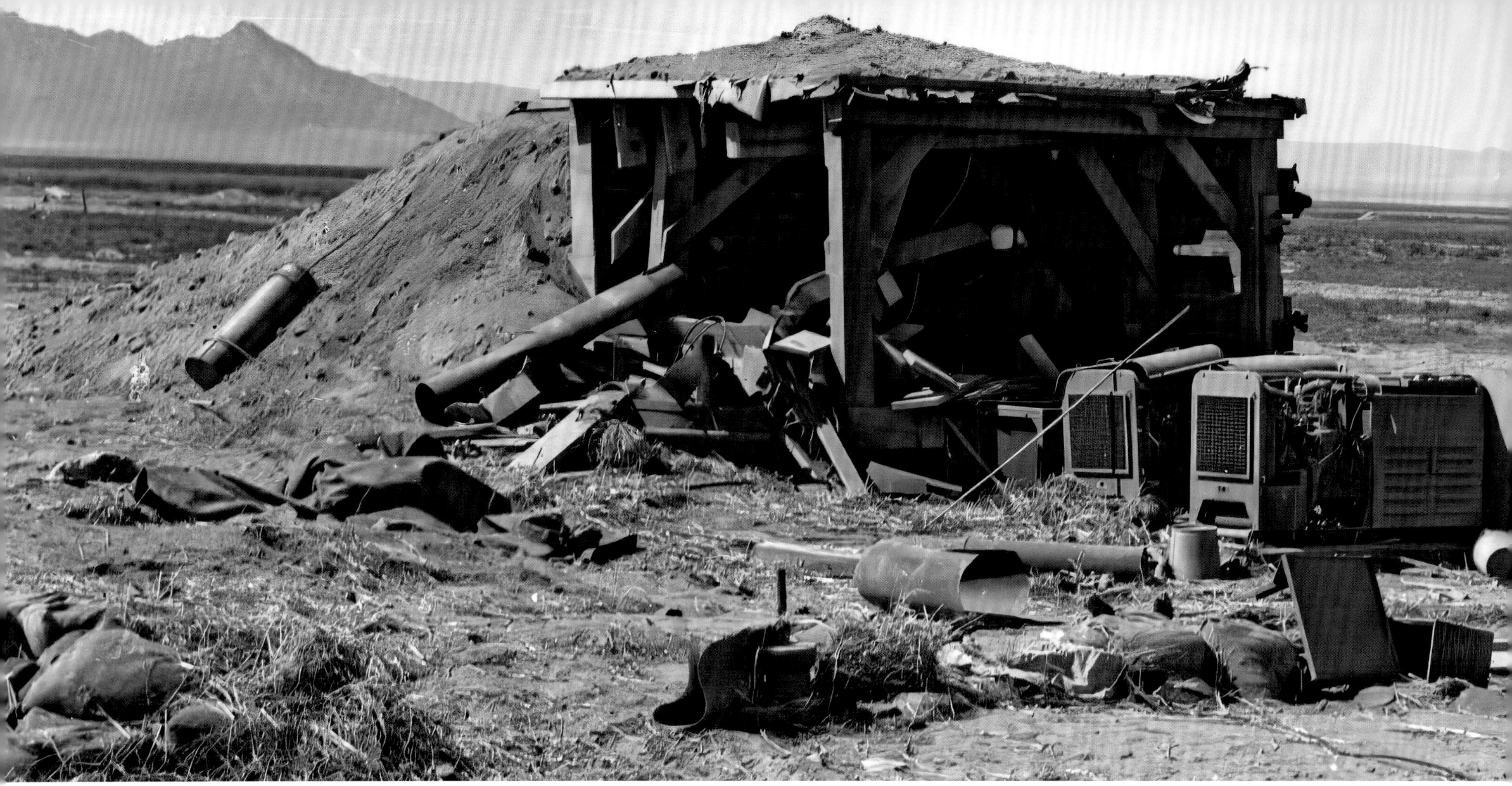

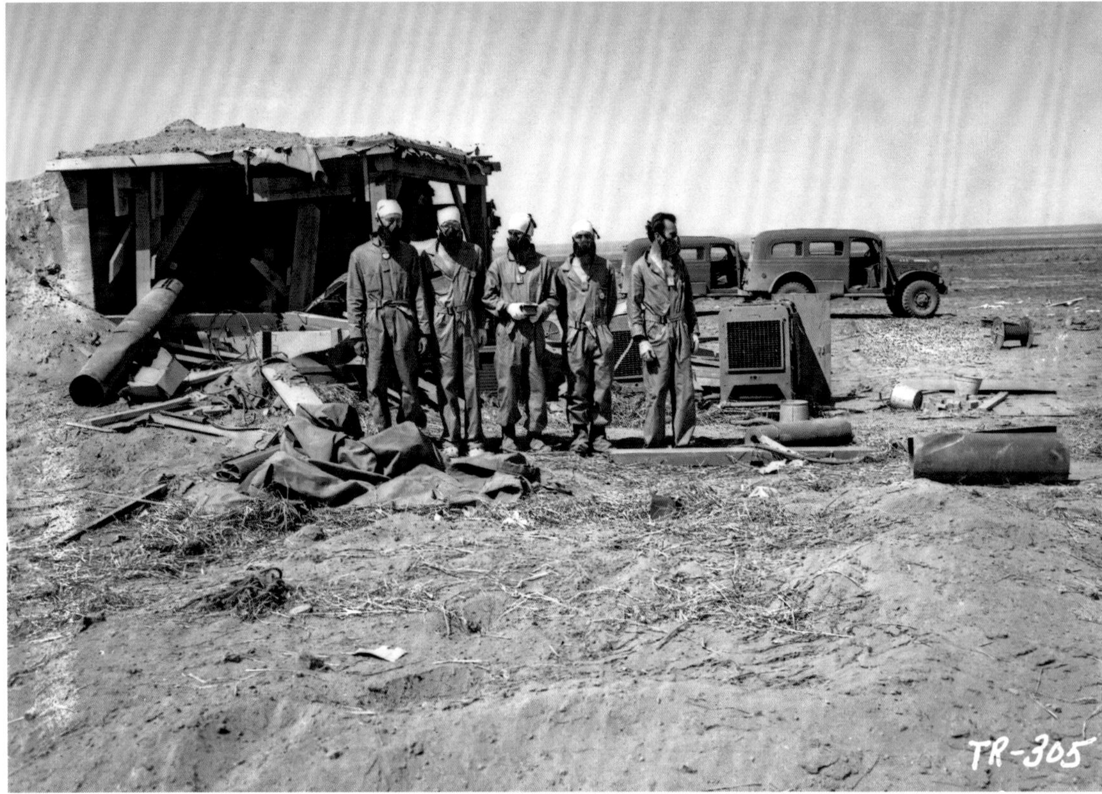

At north 1,000, where loose materials were exposed to the blast, going-in parties found the wooden bunker heavily damaged and a chaotic scene around it.

Nearby, pictured above and at right behind a huge, twisted flagpole, they also found a familiar figure standing tall amid the destruction.

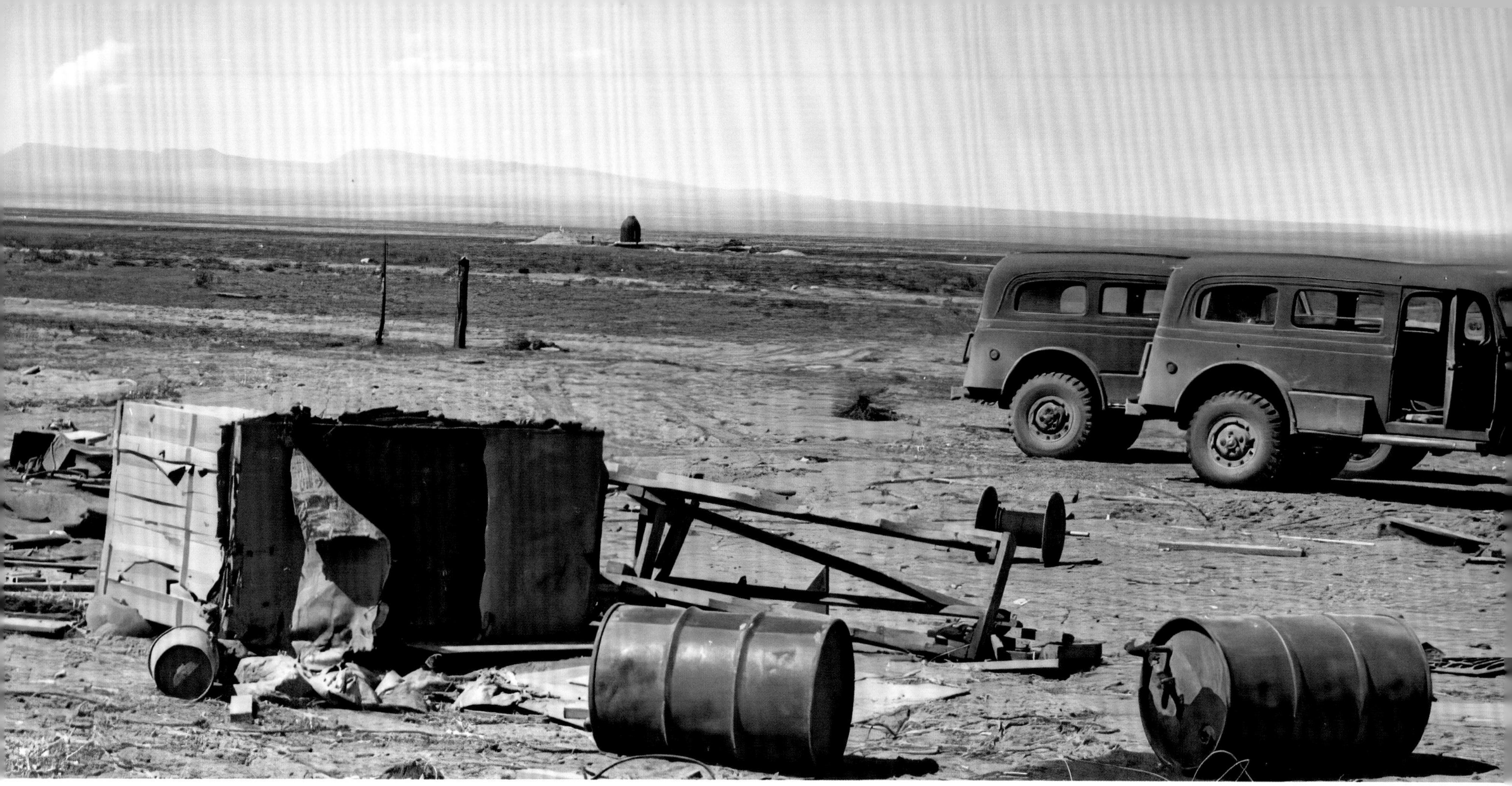

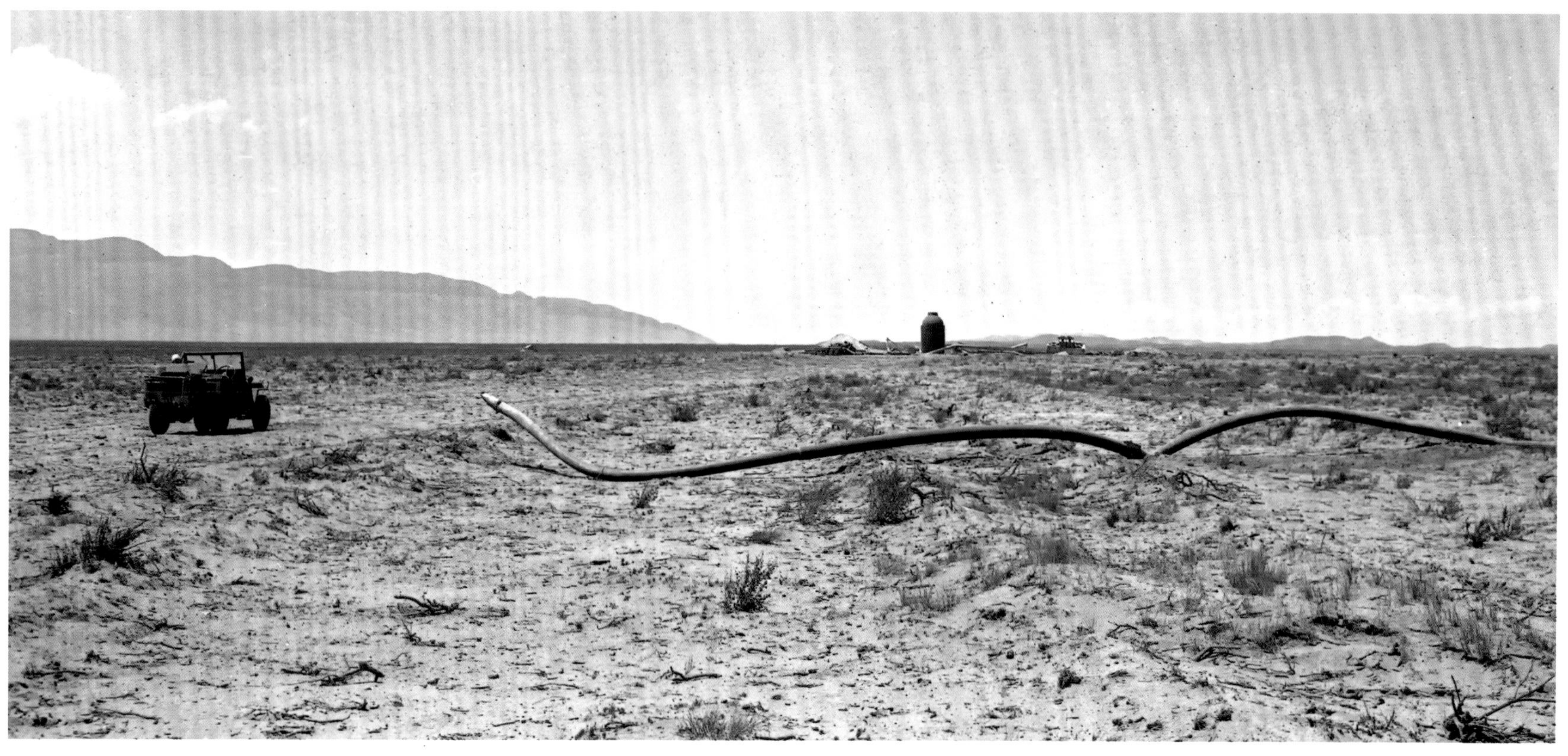

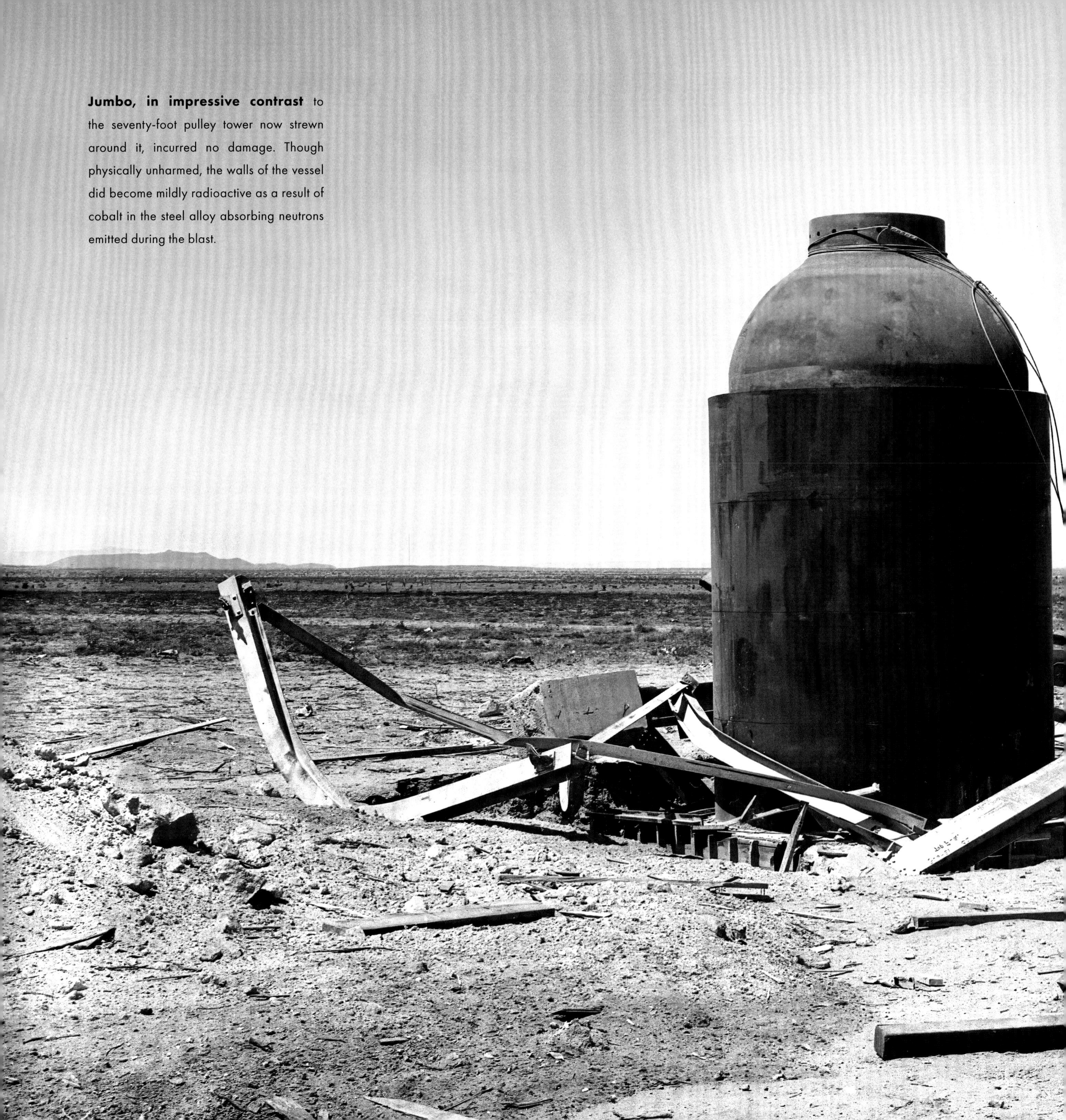

Jumbo, in impressive contrast to the seventy-foot pulley tower now strewn around it, incurred no damage. Though physically unharmed, the walls of the vessel did become mildly radioactive as a result of cobalt in the steel alloy absorbing neutrons emitted during the blast.

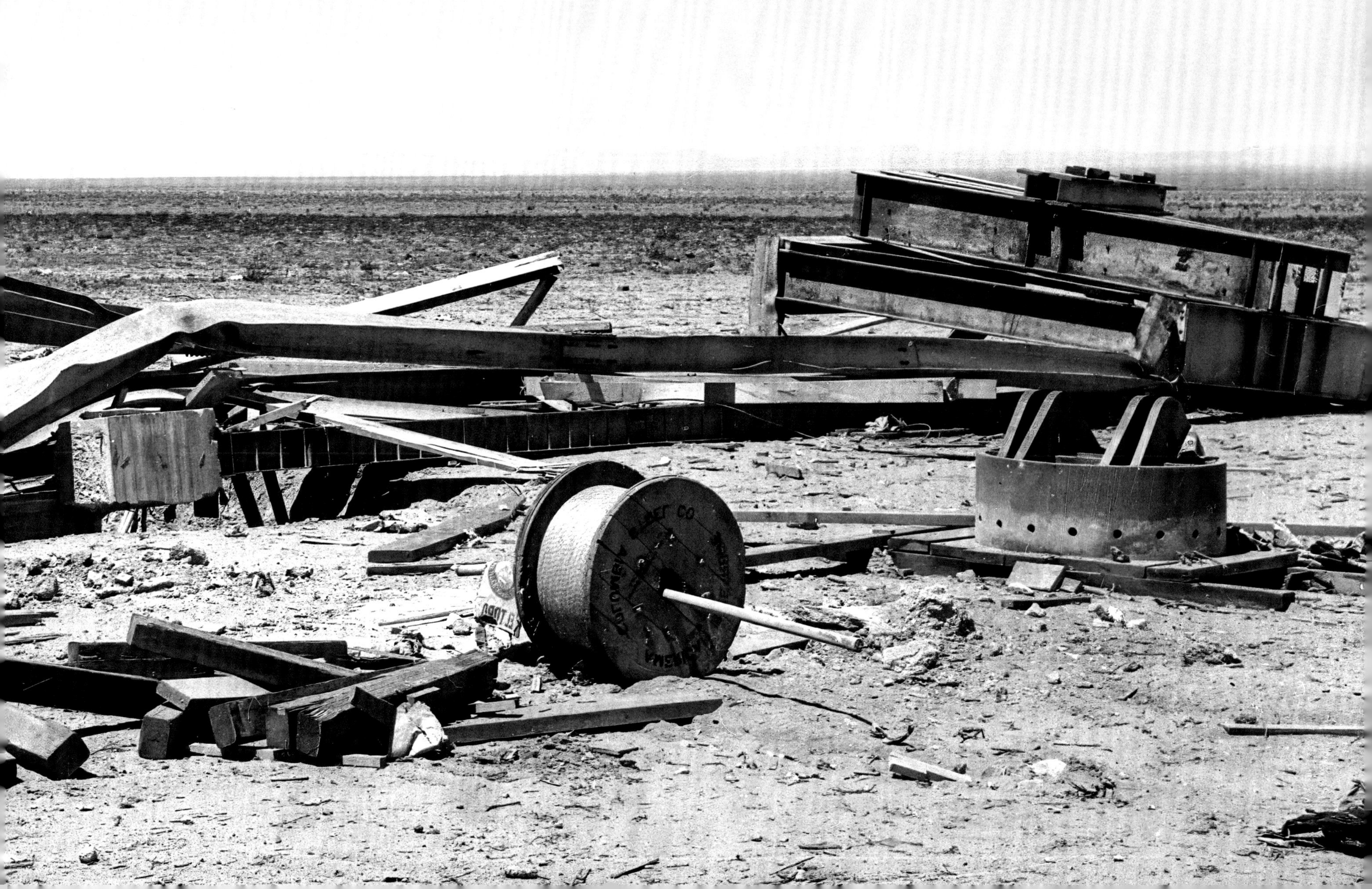

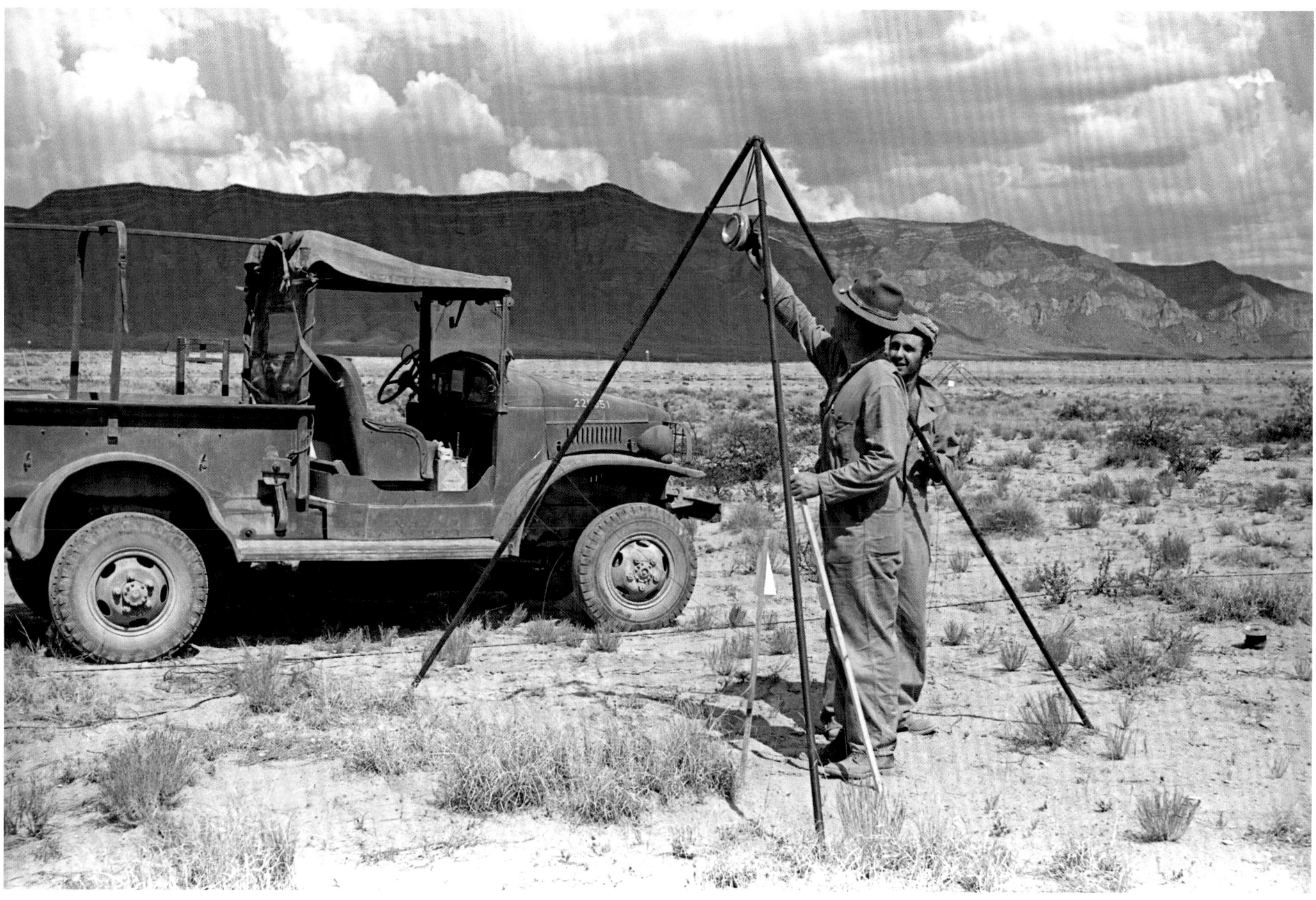

Many of the unsheltered experiments were neither fortified nor calibrated for a blast as powerful as what befell them. Those that were not destroyed operated with mixed results. Among the setups that failed completely were the shock switch experiments tended by Group TR-5 members Ben Benjamin and George Economou (*above*). The torpex and Primacord explosives were set off by thermal radiation almost immediately instead of by the shock wave, as planned. By contrast, the excess velocity microphones, which also detected the passing shock wave, worked quite successfully—performing even better for the Trinity test than they had in the 100-ton dress rehearsal.

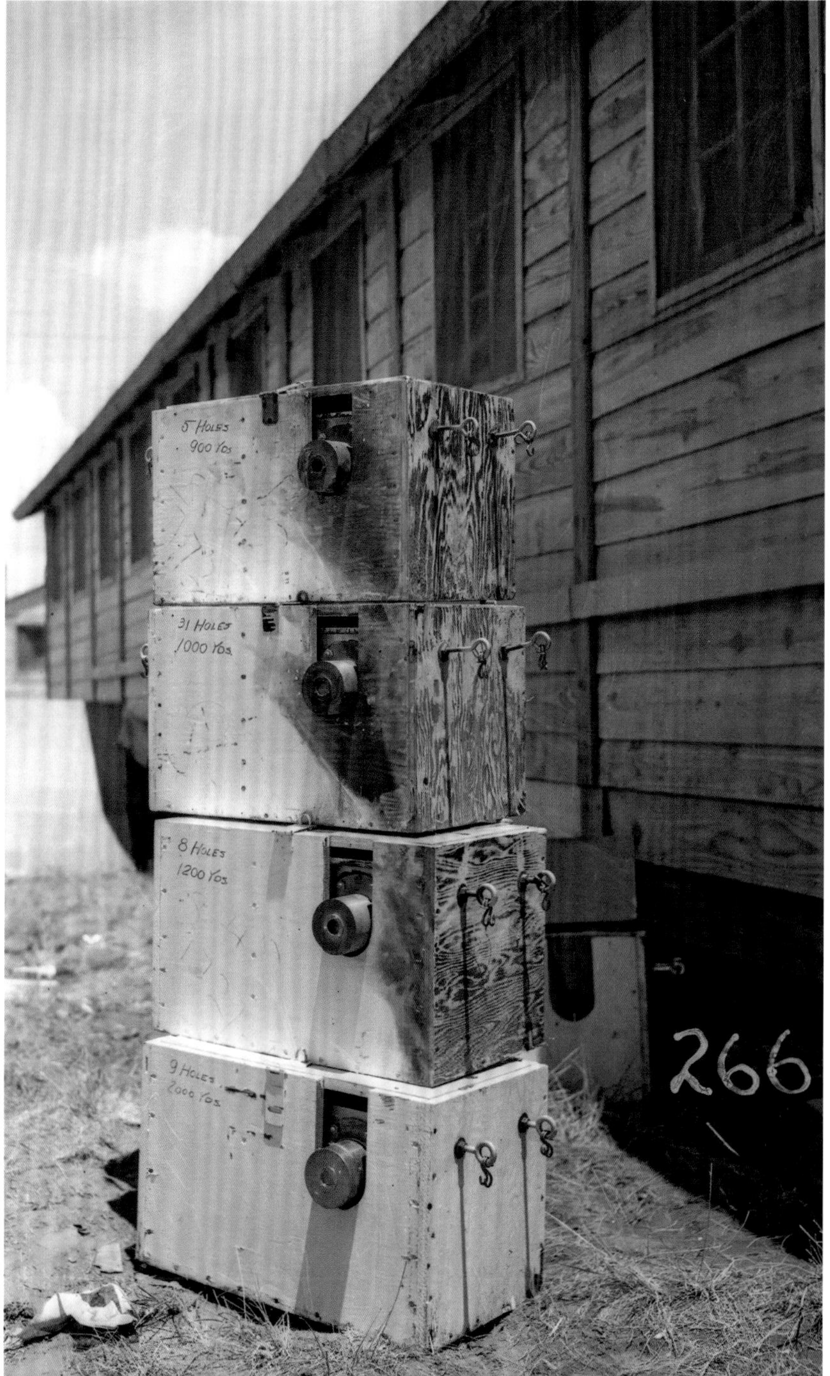

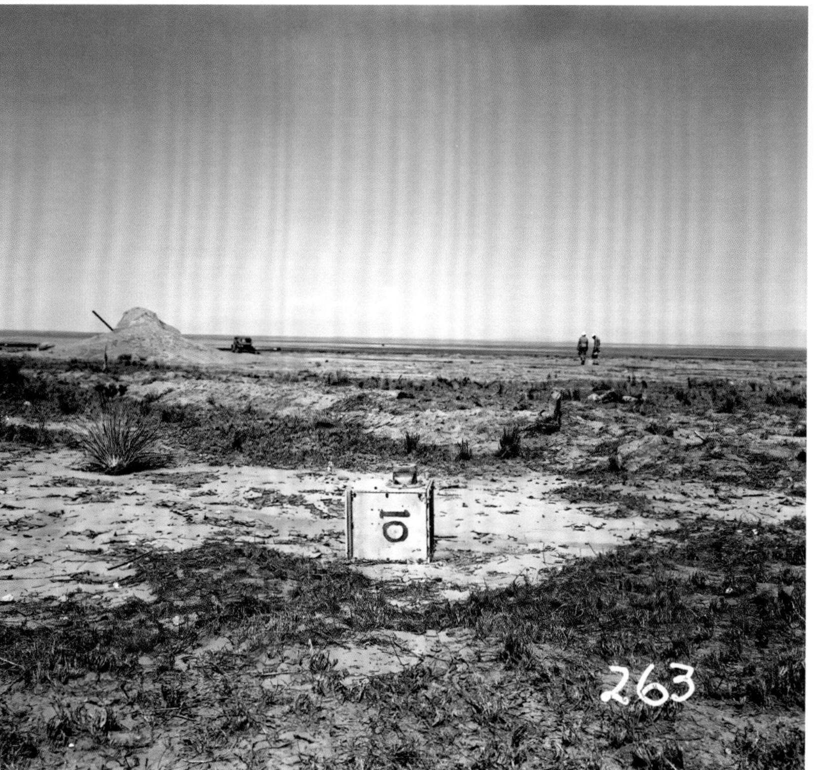

Eight mechanical impulse gauges were recovered unharmed, although only half produced usable results. Three others were found damaged, and one was presumed to have been completely destroyed. The water-filled pistons inside the devices failed to function properly across the full range of pressure inputs they received, and the highest yield indicated by any of the devices was about 10,000 tons of TNT.

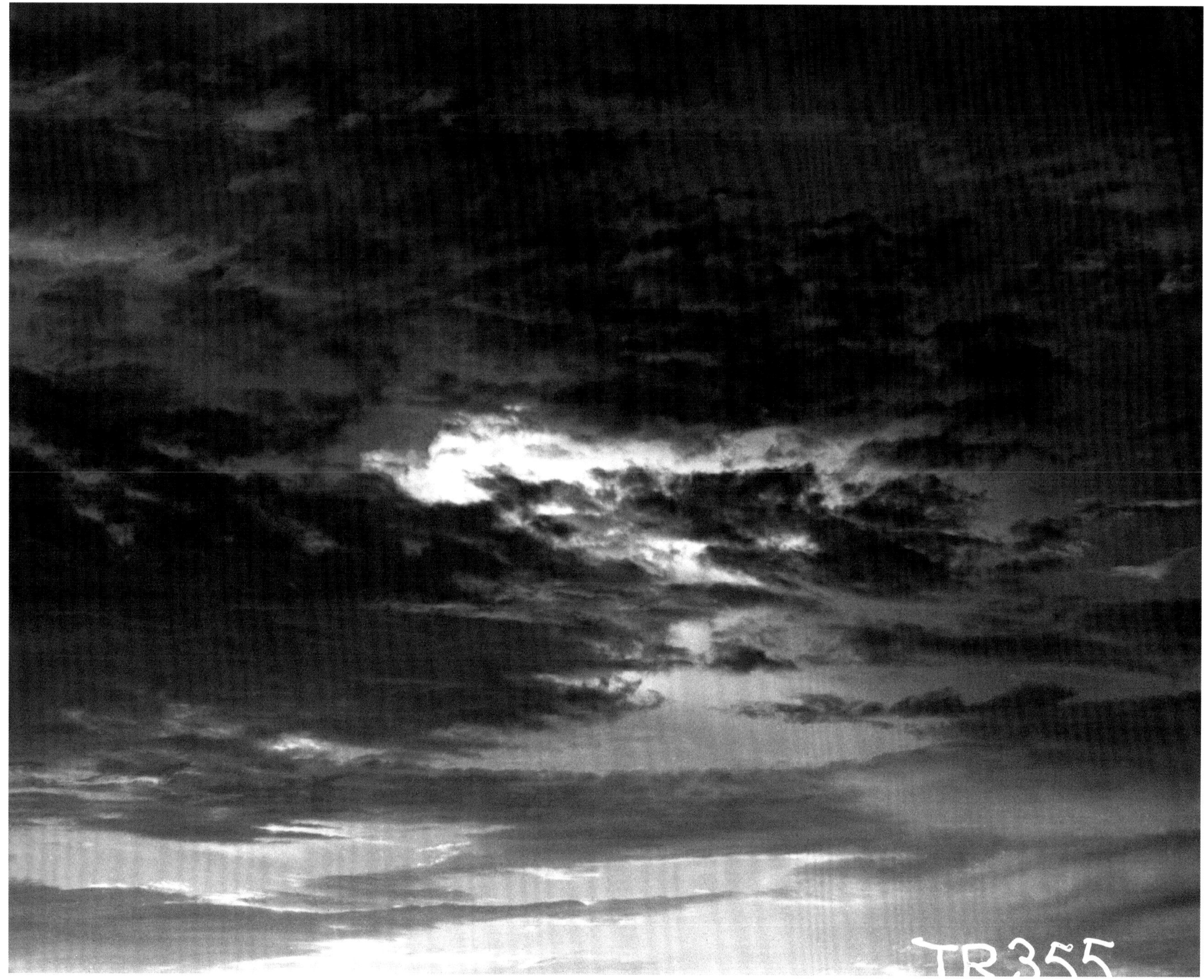

Most of the force associated with the explosion manifested as the airborne shock wave, as the project's seismologist had predicted, although seismographs did detect a small ground tremor in San Antonio, New Mexico, twenty-eight miles away. The other potential danger to surrounding communities—nuclear fallout from the radioactive cloud—would prove to have a longer reach.

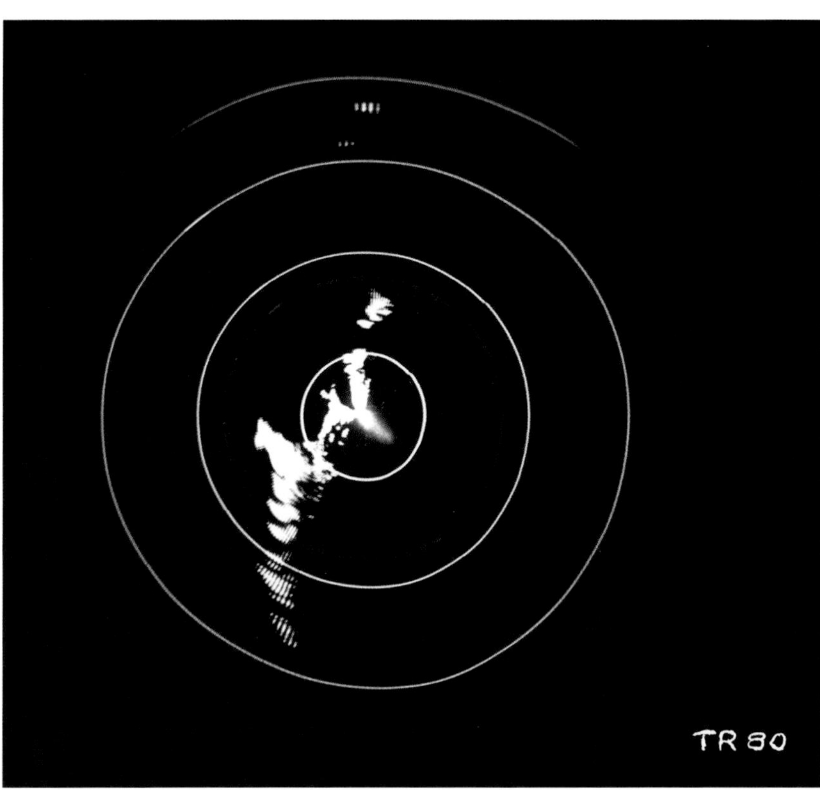

Detonating the Gadget fairly close to the ground resulted in many tons of earth being consumed by the fireball, adding to the radiation load of the cloud as it rose into the sky and began to drift away. The heavier of these particles settled out first and nearest to ground zero, while lighter particles remained aloft as smoke, forming a long, lateral tail in the prevailing winds.

Radar teams at north, south, and west 10,000 were able to track the dissipating cloud for eighteen miles, while pairs of technicians with detection equipment followed its general direction by car to verify that radioactivity at ground level remained below a tolerance threshold established by the Health Group. The expectation was that irradiated material would disperse high in the atmosphere and fall over a huge area in negligible quantities.

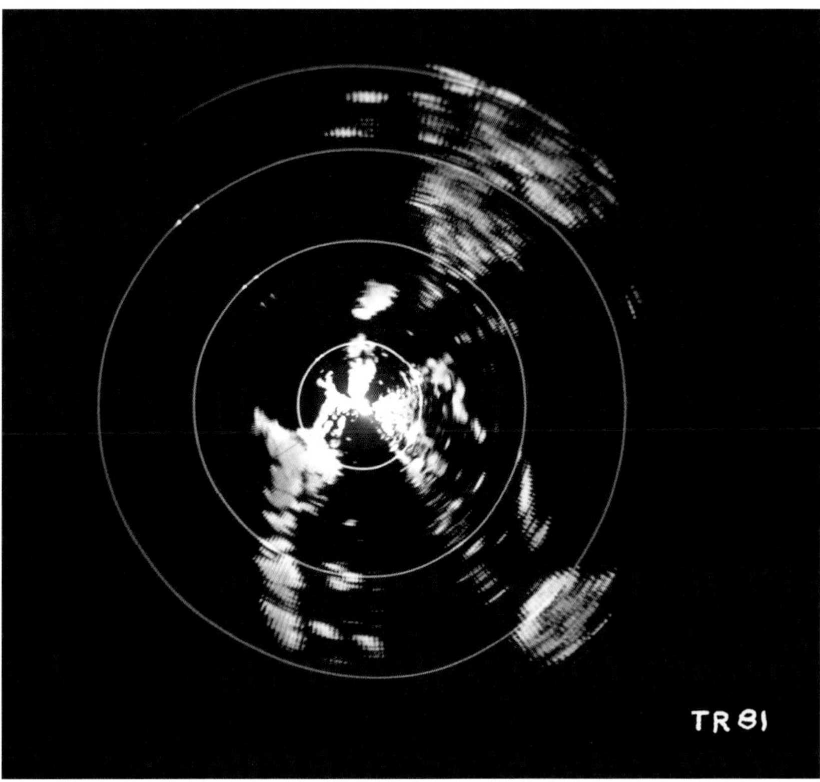

As they traveled along prescribed routes covering hundreds of square miles, while instructions came in from the radar stations, most of the two-person monitoring teams checked in hourly with Lt. Col. Hymer Friedell, a physician based in Albuquerque. Those working near the towns of Carrizozo and Socorro—areas considered most at risk because of their proximity to ground zero—checked in every thirty minutes. The detailed safety plan also tasked specific monitoring teams with the responsibility of evacuating specific families should the need arise.

Comments on the Administration of Town Monitoring.

1. Headquarters in Albuquerque apparently failed in its function to direct monitors. Anderson came to Bingham from Carrizozo for lack of instructions.

2. The police had two contradictory orders about letting me and Palmer in to Guard Tower #2.

3. Hirschfelder violated the monitoring program at Bingham by telling Palmer and the monitors it was a waste of time.

4. As a result of Hirschfelder's argument Leonard went off and left the program at Bingham without instructions

5. Magee's method of reading meters was inadequate. He also disposed of his meter and went home, later to pick up the contaminated meter and handle it in the Tech Area

6. Provisions for testing whether the radiation was coming from the ground was inadequate. Setting a meter on the ground contaminates it.

7. Since cars became easily contaminated special provisions should have been made for keeping certain cars clean.

8. Instruments such as the bullet counter become useless once contaminated. Sensitive meters such as the GM Survey meters also become useless.

For all the planning and resources that were dedicated to public safety, numerous problems were encountered with the monitoring effort, including—as detailed in the critique at left by an unknown author—spotty radio communication, contaminated vehicles and instruments, inconsistent methodology, and occasionally illegible handwriting.

The copious notes taken and meter readings obtained nonetheless provide valuable insight into the scale of Trinity's fallout. Within the first ten to fifteen miles from ground zero, little radioactive material was detected. Beyond this immediate vicinity, varying amounts were recorded in an area approximately one-hundred miles long and thirty miles wide, running northeast of ground zero.

Observed levels of radioactivity surpassed the established dosage limit only in places believed to be uninhabited based on reconnaissance conducted before the test. With no civilians having been relocated by early afternoon on July 16, the evacuation camp set up to receive them disbanded.

Table IVa: Summary of examples of town monitoring: Gamma roentgen doses.

Location	Geometrical Dose, Dg	Integrated Dose, Dn	Total dose after 336 hrs, Dg + Dn	Increment of dose during 2 week intervals after the shot: 3–4 weeks	5–6 weeks	7–8 weeks	19–20 weeks
White							
a) Dose on ground*	8.4 R	21.8 R	30.2 R	1.5 R	0.9 R	0.54 R	0.087 R
b) Dose on ground corrected for house wall effect**	8.4	11.8	20.2	0.81	0.48	0.29	0.045
c) Dose (b) corrected to torso level dose***	4.2	5.8	10	0.81	0.48	0.29	0.045
Bingham							
a) Dose on ground*	3.3	24	27.3	1.7	1	0.6	0.096
b) Dose on ground corrected for house wall effect**	3.3	13	17.3	0.91	0.54	0.32	0.052
c) Dose (b) corrected to torso level dose***	1.7	6.5	8.1	0.91	0.54	0.32	0.052
Hot Canyon							
a) Dose on ground*	24	115	139	8	4.8	2.8	0.46
b) Dose on ground corrected for house wall effect**	24	62	88	4.3	2.6	1.5	0.25
c) Dose (b) corrected to torso level dose.	15	41	56	4.3	2.6	1.5	0.25

* 10 cm above ground. ** adobe house, reduces gamma dosage to 54%

*** Dose at torso level is 50% of that at 10 cm above ground level. applies only to the dosage during the first 336 hrs after the shot. after that time ground and torso doses are the same, see section II d.

The expectation of Trinity personnel and the purpose of the carefully planned evacuation protocol was that no civilians would be harmed, other than possibly experiencing a brief displacement from their homes. Trinity was sobering in many respects—not least of which is that this expectation relied on the radiation safety standards of the 1940s, which preceded long-term studies informing the greater precautions in place today. Through a modern lens, it is evident that some New Mexicans living northeast of the explosion were potentially exposed to dangerous levels of radiation in the aftermath of the test shot.

ALBUQUERQUE JOURNAL
NEW MEXICO'S LEADING NEWSPAPER
65th Year Volume 265 Number 17 Entered as second class matter Albuquerque, N. M., postoffice under act of Congress 1879 Tuesday Morning, July 17, 1945 Published Every Morning

Alamogordo Base Explosives Blast Jolts Wide Area

Windows at Gallup, 235 Miles Away Rattle; No Loss of Lives

By the Associated Press

Following a blast felt over hundreds of miles Monday morning, explosion of "a considerable amount of high explosive and pyrotechnics" in a remote area of the Alamogordo air base reservation was reported by Col. William O. Eareckson, commandant.

Although the blast rattled windows 235 miles away at Gallup in northwestern New Mexico, Col. Eareckson said there were no loss of life or injury to anyone.

"Property damage outside of the explosives magazine itself was negligible," the commandant reported.

Variety of Reports

Reports from over the state listed the blast variously as an earthquake, meteor and air plane crash.

Members of the crew and passengers aboard a Santa Fe railway train near Mountainair thought they saw a bomber explode and burn in the sky.

So brilliant was the flash from the explosion Miss Georgia Green of Socorro, blind University of New Mexico student, exclaimed "What's that."

She was being driven to Albuquerque by her brother-in-law Joe Wills, Socorro theater operator.

Brightens Sky

The flash "lighted up the sky like the sun," Willis said. "The light lasted several moments, followed by a large crimson light to the southeast. We drove down the road several minutes before we heard the explosion."

Continued on Page Two

Continued from Page One

Albuquerque is 150 miles from Alamogordo.

Gallup residents reported their windows were rattled by two explosions at about 5:45 a. m. Officials at the nearby Wingate ordnance depot reported they knew of no blast there.

At Silver City, 135 miles west of Alamogordo, the windows rattled and at a tower on Lookout Mountain near Beaverhead, northwest of Silver City, Forest Ranger Ray Smith said he saw a flash of fire followed by a violent explosion and smoke in the direction of Alamogordo.

Smaller Explosions

He said there were several smaller explosions and placed the time at 5:30 a. m.

In Alamogordo, 10 miles from the base, Mrs. Tom Charles said the explosion had caused no particular comment and no damage so far as she knew.

Eareckson's statement said "weather conditions affecting the content of gas shells exploded by the blast may make it desirable for the Army to evacuate temporarily a few civilians from their homes."

There is a civilian area on the Alamogordo reservation.

Despite efforts to maintain safety and secrecy, on July 16 the effects of the explosion traveled well beyond the boundaries of the test site. Windows shook over 100 miles away. Early risers across the state saw a brilliant light fill the predawn sky. And an ominous cloud having streamed up tens of thousands of feet over the Jornada del Muerto drifted northeast into the distance.

The strange phenomena were the talk of the day in surrounding towns, despite news outlets reporting dully that an accidental explosion had occurred on the army base. Word traveled swiftly through official channels as well. Groves sent encoded notice to Washington that the weapons engineering arm of the Manhattan Project had accomplished its objective. To what end, the Trinity technicians and scientists already knew. The military reality of the atomic age weighed heavily as they completed work at the test site, departing for Los Alamos throughout the remainder of July.

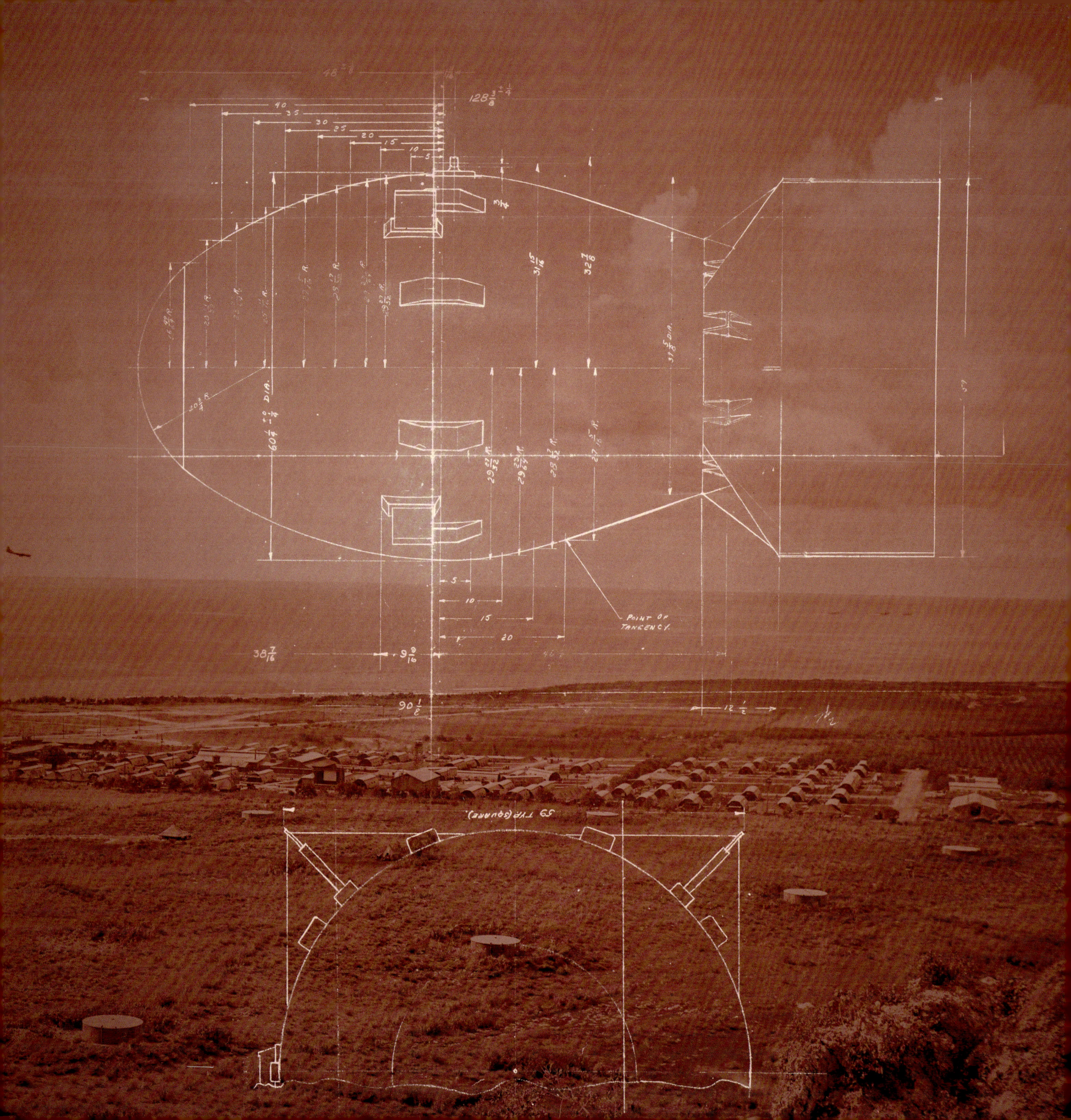

POINT OF
TANGENCY
59 TYP (SQUARE)

08 Trinity to Tokyo Bay

Although the test at Trinity marked the Manhattan Project's technical crescendo, much work remained for the Los Alamos staff supporting military deployment of the weaponized version of the Gadget—Fat Man—and its uranium-fueled counterpart, the gun-type bomb called Little Boy.

The strikes were to be executed by the elite 509th Composite Group, which had begun special training in Wendover, Utah, in December 1944—just before the Trinity site was occupied and long before any proof existed that an atomic bomb would work. Led by Col. Paul W. Tibbets Jr., the 1,800 members of the 509th had prepared intensely for a secret mission—secret even to them—learning to load and drop extremely heavy, oddly shaped bombs with precision from approximately six miles in the air. The group departed to the United States' Pacific air base at Tinian Island in the summer of 1945 with a fleet of B-29 Superfortress bombers specially modified to carry atomic weapons.

The 509th was joined in its mission by military and civilian personnel from Los Alamos organized in March under the codename Project Alberta. The group, led by Navy Captain William "Deak" Parsons, supplied expertise in bomb assembly and shot diagnostics, preparing each of the weapons for deployment at Tinian as well as flying in the strike missions. According to physicist Norman Ramsey, scientific and technical deputy to Parsons, their objective was "to assure the successful combat use of an atomic bomb at the earliest possible date after a field test of an atomic explosion and after the availability of the necessary nuclear material."

Owing to the preparedness of Project Alberta members and the 509th Composite Group, the first use of the new weapon occurred only seventeen days after the test at Trinity. The timing was ultimately dictated by the suitability of weather conditions, in combination with a directive from President Truman to allow Japan one week to accept the terms of surrender outlined in the Allies' July 26 Potsdam Proclamation.

On August 6, the uranium-fueled Little Boy atomic bomb was detonated above Hiroshima, Japan—a military headquarters as well as a major city populated by approximately 255,000 people. Though Little Boy was 6,000 tons smaller in yield than the implosion device tested at Trinity, the destruction and death it left behind were catastrophic. The War Department issued a press release that included the following announcement: "The energy of the atom has been harnessed to produce the deadliest weapon ever devised, the atomic bomb . . . [with] an explosive force such as to stagger the imagination."

Although a common reaction after the Little Boy strike and even the Trinity test was some form of belief that the war was over, Groves and others had long believed that at least two strikes would be necessary to convince the enemy that nuclear weapons posed a repeatable threat. This suspicion had underpinned the importance of creating a working plutonium weapon; of the two designs, Fat Man alone was quickly reproducible. It would have been months before enough uranium for another Little Boy unit was available.

The order from Washington had been to commence bombing of the four identified targets: Hiroshima, Kokura, Niigata, and Nagasaki. Accordingly, hearing no response from the Japanese government, another strike mission proceeded as soon as the plutonium weapon could be readied. On August 9, the second atomic bomb, Fat Man, was detonated, yielding 21,000 tons, high above the port city of Nagasaki—a hub of ship and arms manufacturing and home to an estimated 195,000 people. Again, the impact was devastating and the casualties enormous.

Facing an abrupt and calamitous change in circumstances, Japan's Supreme War Council debated into the early morning hours of August 10 on how to proceed. Adding to the futility of their situation, the Soviet Union had declared war against the Empire and launched an invasion of Japanese-occupied Manchuria just hours before the bombing of Nagasaki. Still, a consensus could not be reached. Ultimately, Prime Minister Suzuki asked the emperor for *go-seidan:* a sacred decision. In his short address to the council that morning, Emperor Hirohito conceded: "In this situation, there is no prospect of victory over the American and British forces with such technological power."

On August 14, Hirohito publicly announced that his nation had accepted the terms dictated from Potsdam, citing among the reasons the use of "a new and most cruel bomb." On September 2, the formal surrender was signed aboard the USS *Missouri* in Tokyo Bay: at last, the long violence of a war that left tens of millions dead was over. At what cost, and with what implications for the future, remained to be learned.

* * * * * * *

During the two-week meeting at Potsdam, which convened the day after the Trinity test, the leaders of the United States, Great Britain, and the Republic of China drafted and issued terms of surrender for the Empire of Japan. With the atomic bomb now a reality and components of the uranium weapon already on its way to an air base in the Pacific, the call for surrender concluded with a final warning: "The alternative for Japan is prompt and utter destruction." Alongside the ultimatum, President Truman ordered that the new weapons were not to be used right away, in case Japan chose to accept defeat. The official directive, issued to the Strategic Air Forces by the secretary of war and the president's chief of staff, is shown at right. (National Archives Catalog, 542193.)

Instead, attacks were to begin as soon as possible "after about 3 August," with weather conditions to dictate the dates of the strikes. Good visual bombing weather was difficult to come by in Japan. Sometimes weeks would pass without a break in the constant clouds. The worst month was thought to be June, with conditions improving in July and August before worsening again in September.

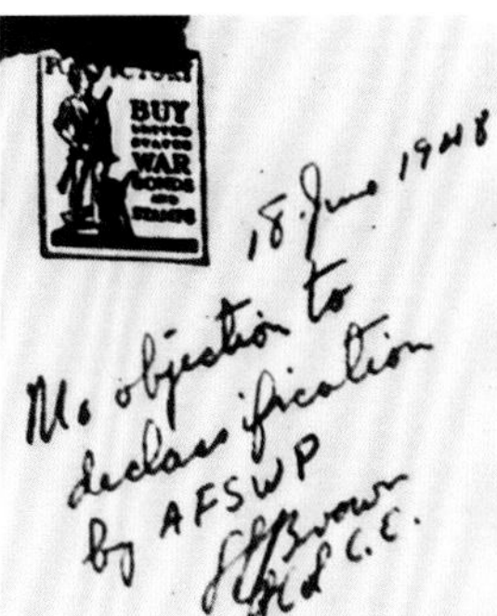

WAR DEPARTMENT
OFFICE OF THE CHIEF OF STAFF
WASHINGTON 25, D. C.

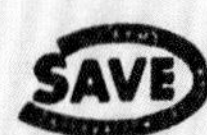

25 July 1945

TO: General Carl Spaatz
Commanding General
United States Army Strategic Air Forces

1. The 509 Composite Group, 20th Air Force will deliver its first special bomb as soon as weather will permit visual bombing after about 3 August 1945 on one of the targets: Hiroshima, Kokura, Niigata and Nagasaki. To carry military and civilian scientific personnel from the War Department to observe and record the effects of the explosion of the bomb, additional aircraft will accompany the airplane carrying the bomb. The observing planes will stay several miles distant from the point of impact of the bomb.

2. Additional bombs will be delivered on the above targets as soon as made ready by the project staff. Further instructions will be issued concerning targets other than those listed above.

3. Dissemination of any and all information concerning the use of the weapon against Japan is reserved to the Secretary of War and the President of the United States. No communiques on the subject or releases of information will be issued by Commanders in the field without specific prior authority. Any news stories will be sent to the War Department for special clearance.

4. The foregoing directive is issued to you by direction and with the approval of the Secretary of War and of the Chief of Staff, USA. It is desired that you personally deliver one copy of this directive to General MacArthur and one copy to Admiral Nimitz for their information.

Thos T Handy
THOS. T. HANDY
General, G.S.C.
Acting Chief of Staff

Incl #1

UNCLASSIFIED

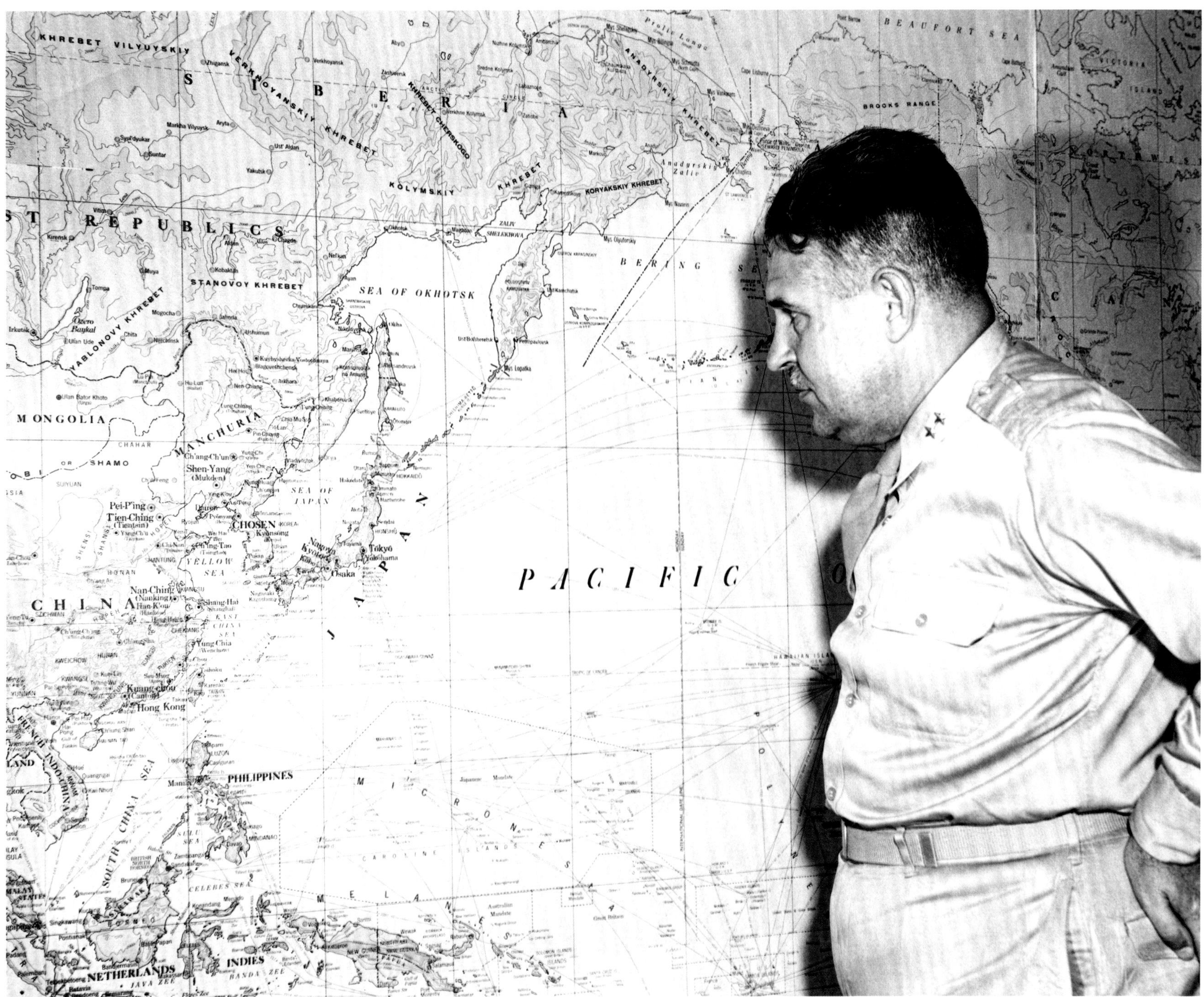

The target cities had been recommended in May of 1945 by a Target Committee, chaired by Groves, based on military value to the Japanese and the psychological effect that could be achieved by their destruction. It was also deemed important that the cities be undamaged so proper assessment could be made of the effects of the new weapons. The Twentieth Air Force was asked to exclude the chosen targets from their campaign to eliminate all of Japan's major cities by January of 1946 via conventional bombing. The originally proposed targets were the Kokura Arsenal and surroundings; Hiroshima, which was the second-largest Japanese city untouched by American bombing raids; Kyoto, which was the largest and was stricken from the original list because of its cultural significance; and Niigata, a city growing in military importance as other ports of embarkation were destroyed. The port of Nagasaki was later added in place of Kyoto. (Photo by Ed Westcott, courtesy of US Department of Energy.)

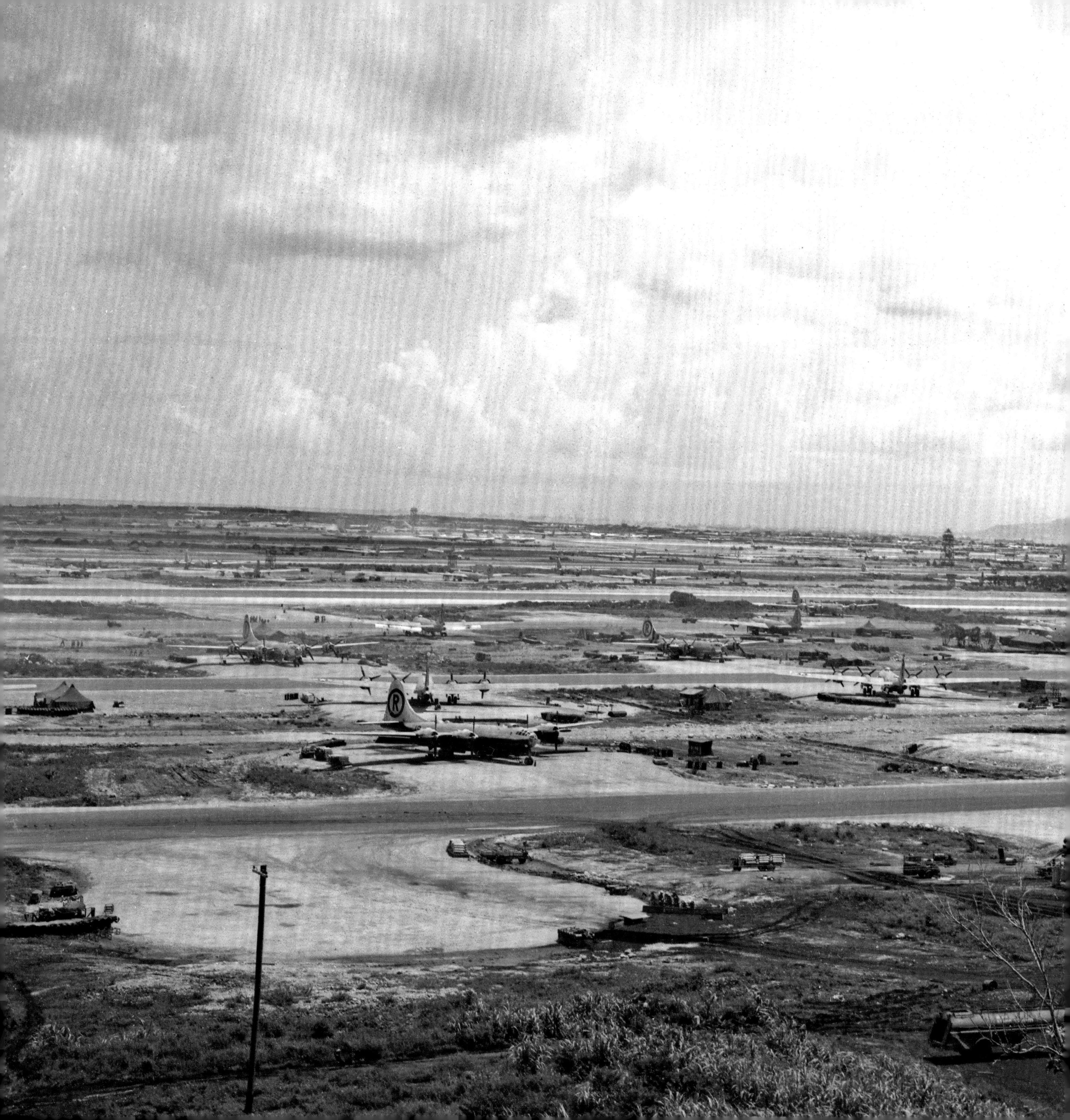

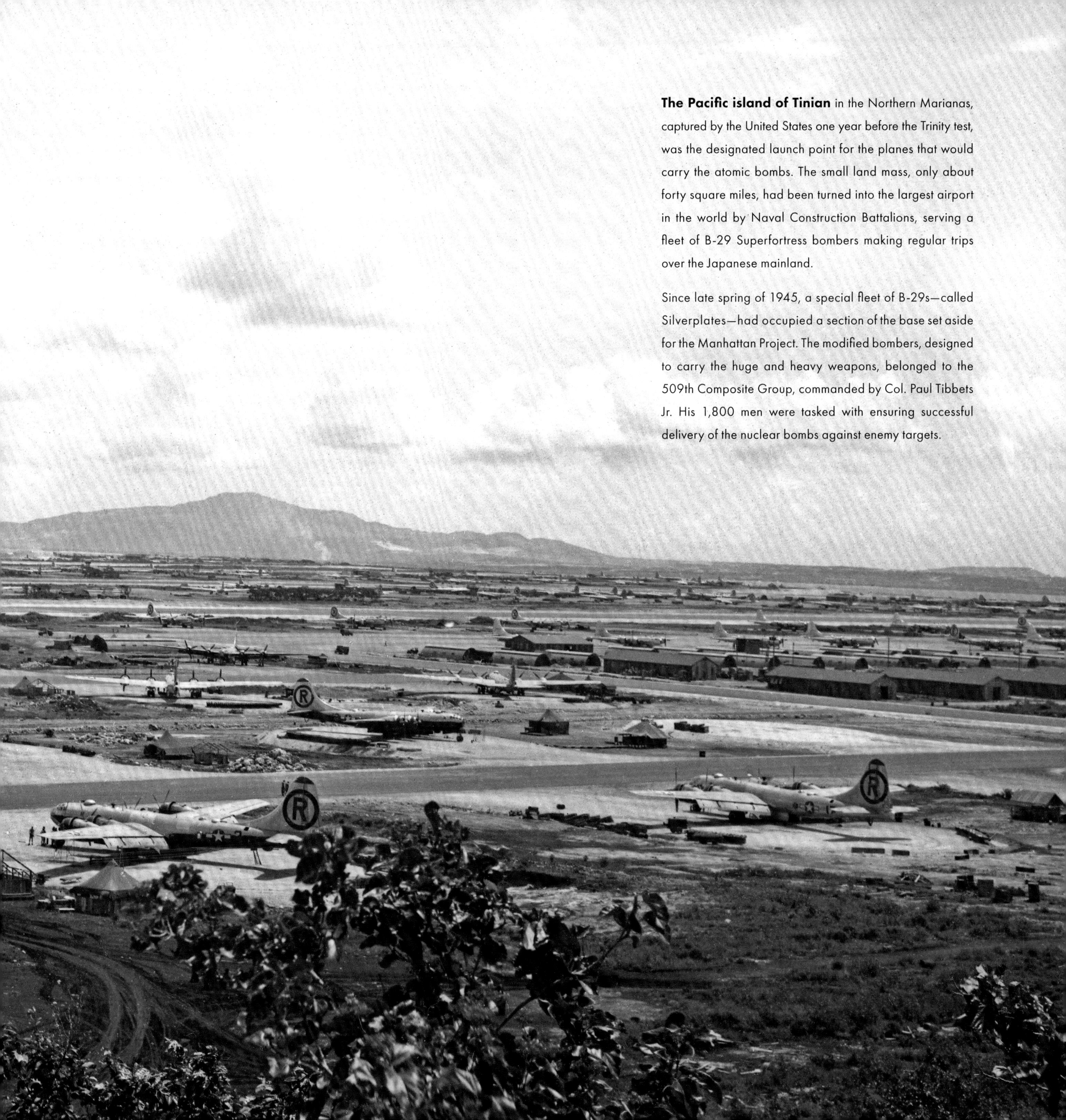

The Pacific island of Tinian in the Northern Marianas, captured by the United States one year before the Trinity test, was the designated launch point for the planes that would carry the atomic bombs. The small land mass, only about forty square miles, had been turned into the largest airport in the world by Naval Construction Battalions, serving a fleet of B-29 Superfortress bombers making regular trips over the Japanese mainland.

Since late spring of 1945, a special fleet of B-29s—called Silverplates—had occupied a section of the base set aside for the Manhattan Project. The modified bombers, designed to carry the huge and heavy weapons, belonged to the 509th Composite Group, commanded by Col. Paul Tibbets Jr. His 1,800 men were tasked with ensuring successful delivery of the nuclear bombs against enemy targets.

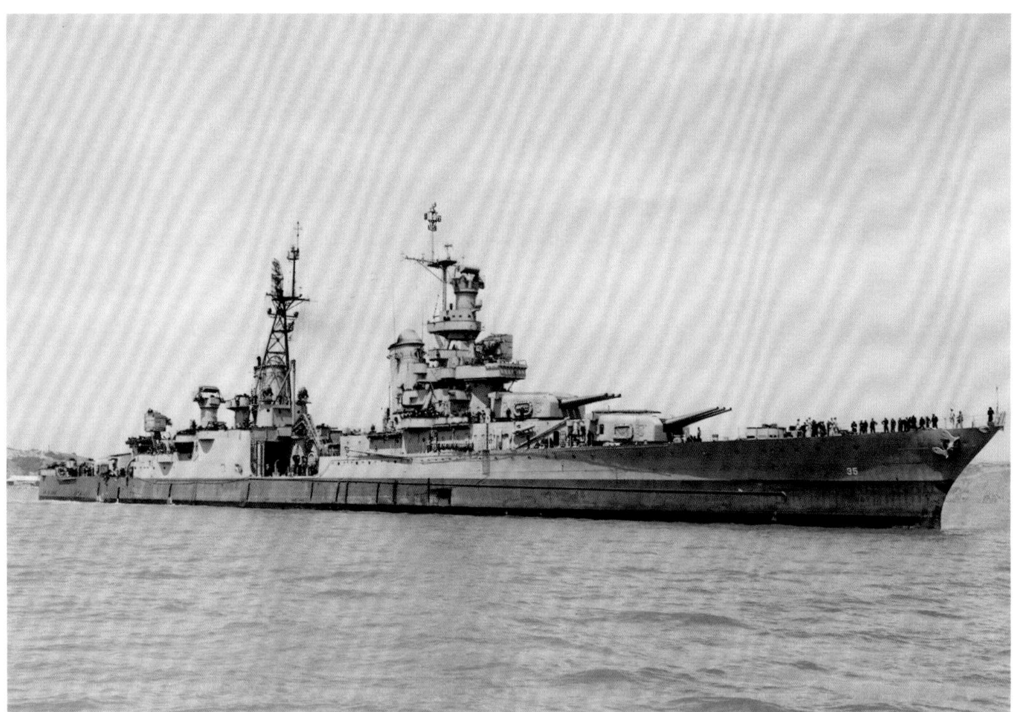

The first pieces of the uranium gun-type bomb, Little Boy, arrived on Tinian the day of the Potsdam Declaration, having set sail immediately after word arrived of the successful Trinity test. The courier, a warship, had only just finished receiving repairs after taking heavy damage during a kamikaze attack in the battle for Okinawa. The USS *Indianapolis* rushed across the ocean in a record-setting ten days, refueling at Pearl Harbor and reaching its destination with the Little Boy ballistic case and a portion of its uranium fuel on July 26, 1945. (National Archives Catalog, 148727930.)

Transport through Pacific airspace and waters was not without risk, and a tragic reminder was soon in coming: en route to the Philippines from the Marianas, only days after delivering its precious cargo, the *Indianapolis* was sunk by torpedoes from a Japanese submarine, resulting in the deaths of nearly 900 of the vessel's sailors.

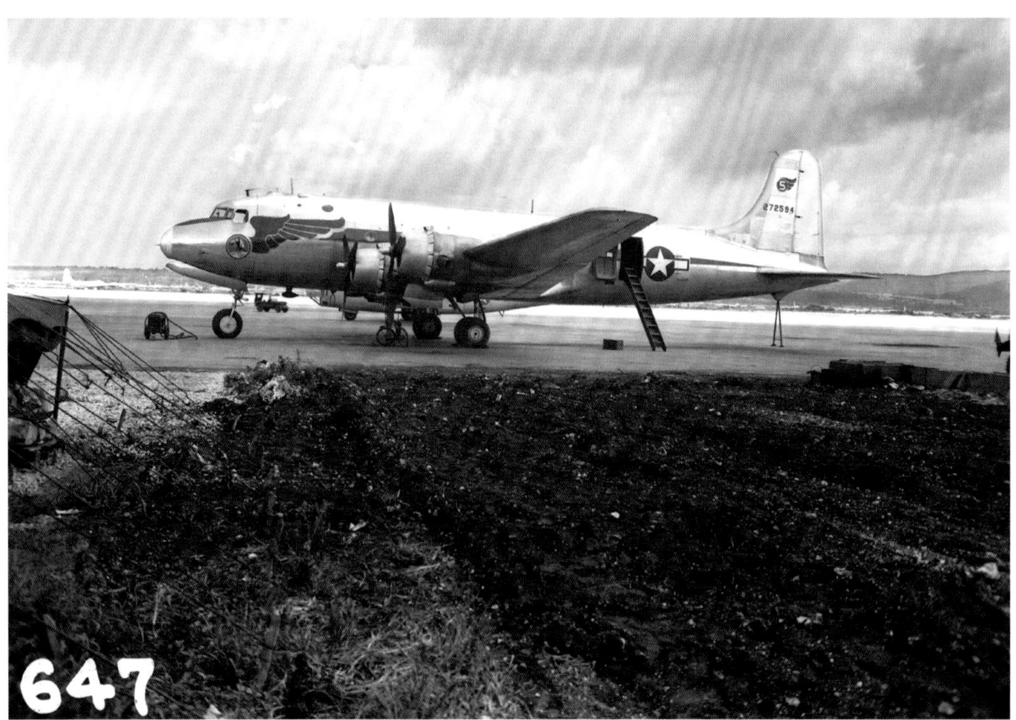

In the days before the *Indianapolis* was lost, the remainder of Little Boy's uranium, along with Fat Man's plutonium core, had arrived safely on Tinian aboard Douglas C-54 Skymasters (*left*), part of a fleet of five Manhattan Project transport planes known as the Green Hornet Line.

The non-nuclear components of the plutonium weapon traveled to Tinian by air as well, with three high-explosives spheres (like that used in the Trinity test) the last to arrive of all the weapons cargo. Because of their size and weight, the spheres had to make the overseas journey in the bomb bays of Silverplate B-29s, boarding the aircraft on July 28 for departure from Albuquerque, New Mexico.

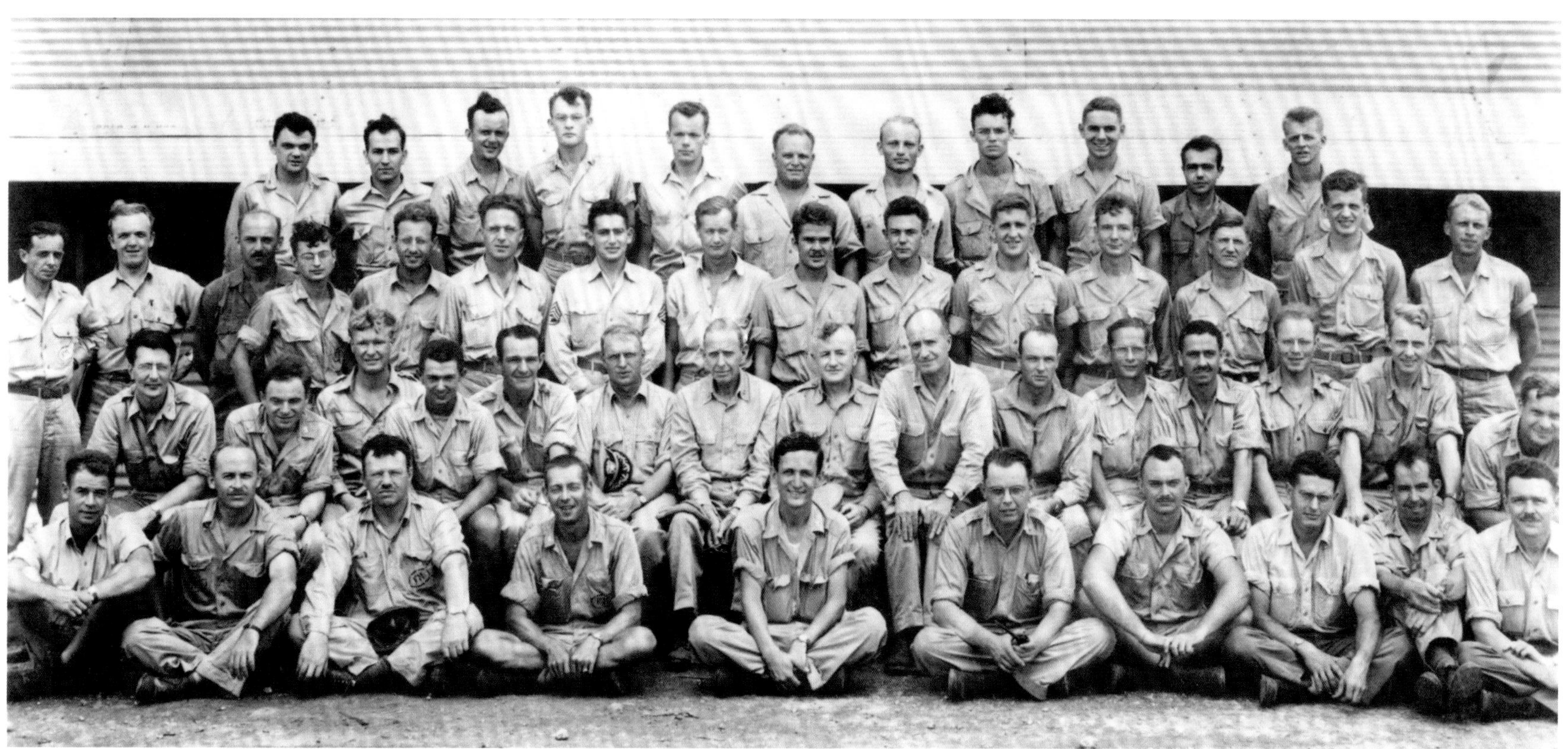

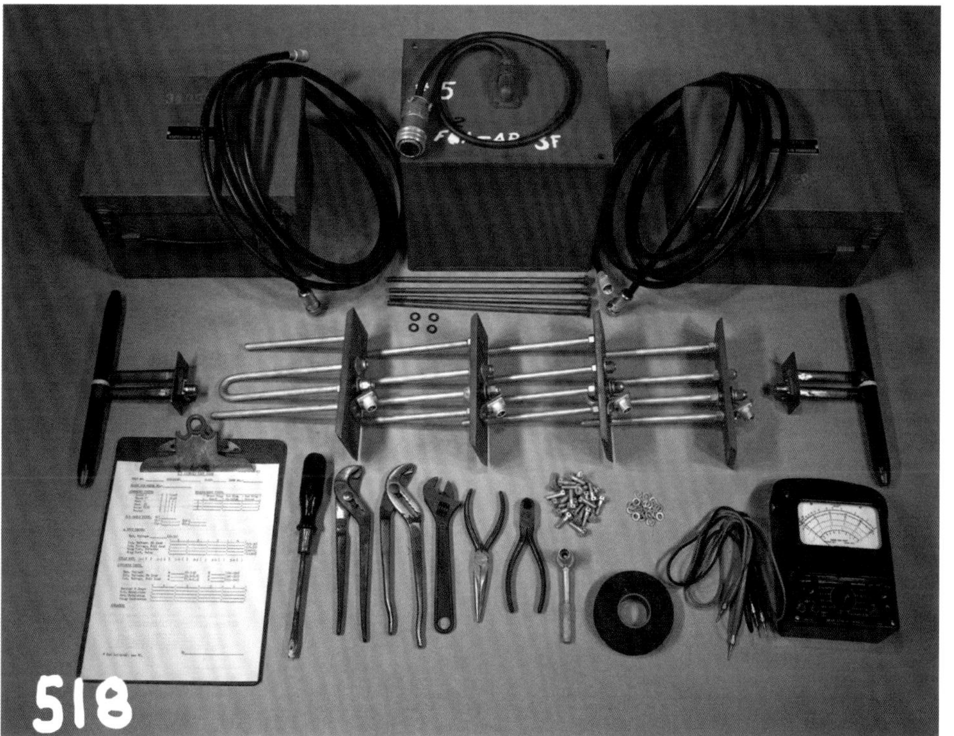

In between arrival at Tinian and deployment over Japan, Little Boy and Fat Man would need to be prepared for detonation by trained experts familiar with the two weapons. To complete the delivery phase of the Manhattan Project, a team of scientists, engineers, and technicians (*above*) voluntarily traveled to the Pacific to work with the 509th. In addition to conducting the assembly processes for both weapons, some of these men would fulfill critical roles in the two upcoming strike missions, joining the B-29 crews to complete in-flight arming, monitoring, and diagnostic procedures.

Assembly teams used special kits, like that at left, shipped from Los Alamos that contained all the tools needed for a particular task. They had trained with these very toolkits in buildings designed as facsimiles of those on Tinian in order to replicate field conditions as closely as possible.

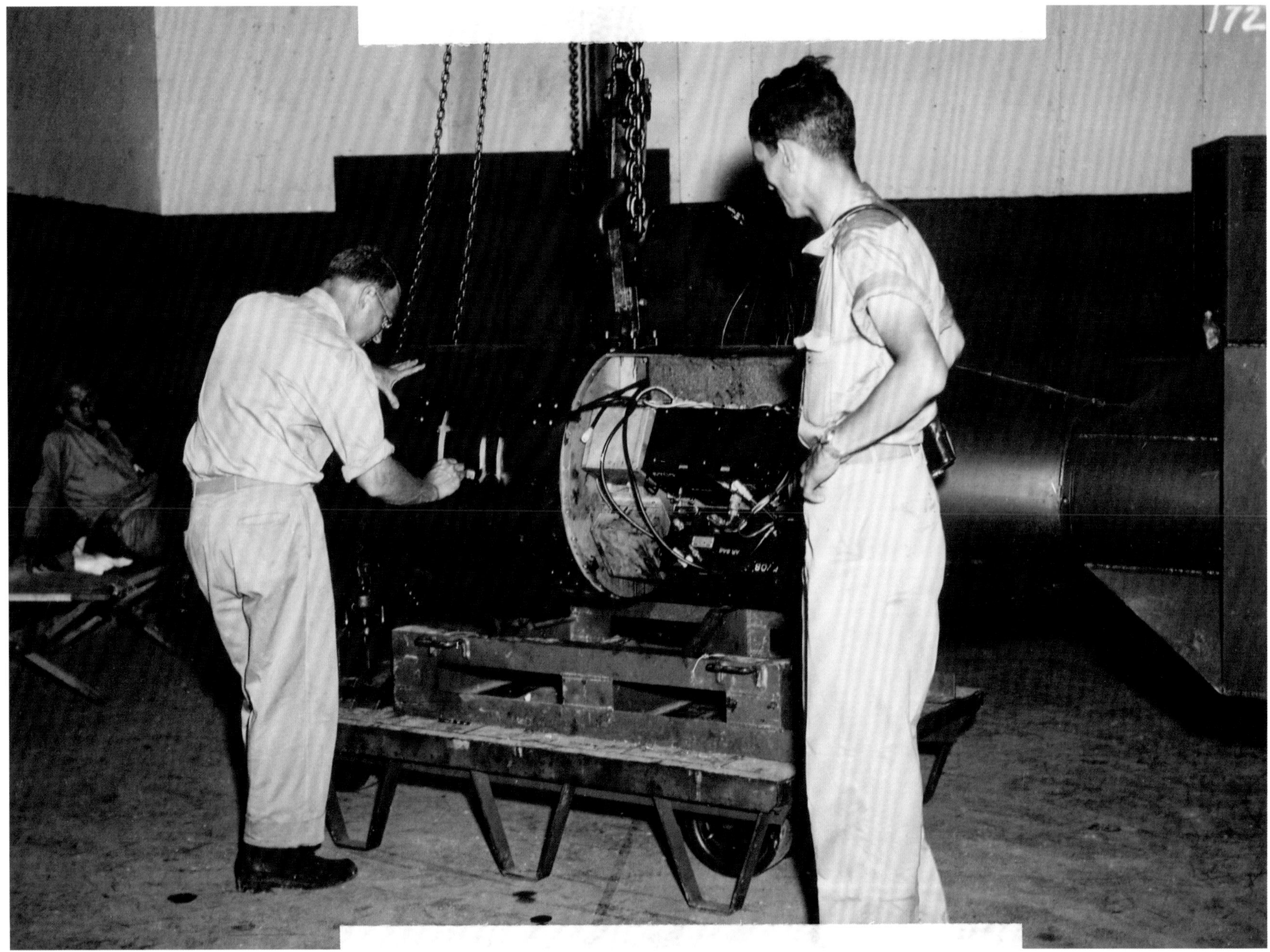

By August 2, with the waiting period set to expire the next day, all components of both weapons had arrived on Tinian and Little Boy was ready for deployment. Above, with Project Alberta deputy Norman Ramsey looking on, Lt. Cmdr. Francis Birch labels the unit designated for the first nuclear attack as L-11. Carbon copies of Little Boy—"L" units bearing numbers one through ten that lacked only nuclear material—had been expended in practice drops and tests of the firing systems to ensure no surprises or mishaps on the day of the mission. With procedures down to a science and the weapon prepared, the date of the first atomic strike now depended only on closely monitored atmospheric conditions over the four targets. After a period of poor weather, on the morning of August 5, a favorable forecast for the next day was finally received. (National Archives Catalog, 519395.)

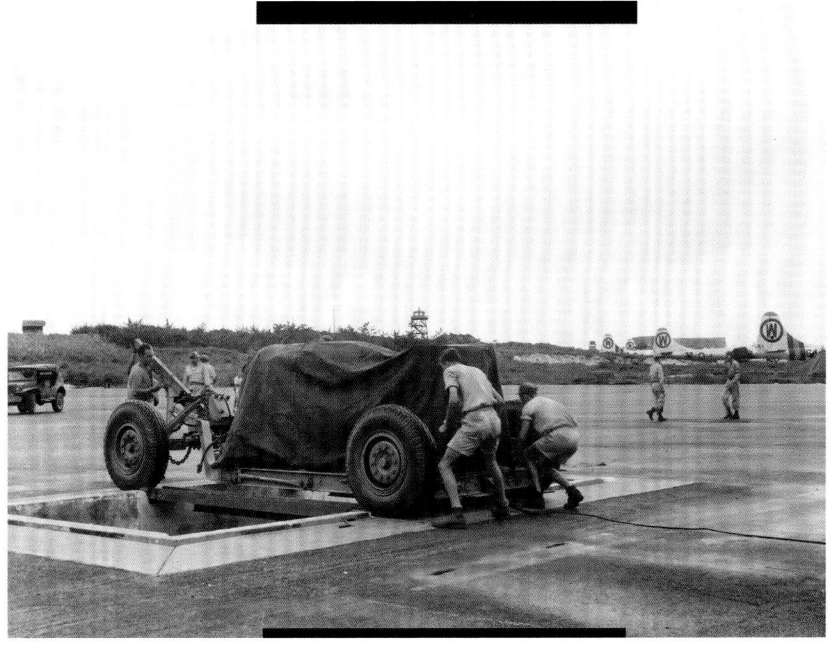

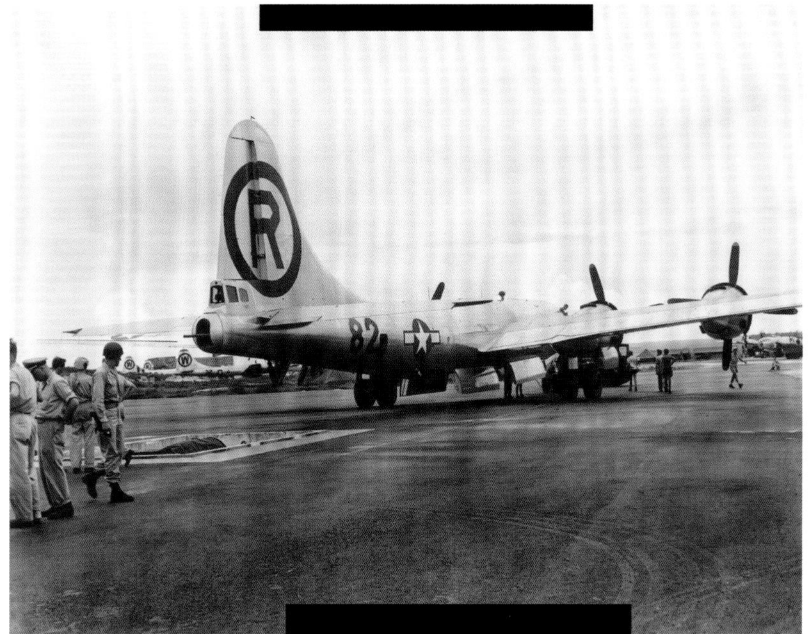

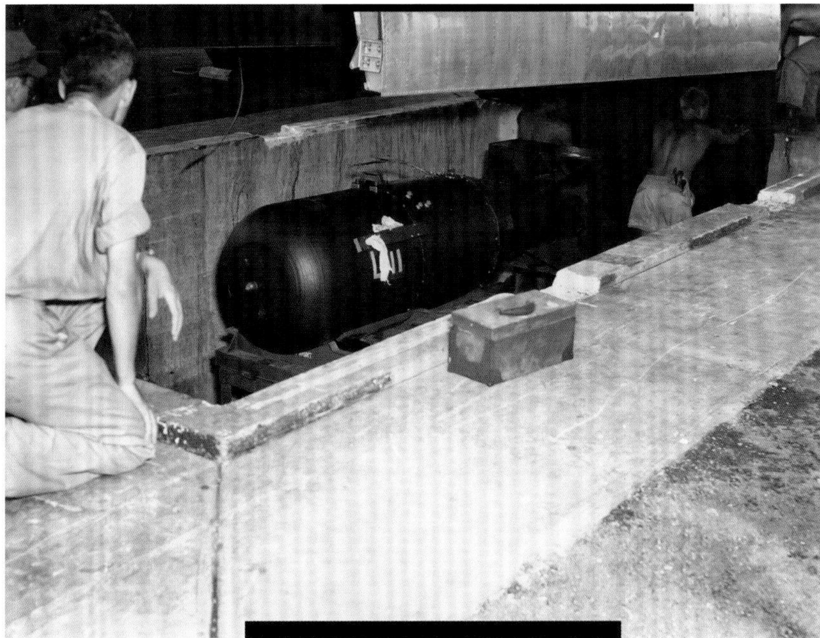

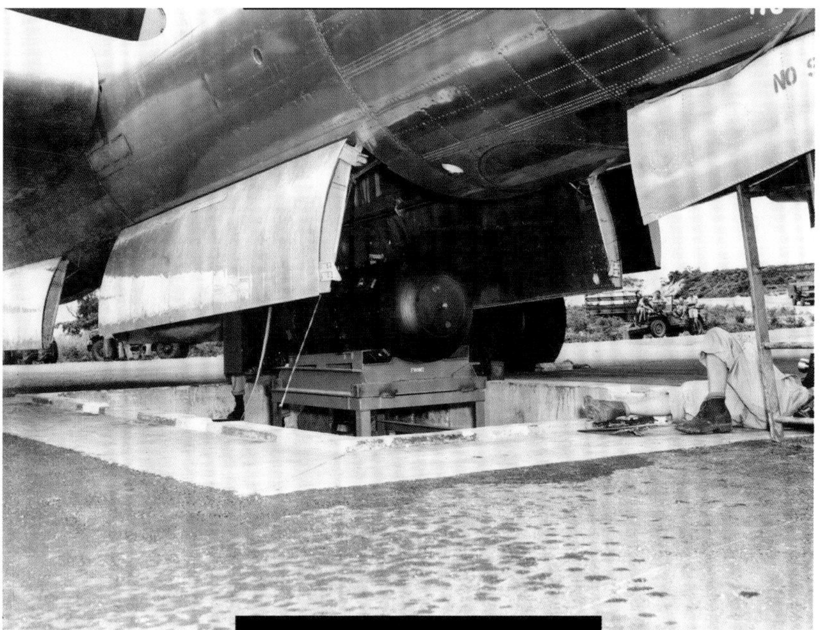

That evening, Little Boy was loaded into the bomb bay of a Silverplate B-29. The planes modified for the 509th had been stripped of heavy, dispensable equipment typically aboard the aircraft, including most of the armor and gun turrets. Other modifications included adding reverse pitch propellers, which provided backward thrust to help slow down the plane during landings; fuel-injected engines with an improved cooling system; pneumatic bomb bay doors; and the British attachment and release systems used to carry the Royal Air Force's 12,000-pound Tallboy bomb. (National Archives Catalog, nos. 76048669, 76048688, 76048710, and 76048660.)

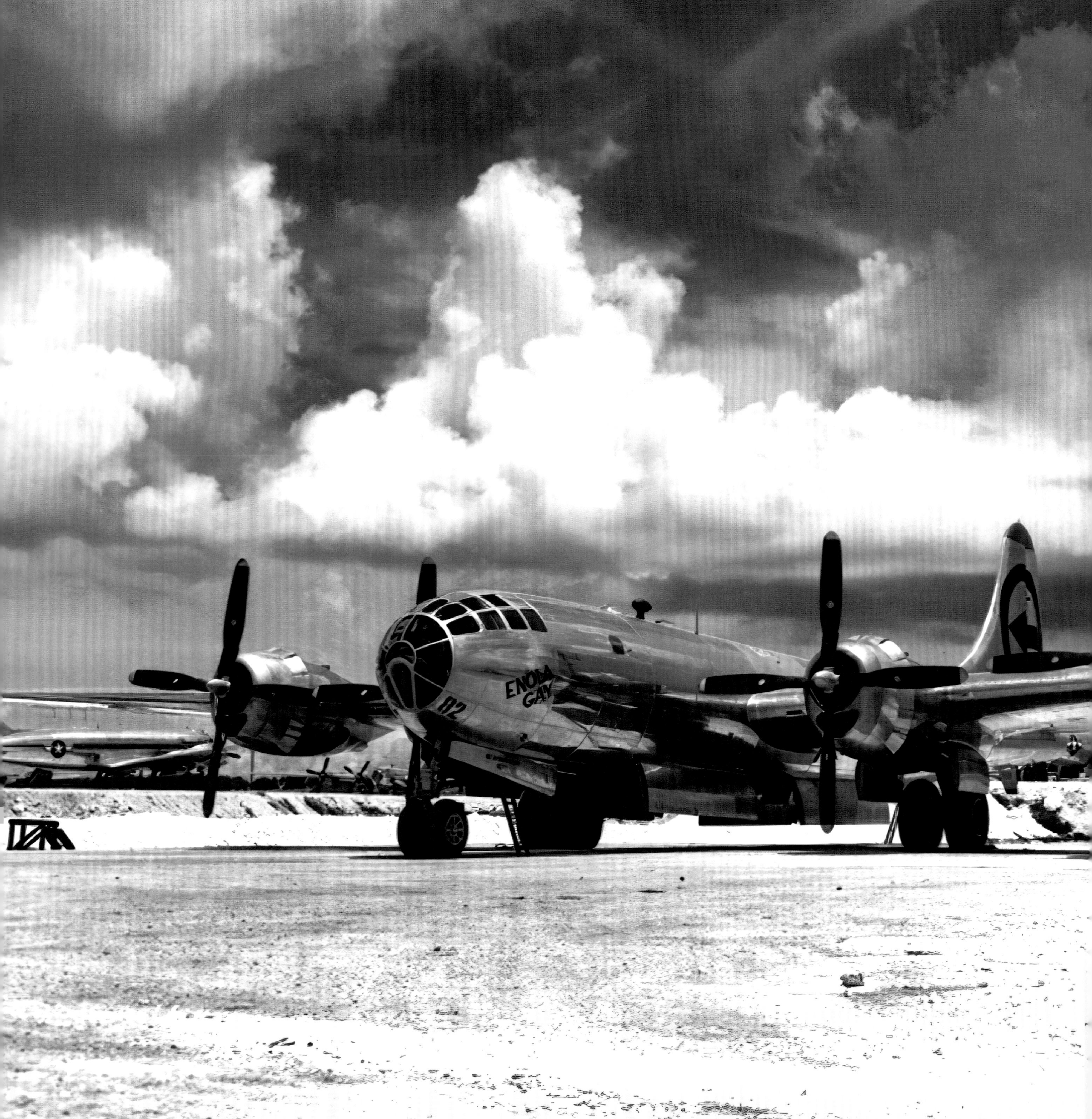
ENOLA GAY
82

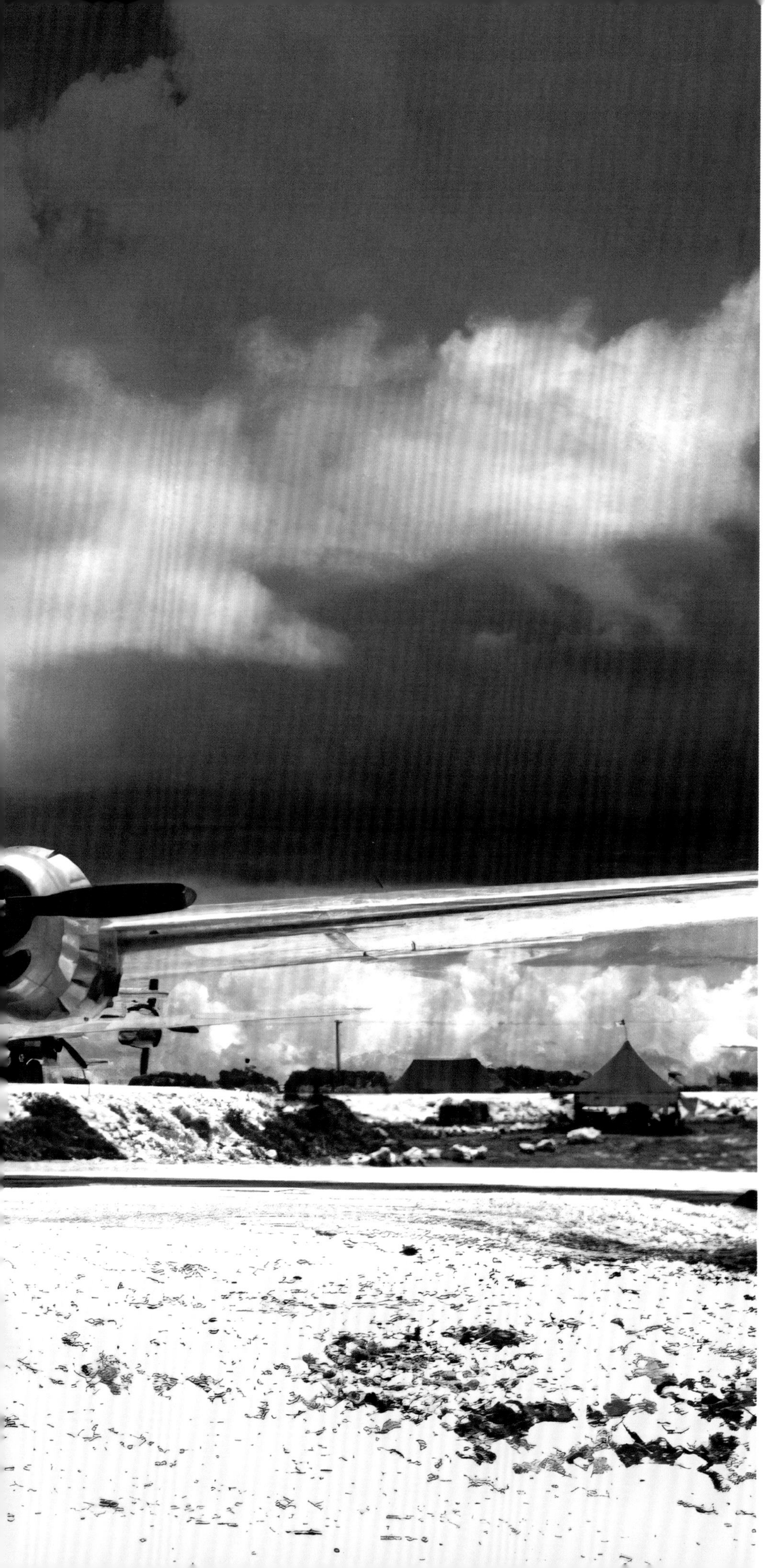

The captain and pilot of the strike plane was 509th Composite Group commander Paul Tibbets. He named the plane after his mother, Enola Gay. In the photo below, taken at a pre-strike briefing, Colonel Tibbets is pictured to the right of Navy Captain William "Deak" Parsons, the head of Project Alberta.

Parsons served as one of Oppenheimer's associate directors at Los Alamos and was the weaponeer aboard the *Enola Gay*, meaning he would arm Little Boy in flight, assisted by Morris Jeppson. Parsons was chosen for the role because orders called for a senior officer present on the plane who had familiarity with the design, development, and tactical features of the bomb. This person was to render final judgment in case an emergency required deviation from the plan.

In briefing those who would take part in the Hiroshima mission, Captain Parsons informed the flight crews of what effects could be expected from the weapon and the anticipated impact on the target city. There was a general awareness that they were dealing with no ordinary bomb, but its capacity for destruction was shocking news to many. (Photo at left by Leon Smith, courtesy of Brad Smith.)

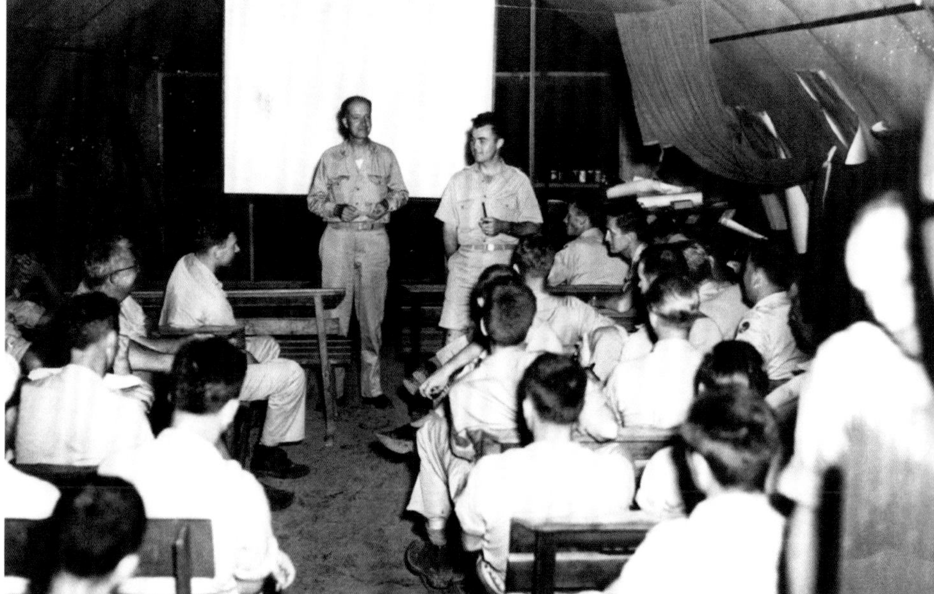

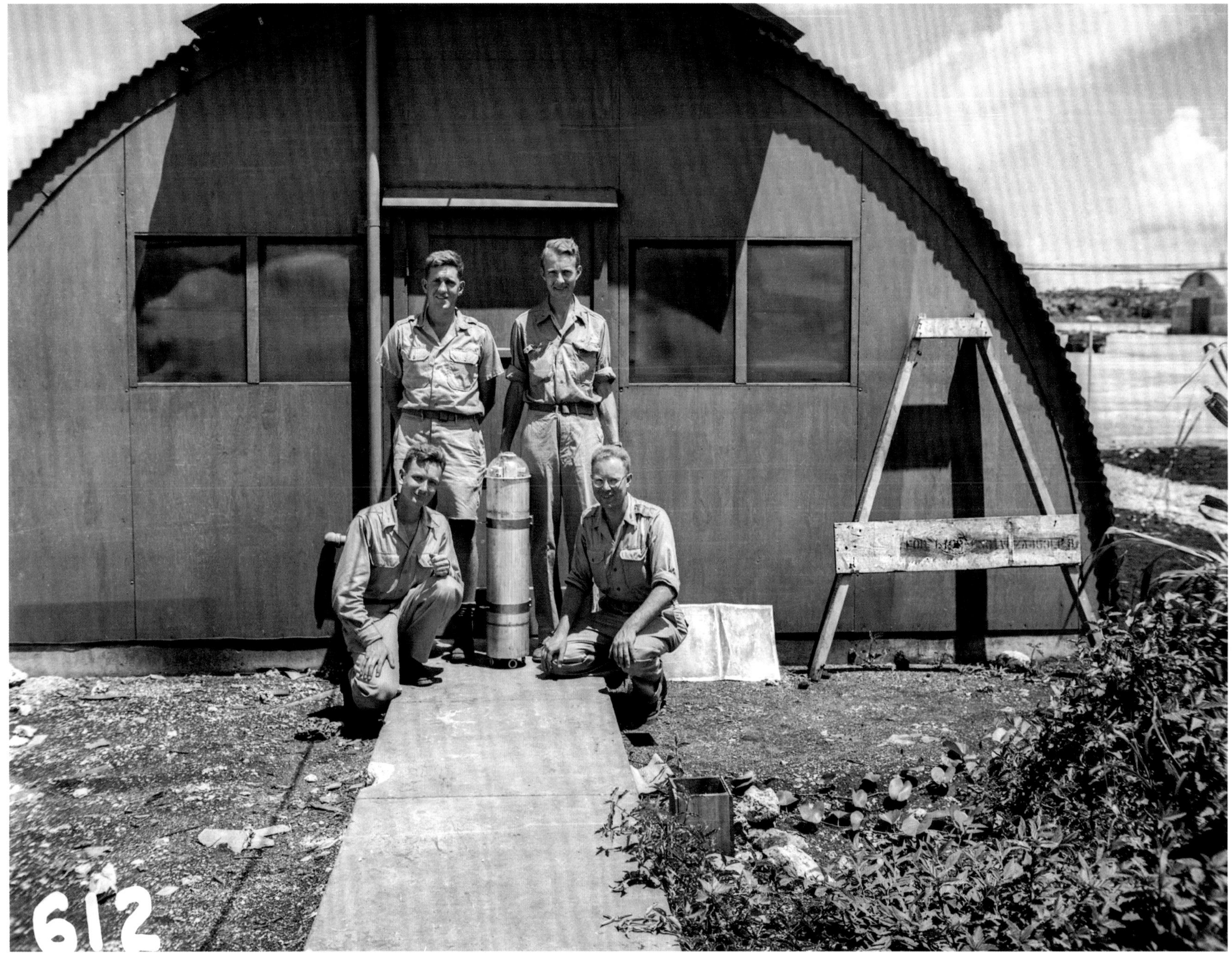

The *Enola Gay* took off just after two o'clock in the morning on August 6, joined in flight by two observation planes. The crew of the instrument plane, *The Great Artiste*, included Los Alamos scientists Harold Agnew (standing, left), Luis Alvarez (right), and (kneeling, left) Lawrence Johnston, inventor of the exploding bridge wire detonator. Having been present for Trinity, Johnston would become the only person to witness the first three nuclear detonations. The fourth man pictured is Bernie Waldman, who rode in the camera plane *Necessary Evil*. The group is posing next to a canister housing gauges for measuring the force of the blast. Dropped by parachute, the canister would transmit data back to *The Great Artiste* by radio.

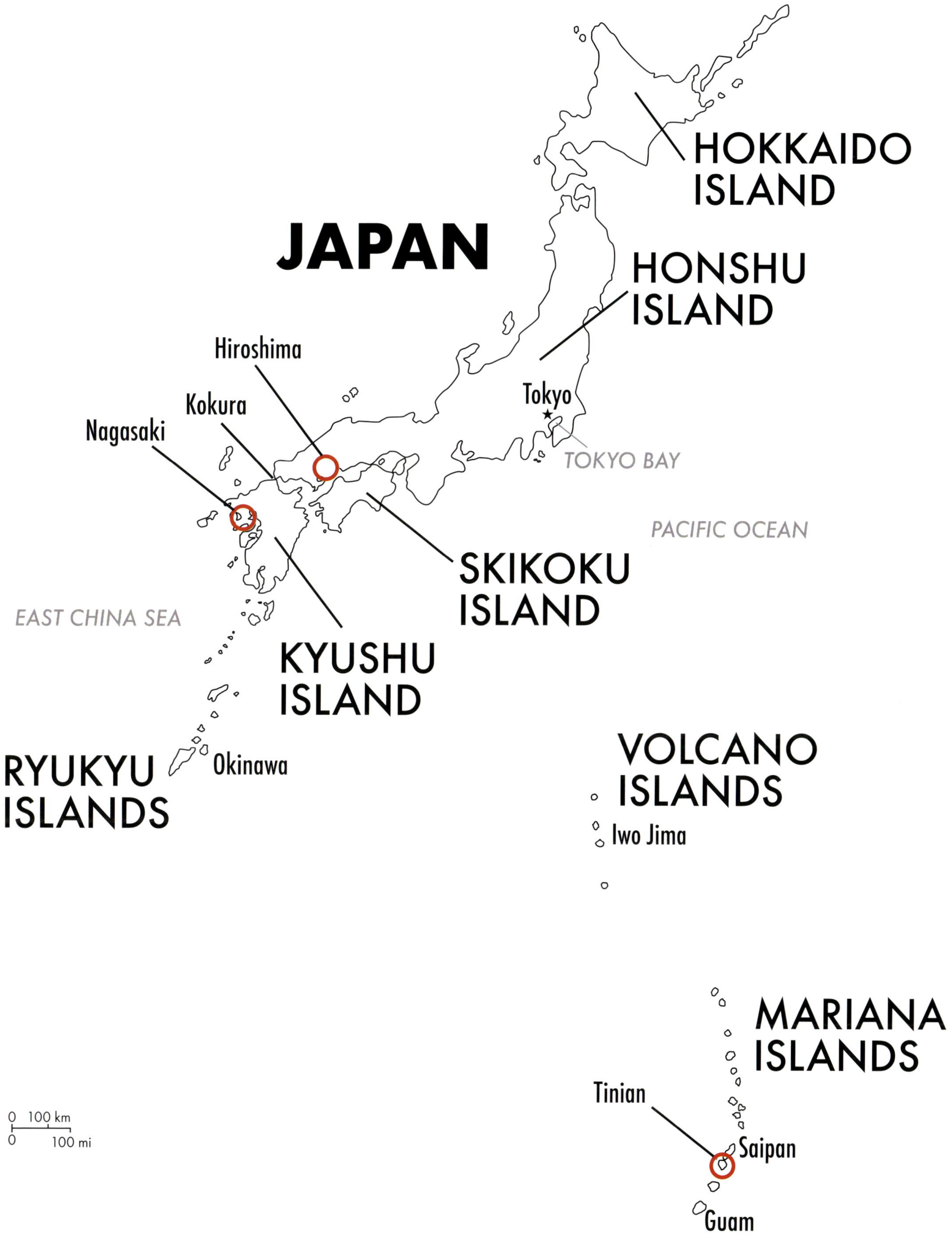

HOKKAIDO ISLAND
JAPAN
HONSHU ISLAND
Hiroshima
Tokyo
Kokura
Nagasaki
TOKYO BAY
PACIFIC OCEAN
SKIKOKU ISLAND
EAST CHINA SEA
KYUSHU ISLAND
RYUKYU ISLANDS
Okinawa
VOLCANO ISLANDS
Iwo Jima

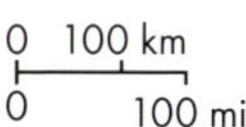
0 100 km
0 100 mi

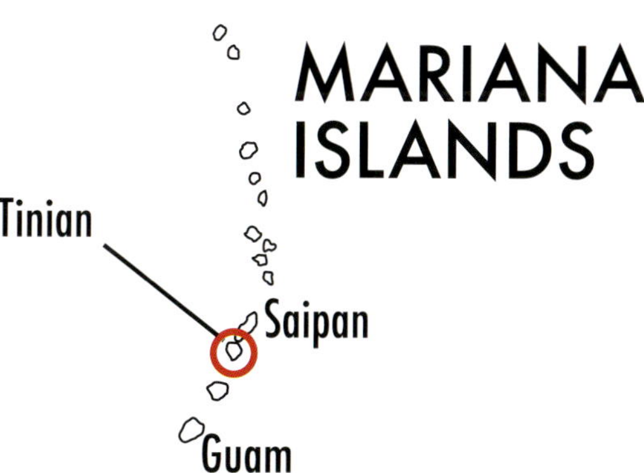
MARIANA ISLANDS
Tinian
Saipan
Guam

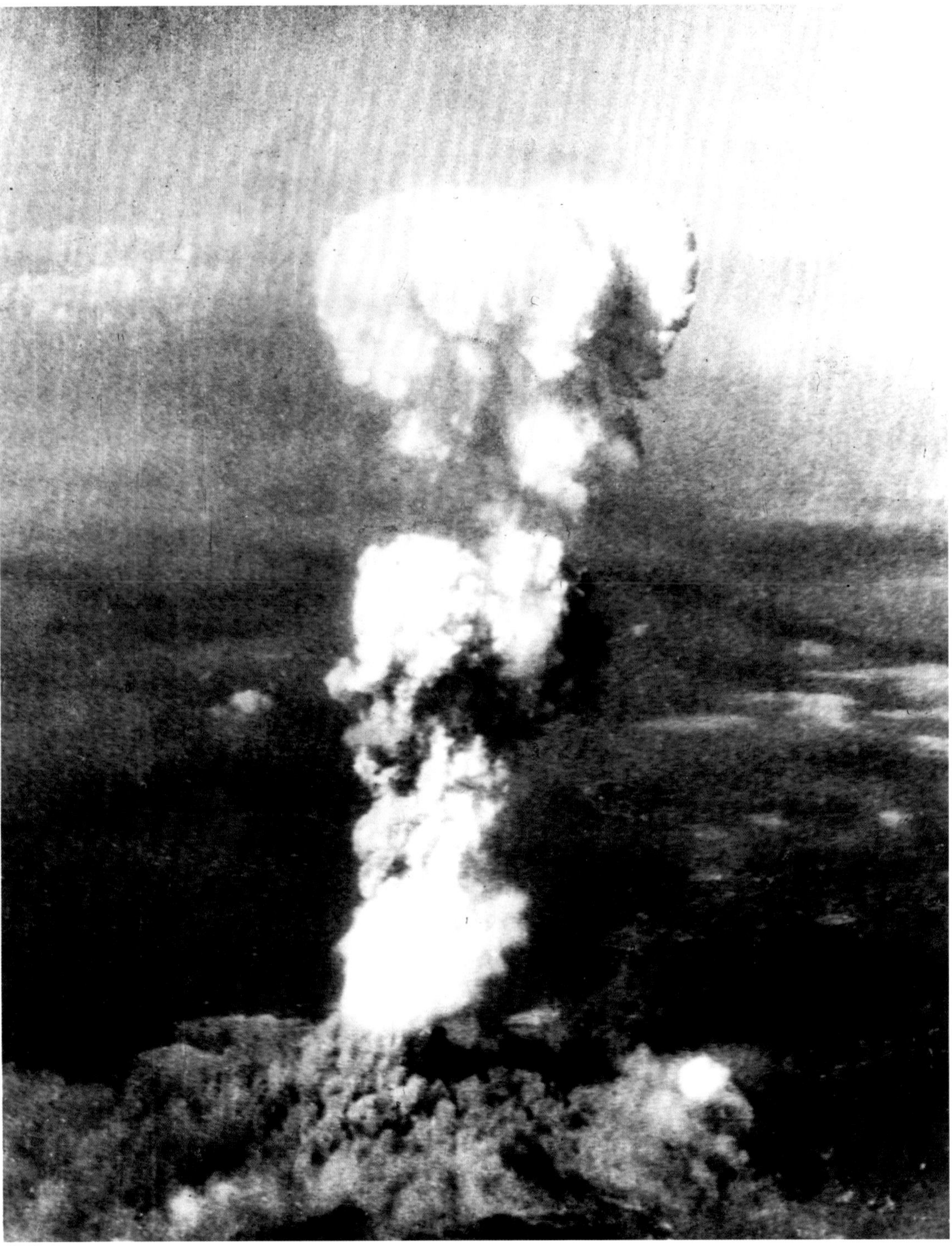

The target, Hiroshima, was located 2,500 miles northwest of Tinian. According to Parsons's flight log, the city came into view at 8:09 a.m. The aiming point was the Aioi Bridge, a unique T-shaped structure near the center of the city that would be easy to see from the air. The operation went as smoothly as it could have. No defensive fire was encountered, the target was sighted easily, and the bomb was released at 8:15 a.m. local time—the exact minute for which the drop had been scheduled.

As Little Boy free fell from over 30,000 feet to its preset detonation height, the *Enola Gay* was banking away in a maneuver Tibbets had practiced many times, gaining as much distance as possible from point zero in the forty-three seconds until detonation. The shock wave rocked the plane nearly a minute after the flash of the explosion, followed by a second wave reflected off the ground.

From a side window of *Necessary Evil*, 2nd Lt. Russell Gackenbach captured the still photograph at left, approximately two minutes after detonation. Those who saw the cloud described it as mushrooming at the top, breaking away from the column, and mushrooming again. Tibbets later wrote that "it was a frightening sight, and even though we were several miles away, it gave the appearance of something that was about to engulf us. Even more fearsome was the sight on the ground below: At the base of the cloud, fires were springing up everywhere amid a turbulent mass of smoke that had the appearance of bubbling hot tar."

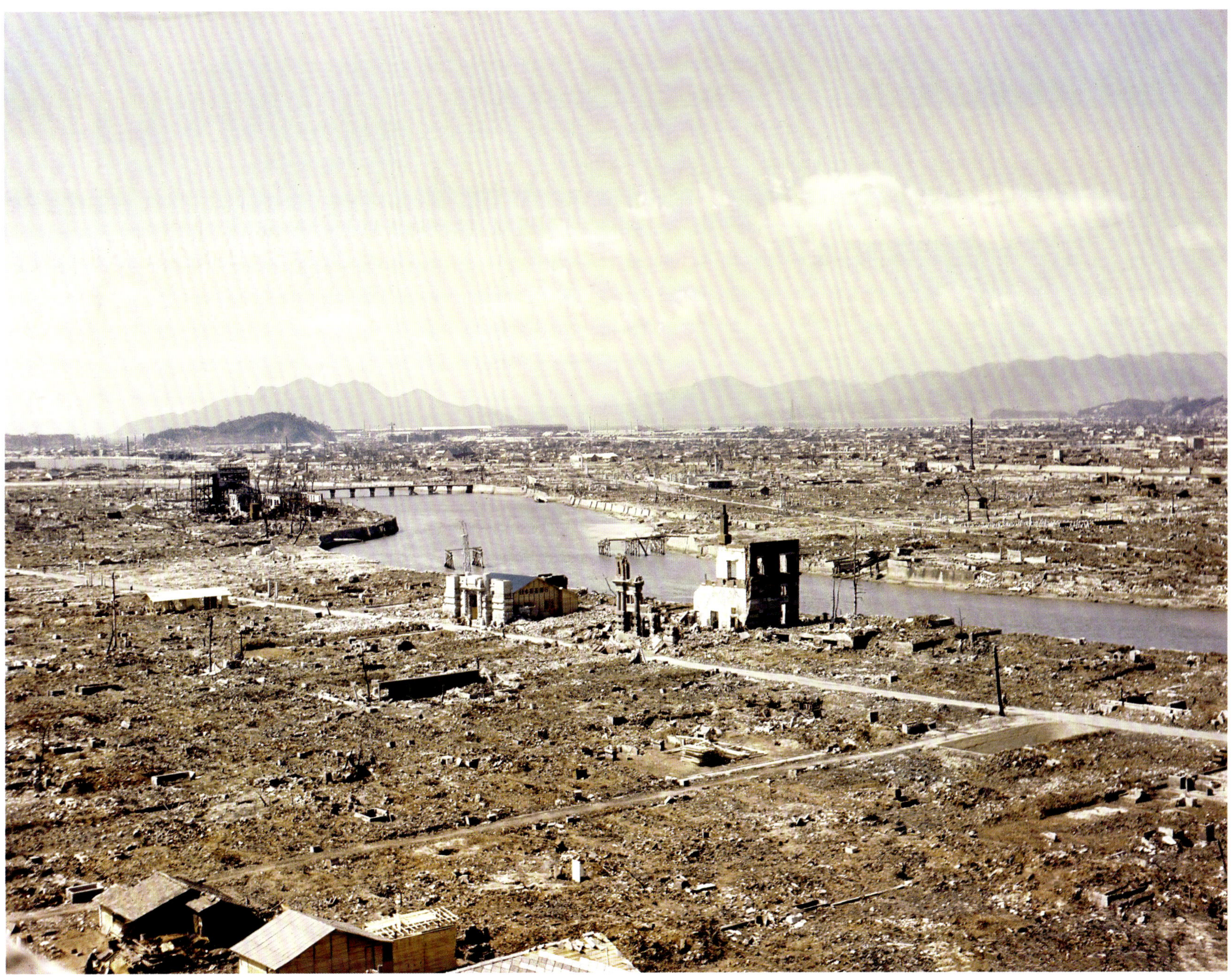

Four hours after the strike, observation planes were unable to take photographs of the city beneath a thick cloud of smoke and dust. Many structures not destroyed by the blast were consumed by a horrific firestorm, which turned large parts of Hiroshima to ashes. Aid was slow to arrive in the ensuing confusion, as Japanese officials investigated a sudden loss of contact with the city and hours later came upon the same scene as the American observation planes. It was not until the next day that the extent of the devastation could be seen. The extent of human suffering and the horrors that took place before the quiet in the scene above would be revealed much later. (National Archives Catalog, 148728174.)

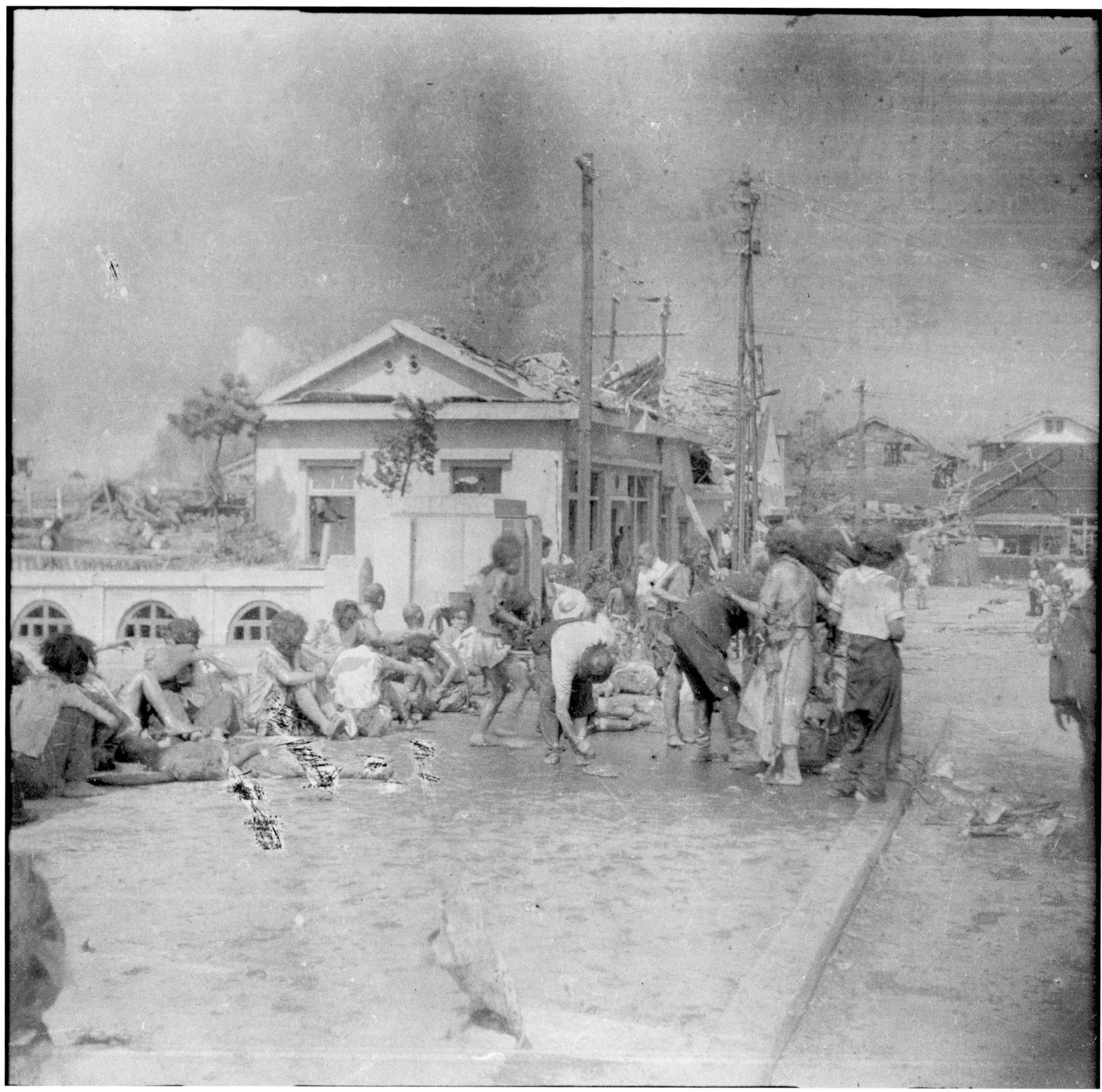

Only five pictures are known to exist of the immediate aftermath in Hiroshima. They were captured by Yoshito Matsushige, a staff photographer at the regional newspaper *Chugoku Shimbun*. Matsushige was at his home 1.7 miles from the hypocenter—the point on the ground directly below the bomb—when Little Boy exploded. "It was such a cruel sight that I couldn't bring myself to press the shutter," he recalled of the first twenty minutes outside his house. He eventually did, and Matsushige later described grappling internally over his actions but feeling duty-bound as a photojournalist to capture the suffering around him. The photograph above was the first one taken, near the west end of the Miyuki Bridge, where burn victims were being treated with cooking oil. (Photos courtesy of *Chugoku Shimbun*/Kyodo News Images.)

Several people have come forward to identify themselves in the photographs since they were published in 1952.

Standing in the back of the crowd in a white shirt (*left*) is Yasutaro Matsubayashi, a physician who was injured when his home collapsed but who treated other wounded people with supplies in his emergency bag. Police sergeant Katsuyuki Fujikawa, in uniform (*right*), was injured by shattered glass; he later wrote in a memoir: "I applied oil and bandages as treatment for the burns of the wounded. . . . Some collapsed on the spot and did not move again. I could do absolutely nothing to help them." Mitsuko Sakamoto, standing closest to the camera, was at school when the bomb went off, throwing her against a wall and knocking her unconscious. She recognized herself in these photographs in 1973 while visiting an exhibit at the Hiroshima Peace Memorial Museum.

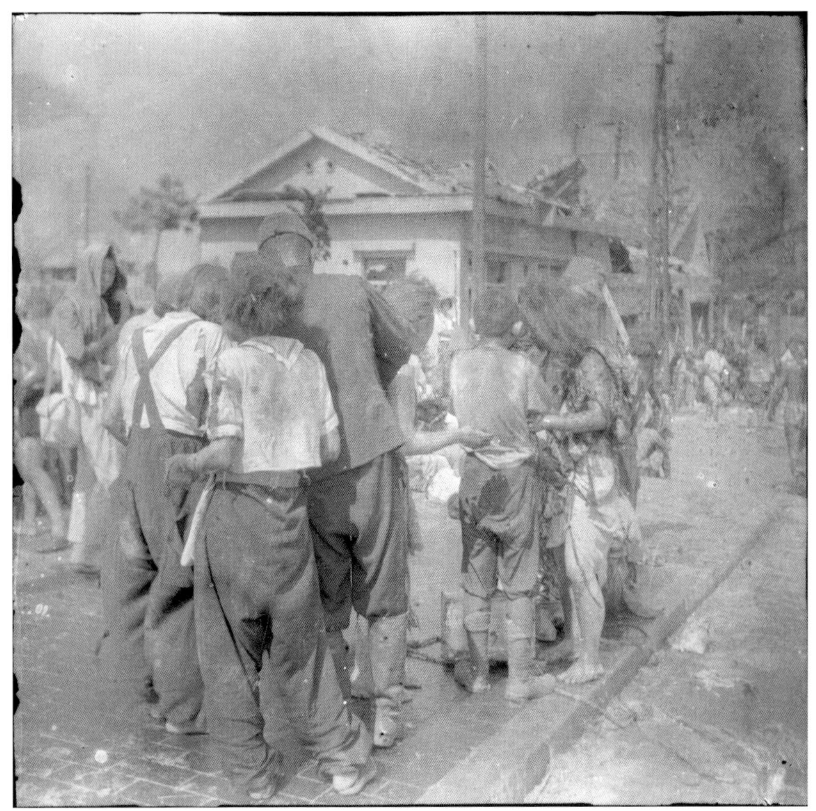

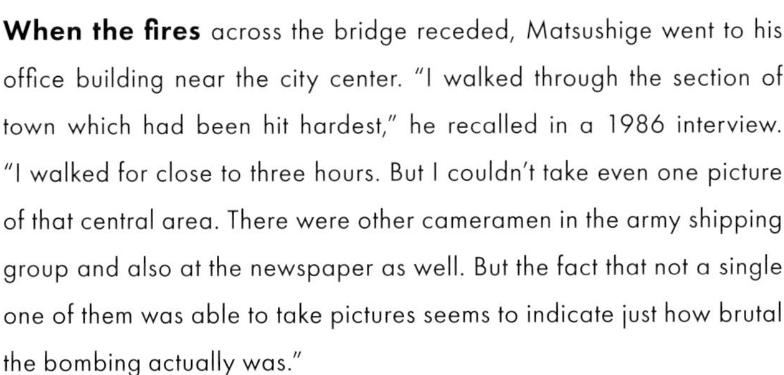

When the fires across the bridge receded, Matsushige went to his office building near the city center. "I walked through the section of town which had been hit hardest," he recalled in a 1986 interview. "I walked for close to three hours. But I couldn't take even one picture of that central area. There were other cameramen in the army shipping group and also at the newspaper as well. But the fact that not a single one of them was able to take pictures seems to indicate just how brutal the bombing actually was."

On his way home, Matsushige passed police officer Tokuo Fujita writing aid tickets near the east end of the Miyuki Bridge (*right*).

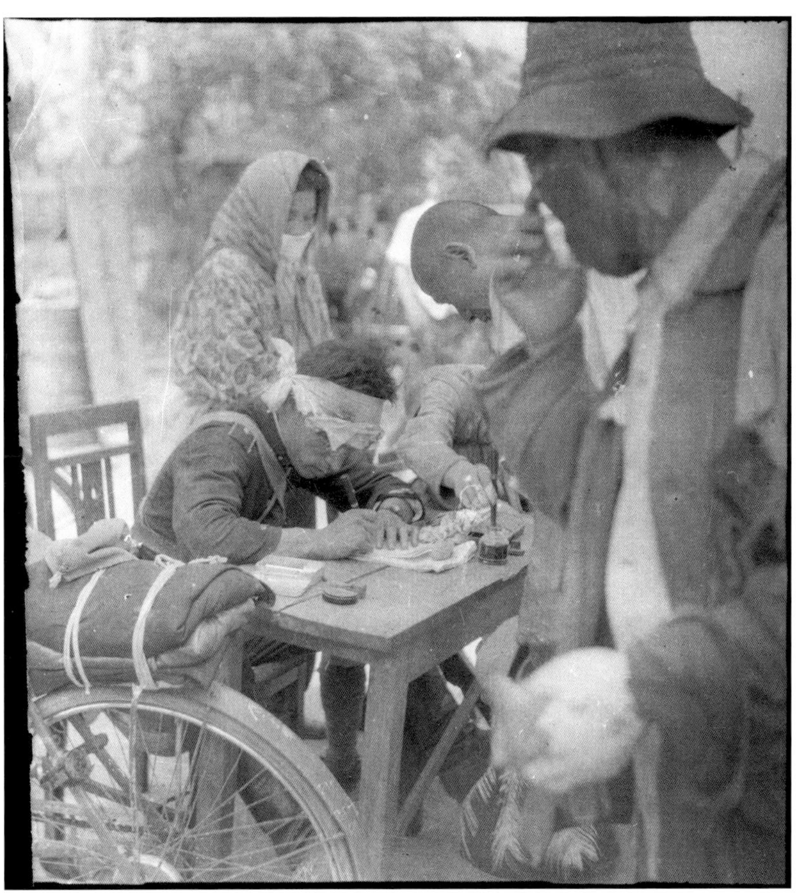

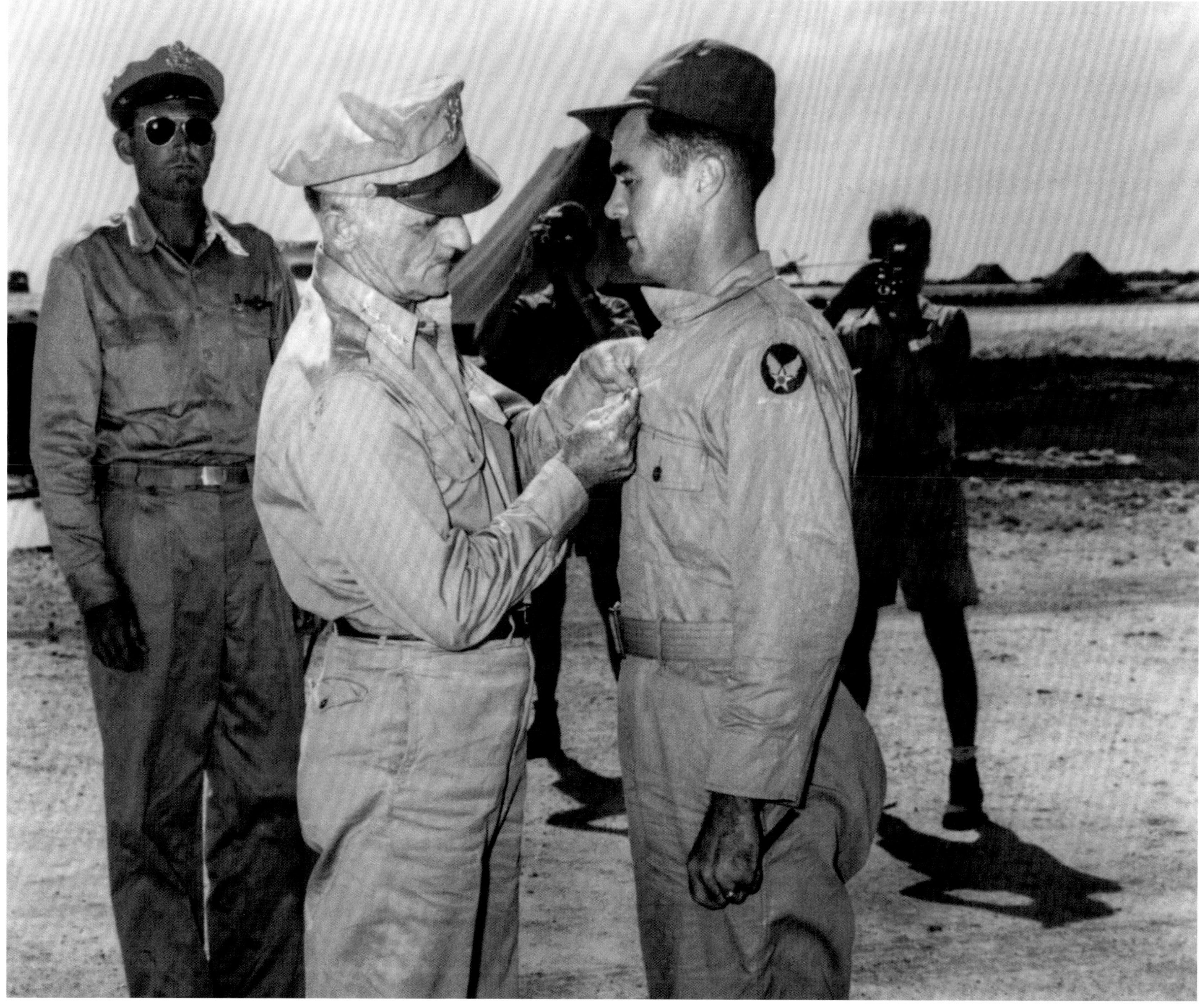

When the *Enola Gay* returned to Tinian, twelve hours after releasing the bomb, Gen. Carl Spaatz, commander of the United States Strategic Air Forces in the Pacific, decorated Colonel Tibbets with the Distinguished Service Cross. The mission, under his command, had been a success. From the military's perspective, an entire city had been rendered unusable as a machine of war. In hastening the end of the fighting, however, a lone weapon had not produced the desired result. As after the Potsdam Declaration, which threatened the "complete and utter destruction" now carried out in Hiroshima, no move toward surrender came.

THIS IS A WARNING TO THE JAPANESE PEOPLE!
LEAVE THIS CITY IMMEDIATELY!

The contents of this flyer are very important. The Japanese people are facing a very significant change. Your Military rulers were presented the opportunity to stop this pointless war in the Thirteen Articles of the Joint Resolution. Your Military Rulers refused. For this reason the Soviet Union has declared war on Japan. Further, the United States has invented and tested a most formidable weapon, the atomic bomb, even though it was thought impossible. This atomic bomb, alone, is as destructive as the usual bomb load of two-thousand B-29's. (end of first page)

You will know this is true when you have seen the devestation caused by only one atomic bomb dropped on Hiroshima. The Japanese military is causing this pointless war to continue, so we are going to destroy them with this fearsome weapon. Before the United States uses many of these atomic bombs on Japan we wish the Japanese people would petition the Emporer to stop this war.

The President of the United States has announced the Thirteen Articles of the Joint Declaration and hopes the Japanese people will accept these Articles very soon, become a peace loving people and build a new Japan. The Japanese people must stop their military resistance now. If you do not stop this war the United States will be forced to use the atomic bomb and other superior military weapons.

LEAVE THIS CITY IMMEDIATELY!

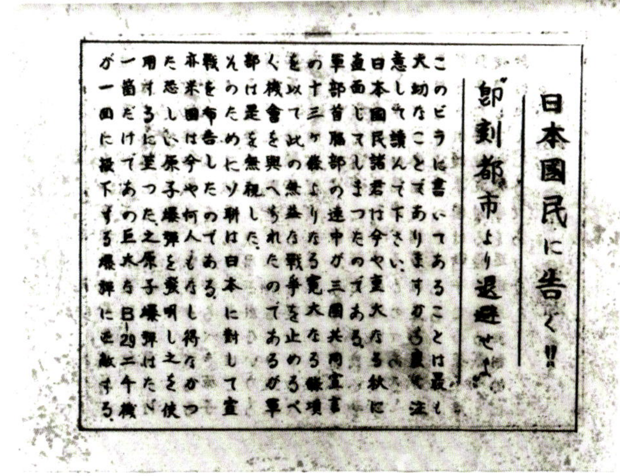
日本国民に告ぐ!!
即刻都市より退避せよ

Preparations continued, in accordance with Truman's July 25 order, to deploy Fat Man over the second target city. With more implosion weapons not far behind in production, a warning message was communicated directly to the people of Japan. The message was read on Radio Saipan every fifteen minutes and distributed to cities of more than 100,000 people in the form of leaflets (like that shown above) dropped by American bombers. The leaflets warned that the atomic bomb, which had devastated Hiroshima, was "as destructive as the usual bomb load of two thousand B-29s." A full translation is at left.

Printing the millions of copies took time, however, and the residents of Nagasaki did not receive the warning before the second and final nuclear bomb would be detonated over their city.

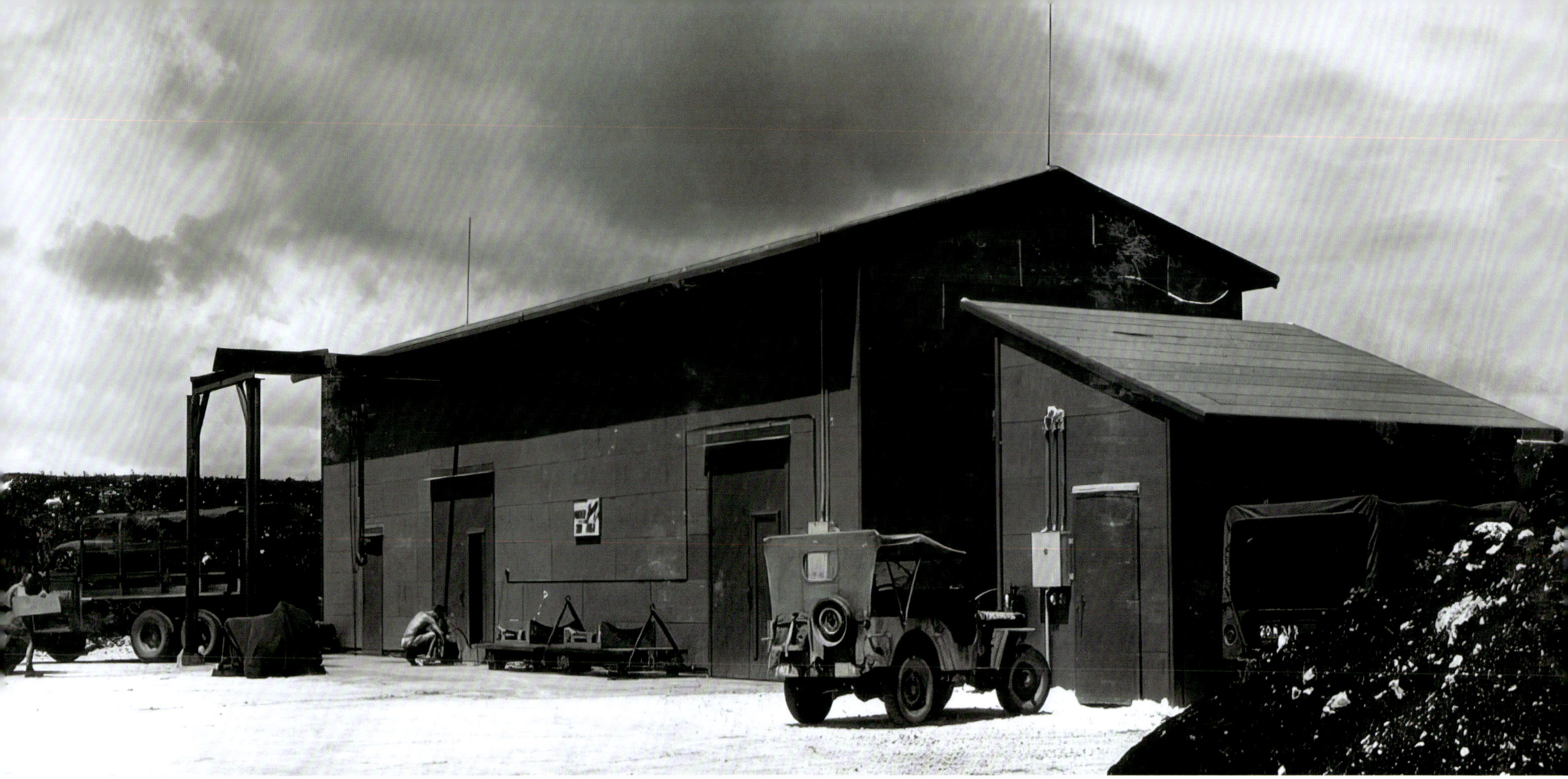

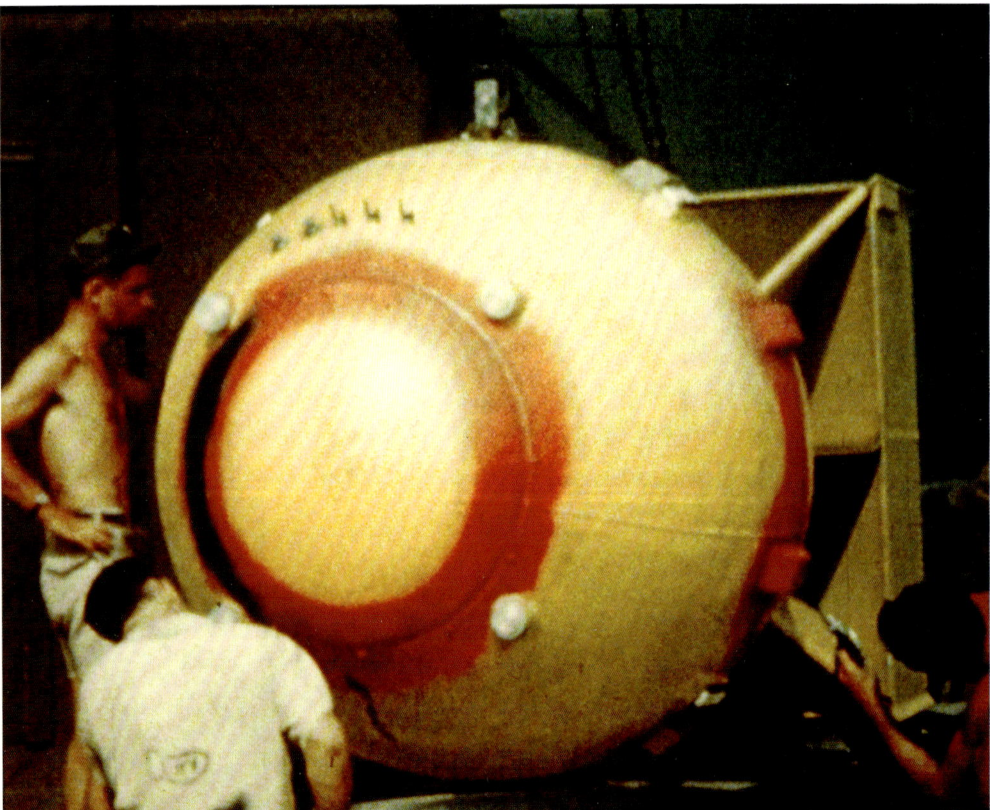

Although scheduled for August 11, the Fat Man strike was advanced to August 9 due to incoming bad weather that would have pushed the mission back by several days. A scramble ensued among Project Alberta workers to assemble the device on a shortened timeline.

Of the three high-explosives spheres that had been delivered to Tinian, one was badly cracked and unusable; another was earmarked for a dry run of combat drop procedures that was successfully conducted on August 8; and the third, assessed to have the highest-quality explosive lenses, was designated to receive the lone plutonium core on the island. This unit, called F-31, was prepared for detonation at the building above. (National Archives Catalog, 204976346.)

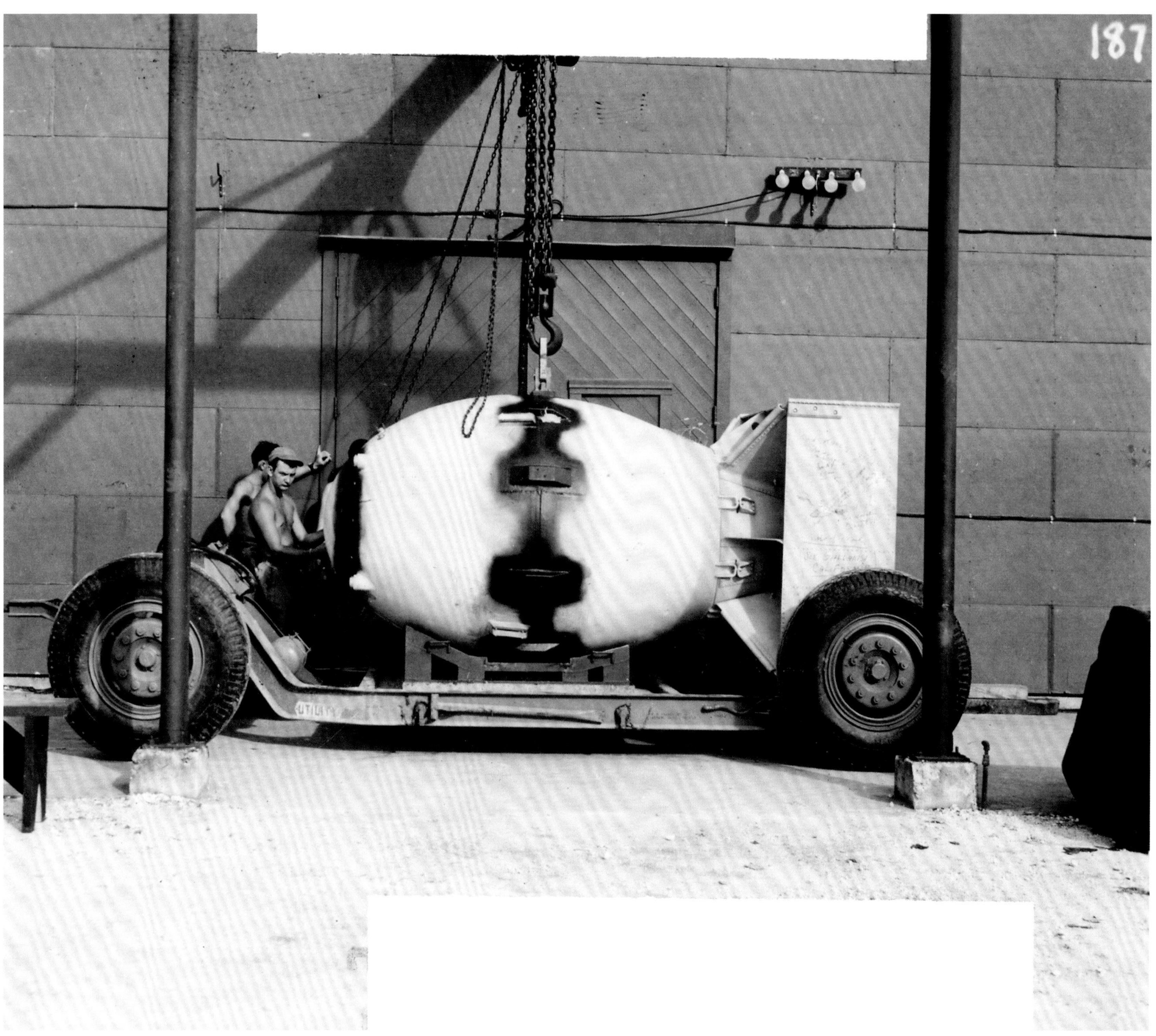

As crew members returning from the first mission readied themselves for another flight, F-31 was sealed at the seams with a waterproofing resin and then placed on a special trailer that would carry it from the hoist at the assembly building to the runway. (National Archives Catalog, 519397.)

Physicist Luis Alvarez (standing on the right, holding Fat Man's core), who rode in an observation plane during the first mission, had written a letter to his son while the plane returned to Tinian, as he reflected on his role in the bomb's creation and now its use. With a second strike approaching, Alvarez penned another letter, with help from fellow Project Alberta members Robert Serber and Phillip Morrison. The message, which they placed in a blast gauge canister that would be dropped with Fat Man, was addressed to their former colleague at the University of California, Japanese nuclear physicist Ryokichi Sagane. It reads, in part:

"You have known for several years that an atomic bomb could be built if a nation were willing to pay the enormous cost of preparing the necessary material. Now that you have seen that we have constructed the production plants, there can be no doubt in your mind that all the output of these factories, working 24 hours a day, will be exploded on your homeland.

Within the space of three weeks, we have proof-fired one bomb in the American desert, exploded one in Hiroshima, and fired the third this morning.

We implore you to confirm these facts to your leaders, and to do your utmost to stop the destruction and waste of life which can only result in the total annihilation of all your cities, if continued. As scientists, we deplore the use to which a beautiful discovery has been put, but we can assure you that unless Japan surrenders at once, this rain of atomic bombs will increase manyfold in fury."

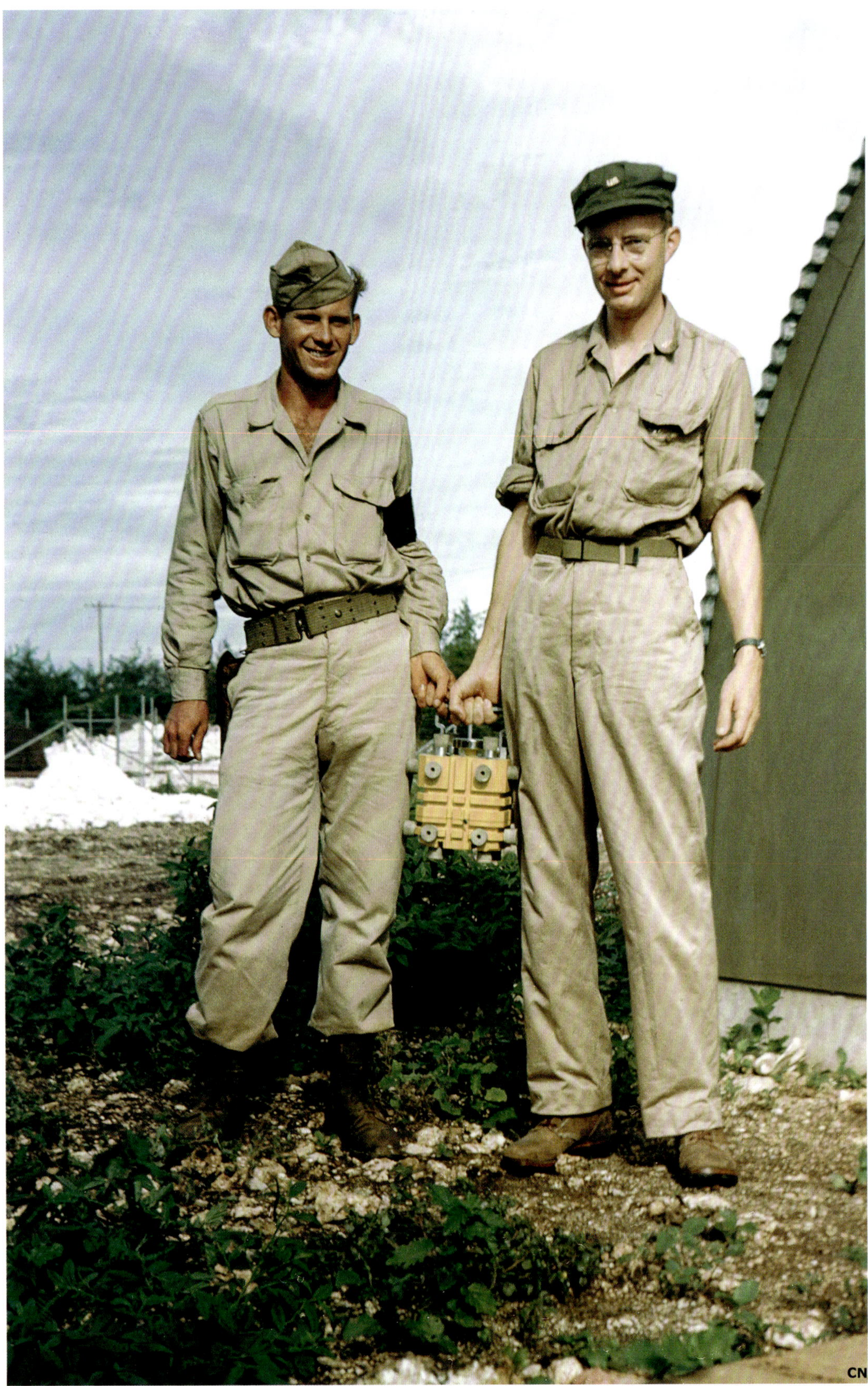
CN

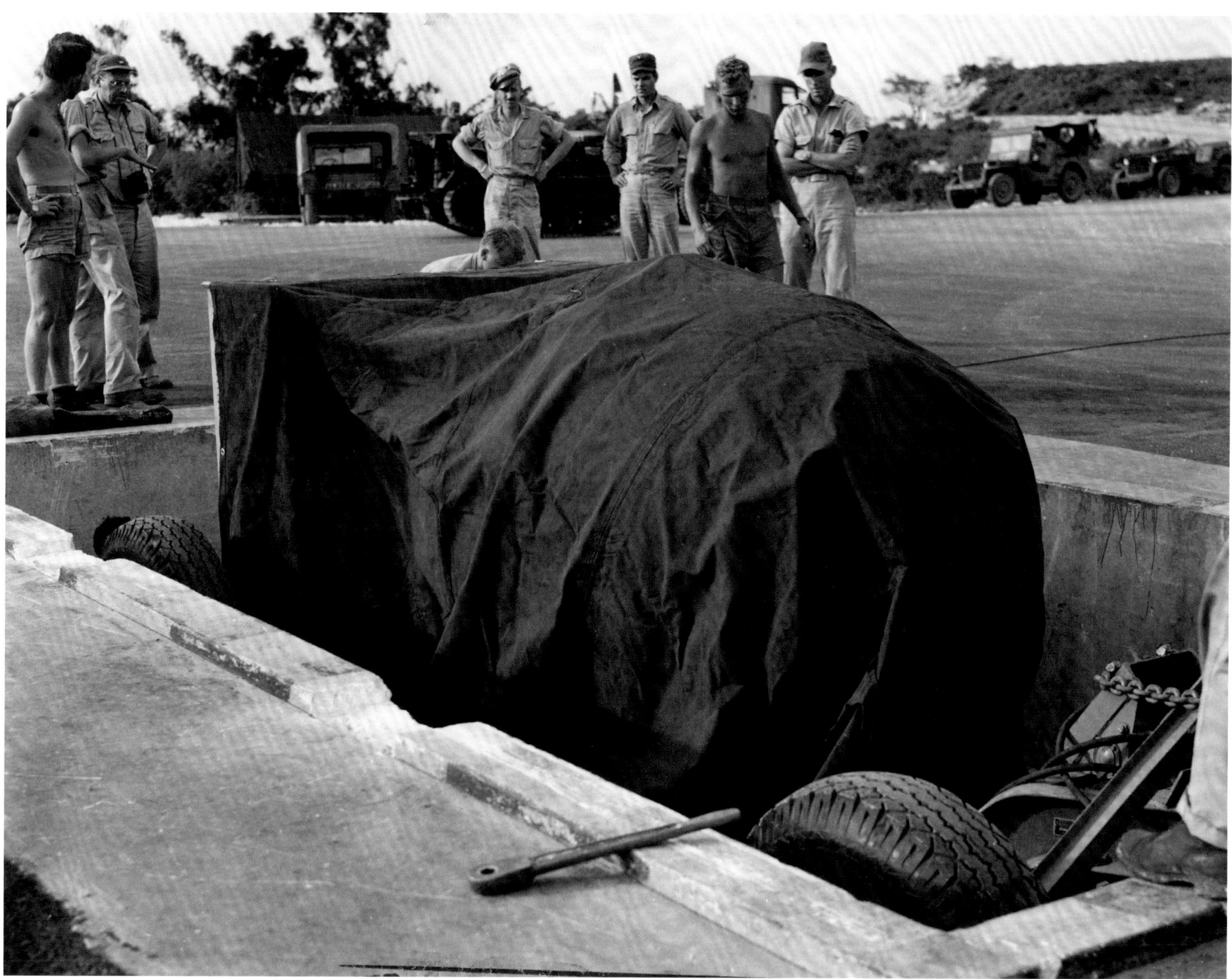

By 10:00 p.m. on August 8, the implosion bomb had been raised from the loading pit and rested in the forward bay of the strike plane, *Bockscar.* The crew for the mission included aircraft commander Charles W. Sweeney, pilot Charles D. Albury, bombardier Kermit K. Beahan, and Frederick Ashworth fulfilling the role of weaponeer. The usual crew of *Bockscar,* led by aircraft commander Frederick Bock, for whom the plane was named, flew *The Great Artiste* on observation duty. The *Enola Gay* was assigned to weather reconnaissance. (National Archives Catalog, 76048738; NND 730339.)

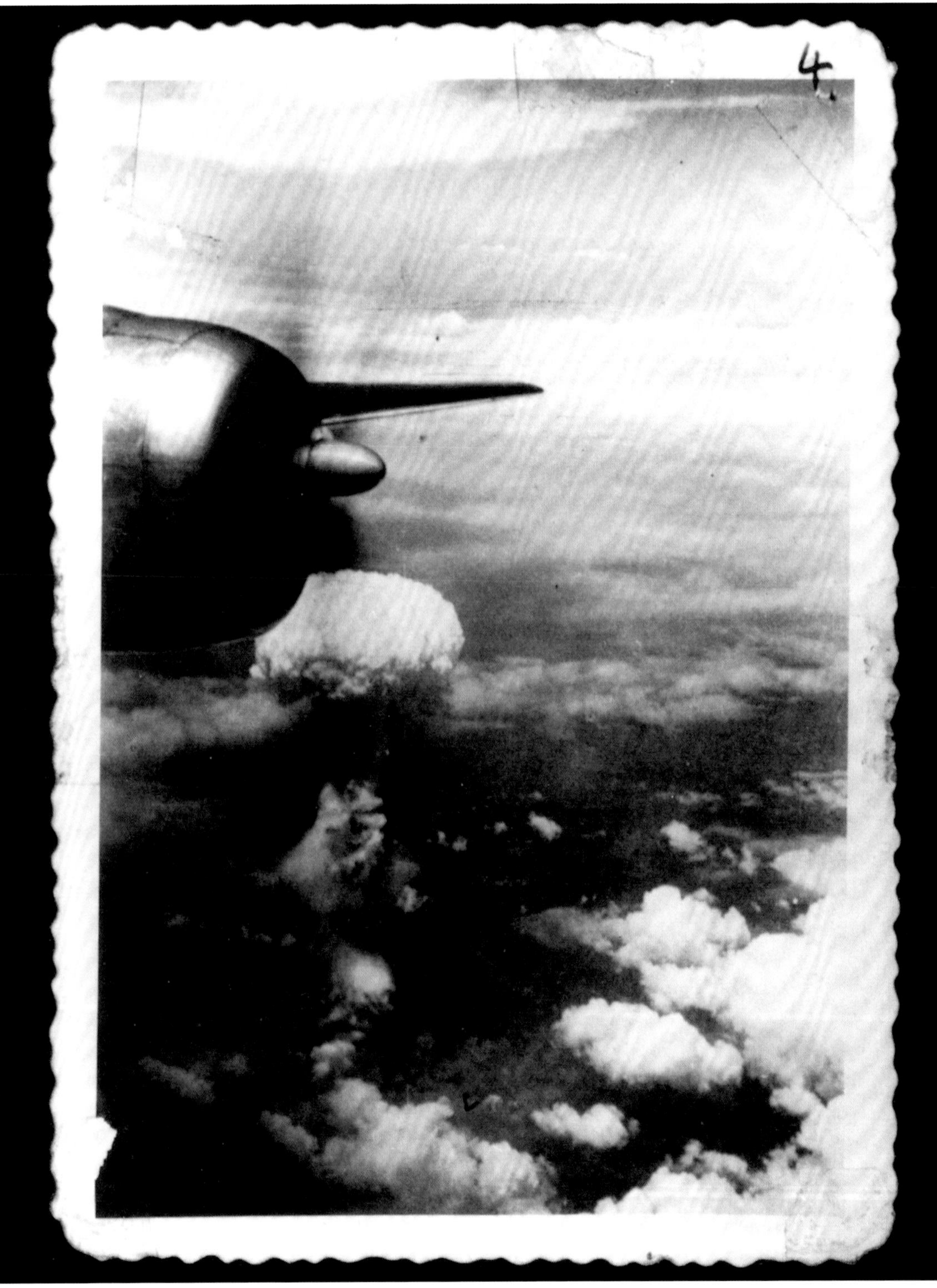

Unlike the previous mission, the flight of *Bockscar* was fraught with challenges. Before takeoff, it was discovered that a broken pump would render several hundred gallons of fuel on the plane inaccessible. With poor weather expected in the days ahead, it was decided to proceed with the mission. Then, the crew was unable to make contact with one of the observation planes at the rendezvous point. Again, the decision was to proceed.

The pilot made three runs over the primary target, Kokura, spanning forty-five minutes, but the ground was obscured by a smoky haze. Anti-aircraft fire had begun as well. The only backup target on this mission was the port city of Nagasaki about 100 miles away. The crew calculated that they did not have enough fuel to return to a safe landing spot off the mainland with the heavy bomb on board. If they were able to fly to Nagasaki and release it, however, the 10,000-pound weight reduction would allow them to reach the nearest island base of Okinawa.

Bockscar approached Nagasaki entirely by radar, and it seemed that, again, visual bombing would not be possible. Then, through a break in the clouds, Beahan spotted part of the city below. The plane lurched upward as he released the bomb.

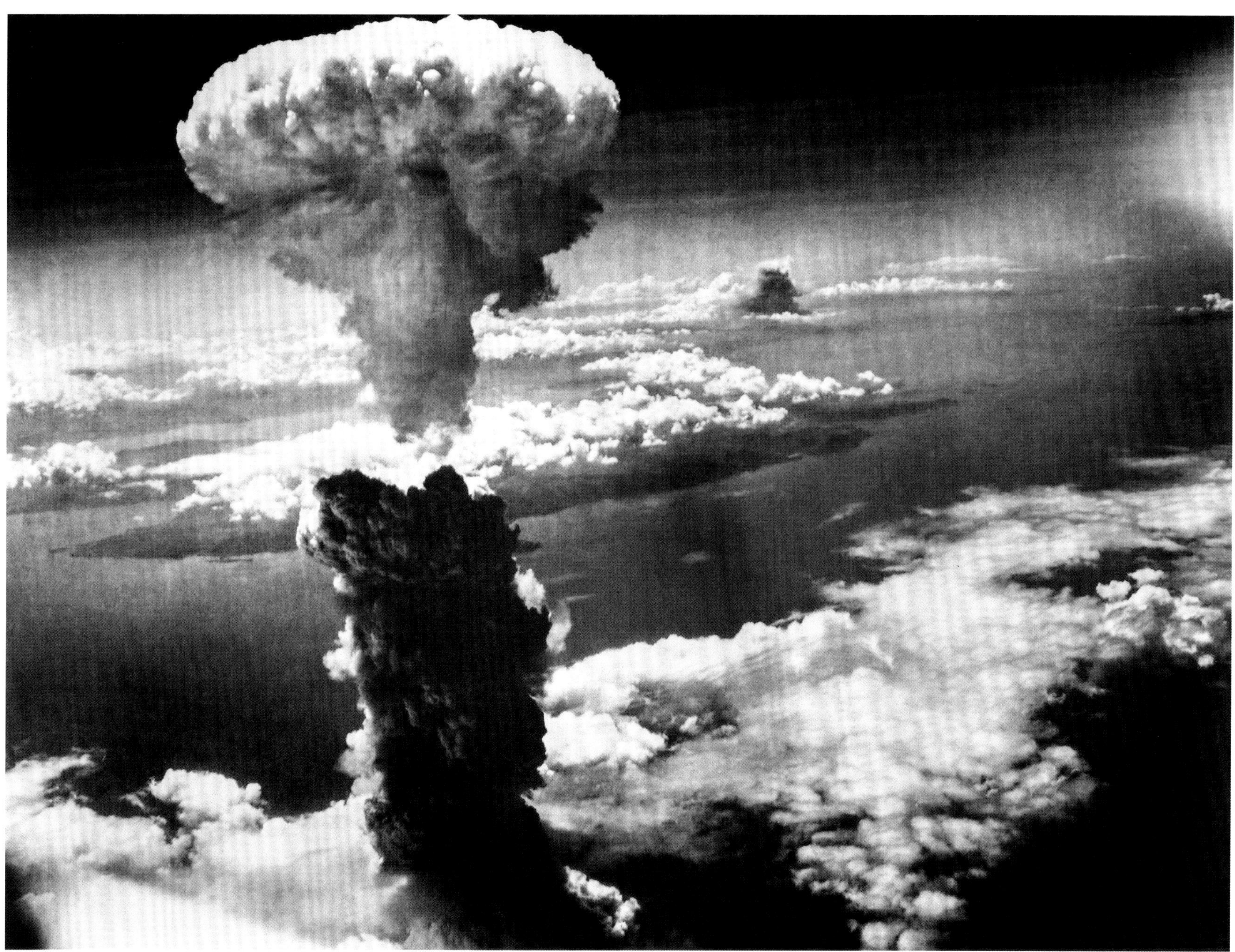

As the crew of *Bockscar* banked away from the third and final atomic explosion of 1945, the nightmare experienced by those in Hiroshima began to repeat itself below, obscured beneath the overcast sky and a new, man-made cloud that mushroomed over the city.

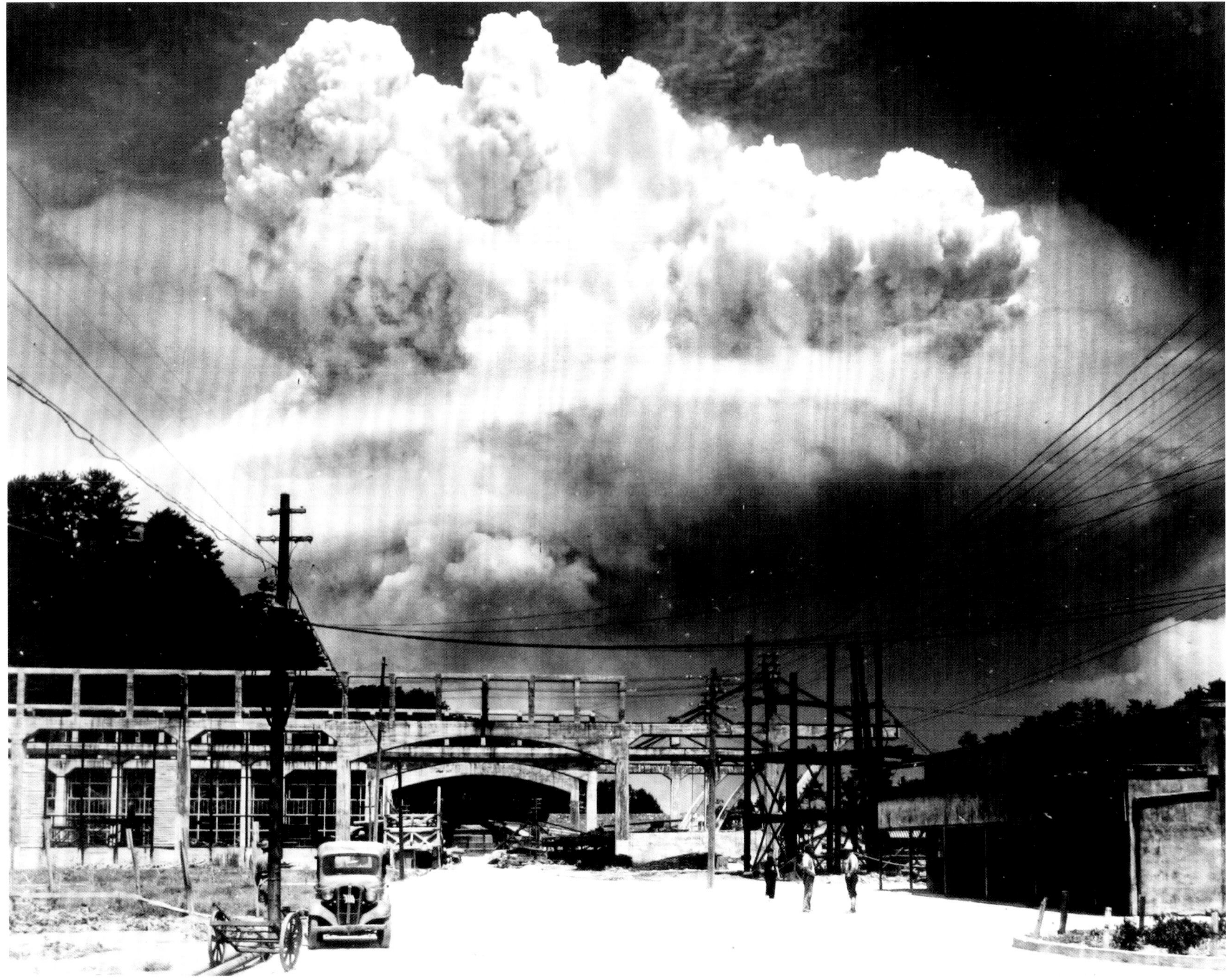

The earliest known image captured from the ground was taken by Hiromichi Matsuda, a photographic technician working at Kawanami Brothers shipyard on the small island of Kōyagi, approximately five miles southwest of the city center. It shows a cloud rising over Nagasaki fifteen minutes after detonation. Later that day, the Japanese Army dispatched a group of three men to survey the impact of the bombing: a painter, Eiji Yamada; a writer, Jun Higashi; and a photographer, Yosuke Yamahata. They arrived early the next morning to a city in darkness, making their way to the military police headquarters. As the sun rose on August 10, the group began to draw, write, and photograph, walking through what Yamahata later described as streetless, ashen ruins.

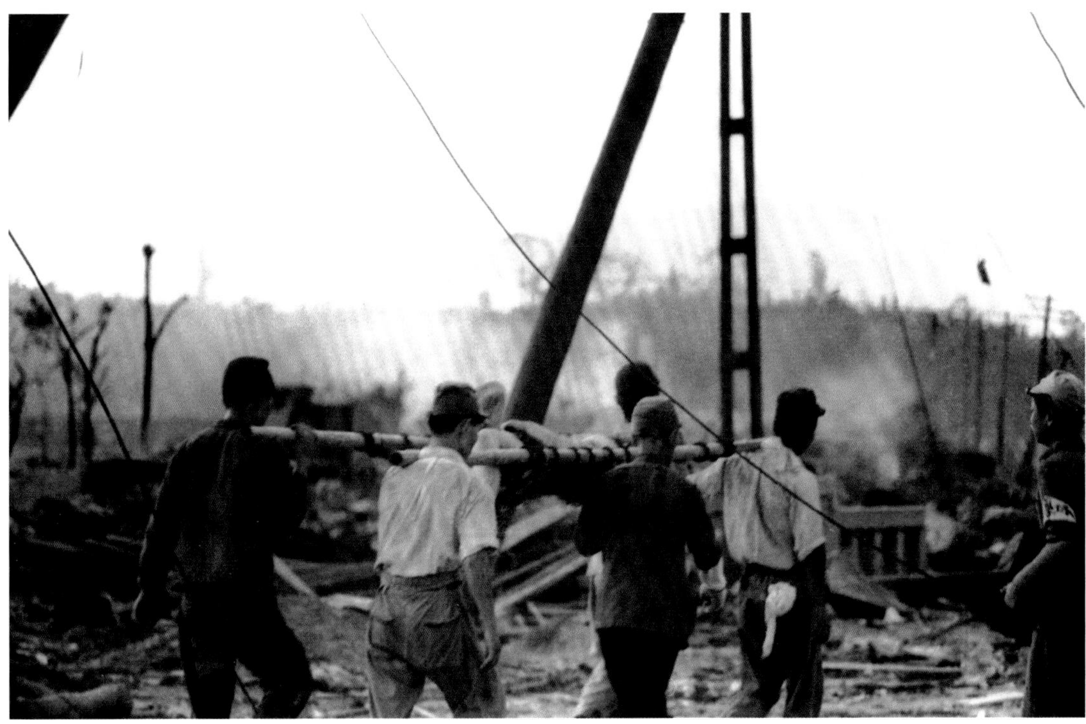

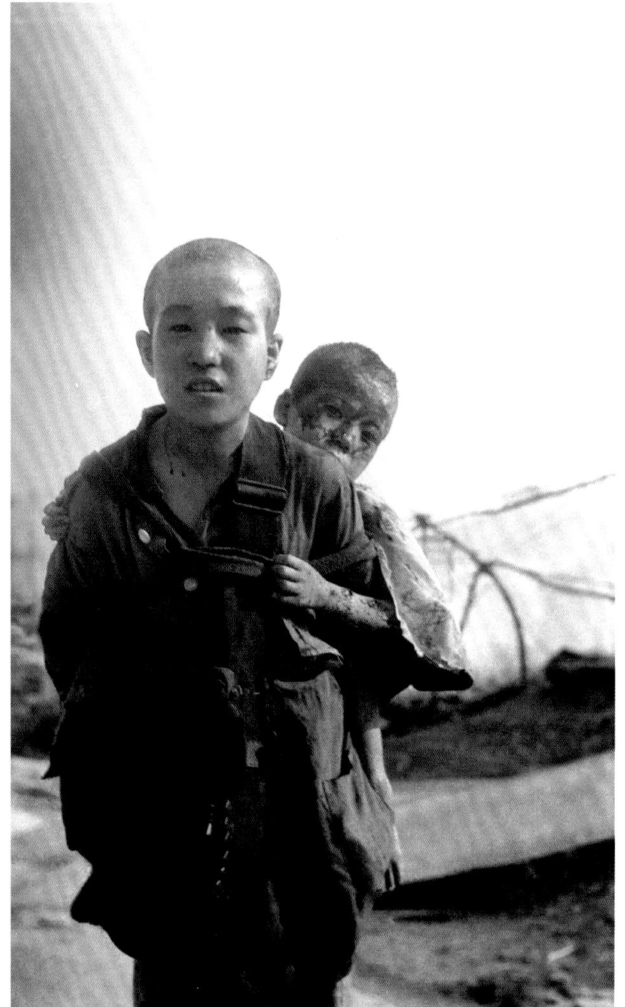

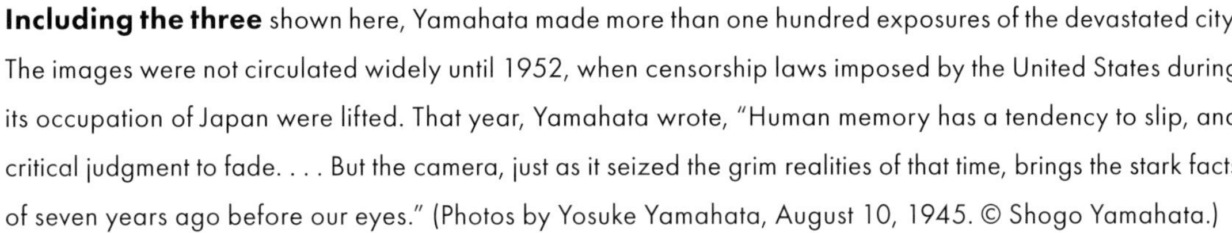

Including the three shown here, Yamahata made more than one hundred exposures of the devastated city. The images were not circulated widely until 1952, when censorship laws imposed by the United States during its occupation of Japan were lifted. That year, Yamahata wrote, "Human memory has a tendency to slip, and critical judgment to fade. . . . But the camera, just as it seized the grim realities of that time, brings the stark facts of seven years ago before our eyes." (Photos by Yosuke Yamahata, August 10, 1945. © Shogo Yamahata.)

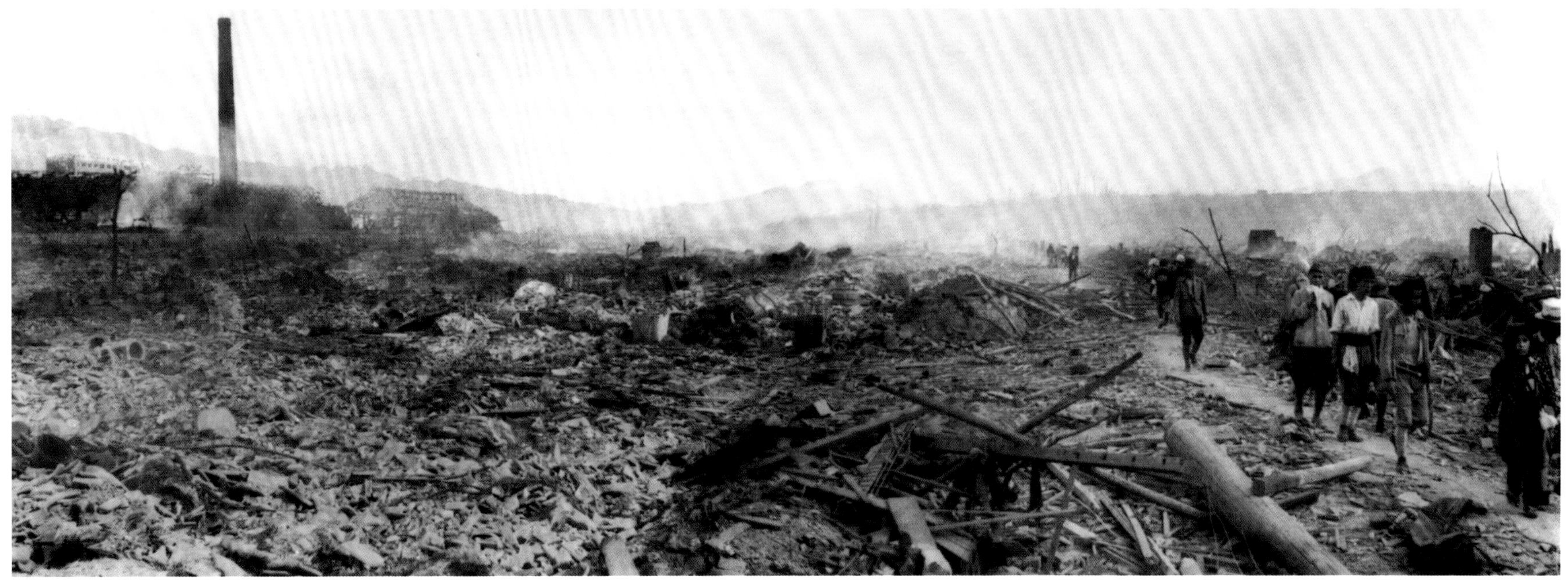

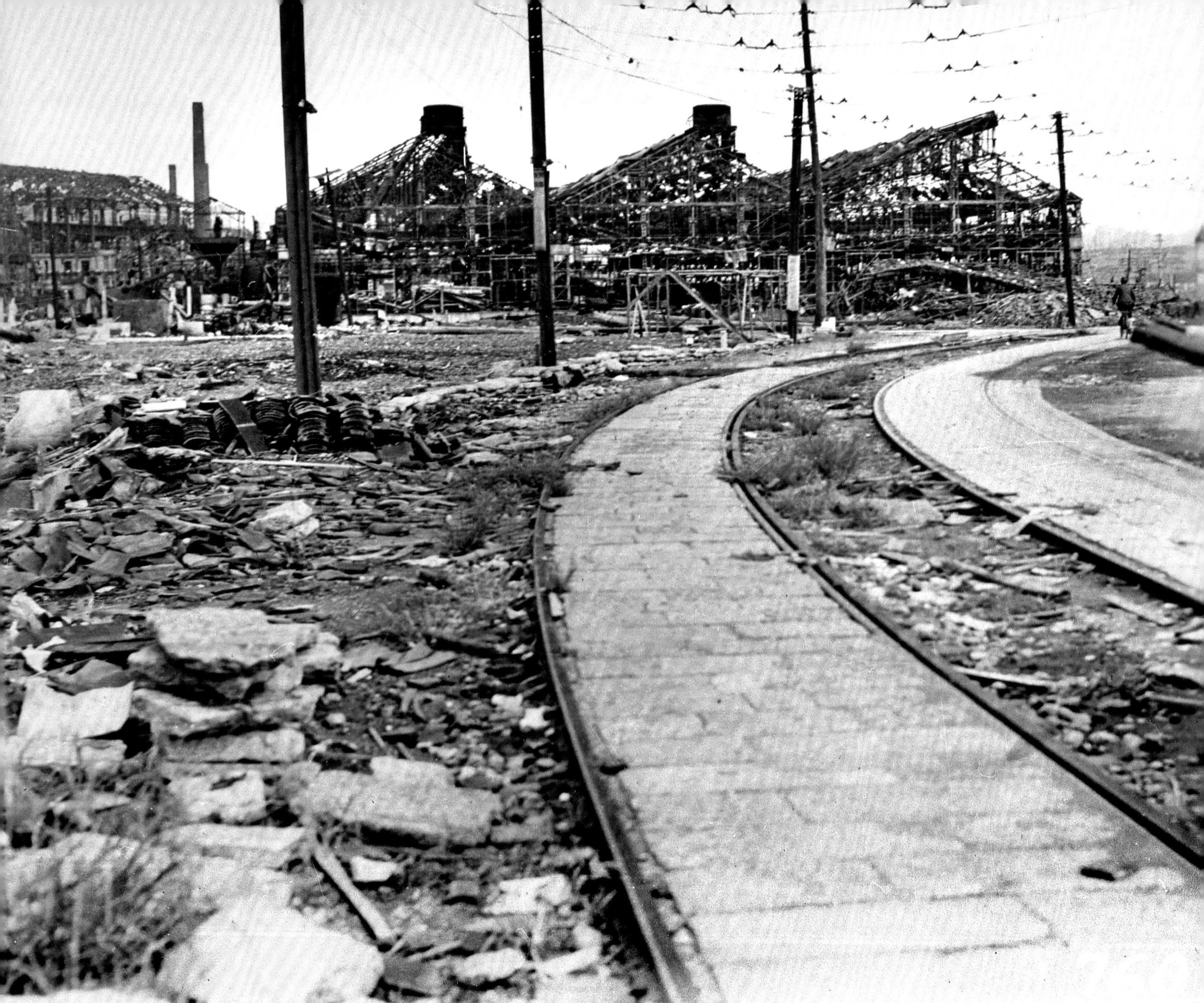

A month after the attacks, in September 1945, a group of soldiers, doctors, and scientists made an official visit to the two Japanese cities with Groves's deputy, Brig. Gen. Thomas Farrell, to document the effects of the weapons. Among the surveyors was Col. Stafford Warren, the chief medical officer of the Manhattan Project, who was also at the Trinity test. The group assessed loss of life, loss of structures, and lingering radiation; the latter, Warren reported, was not detected in appreciable amounts. They captured many photographs, including the above image of the Mitsubishi Arms Works, 1,500 yards from the hypocenter in Nagasaki.

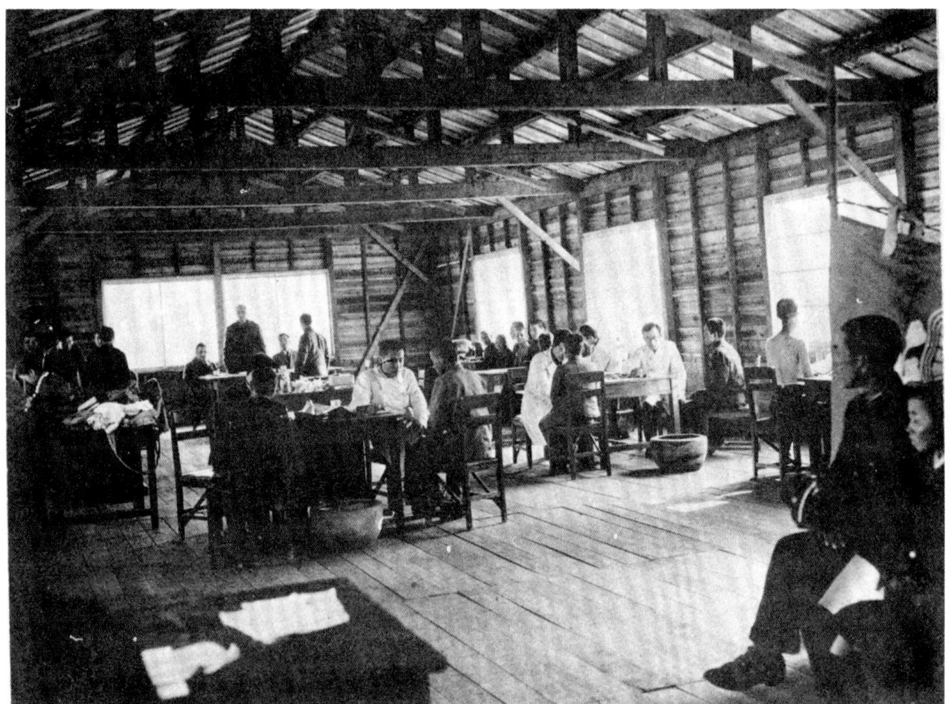

A Joint Commission for the Investigation of the Atomic Bomb, consisting of both American and Japanese researchers, later conducted a more thorough assessment of the number and nature of casualties resulting from the attacks. They analyzed medical effects of the nuclear blasts through questionnaires, interviews, clinical data, autopsies, and case studies. One of the commission's clinics is shown at left, inside the Hiroshima Central Railroad Station.

Most injuries and deaths in both cities were found to have resulted from thermal radiation—a hot burst of light so intense that it caused significant burns to exposed skin within a radius of over two miles. The woman pictured below is identified as Akamatsu in the commission's 1951 report. She was 1,500 yards from the hypocenter in Hiroshima, standing outdoors in the open. Except where the seams and straps of her clothing provided protection, her back and arms were severely burned.

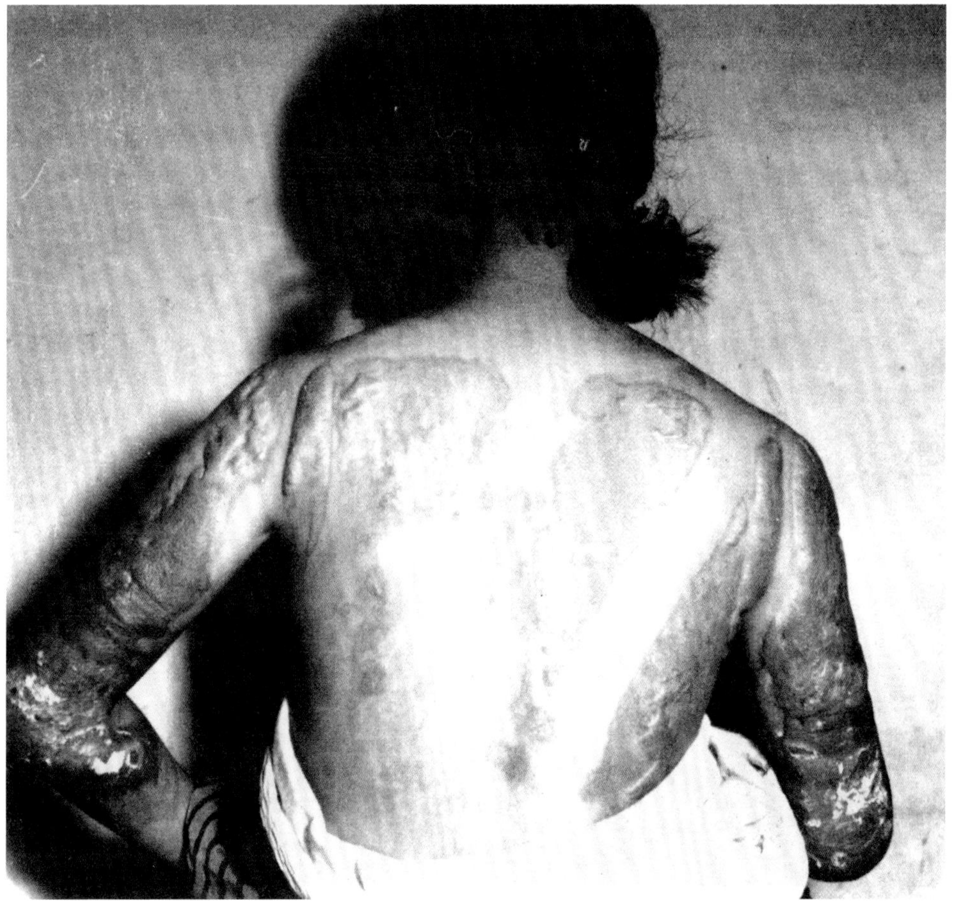

The searing heat lasted only a few moments, but the force of the explosion continued to cause serious and fatal injuries as it collapsed buildings, created flying debris, and led to secondary fires. In addition to facing these perils, many people were harmed by an invisible effect of the weapon—radiation, in the form of gamma rays and neutrons—the effects of which often did not surface for days or weeks after exposure. Like flash burns, the severity of radiation injuries depended on a person's proximity to the bomb and what stood in between.

The commission's report estimated that 72,000 people were injured in Hiroshima and 64,500 had died by mid-November 1945; in Nagasaki, approximately 25,000 were injured and 39,000 were estimated to have died by the same date. In the late 1970s, an international effort led by Japan produced another comprehensive study, which estimated the number of deaths as being closer to 140,000 in Hiroshima and 70,000 in Nagasaki by the end of 1945. Both groups of investigators acknowledge a high degree of uncertainty in their findings and were hindered by gaps and contradictions in the available data: the cities' populations at the time of the bombings remain unknown, and the records of dead and wounded that were made in the aftermath vary widely. Both assessments conclude that most fatalities occurred on the days of the bombings and nearly all within the next several months.

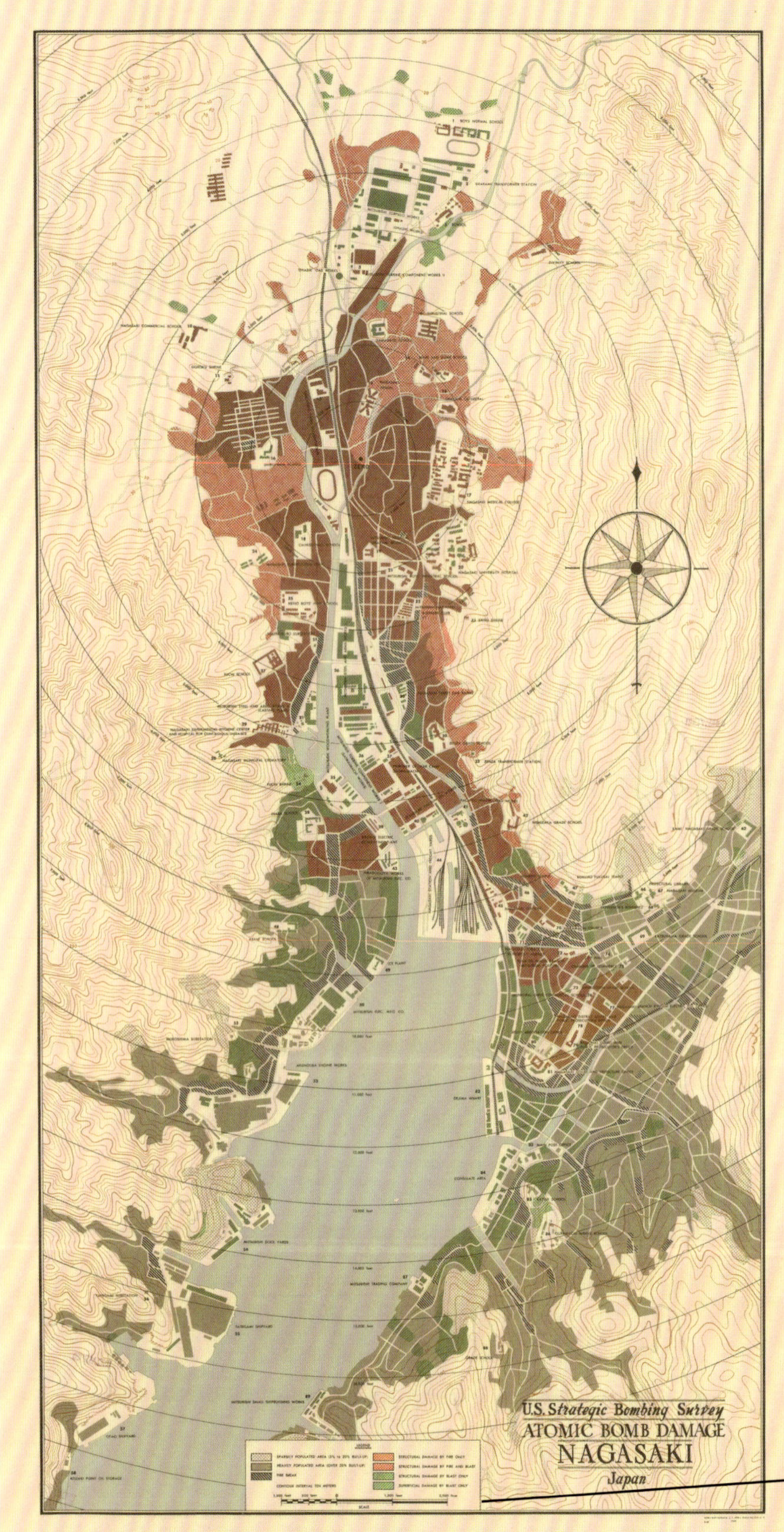

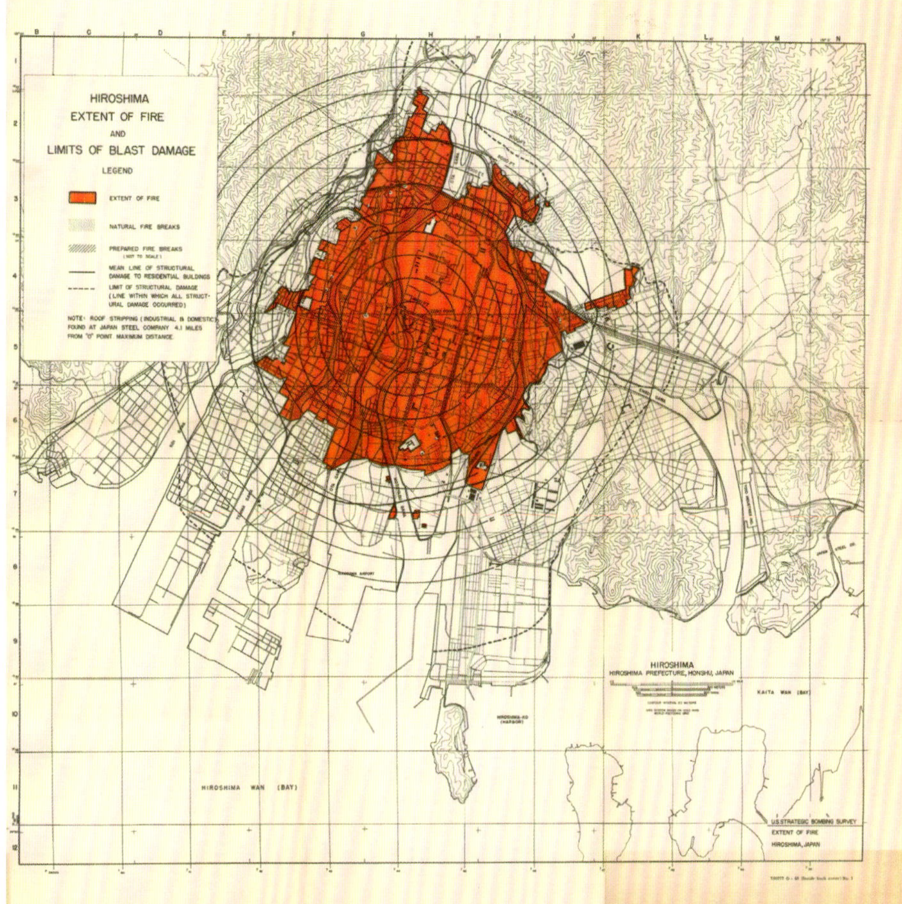

These two 1946 maps, produced by the US Strategic Bombing Survey, illustrate the extent of damage in both cities. The map of Nagasaki includes population density around the hypocenter; the largely industrial Urakami Valley was directly hit, and portions of the southern, residential portion of the city were shielded from heat and radiation because of its geographic layout. Even so, the blast destroyed approximately 40 percent of Nagasaki, affecting forty-three square miles.

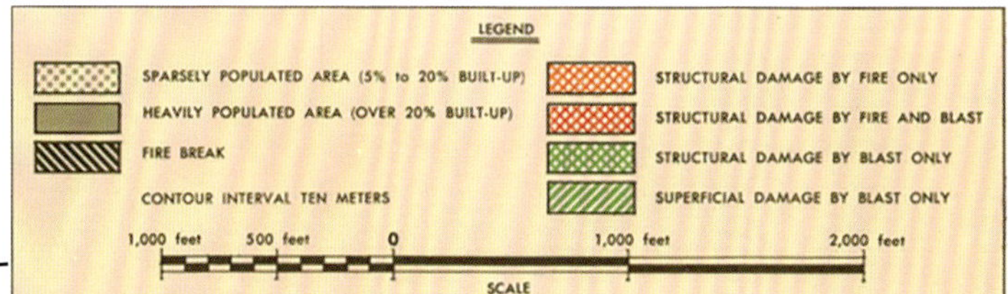

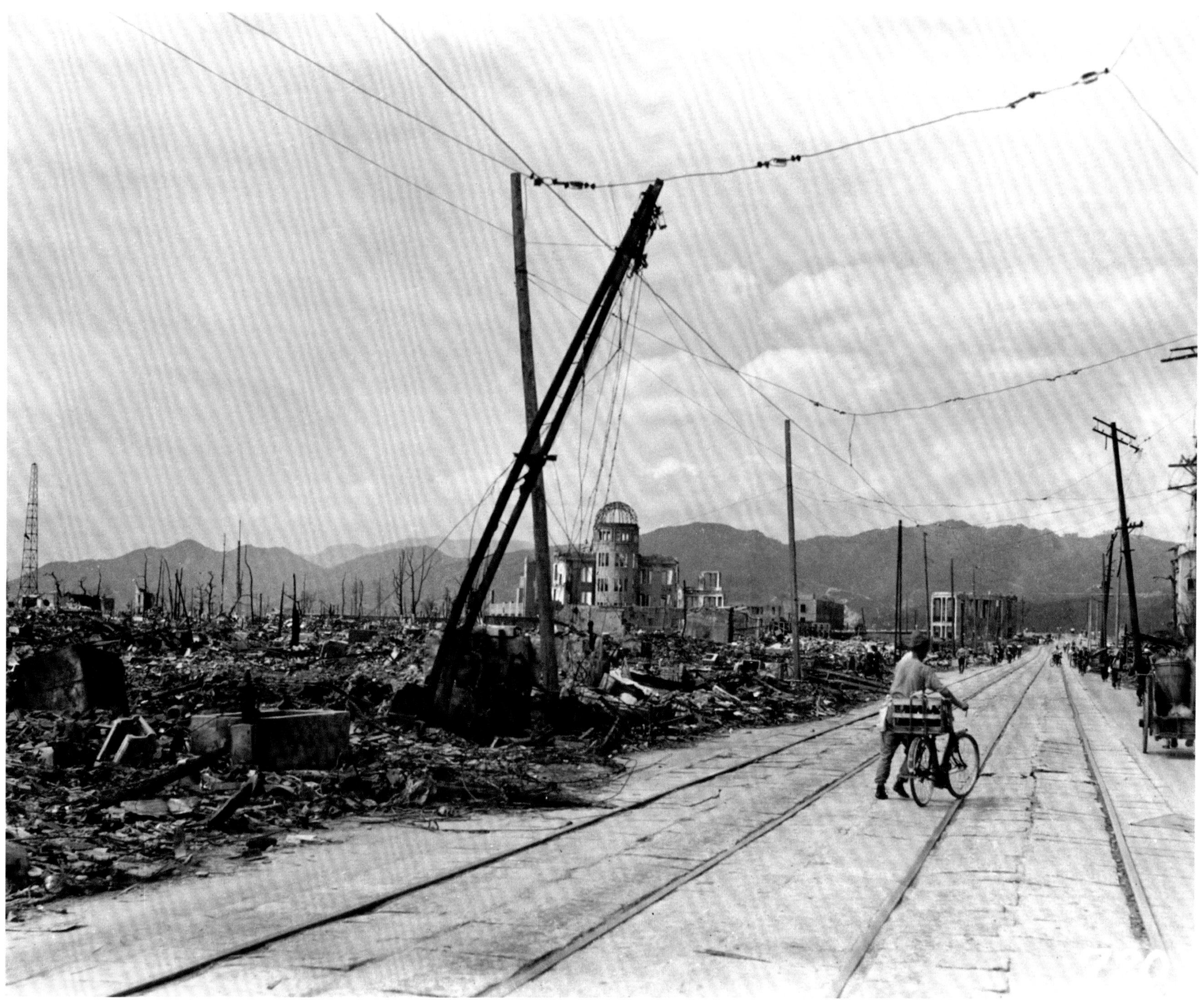

Although the yield of the implosion bomb, Fat Man, was higher than that of Little Boy, and its radius of peak destruction was larger, the human toll in Hiroshima was significantly higher because the bomb exploded directly over the center of the concentrated city. The image above shows one of the few buildings left standing in downtown Hiroshima, the Prefectural Industrial Promotion Hall. It is today a World Heritage Site known as the Atomic Bomb Dome. At the center of the city's Peace Memorial Park, it serves as a call for world peace and abolishment of nuclear weapons.

Five days after the second bombing, on the fourteenth of August, Japan accepted the terms of the Potsdam Declaration. Emperor Hirohito addressed his nation: "Despite the best that has been done by everyone—the gallant fighting of military and naval forces, the diligence and assiduity of Our servants of the State and the devoted service of Our one hundred million people, the war situation has developed not necessarily to Japan's advantage, while the general trends of the world have all turned against her interest. Moreover, the enemy has begun to employ a new and most cruel bomb, the power of which to do damage is indeed incalculable, taking the toll of many innocent lives. Should We continue to fight, it would not only result in an ultimate collapse and obliteration of the Japanese nation, but also it would lead to the total extinction of human civilization. Such being the case, how are We to save the millions of Our subjects; or to atone Ourselves before the hallowed spirits of Our Imperial Ancestors? This is the reason why We have ordered the acceptance of the provisions of the Joint Declaration of the Powers."

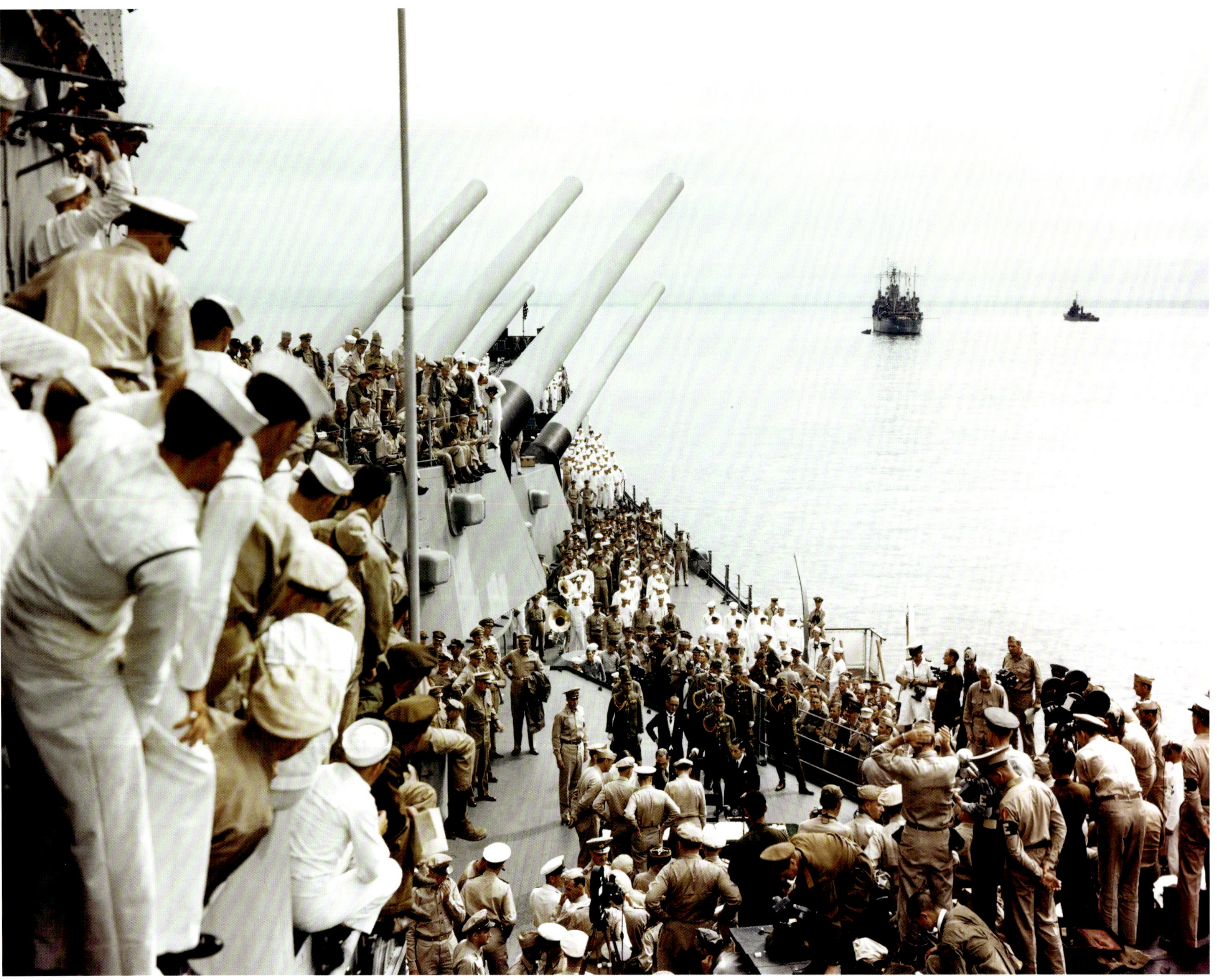

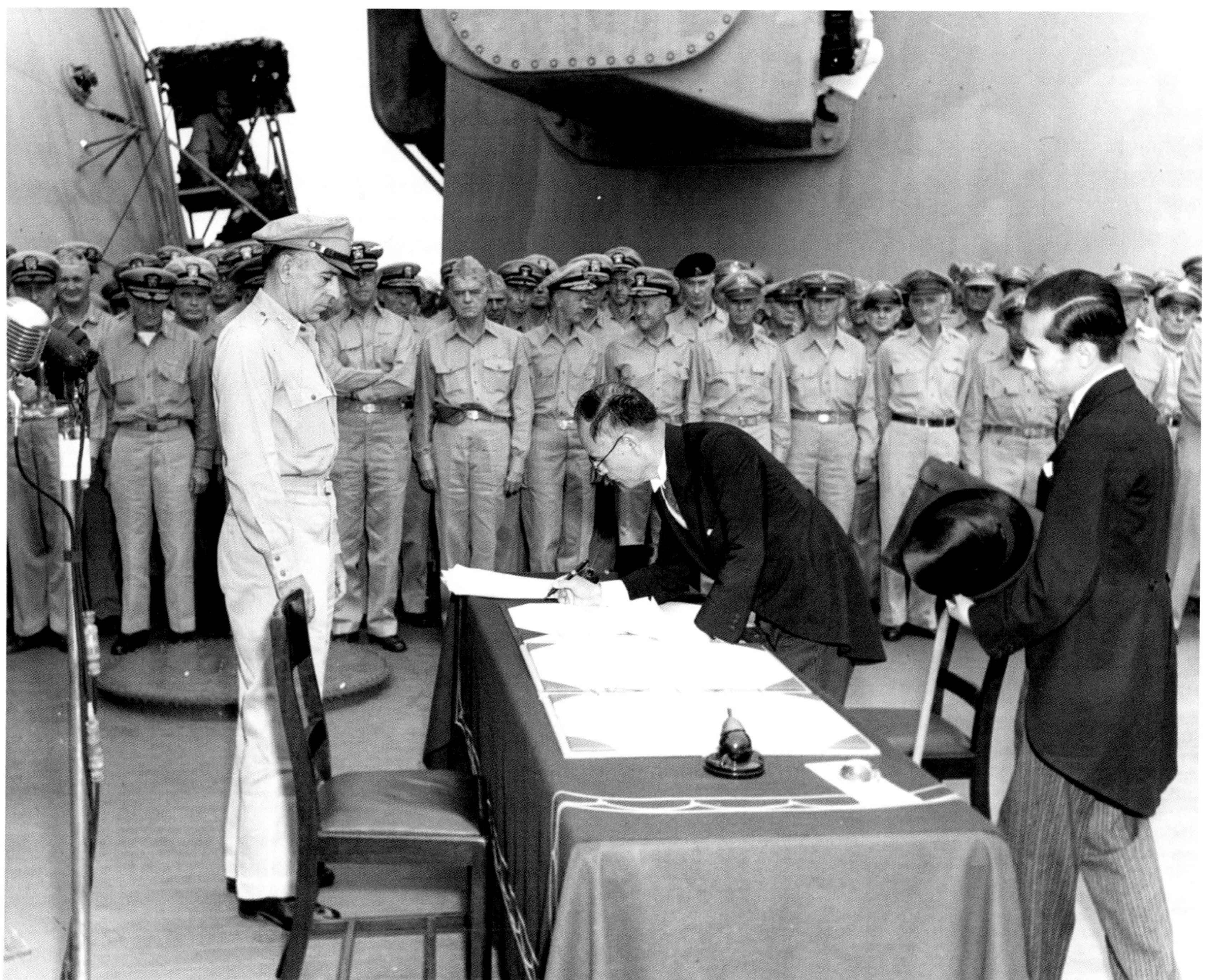

On September 2, 1945, Japanese representatives boarded the USS *Missouri* in Tokyo Bay, where foreign affairs minister Mamoru Shigemitsu and other dignitaries signed the formal Instrument of Surrender. Under the terms of the agreement, the emperor would be stripped of authority, the Japanese military would immediately disarm, and Japan would be occupied. Described by Groves as a somber and thoughtful event, this moment in history would become known as V-J Day, for Victory over Japan. "We resolved the immediate problem of ending the war," he recalled, "but in doing so we raised up many unknowns." (Army Signal Corps Collection, Archives Branch, Naval History and Heritage Command: SC 213700 (*above*), USA C-4626 (*left*).)

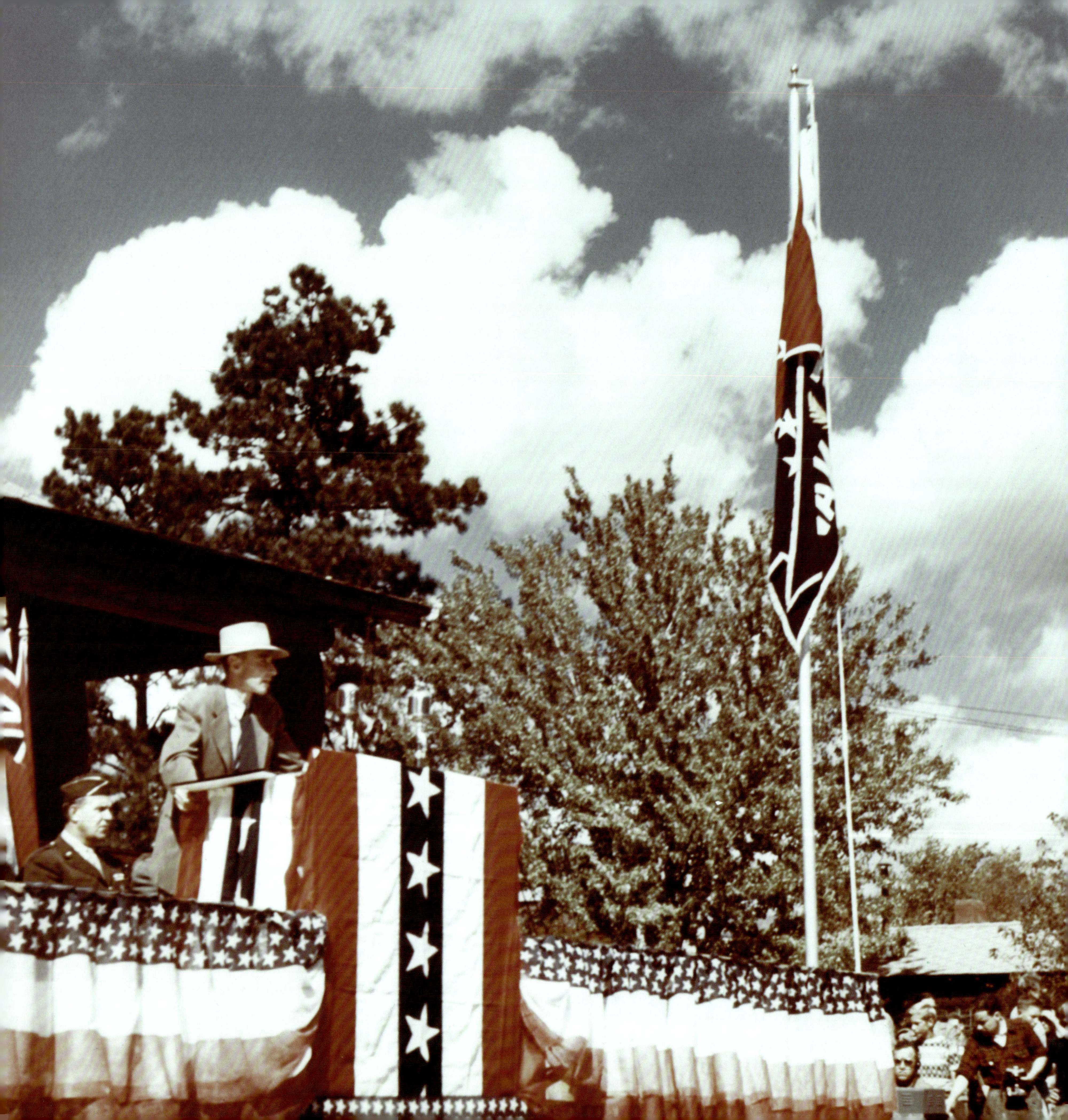

A few days after Japan's surrender, J. Robert Oppenheimer received a letter. The staff of the laboratory at Los Alamos was to be presented with the Army-Navy "Excellence in Production" Award for their outstanding contributions to the war effort. At an October 16 ceremony in New Mexico—exactly three months after the Trinity test—the honor was presented by Groves. As had been true since the discovery of fission in 1938, the excitement surrounding progress in atomic science was darkened by the shadow of war and of the technology's destructive potential. Oppenheimer gave a haunting speech of acceptance:

"It is with appreciation and gratitude that I accept from you this scroll for the Los Alamos Laboratory, for the men and women whose work and whose hearts have made it. It is our hope that in years to come we may look at this scroll, and all that it signifies, with pride.

Today that pride must be tempered by a profound concern. If atomic bombs are to be added as new weapons to the arsenals of a warring world, or to the arsenals of nations preparing for war, then the time will come when mankind will curse the names of Los Alamos and of Hiroshima.

The peoples of this world must unite, or they will perish. This war, that has ravaged so much of the earth, has written these words. The atomic bomb has spelled them out for all men to understand. Other men have spoken them, in other times, of other wars, of other weapons. They have not prevailed. There are some, misled by a false sense of human history, who hold that they will not prevail today. It is not for us to believe that. By our work we are committed, committed to a world united, before this common peril, in law, and in humanity."

SANTA FE NEW MEXICAN

The Oldest Newspaper in the Southwest, Founded in 1849

Vol. 96, No. 211 | MEMBER AUDIT BUREAU OF CIRCULATIONS | SANTA FE, NEW MEXICO, MONDAY, AUGUST 6, 1945 | ASSOCIATED PRESS UNITED PRESS | Price 5c

Los Alamos Secret Disclosed by Truman

ATOMIC BOMBS DROP ON JAPAN

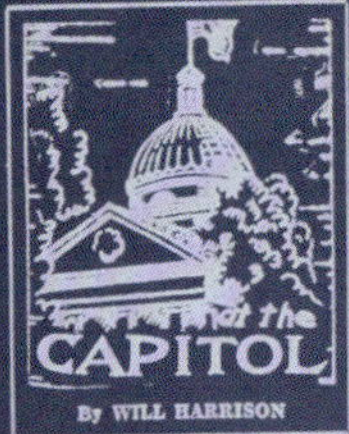

By WILL HARRISON

Deadliest Weapons in World's History Made In Santa Fe Vicinity

Santa Fe learned officially today of a city of 6,000 in its own front yard.

The reverberating announcement of the Los Alamos bomb, with 2,000 times the power of the great Grand-Slammers dropped on Germany, also lifted the secret of the community on the Pajarito Plateau, whose presence Santa Fe has ignored, except in whispers, for more than two years.

Decision to locate the Atomic Bomb Project Laboratory on mesa an hour's drive from Santa Fe, meant that it was necessary for the Army Engineers to construct an entirely new town to house the workers and their families. Primary reason for selection of the isolated site was security.

in mess halls, or in a large cafeteria, was housing. Various types were constructed, to meet different needs. There are three-room prefabricated, individual houses; three-room apartments, eight to a building, and four- and five-room apartments, four in a two-story unit. There are some hutments, Quonset-type huts and government and personally-owned trailers.

Bachelor Dorms

Dormitories have been constructed for unmarried personnel, or persons who do not have their families with them. Rents, for family groups, are based on earnings. Apartments are unfurnished and family groups ordinarily bring their furniture with them, although some items of government furniture have been available.

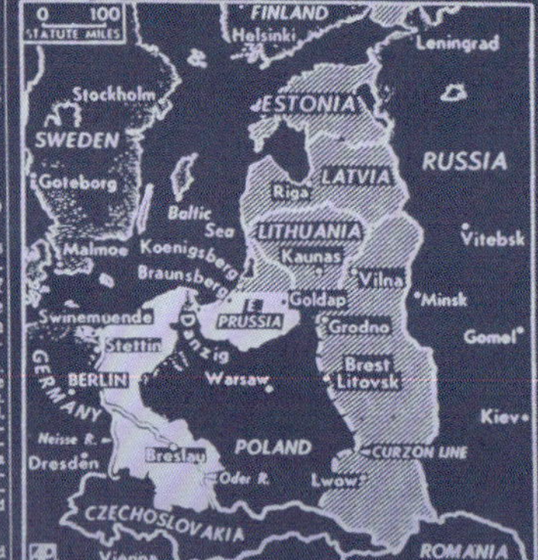

TERRITORIAL CHANGES OF WORLD WAR II—Black areas on map are those parts of Germany which the Big Three proposes will come under Polish rule. Shaded area is territory which Russia has taken control over since the start of hostilities on the continent. Northern East Prussia, proposed as Russian by the Big Three, is the newest addition to Soviet territory. There still remains some question as to final disposition of the port of Stettin. Note the large section of Baltic coastline of the proposed new Poland.

4 More Nippon Cities Now Smoldering Ruins

'Utter Destruction,' Promised in Potsdam Ultimatum, Unleashed; Power Equals 2,000 Superforts

WASHINGTON, Aug. 6 (AP)—The U. S. Army Air Force has released on the Japanese an atomic bomb containing more power than 20,000 tons of TNT.

It produces more than 2,000 times the blast of the largest bomb ever used before.

The announcement of the development was made in a statement by President Truman released by the White House today.

The bomb was dropped 16 hours ago on Hiroshima, an important Japanese army base.

The President said that the bomb has "added a new and revolutionary increase in destruction" on the Japanese.

Mr. Truman added:

"It is an atomic bomb. It is a harnessing of the basic power of the universe. The force from which the sun draws its power has been loosed against those who brought war to the Far East."

The base that was hit is a major quartermaster depot and has large ordnance, machine tool and aircraft plants.

The raid on Hiroshima, located on Honshu Island on the shores of the Inland sea, had not been disclosed

in their war effort but failed. Meantime American and British scientists studied the problem and developed two principal plants and some lesser factories for the production of atomic power.

The President disclosed that more

MADE IN SANTA FE

WASHINGTON, Aug. 6 (AP)—The atomic bomb disclosed by President Truman today was developed at factories in Tennessee, Washington and New Mexico.

WILL SHORTEN WAR

May Be Tool To End Wars; New Era Seen

Mankind's successful transition to a new age, the Atomic Age, was ushered in July 16, 1945, before the eyes of a tense group of renowned scientists and military men gathered in the desertlands of New Mexico to witness the first band results of their $2,000,000,000 effort. Here in a remote section of the Alamogordo Air Base 120 miles southeast of Albuquerque the first man-made atomic explosion, the outstanding achievement of nuclear science, was achieved

09 Legacy

The world first learned of the Manhattan Project as it first learned of the atomic bombings. Reports detailing the widespread destruction in Japan also introduced the public to Leslie Groves, J. Robert Oppenheimer, and the Trinity test. The veil of secrecy concealing Project Y was lifted as well, and in fall 1945 a new name came into use: Los Alamos Scientific Laboratory.

Having accomplished the mission, hundreds of laboratory employees left Los Alamos quickly. Many scientists eagerly transitioned away from weapons development, returning to graduate programs or professorships. Oppenheimer hoped to return to academia himself. Though continuing to serve as a conduit between scientists and policymakers, he resigned as director in October 1945, handing the reins over to Norris Bradbury. The physicist had wanted to resume teaching after the war but reluctantly agreed to lead Los Alamos for six months: he ultimately led the institution for the next quarter century.

Though he possessed an aversion to them, Bradbury believed in the importance of thoroughly understanding nuclear weapons. As such, he stabilized the staffing situation at Los Alamos and directed his scientists to continue refining fission bombs. In the years that followed, continual improvements to the Gadget's descendants strengthened and diversified America's nuclear arsenal. During that time, scientists also explored the possibility of creating weapons hundreds of times more powerful than the bombs that exploded above Hiroshima and Nagasaki: weapons deriving their power from the fusion of hydrogen atoms.

After Trinity, testing of full-scale nuclear weapons resumed in summer 1946, but not in New Mexico. The US Navy had requested test devices and technical assistance from Los Alamos in late 1945 to study the effects of nuclear weapons against naval vessels. The two shots of Operation Crossroads, conducted at Bikini Atoll in the Marshall Islands, demonstrated that radioactive contamination could cripple a fleet of ships even more efficiently than blast damage. The test series proved controversial for several reasons, including the resulting displacement of native populations; some considered it unnecessary, and others feared it might stoke a burgeoning rivalry with the Soviet Union (then three years away from detonating its own nuclear bomb). In tandem with significant infrastructure investments and Bradbury's leadership, Crossroads also helped the laboratory establish footing during the nation's transition to peacetime.

The Manhattan Project formally came to an end at midnight on January 1, 1947. At that moment, the Atomic Energy Commission (AEC), a civilian organization created by Congress, inherited the nation's vast nuclear complex—including Los Alamos. Under the purview of the Department of Defense and the AEC (followed by its successors, which include the Department of Energy), nuclear weapons testing continued for nearly fifty years, building on the diagnostic techniques and data that originated at Trinity.

The first formal international arms control agreement—the Limited Test Ban Treaty—came in 1963, when the United States, the Soviet Union, and Great Britain agreed to discontinue nuclear tests in the atmosphere, outer space, and underwater. Extensive testing continued underground until shortly after the collapse of the Soviet Union in 1991. The United States voluntarily ceased testing, and the United Nations General Assembly later adopted a comprehensive treaty prohibiting all nuclear explosions. Though the United States has not ratified the treaty, it has extended its own moratorium indefinitely. Today, Los Alamos National Laboratory and other institutions use data collected during the testing era to maintain the US nuclear stockpile on behalf of the Department of Energy.

As for Trinity, the military retained operational control of the site and—despite many years of protest from the National Park Service—took a minimalist approach to stewardship, primarily concerned with keeping trespassers away from the crater and tourists away from what was becoming a busy missile testing range. By the time Trinity Site was opened regularly to visitors, in the 1960s, only a few relics remained from the Manhattan Project's flurry of activity in 1945. Today, the shallow crater is barely discernible, and only tiny fragments from the green sea of trinitite remain. The exposed rebar of the tower footings that survived the blast has been cut down to the ground. In the tower's wake, for many years there has been an obelisk bearing a plaque that reads, "Trinity Site: Where the World's First Nuclear Device Was Exploded on July 16, 1945." A second plaque, added in 1975, identifies the site as a National Historic Landmark.

Military activities continue in the area surrounding Trinity, where the army's White Sands Missile Range hosts a variety of non-nuclear tests in the interest of national security. Despite the sensitivity of the work, Trinity Site is opened to the public once a year. Visitors travel from across the globe to the birthplace of the atomic bomb, both to discover its history and, for some, to reconcile their own experiences of its impact upon the world.

* * * * * * *

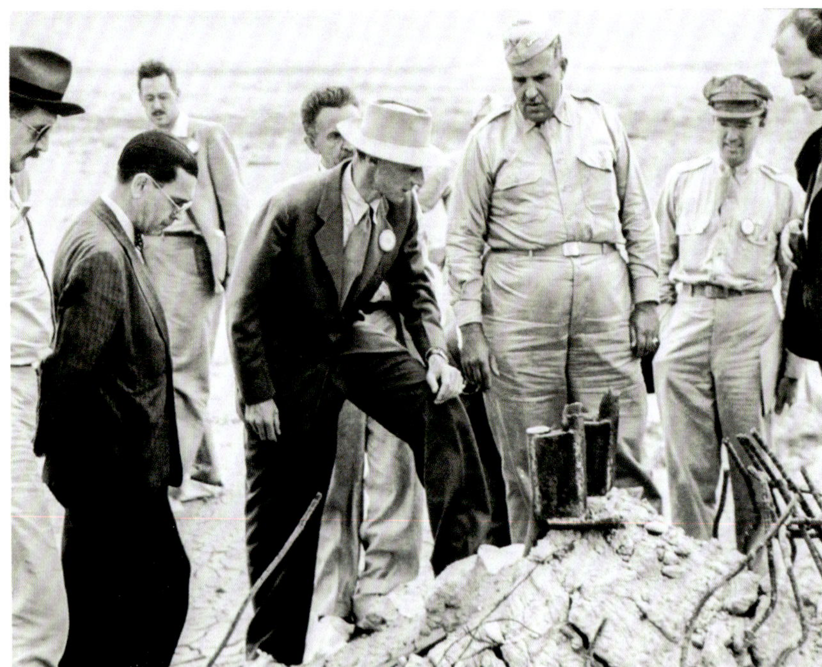

On September 9, 1945, one week after the end of the war, Groves, Oppenheimer, and a small group of test participants hosted members of the press at the Trinity site. The concrete and rebar of the tower footings, visible in these photographs, was cut to the ground in later years; what remains today is shown below alongside Charlie McMillan, the tenth director of Los Alamos.

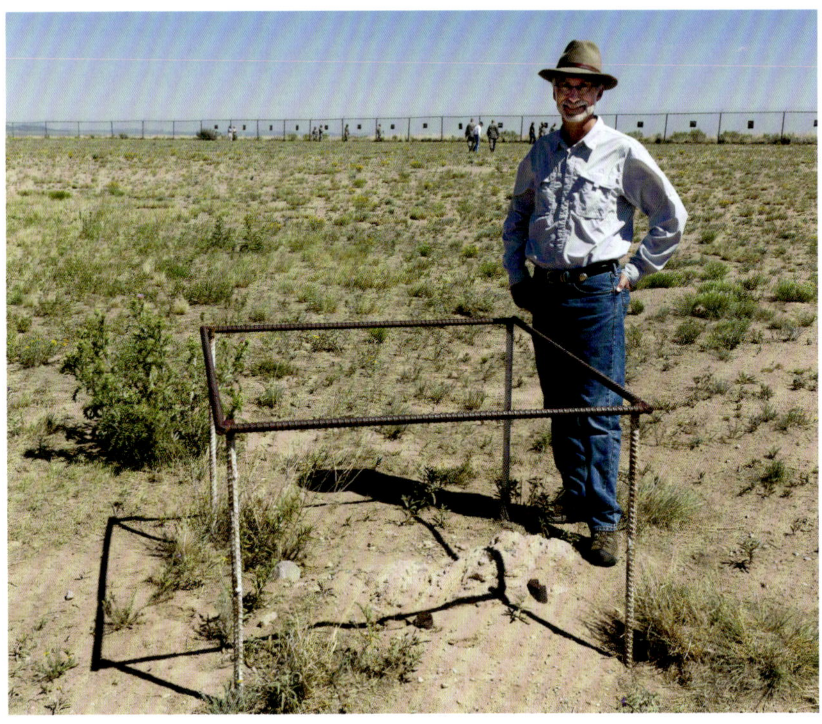

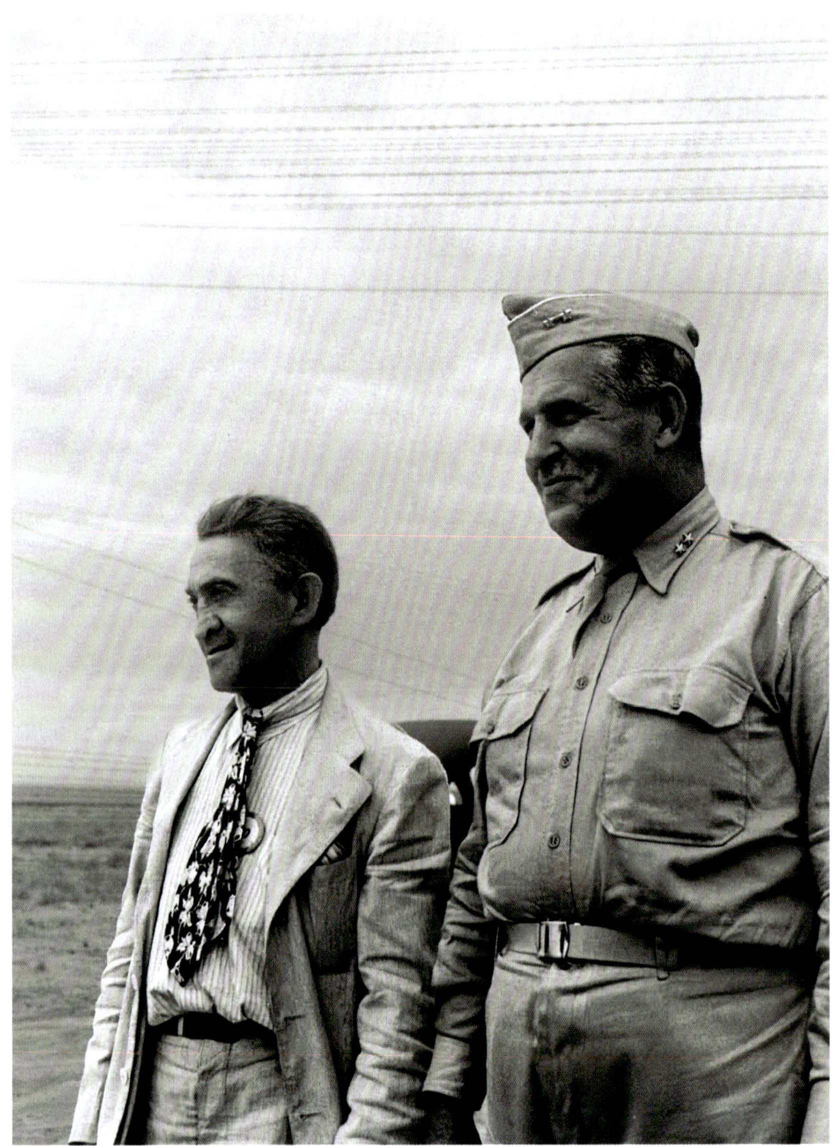

One member of the press, William Laurence (above, with Groves), was making his second visit to the test site. An observer on July 16, he had been recruited years earlier to shadow the Manhattan Project activities at Los Alamos, Trinity, and eventually Tinian. In addition to witnessing the Trinity explosion, Laurence observed takeoff procedures for the Hiroshima strike and was on the instrument plane that flew the Nagasaki mission. (Fritz Goro/The LIFE Picture Collection/Shutterstock.)

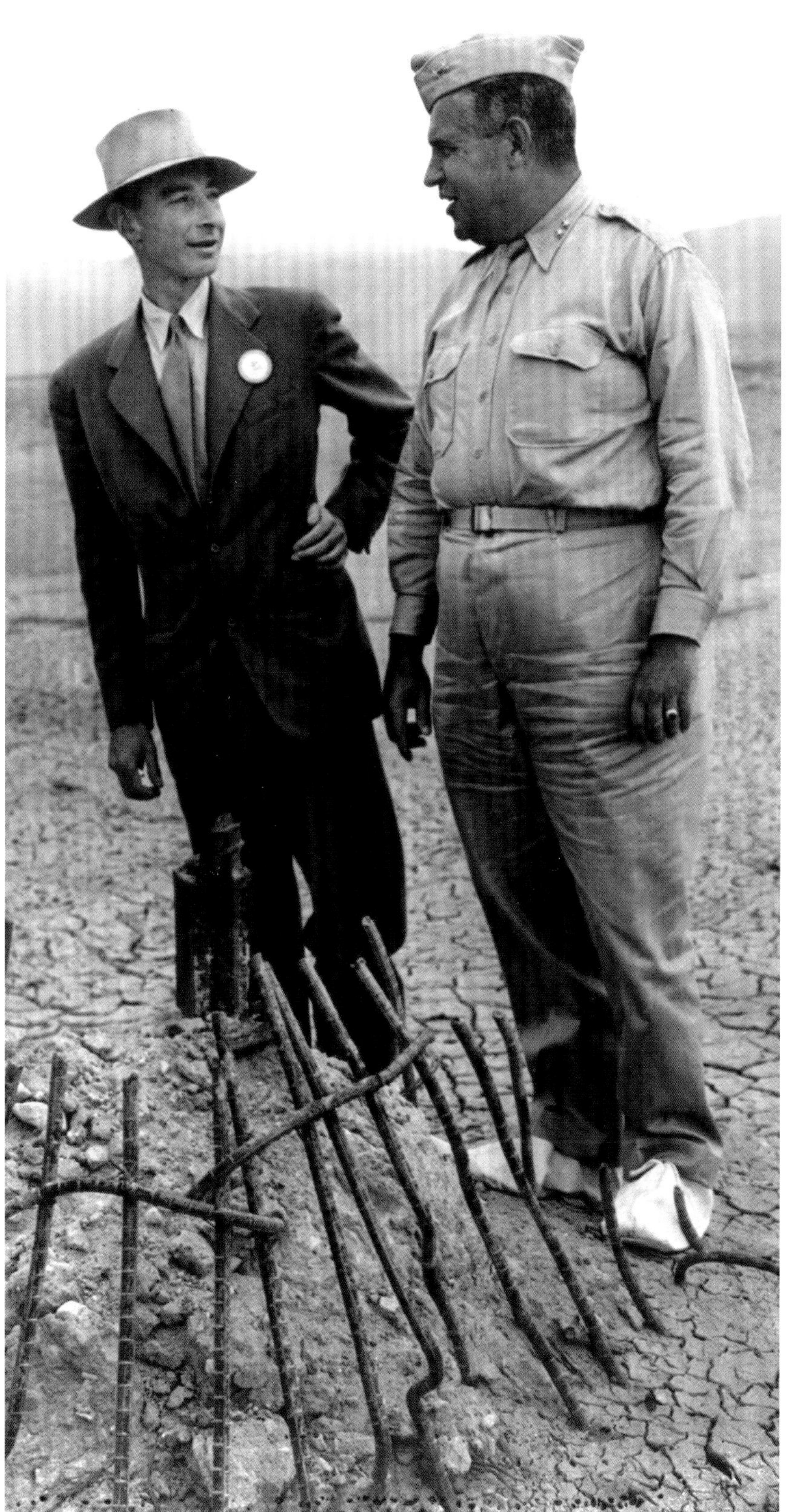

The group wore protective coverings over their shoes to avoid carrying contaminated soil home with them. Trinity personnel who had been exposed to radiation after the test shot also wore a film badge to track their exposure during the press visit.

Certain areas of the crater were still quite radioactive, producing readings as high as seven roentgens per hour. As a precaution, guests were only allowed to remain in the area for thirty minutes. Health Group leader Louis Hempelmann estimated the reporters received an average dose of one roentgen during the visit. This would have been supplemented by the small pieces of trinitite—the radioactive green glass produced during the nuclear explosion—they were allowed to take as souvenirs (after a radiation check, as shown below; Fritz Goro/The LIFE Picture Collection/Shutterstock).

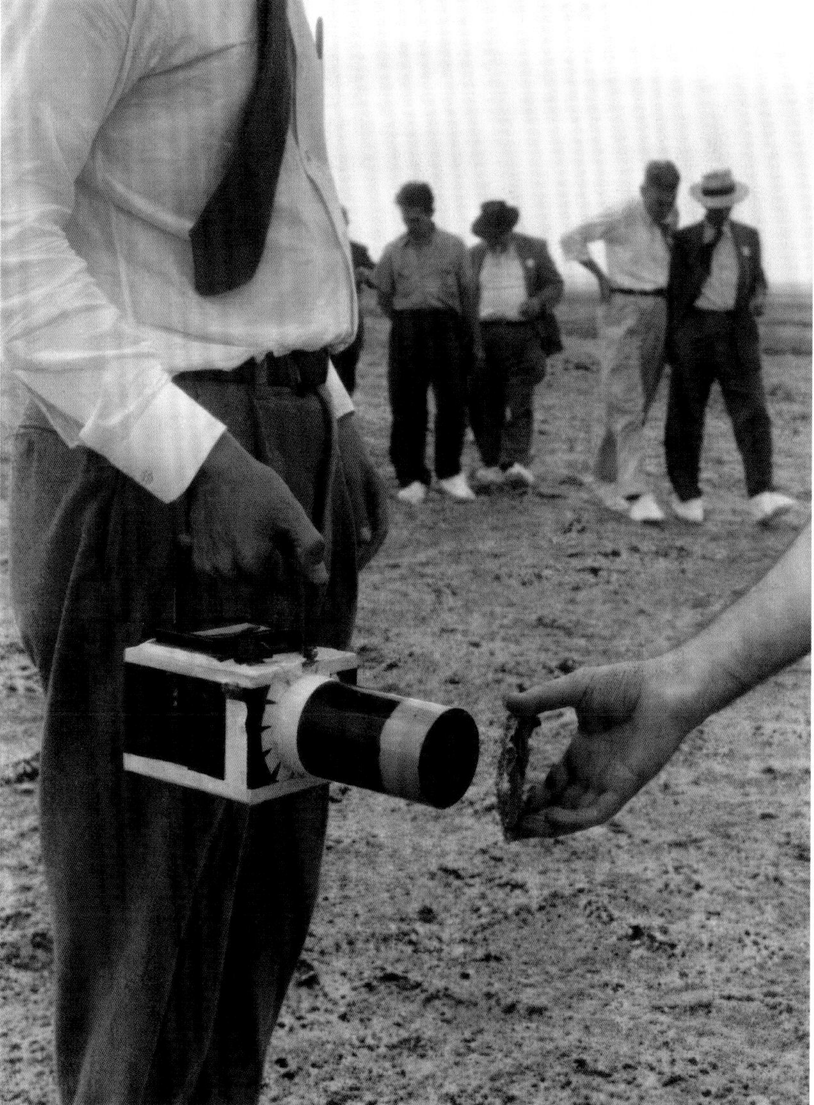

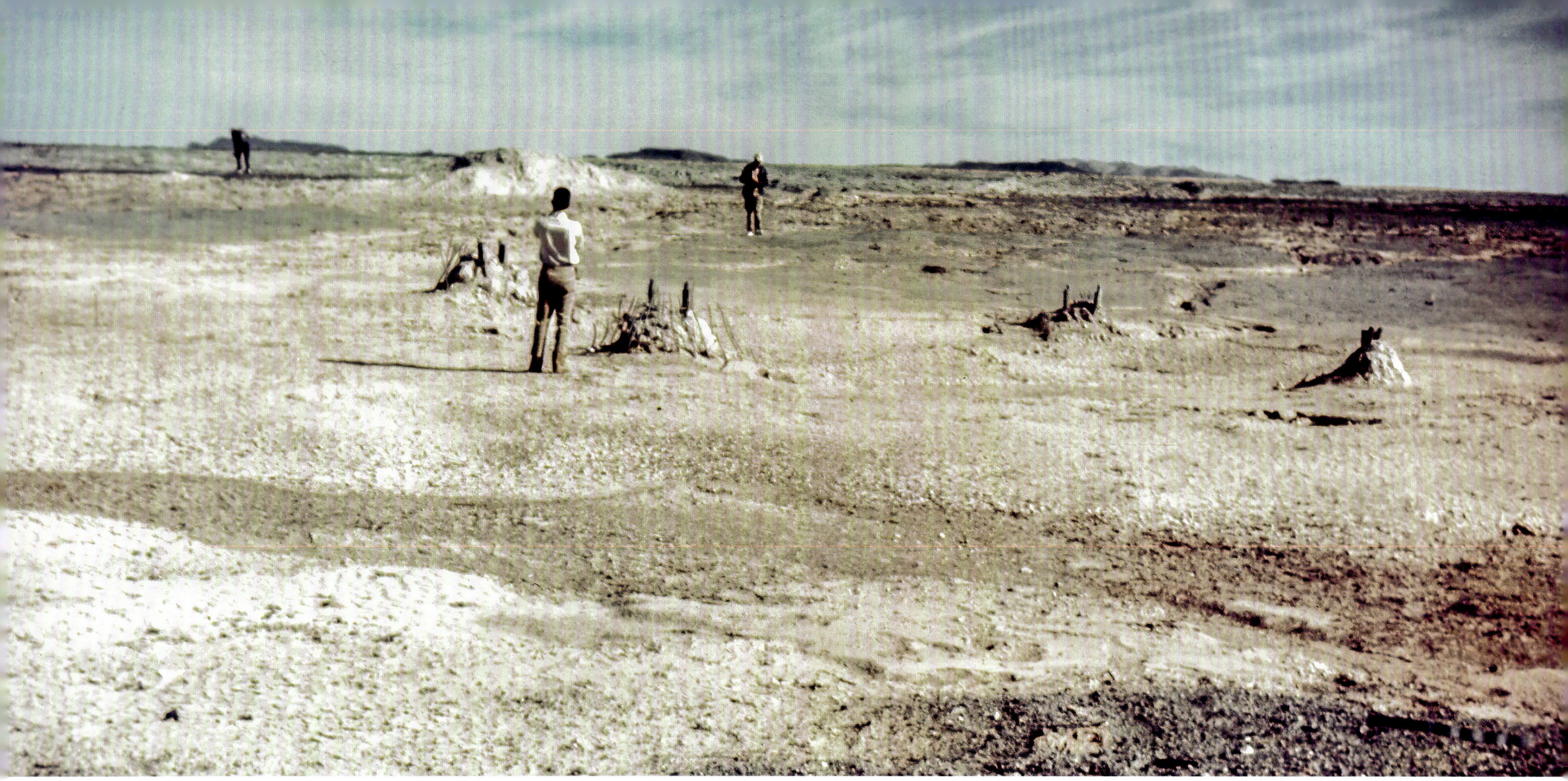

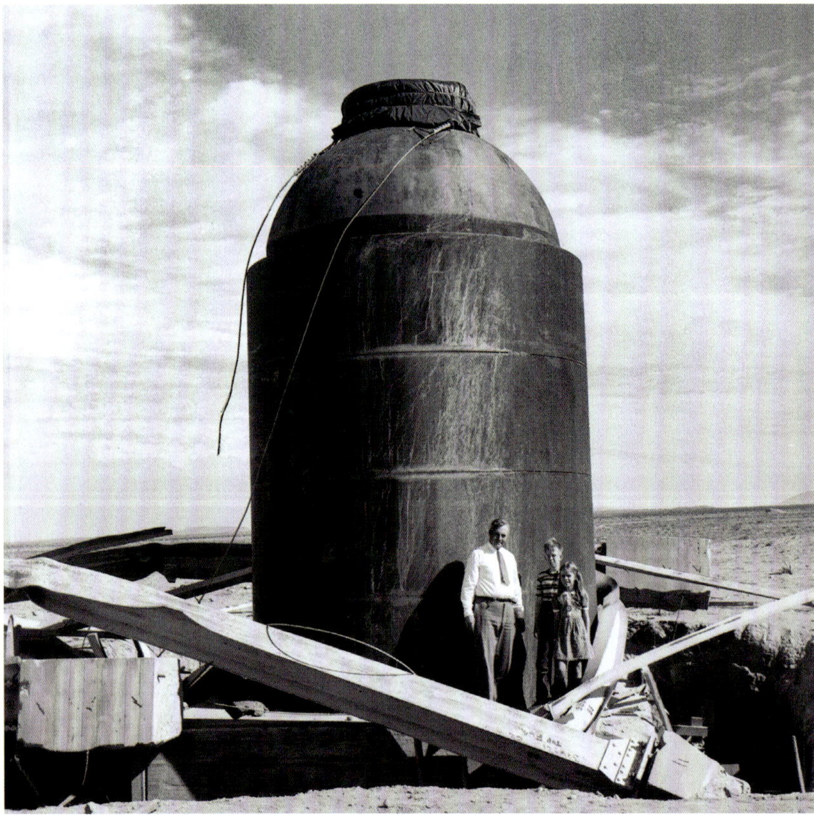

Even among those who had seen the Trinity test, the crater became somewhat of a tourist attraction in the weeks and months that followed. Lt. Jerry Allen, a member of the radiation monitoring team, complained even before the war was over that there were "entirely too many groups entering the contaminated area." Many of these visitors claimed to be recovering equipment, though it appears some may have been more interested in collecting trinitite souvenirs.

A number of Los Alamos scientists returned with personal guests. In the mostly empty desert basin, Jumbo was a popular photo destination. At left, Spectrographic and Photographic Measurements Group leader Julian Mack poses beside the steel bottle with his children.

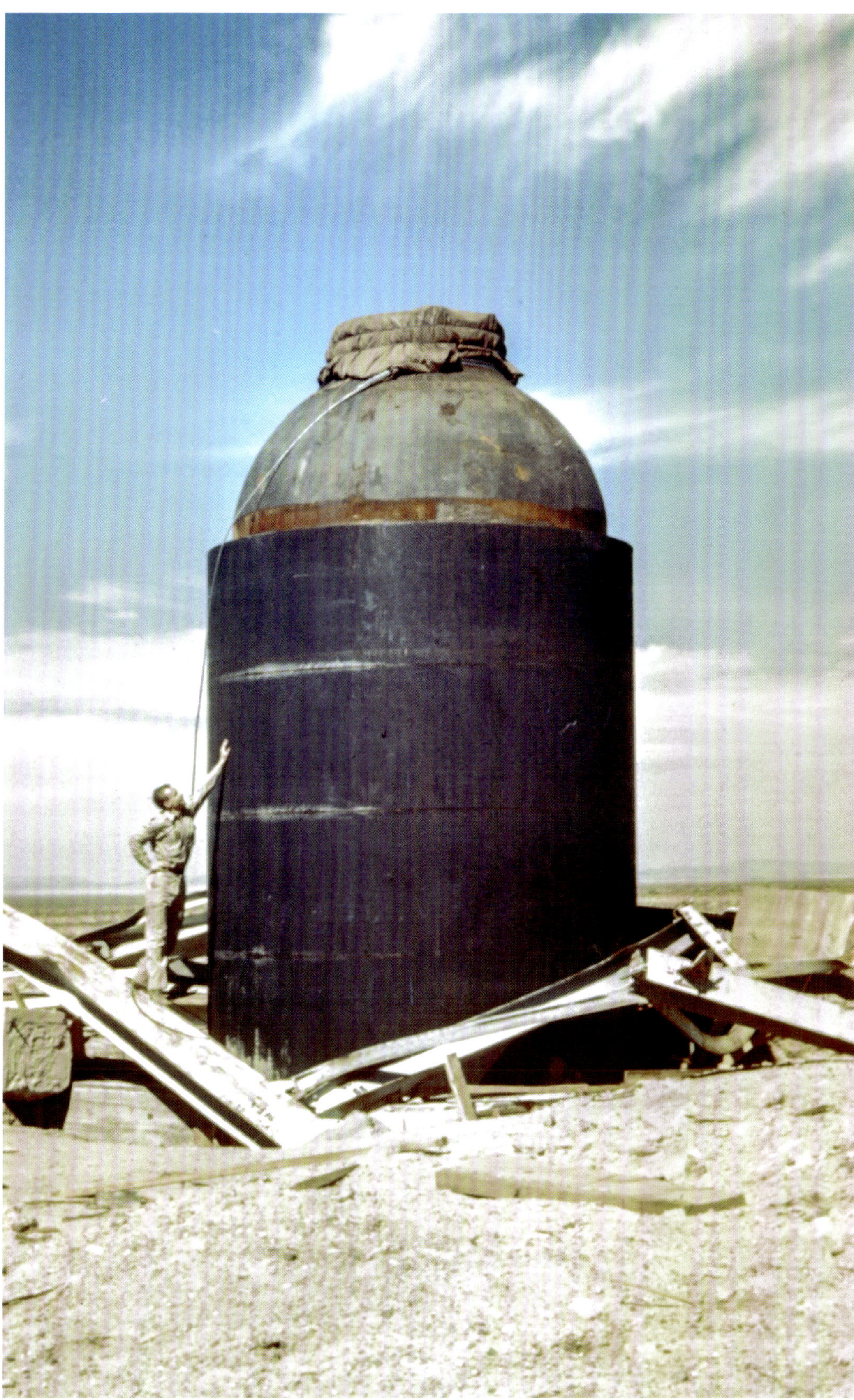

After Trinity, Jumbo's manway had been sealed with a waterproof cover to preserve the interior for future use. In April 1946, the containment vessel received an assignment, to be conducted by members of Z Division—the ordnance arm of the laboratory that had been relocated to Sandia Base in Albuquerque.

After being sidelined for the nuclear test, Jumbo had its first mission: hosting the detonation of eight unserviceable AN-M64 bombs. The explosive force of the 500-pound, general-purpose bombs totaled lower than what the vessel had been designed to withstand. Under the proper conditions, this third and smallest Trinity Site explosion would, in theory, not leave much of a story to tell.

However, the ill-fated operation was not conducted in accordance with Jumbo's design. The explosives were placed at the bottom of the vessel, instead of suspended in its reinforced middle section. And the heavy manway lid was not installed; detonator wires ran through the open top to a two-person firing party set up only 600 feet away.

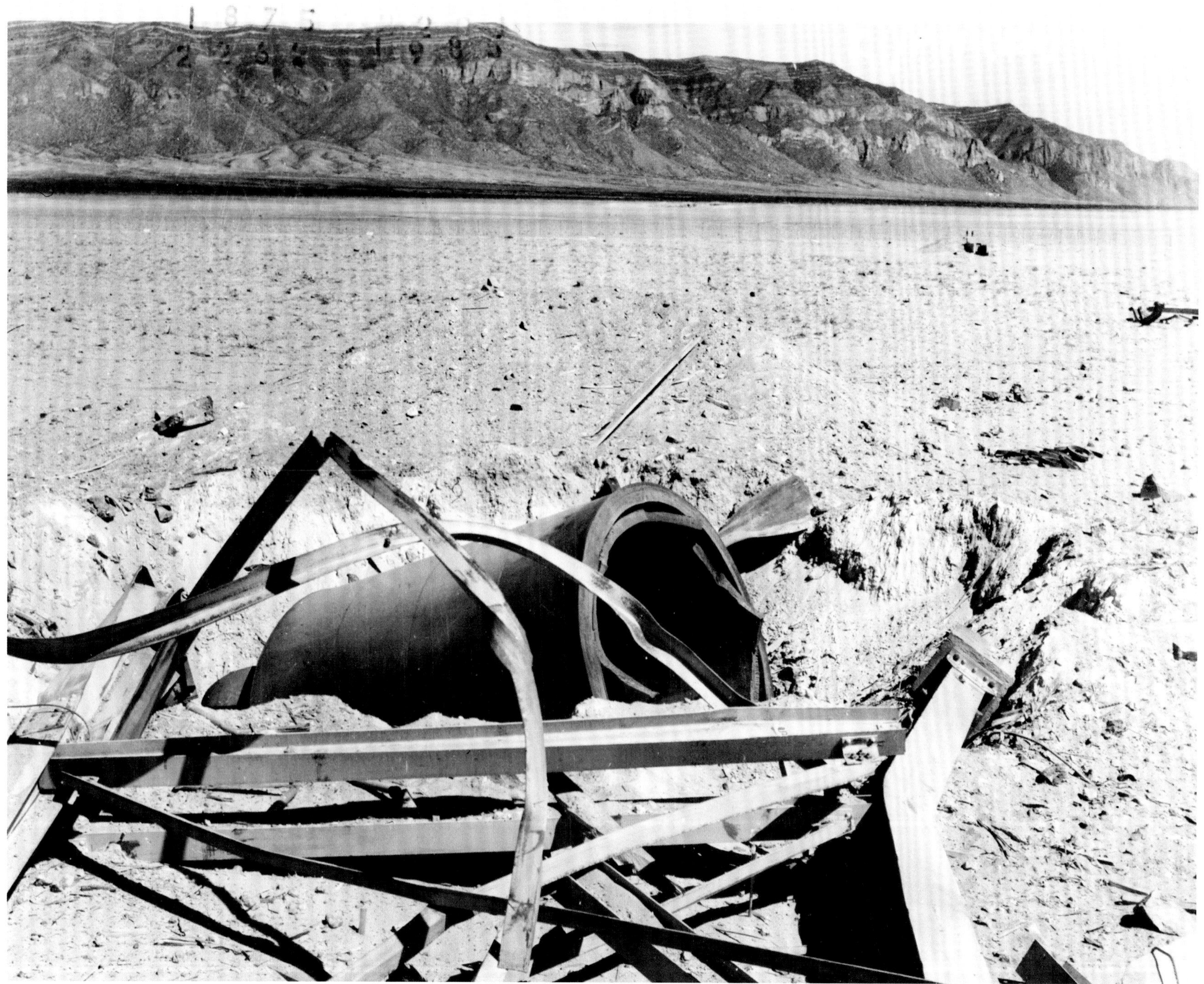

When the charges fired, disaster ensued. Jumbo's concrete foundation erupted and was scattered across the desert floor. Both ends of the steel bottle separated from the center and were catapulted as far as three-quarters of a mile, well beyond the firing party set up at the small instrument bunker visible above. What was left of Jumbo rose off the ground and then fell on its side, coming to rest among the ruins of the pulley tower. The images shown are from a report entitled "Demolition Operation at Trinity" submitted by Lt. Richard Blackburn, a Z Division officer who participated in the event.

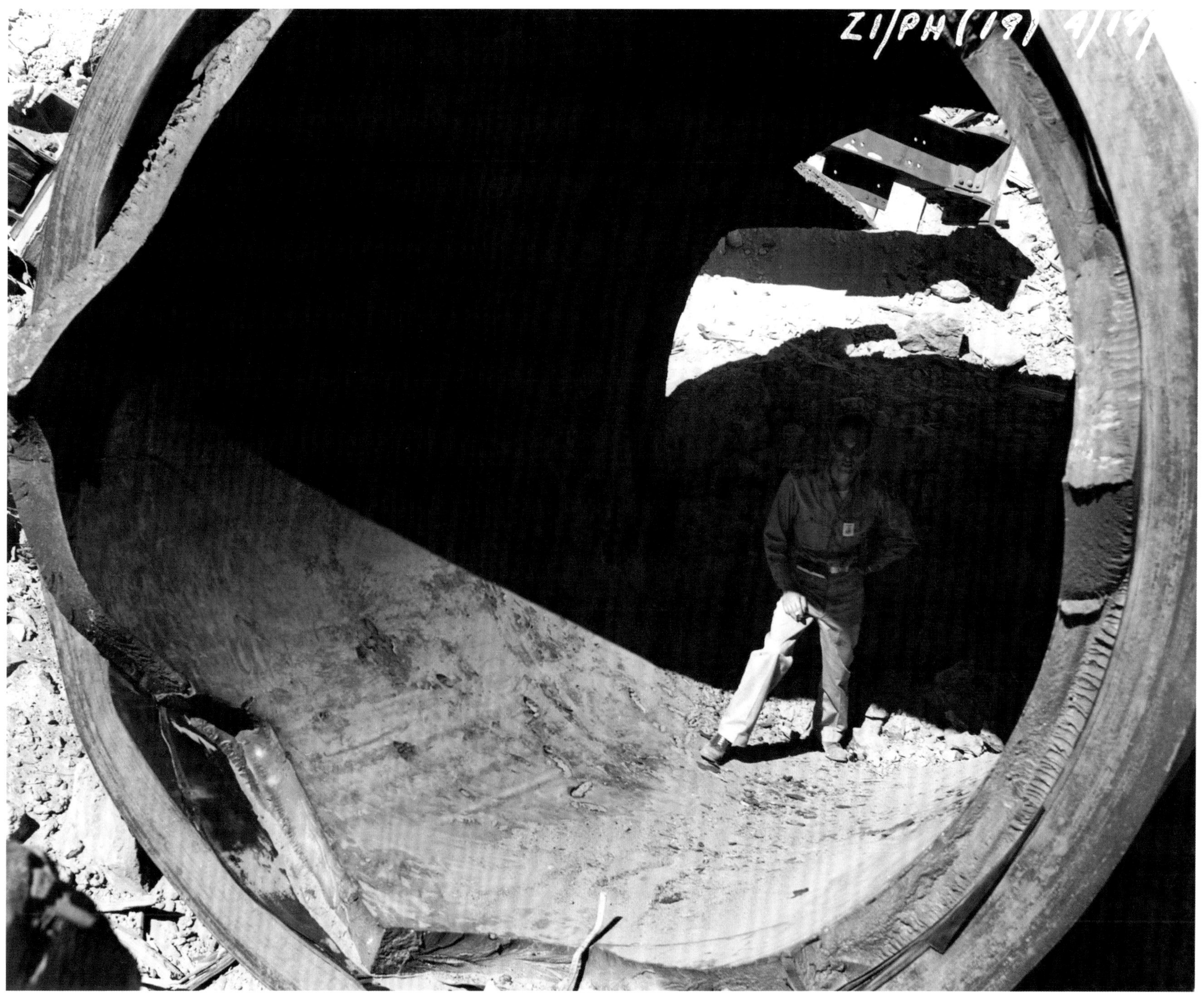

Various elements of the story have led to speculation that Jumbo, rather than the unserviceable bombs, was the target of the "demolition operation." In memos written by Blackburn and his superior, Lt. Col. A. J. Frolich, the April 16 mission is described as "the charge set off to 'deface' Jumbo" and "the operation for the defacing of Jumbo." Frolich also mentions a prior plan to detonate a box of TNT inside of Jumbo—perhaps to give the appearance that the expensive detonation chamber had served some purpose. While it appears true that inflicting some amount of damage was the objective, none of the available documentation suggests an intent to destroy Jumbo. Blackburn reports, in fact, that he was told fragmentation need not be expected—hence the close proximity of the firing party. Whomever ordered the operation, and for whatever reason, it seems no one anticipated that the damage would be so spectacular.

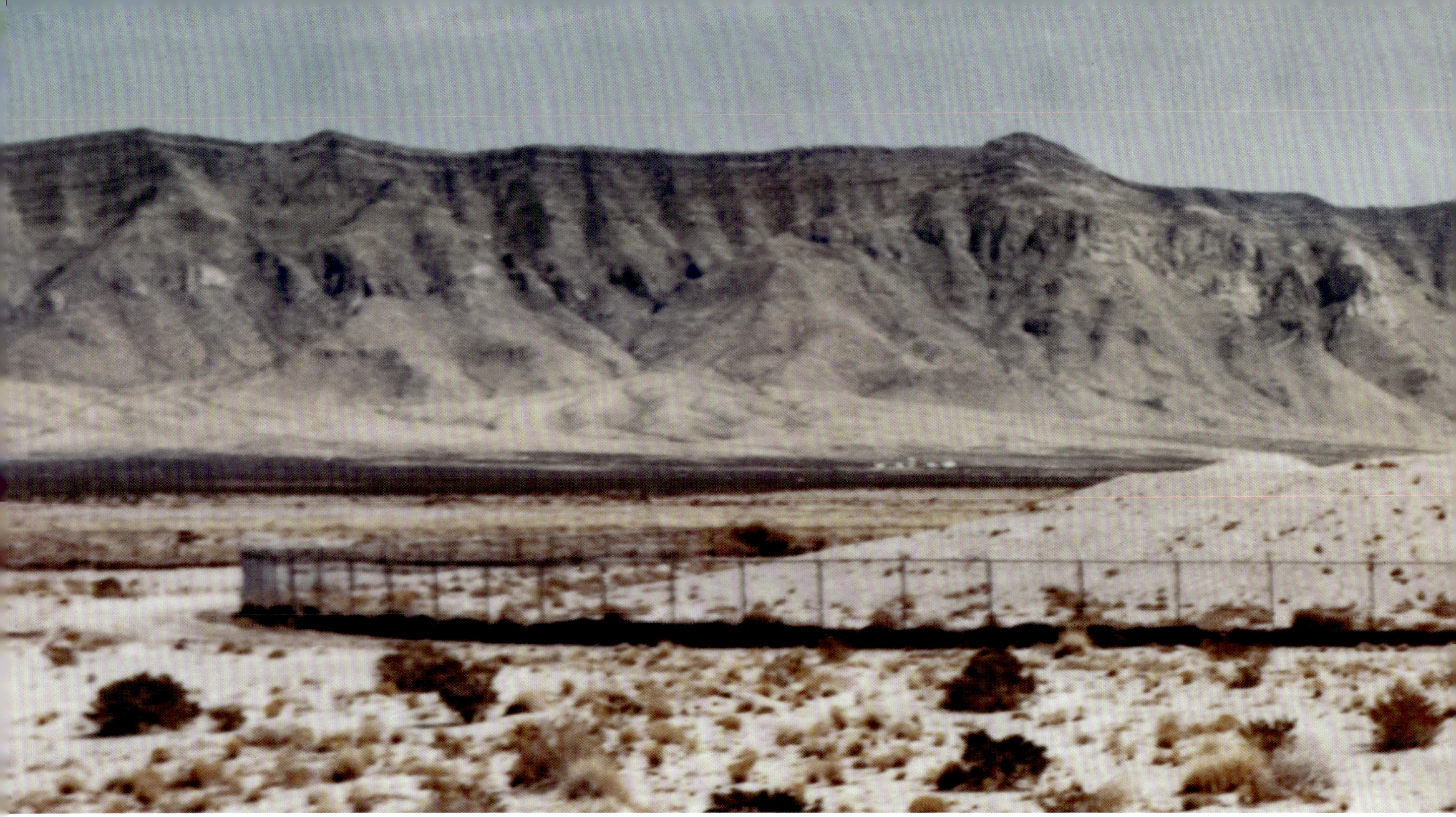

A final operation took place at the Trinity site not long after the destruction of Jumbo. In September of 1946, Los Alamos decided to make use of the isolated area for a series of tests designed to study initiators like that used in the Gadget—devices that inject a burst of neutrons as a nuclear weapon is fired. Since the initiators did contain a small amount of radioactive material, the tests were conducted in bunkers for safety. Perhaps Jumbo would have been involved if still in commission. Instead, three new, underground chambers were built about 530 yards from ground zero.

The concrete bunkers—with walls two feet thick—were buried forty feet underground, covered with a mound of sand for good measure, and thereafter accessible only through a shaft, shown at right. Once a bunker had been prepped for detonation, bulldozers filled in the shaft as well—leaving excavation as the only means of reentry.

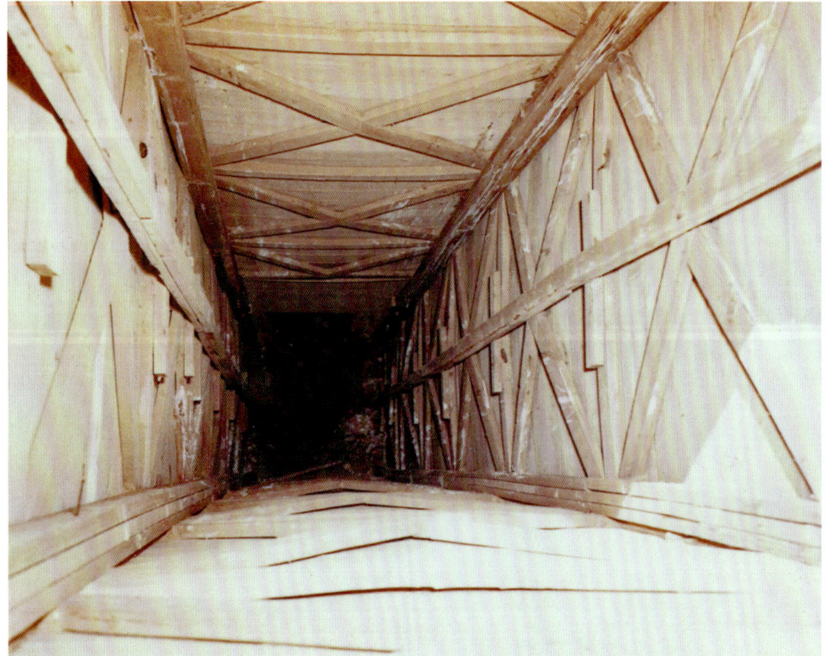

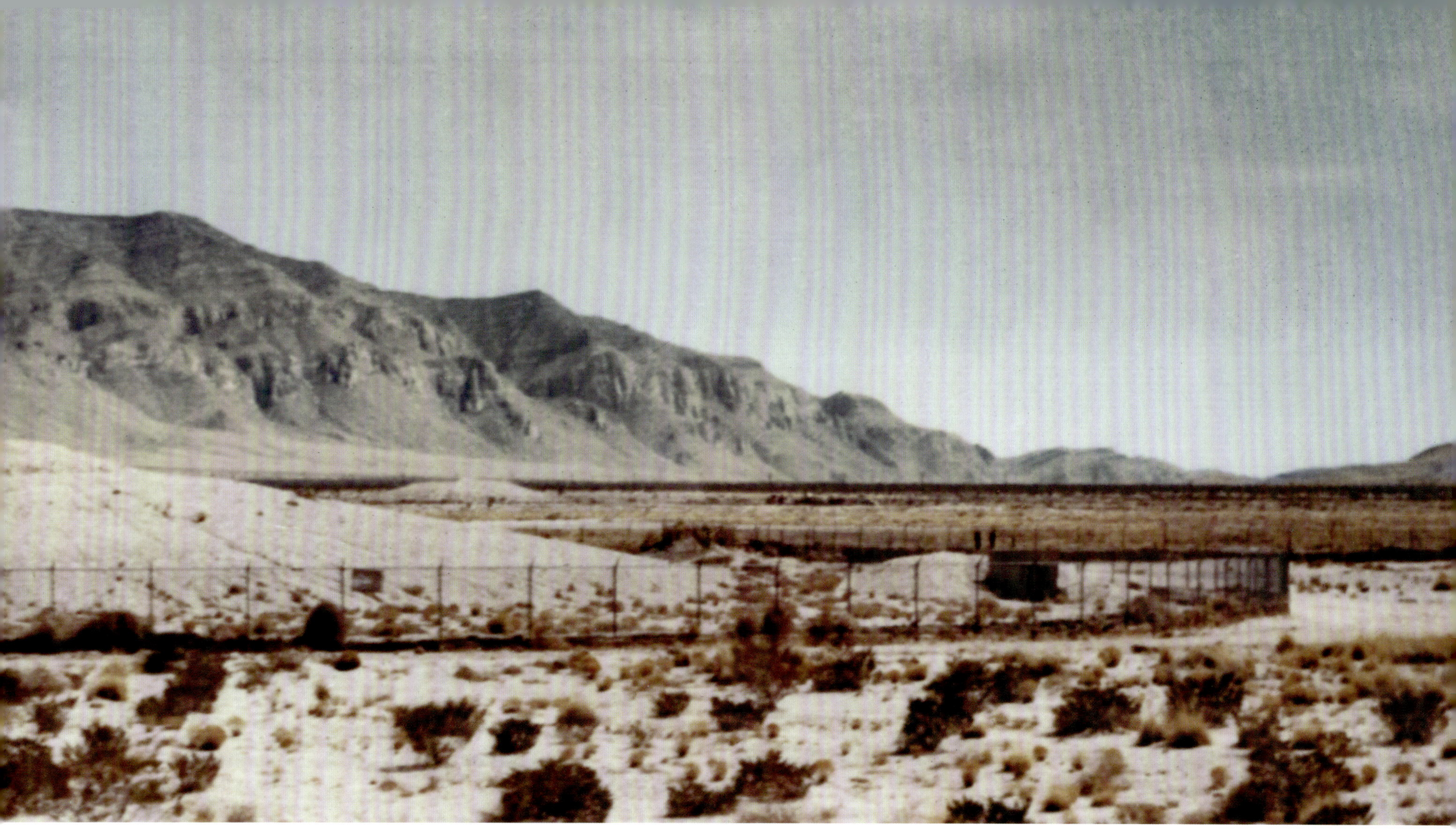

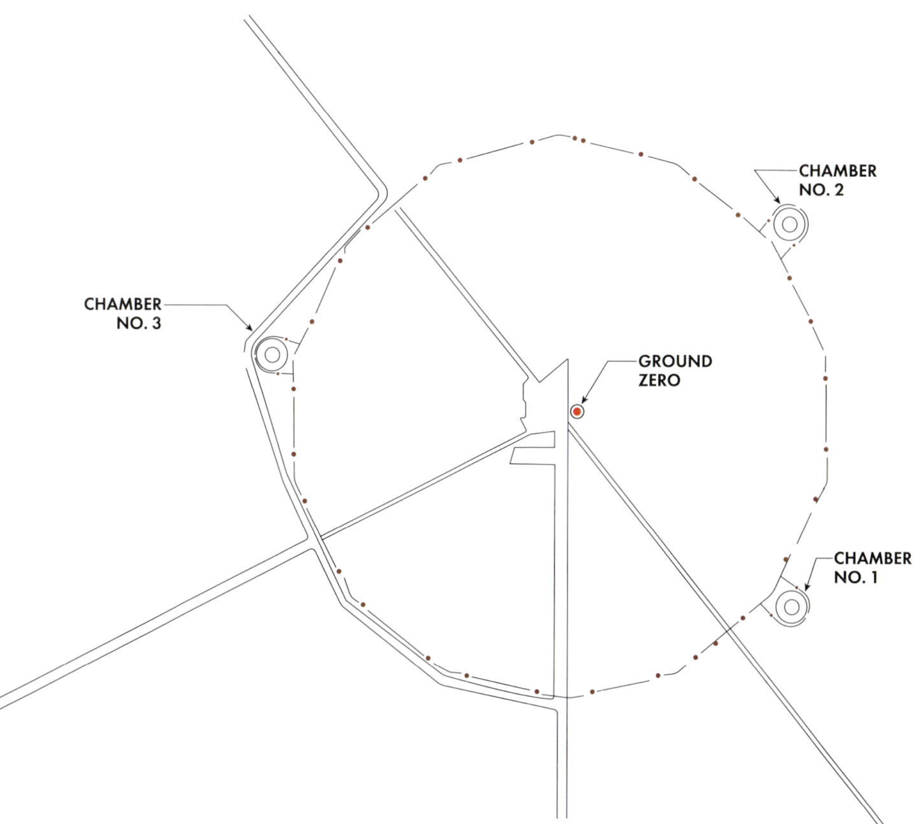

The first test took place in the southeast bunker, shown above. Detonation of a small charge was supposed to trigger a firing switch in the bunker to set off a much larger high-explosive charge. When the small charge was fired, however, monitoring instruments in the bunker showed that the high explosives had not followed suit. If and when they might was anybody's guess. Rather than taking the risk of excavating the bunker filled with live explosives to determine what had happened, they left the experiment in place, where it became known as Sleeping Beauty. An identical test was conducted successfully in the northeast bunker, and the third was not used.

Like Jumbo, just across the crater to the northwest, Sleeping Beauty would be abandoned for decades.

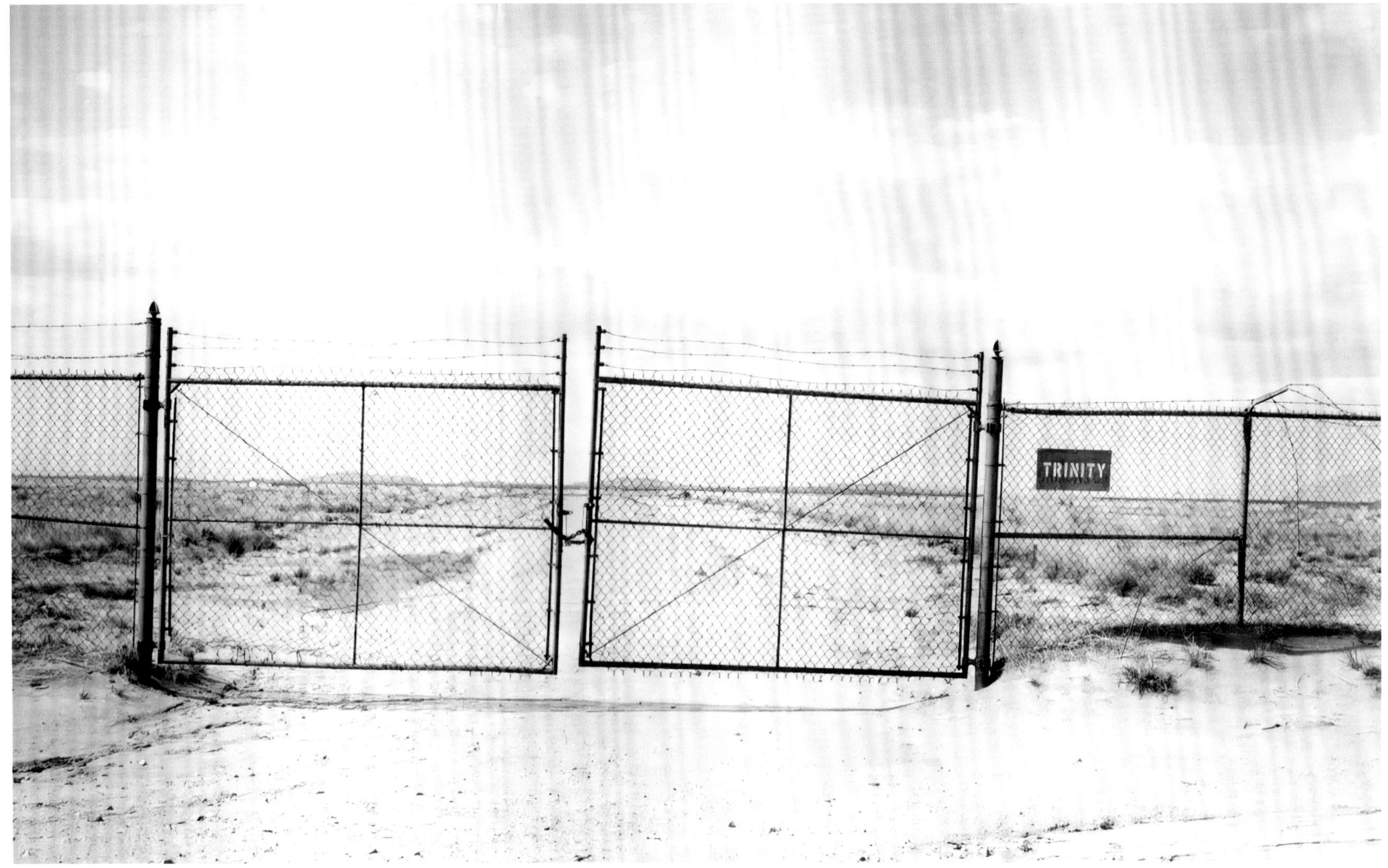

Los Alamos and the Trinity site remained under the purview of the Army via the Manhattan Engineer District until January 1, 1947, when by an act of Congress, the nuclear enterprise transitioned to civilian control under the newly established Atomic Energy Commission (AEC). A small group of Military Police kept guard at the test site as the AEC, the army, and a third stakeholder—the National Park Service—considered how to address the competing issues of security, maintenance, public health, use of the grounds, and historic preservation. (Photo courtesy of White Sands Missile Range Museum.)

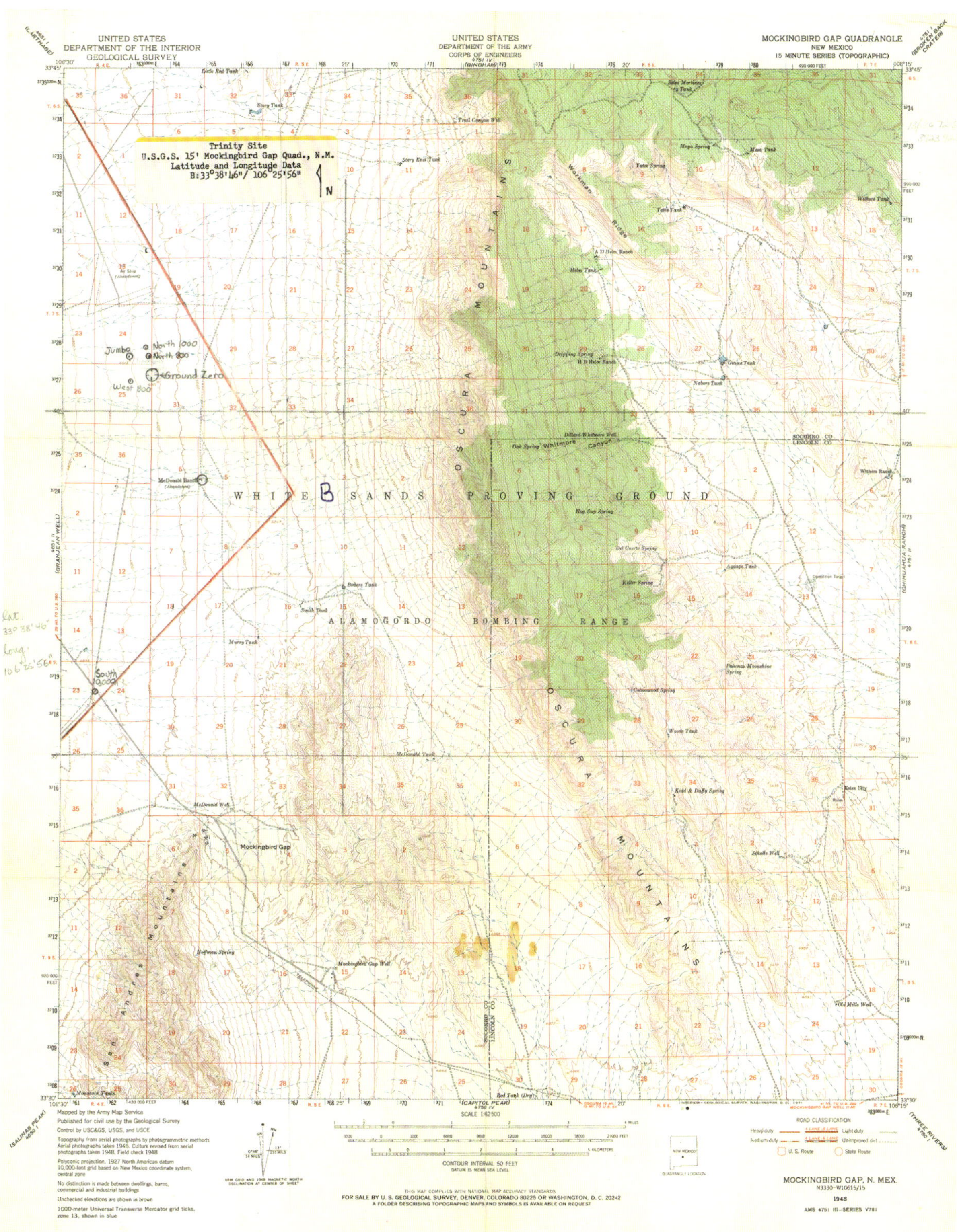

The National Park Service had acted quickly, first expressing its intent to turn the Trinity site into a national monument in January of 1946. Initial momentum, including support from Secretary of War Robert Patterson, eventually ran up against national security concerns related to the military installation that encompassed the site. Along with the old Alamogordo bombing range, Trinity had become part of the new White Sands Proving Ground a week before the July 16, 1945, test shot. The map at left shows the eastern boundary of the Trinity site, outlined in red.

No formal agreement was signed when the Army Air Forces originally transferred a portion of the bombing range to the Manhattan Engineer District in 1944. The AEC thus had no legal claim to the land. They retained responsibility for safety due to the presence of radioactive material, but the Army Air Forces (and later the air force, then the army again) held jurisdiction and operational control over the area.

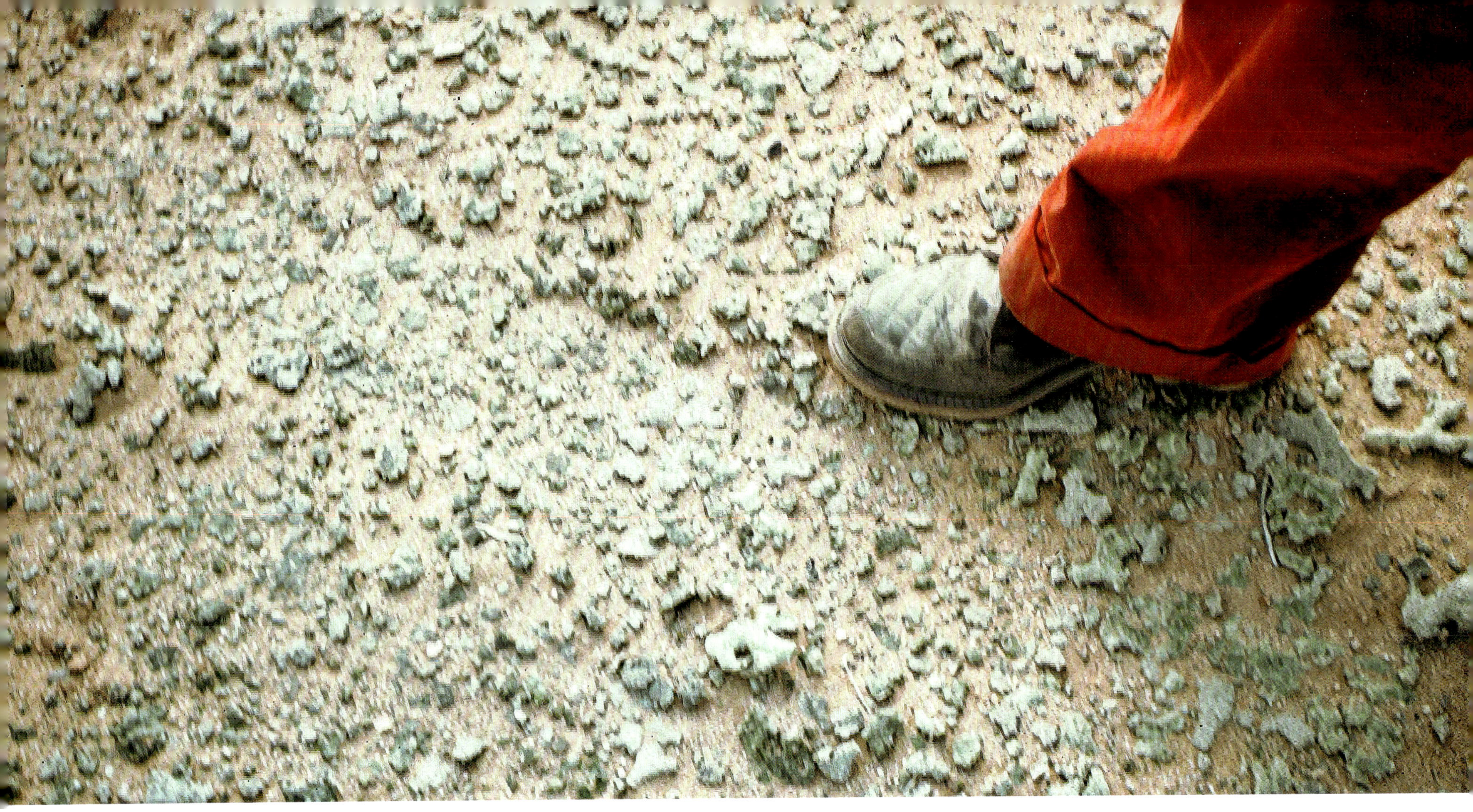

White Sands commanders viewed a national monument as a nonstarter, given ongoing military use of the area around the Trinity site. The AEC had reservations about tourism as well, choosing to extend what was initially a two-year moratorium on National Park Service activity because of the presence of dwindling but still detectable radiation, which rendered the site potentially unsafe.

As years passed, the harsh climate of the Jornada del Muerto steadily wore away the small reminders of human activity at the test site—including the sheet of trinitite, shown above in 1952. On the basis of several radiation surveys taken at the crater, the AEC determined that the green glass, as it slowly broke down and blew away, posed a health hazard—or was a liability in that it gave the appearance of a hazard—and ought to be removed or covered up. (Photo above by Robert Alley, courtesy of Jim Eckles.)

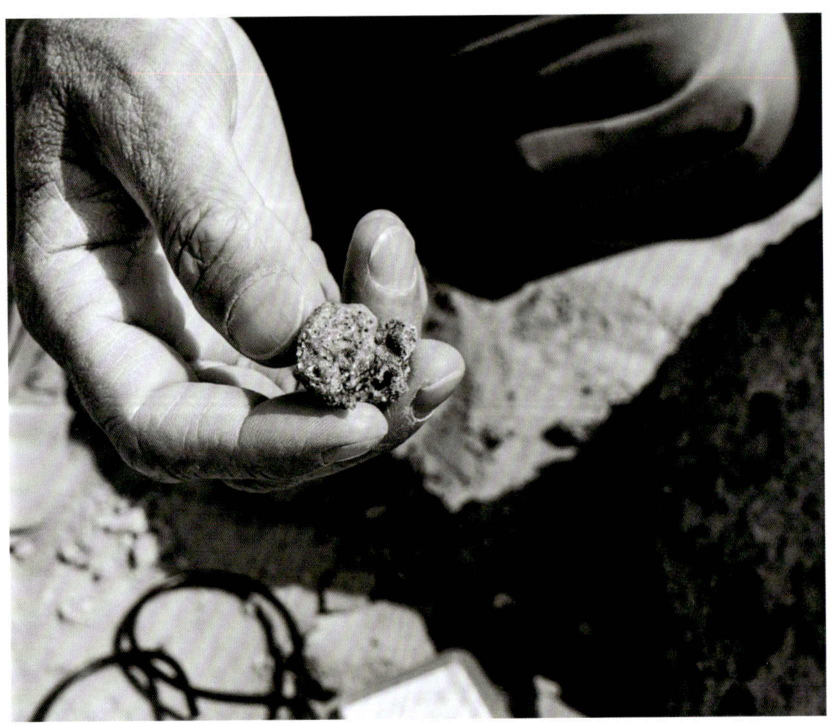

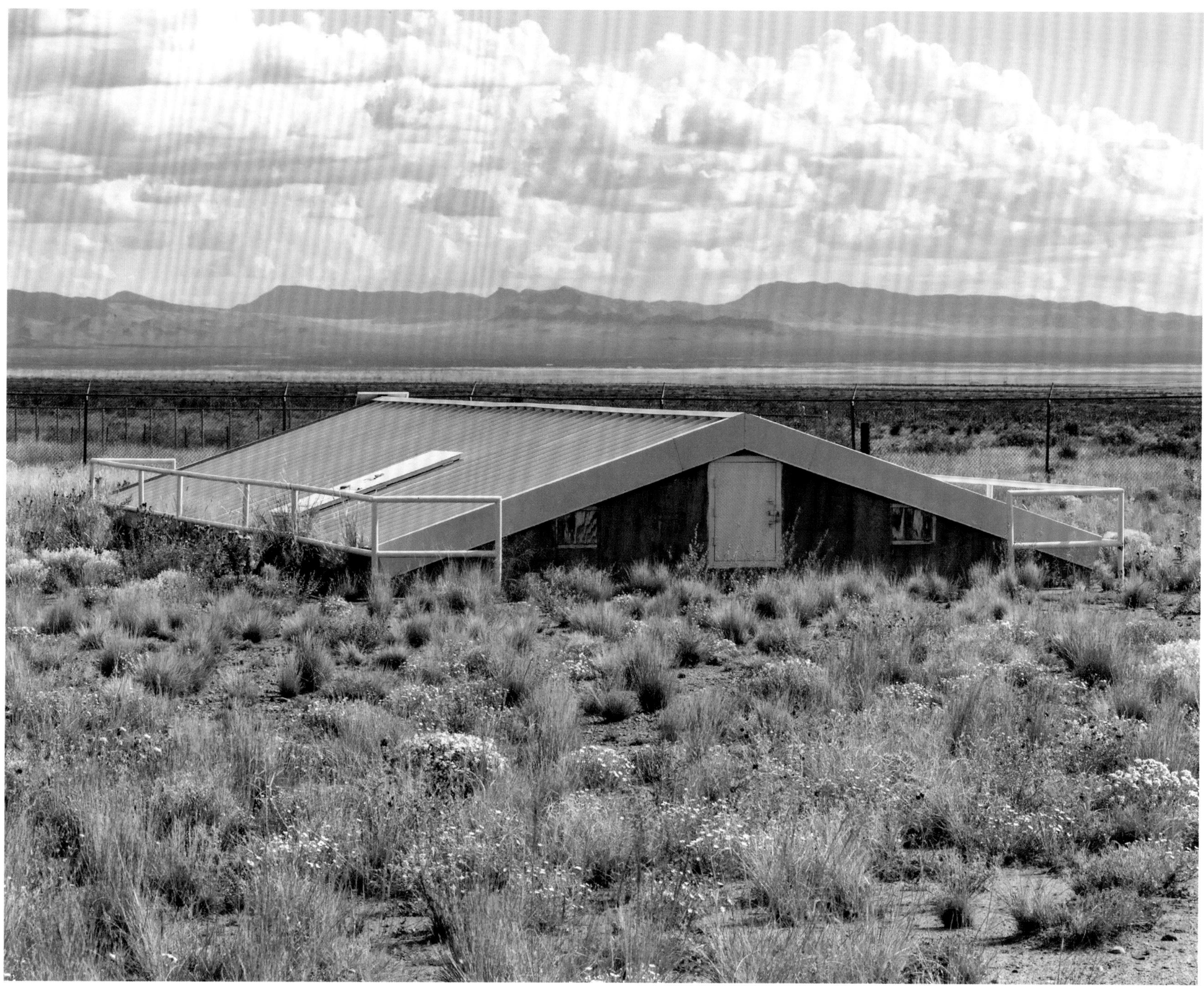

For several years, while the Trinity site lay vacant, plans and contracts to clean up the trinitite kept coming together only to fall apart. Opposition arose at various times from internally within the AEC, from Los Alamos radiation experts who disagreed with the hazard assessment, and from the National Park Service, which wanted the crater preserved in its current state for historical accuracy. At a 1952 conference in Washington, DC, congressional representatives from New Mexico and members of the AEC, the National Park Service, and the Department of Defense agreed on a plan to remove the trinitite while protecting a portion for a future exhibit. There was already a shelter covering a small patch of the material, and it was decided after a field visit to improve this original shelter. (Another renovation in 1984 resulted, ultimately, in the structure shown above.) The parties also unanimously agreed that the site should be turned into a national monument as soon as practicable, with the understanding that this would be postponed as long as the surrounding area remained in active use for missile testing.

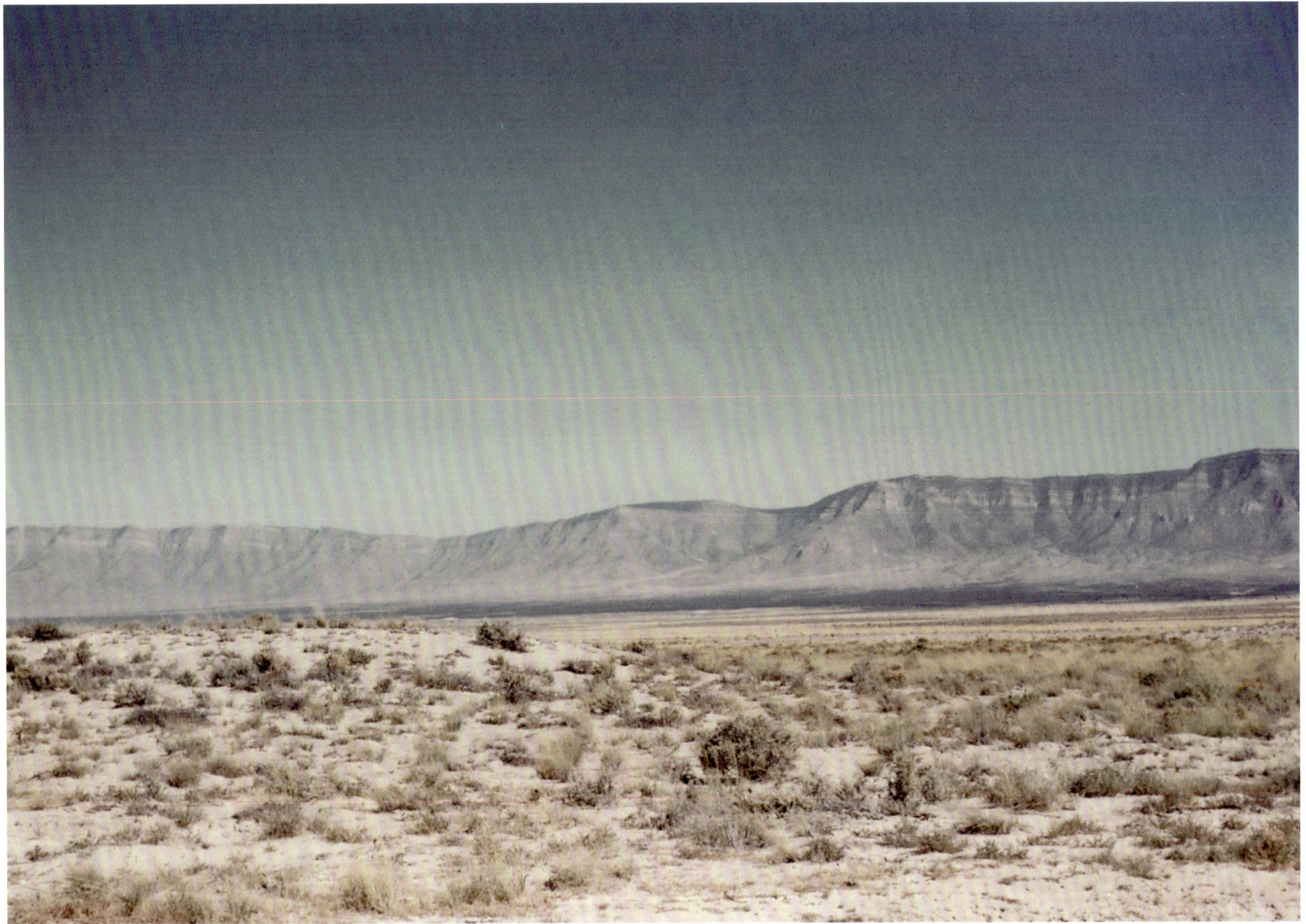

The summer of 1953 saw the trinitite finally scraped off the basin floor. Contractors moved bucketloads of sandy soil, combined with the radioactive glass, to shallow trenches about 1,200 feet from ground zero. That September, with the crater freshly cleared, White Sands hosted the first open house at the Trinity site. Wind had long since leveled off the cratered earth, leaving not much to be seen of ground zero as it would have appeared in 1945. Nevertheless, 650 people attended. Another open house was held in 1955 on the ten-year anniversary of the test, and by 1960 the event was occurring annually. (Photo taken October 30, 1954; courtesy of White Sands Missile Range Museum.)

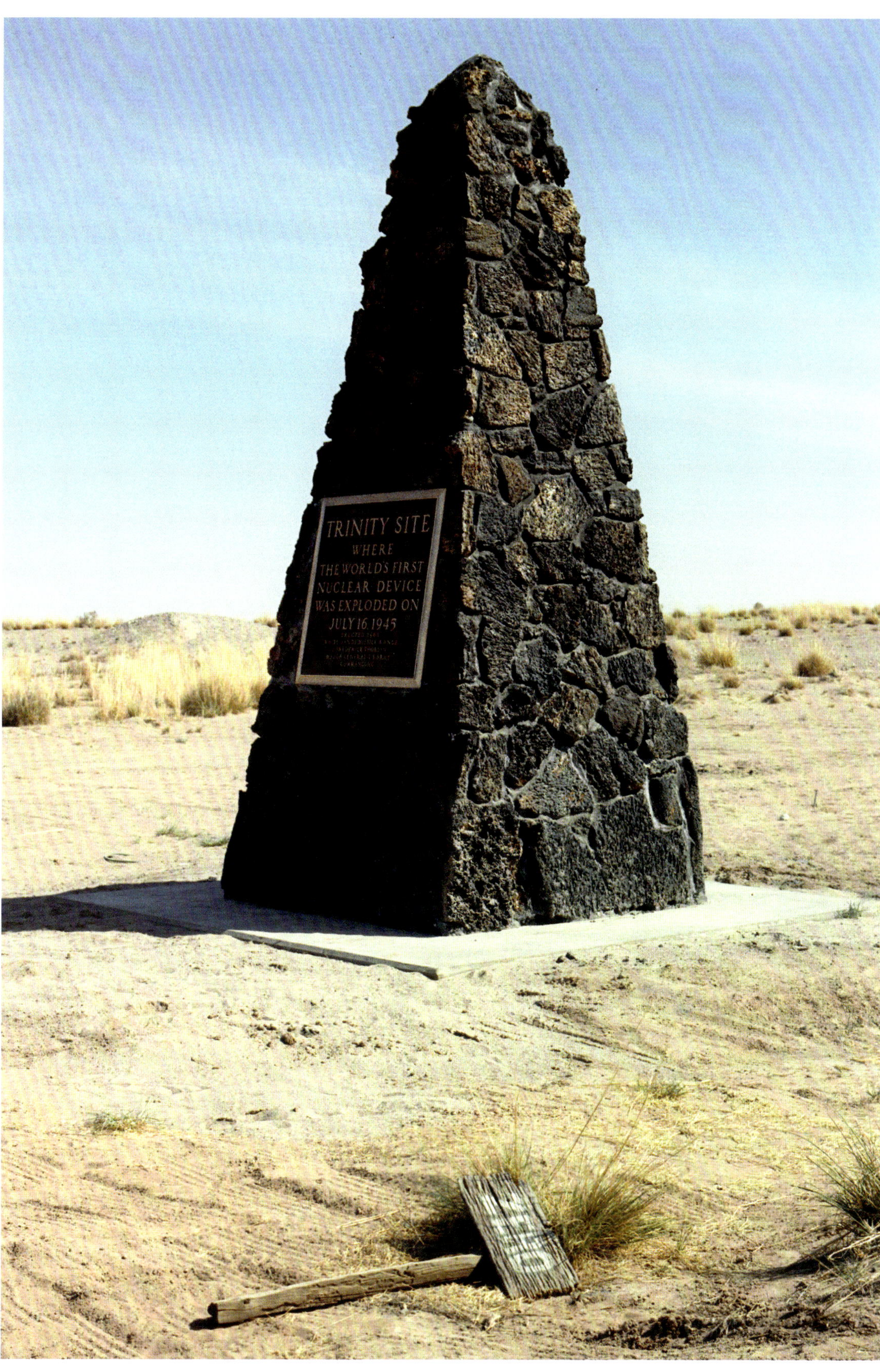

On the twentieth anniversary of the test, a new landmark arrived at ground zero. Visitors in July of 1965 were treated to a monument installed by then-commander of White Sands J. Frederick Throlin. Fashioned from black lava rock harvested on the east side of the Oscura Mountains, the obelisk replaced a simple wooden sign labeled "zero," discarded in the foreground at left.

A plaque on the obelisk identified its location as "Trinity Site: Where the World's First Nuclear Device Was Exploded on July 16, 1945."

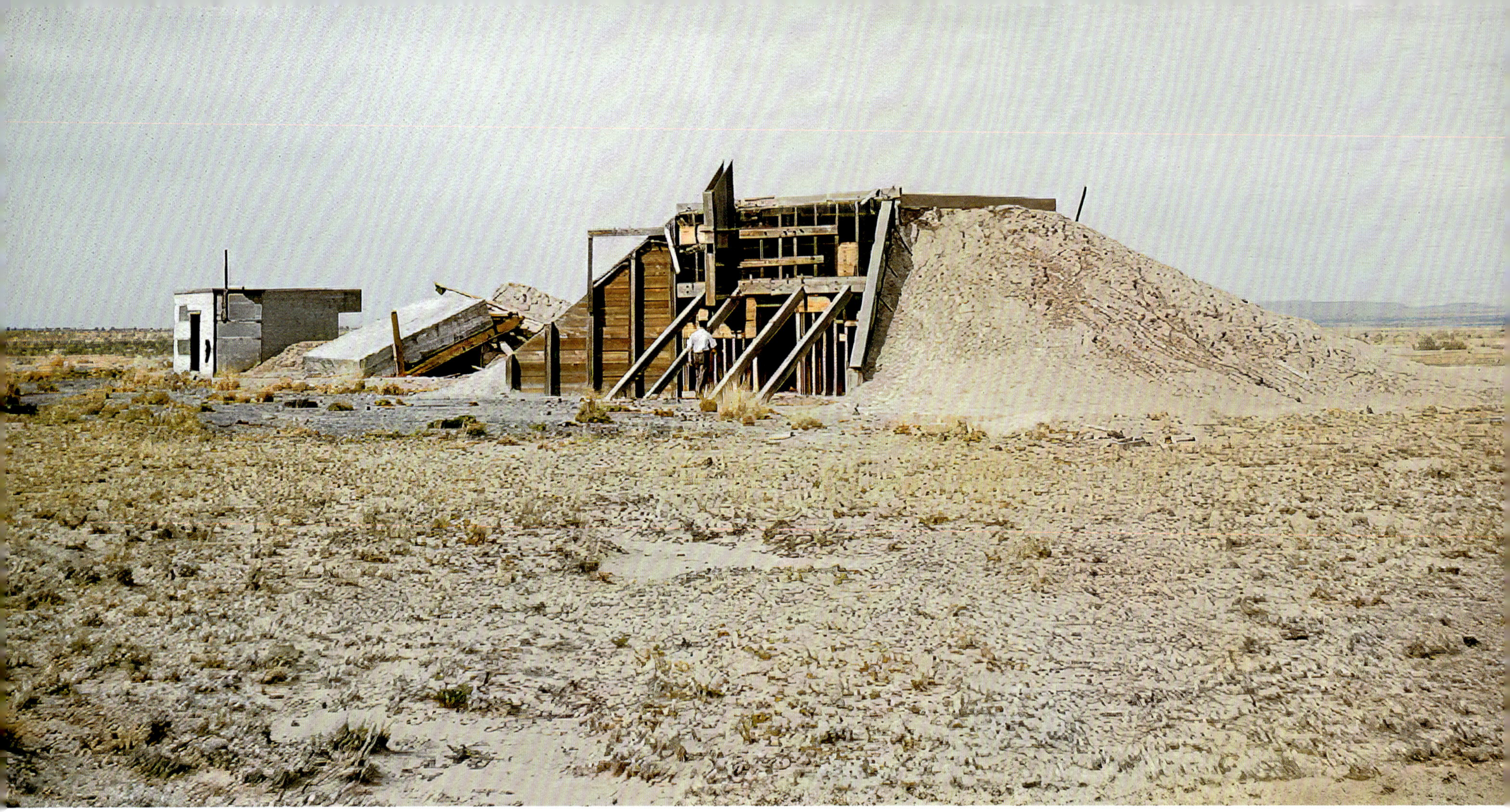

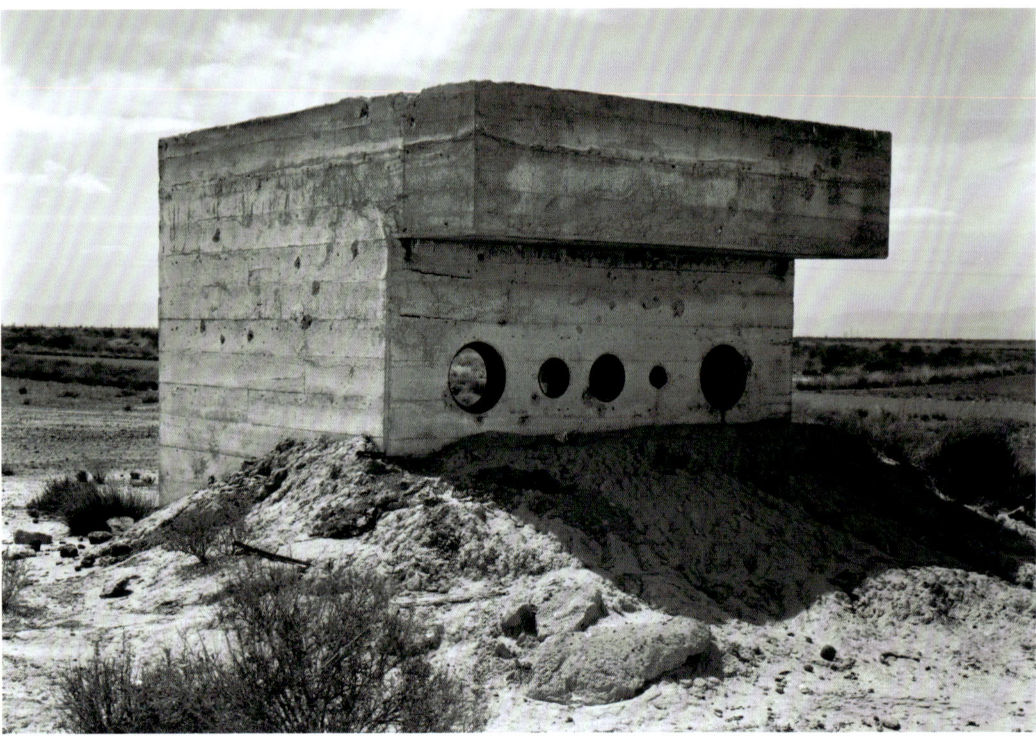

Even as a physical monument was finally emplaced, efforts to create a national monument slowed again to a halt, while the built environment continued to deteriorate. Within only a few years of the Trinity test, some of the shelters had already begun to collapse. West 10,000 is shown above in 1952, with the generator bunker caved in on itself. (Photo by Robert Alley, courtesy of Jim Eckles.)

By 1965, when the obelisk appeared at Trinity, the personnel shelters and adjacent generator bunkers—overtaken by rot and rattlesnakes—had been demolished by the army due to safety concerns. Artifacts like equations and signatures scribbled on beams were lost to the flames. The only structures remaining thereafter at 10,000 yards were the concrete portions of the photography bunkers (*left*); their wooden additions, which housed overflow equipment and included rooftop holes for manned camera turrets, disappeared sometime after the 1952 image was taken.

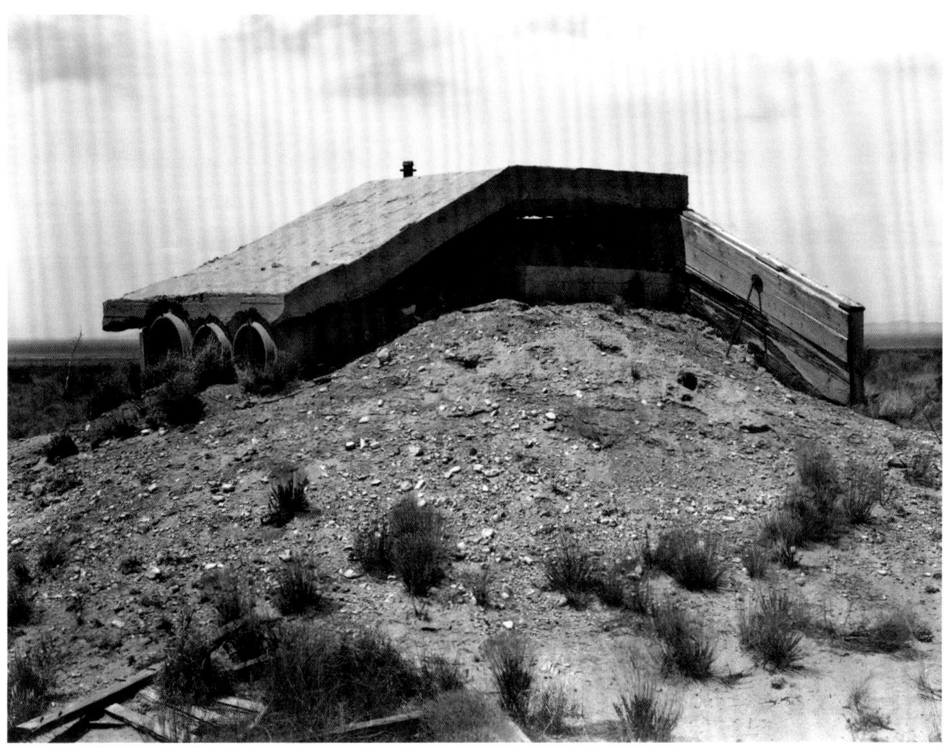

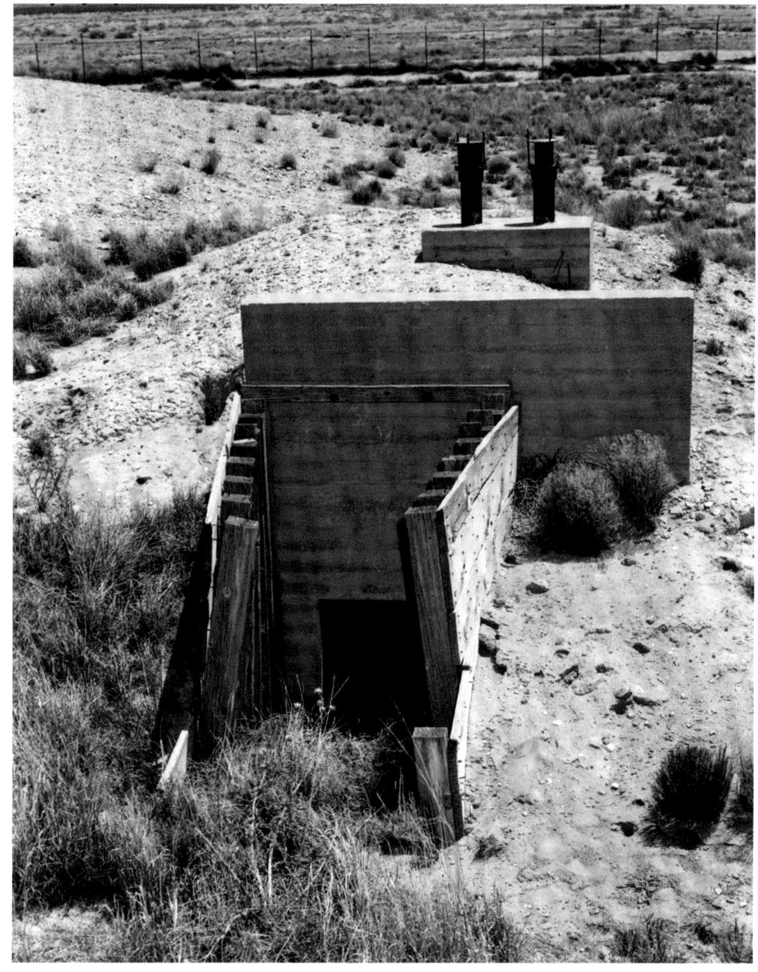

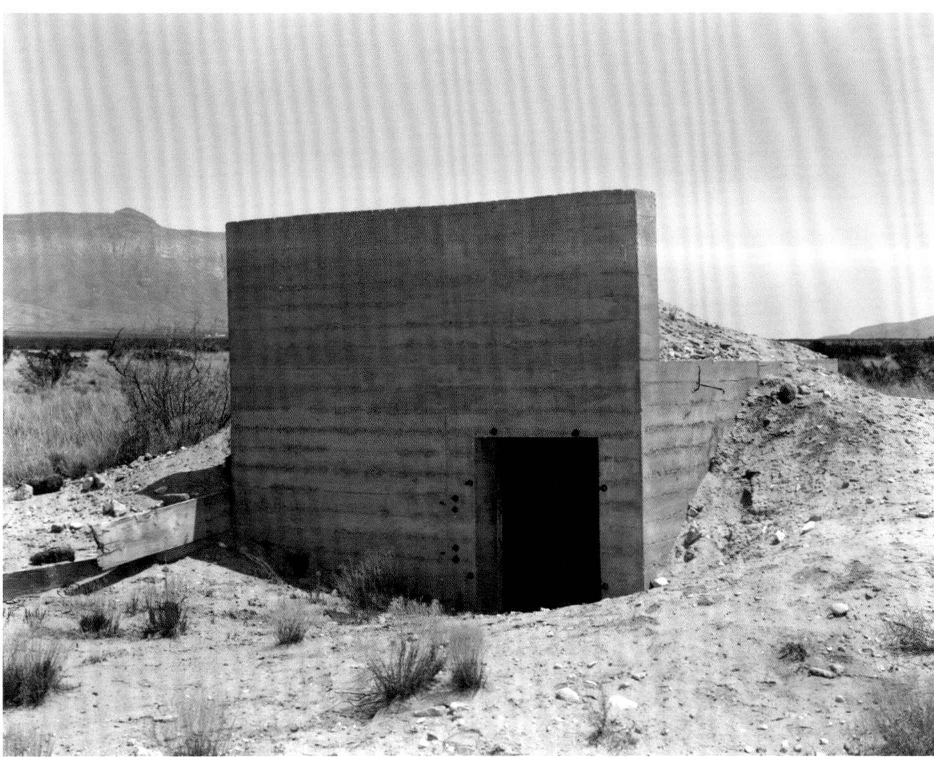

Both 800-yard photography bunkers have survived to the present day (the west bunker is at top left), along with the instrument bunkers at north 1,000 (*left*) and northwest 600 (*above*). The photographs shown were taken in the early 1980s for a Historic American Buildings Survey. (Library of Congress, Prints & Photographs Division, HAER NM-1-A.)

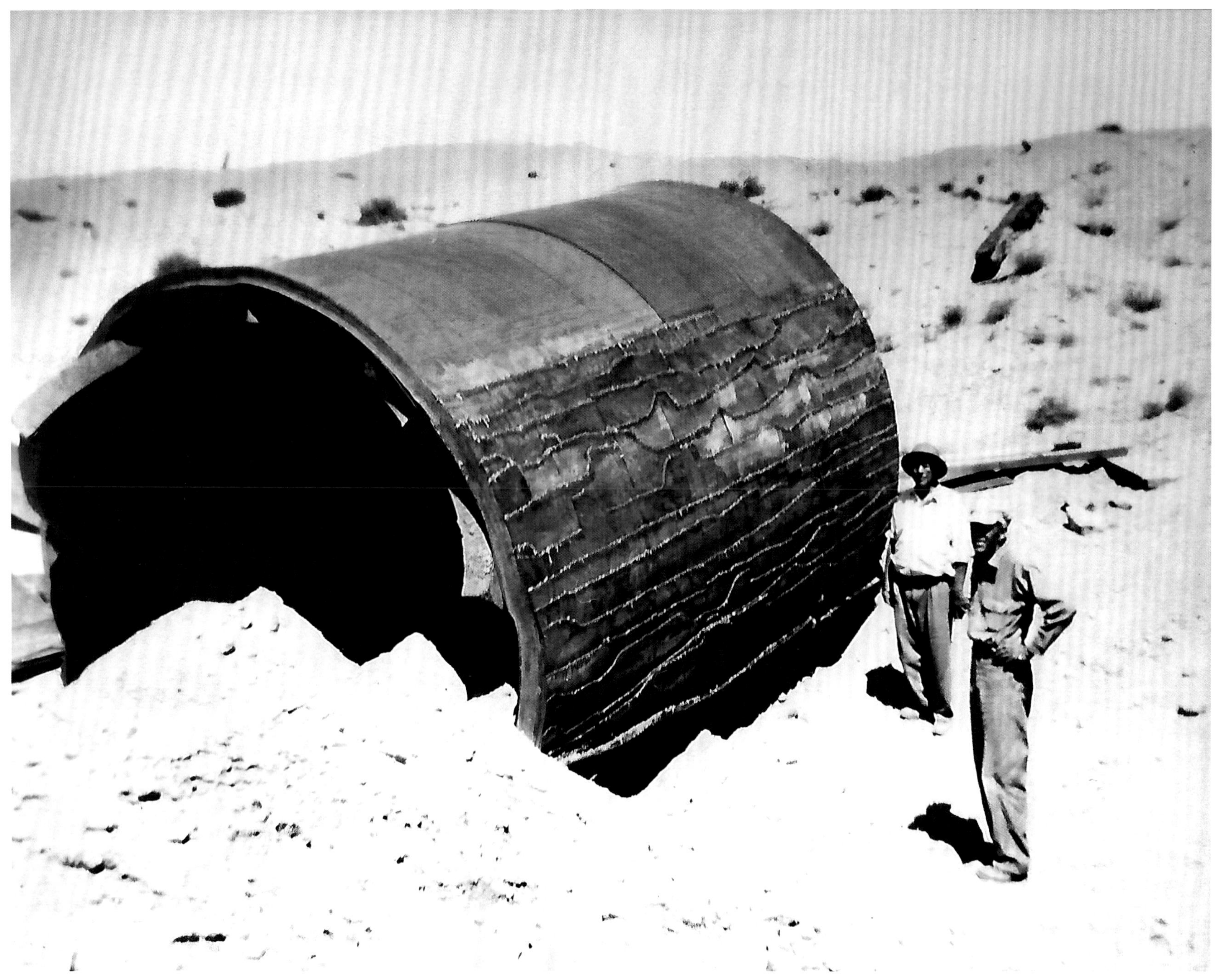

With the exception of resilient concrete structures, most points of interest around the Trinity site fared poorly in the decades after the test. The remnants of Jumbo's tower disappeared sometime in the 1950s, and by 1965 several tons of steel banding had been stripped off the broken vessel in what may have been a failed attempt to make it light enough to relocate. (Photo courtesy of William S. Loring.)

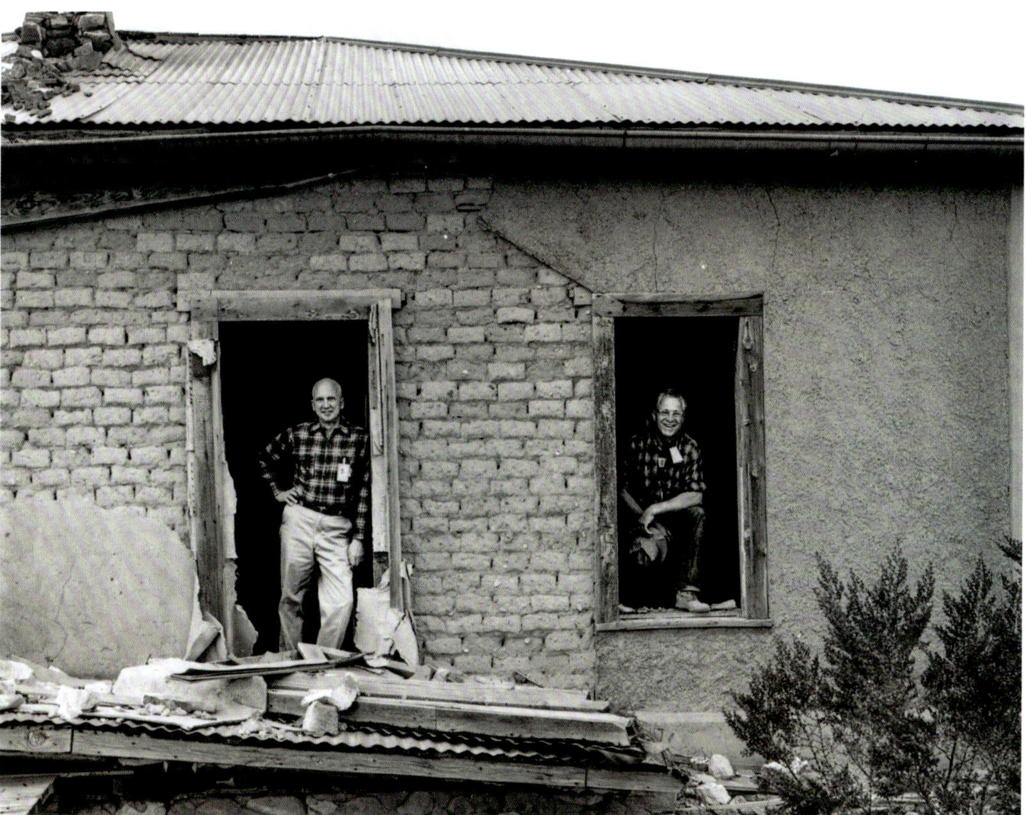

The Schmidt-McDonald Ranch House, although largely undamaged by the Trinity explosion, had fallen apart over many years of neglect and intrusion by the elements.

Trinity test veterans Norris Bradbury (*left*) and Raemer Schreiber visited in the 1960s, well into Bradbury's tenure as the second director of the laboratory. Here, they pose near the collapsed porch roof, inside the clean room where the Gadget's plutonium core was assembled.

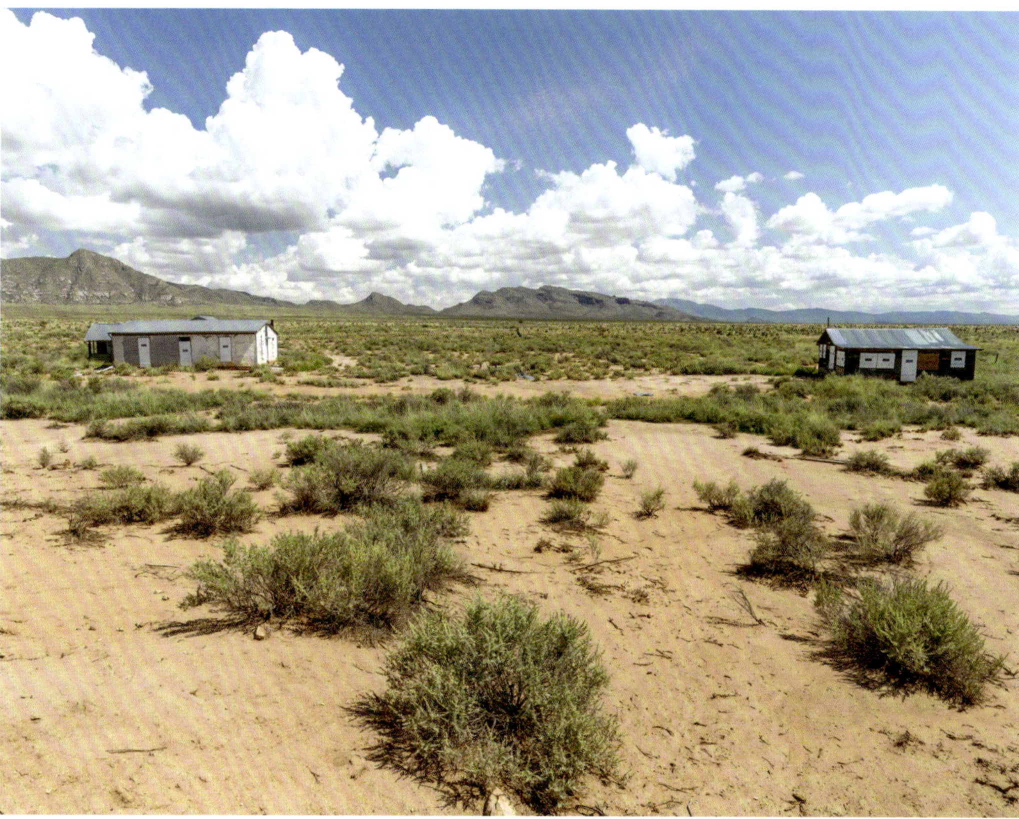

At base camp, where hundreds of soldiers and scientists had lived in the lead-up to the test, only the original ranch structures remain. Available records do not indicate when the government-erected buildings were removed, but it is well documented that within a year or two after the test, when the on-site population had dwindled to only Military Police and the occasional researcher, most buildings fell quickly into disrepair.

The photo at left, taken in July of 2024, recreates the angle of the panoramic image on pages 18–19.

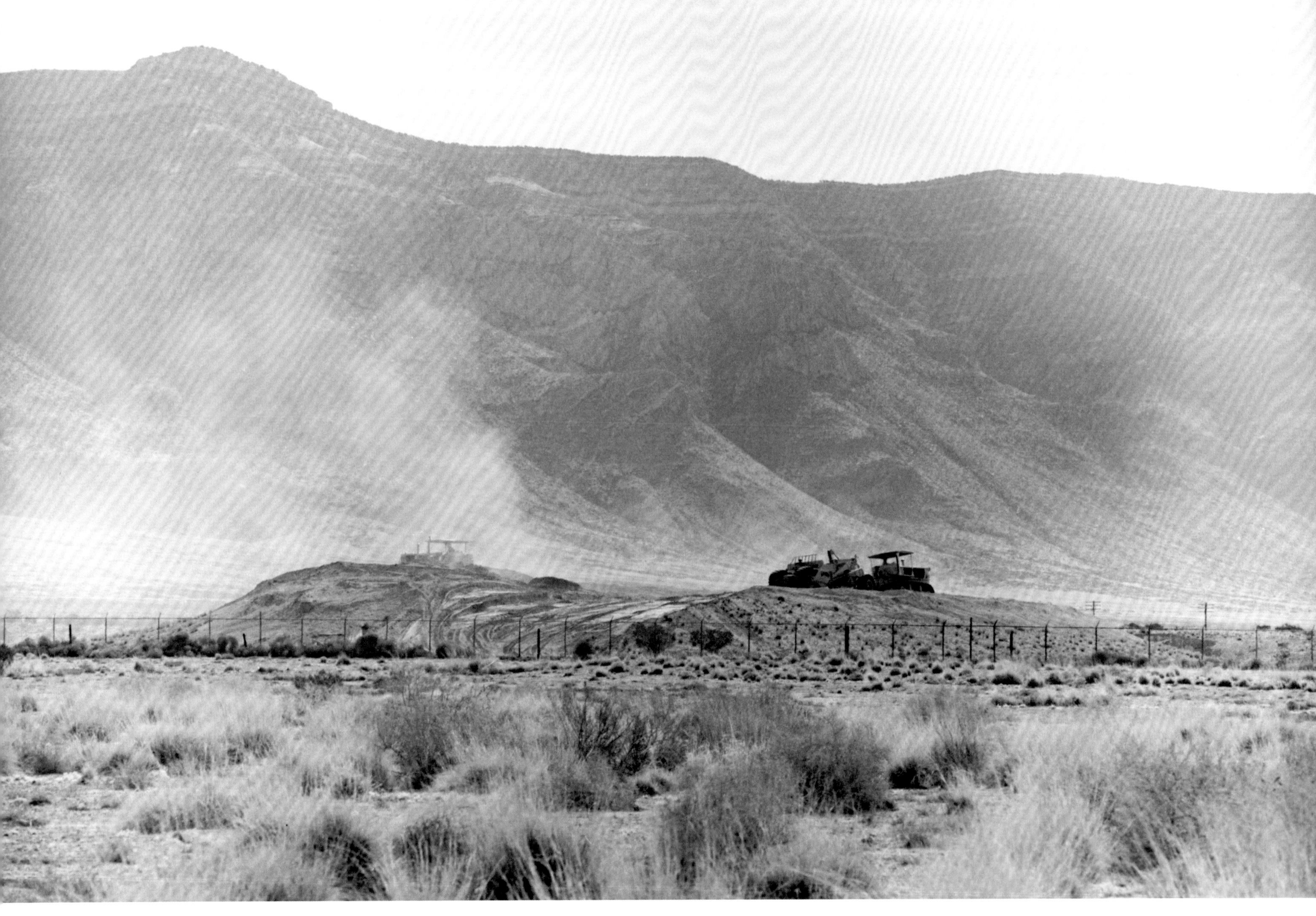

The National Park Service lobbied hard in 1966 for a bill that would mandate proper stewardship of the Trinity site while it remained under care of the Department of Defense. The legislation dissolved, but in light of its near success and rising public interest in Trinity, the AEC and White Sands decided in 1967 to revisit the lingering safety hazards at the test site: trinitite and Sleeping Beauty.

Accessing Sleeping Beauty's underground chamber required heavy equipment. This had kept the explosives and radioactive initiator comfortably out of reach to the casual trespasser, but for the same reason, getting down to the bunker was a major operation. Two Los Alamos scientists who had worked on the original initiator test—Don MacMillan and Robert Lanter—joined the 1967 expedition to do so, hoping to find out why the explosives had failed to detonate all that time ago.

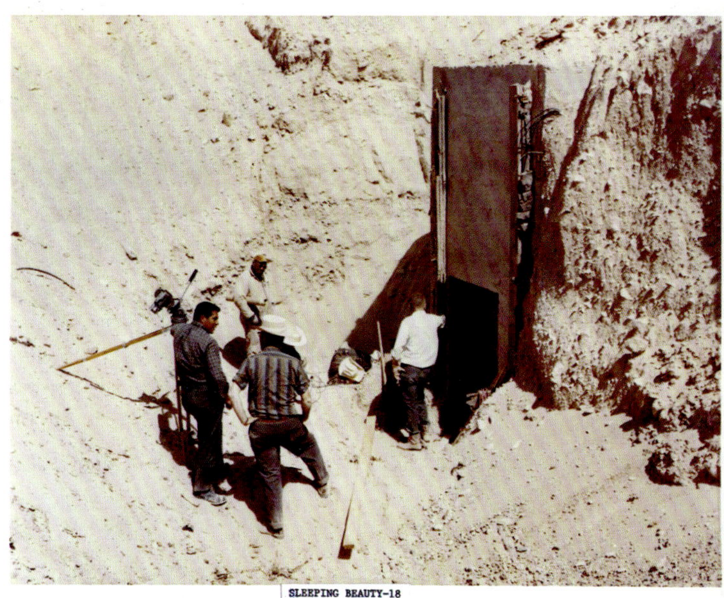

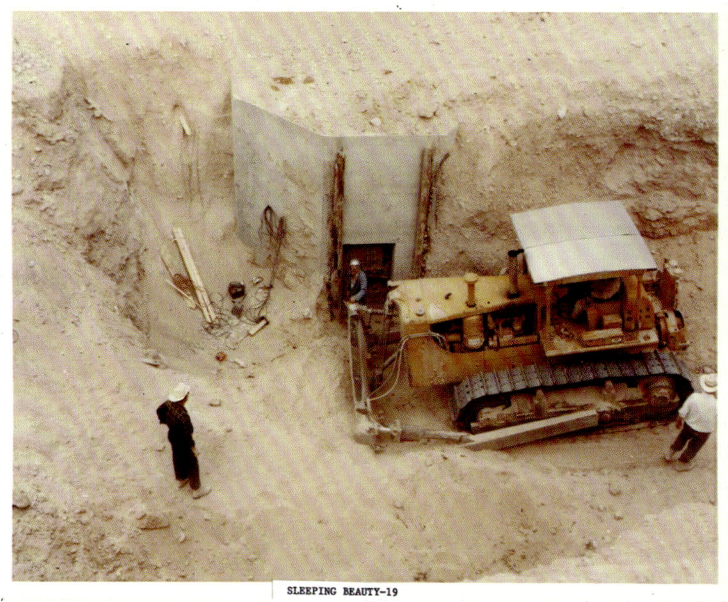

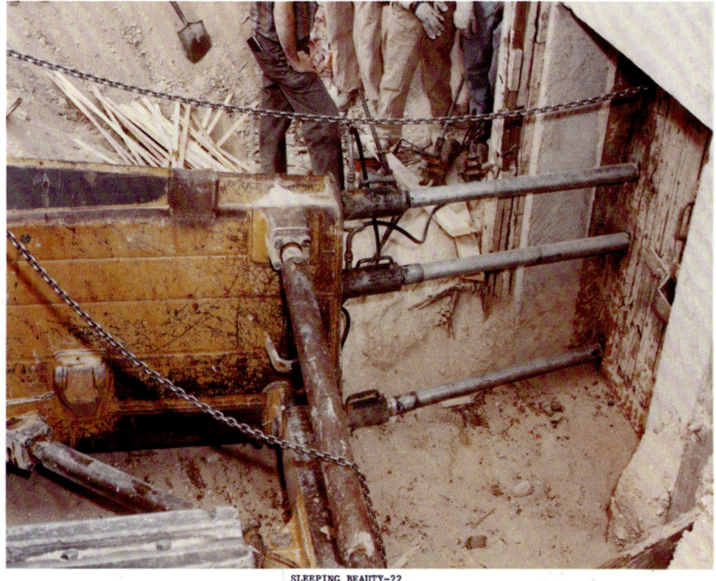

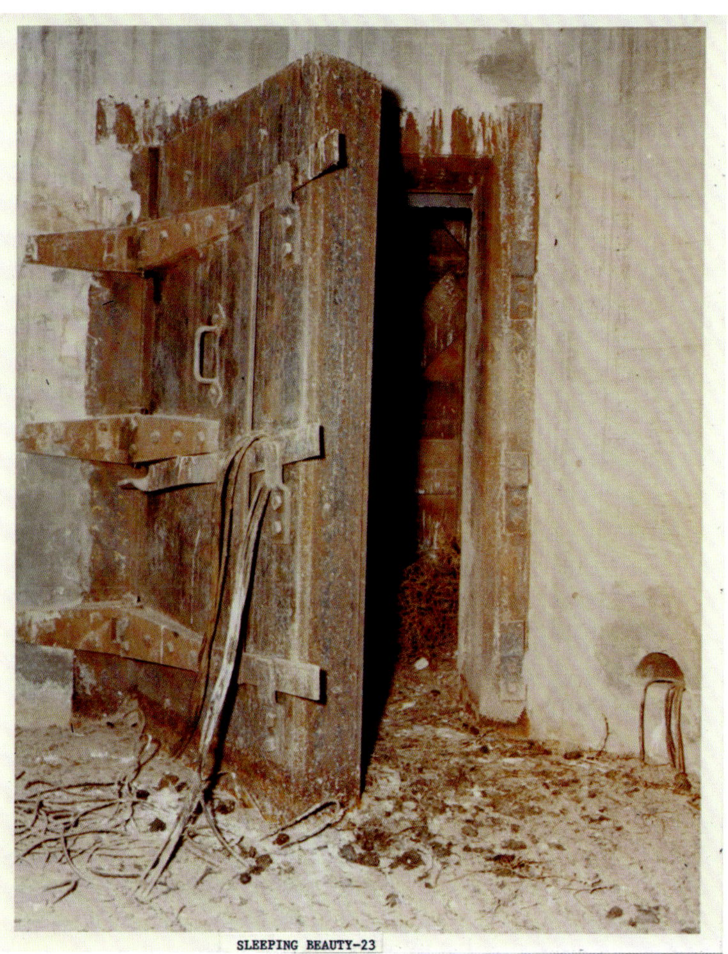

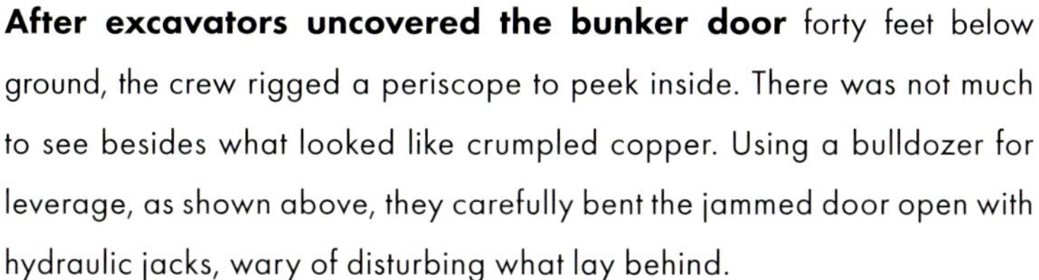

After excavators uncovered the bunker door forty feet below ground, the crew rigged a periscope to peek inside. There was not much to see besides what looked like crumpled copper. Using a bulldozer for leverage, as shown above, they carefully bent the jammed door open with hydraulic jacks, wary of disturbing what lay behind.

A mother barn owl greeted the surprised crew members. She had apparently been entering and leaving the buried safehouse through the forty-foot shaft and a small exhaust pipe designed to let out gases resulting from a successful detonation.

Further inside, the explosive charges that had failed to go off in September 1946 were found intact, still surrounded by neutron-counting devices. The crumpled copper (*far right*) turned out to be the remains of an electrical box containing the faulty firing switch. A new, 100-pound charge made short work of the bunker contents—once the owl had been moved to safety—and bulldozers re-covered the concrete shell, smoothing out the mound of earth that once marked its location.

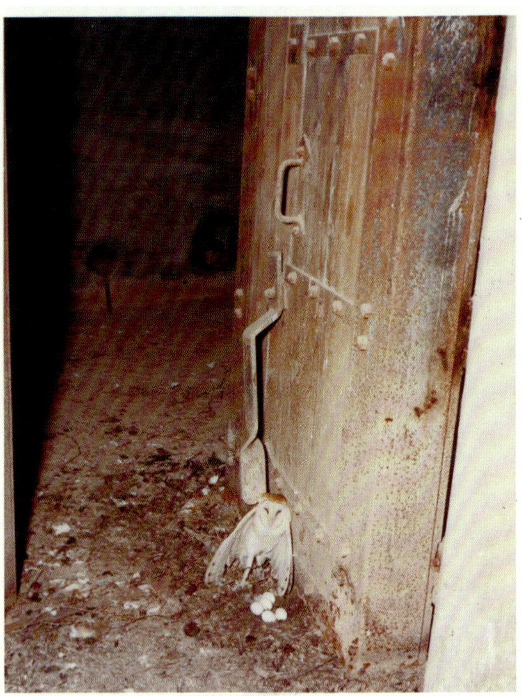

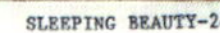

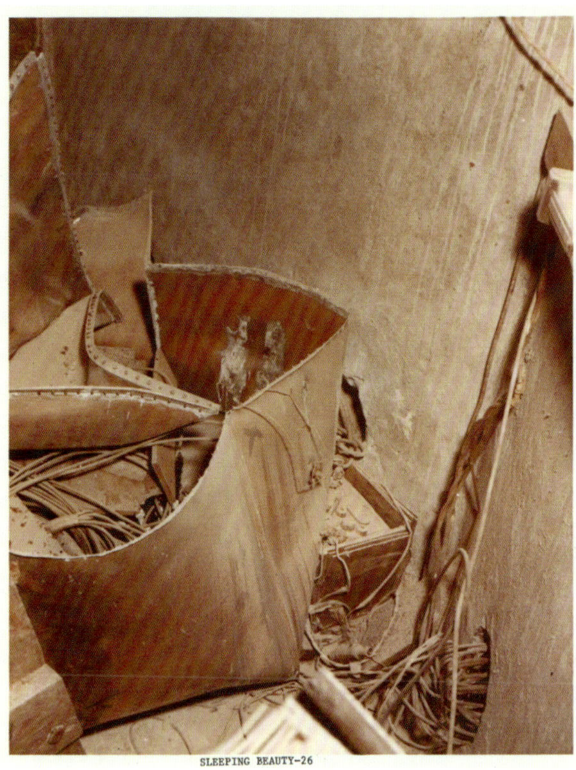

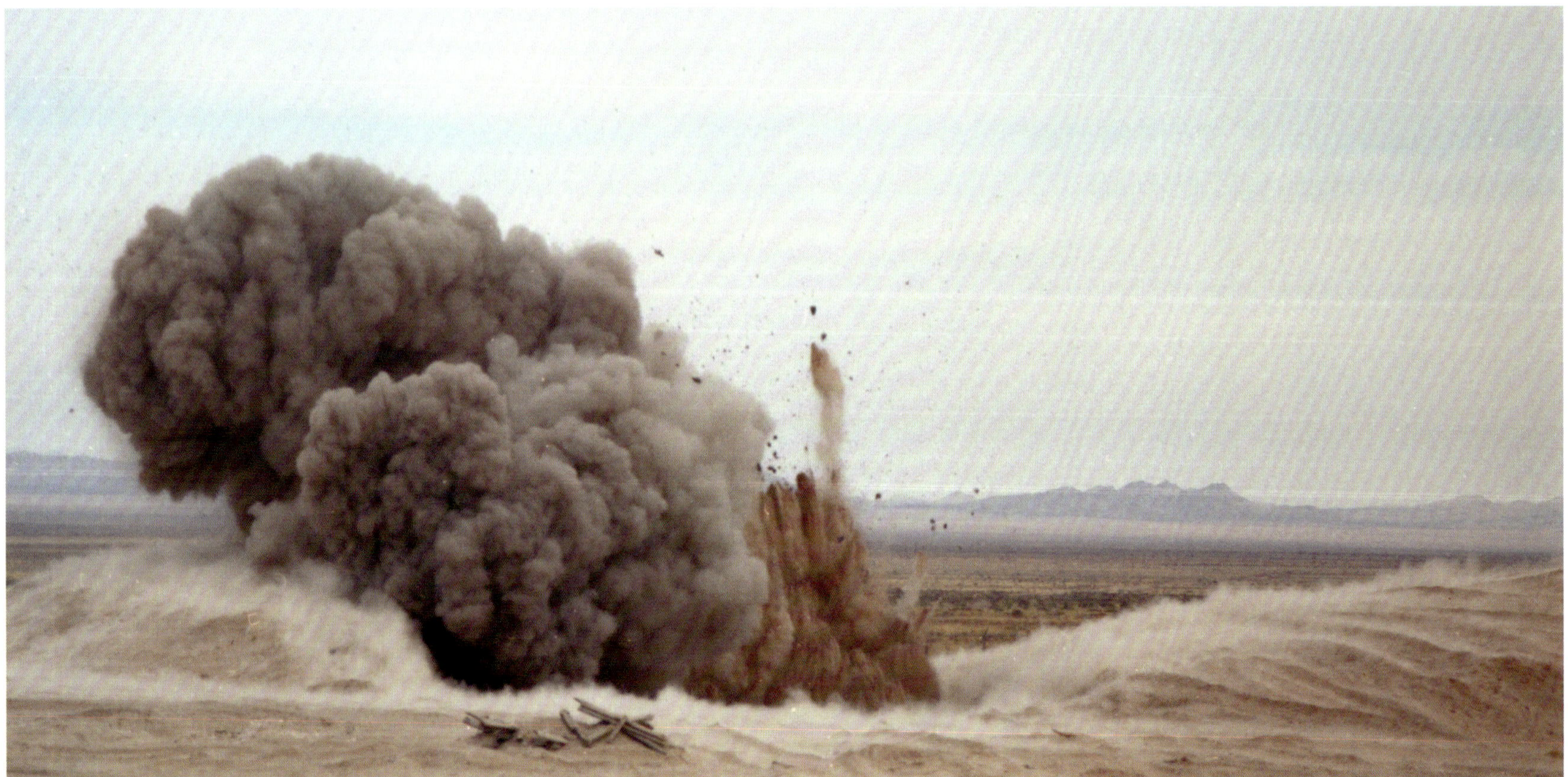

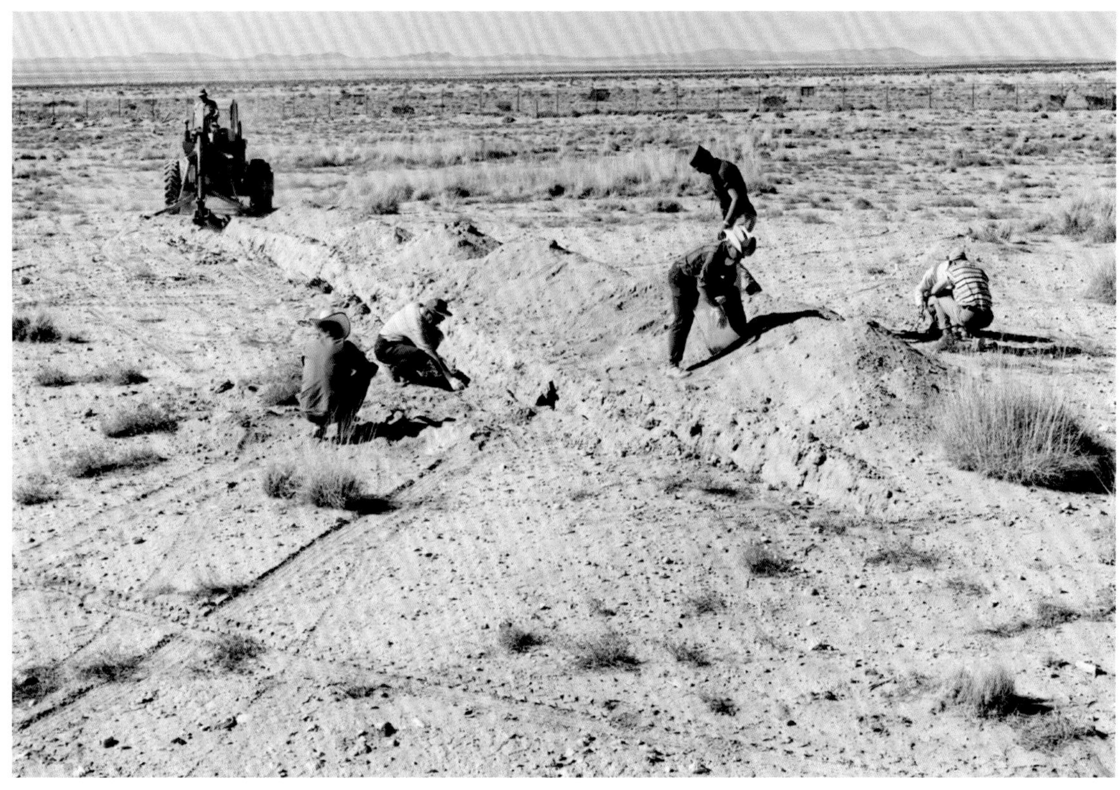

As for the trinitite, while small bits of the green glass still dotted the landscape, the material was mostly confined to a ring of trenches where the AEC had buried the top few inches of crater soil during the 1953 cleanup.

The crew took systematic measurements of radiation levels around the site, a process that included digging out the covered trenches, as shown at left. They found a very low percentage of trinitite amid what was mostly desert sand.

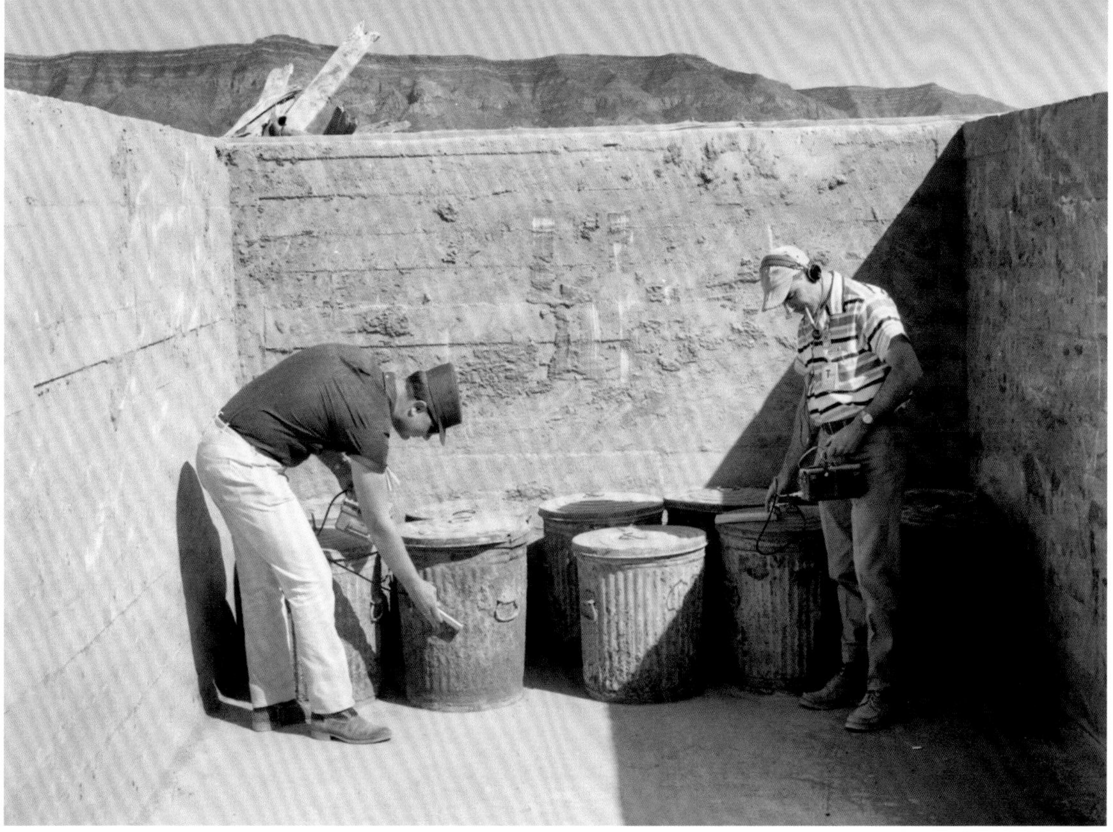

Records also indicated, however, that a buried bunker contained trash cans that had been filled with the most radioactive trinitite shortly after the test shot. Workers uncovered the bunker using a backhoe and sealed the cans inside metal drums, which were taken to Los Alamos for proper disposal.

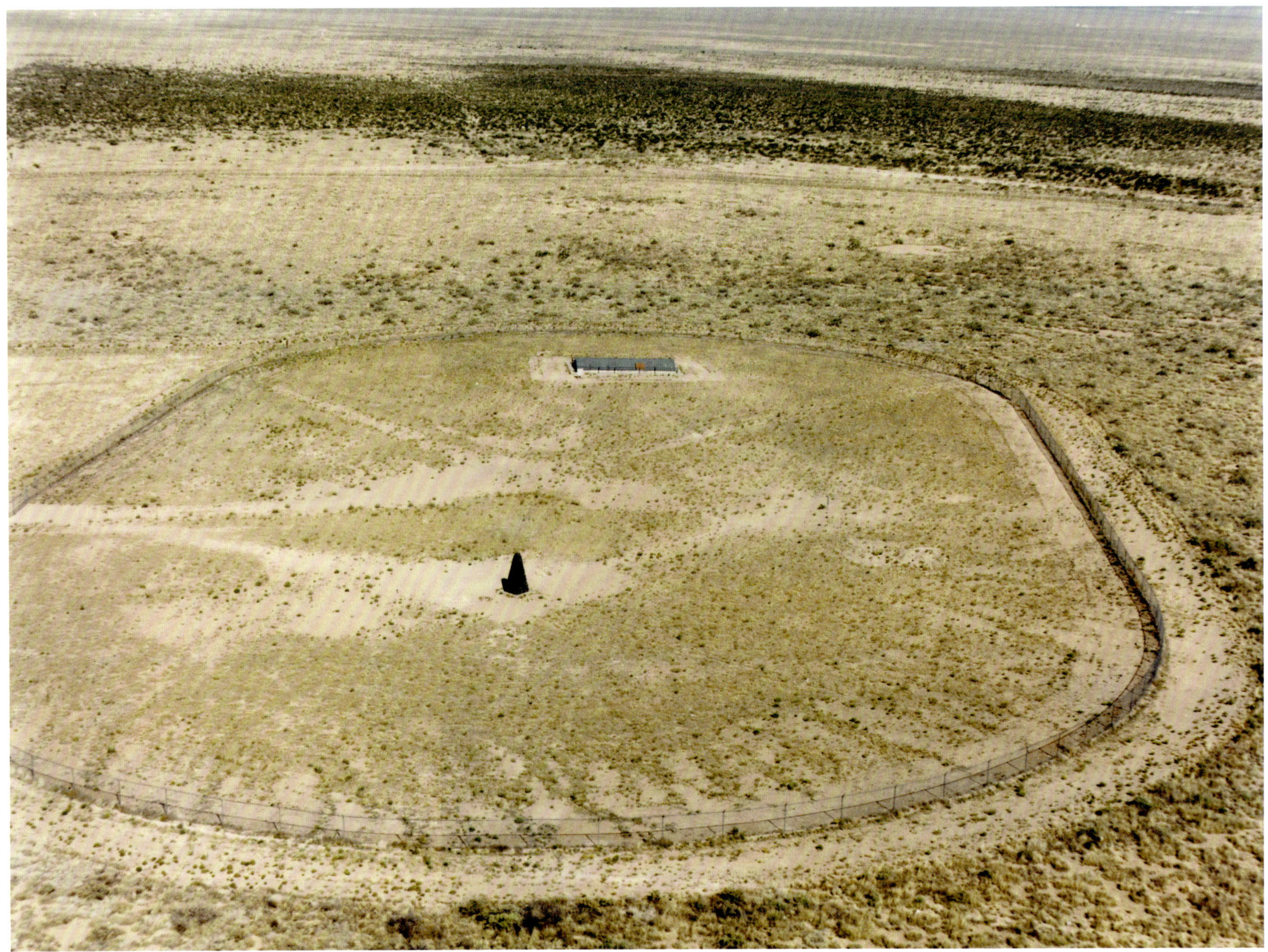

Results from the 1967 survey showed that most of the test site fell within the normal range of background radiation for south-central New Mexico. A few pockets, however, still exceeded the national safety standard for the general public. At the suggestion of the AEC, White Sands erected a fence around the crater to regulate time spent by visitors in the zone with greatest lingering radioactivity. A corridor was later built to shepherd guests from the original, circular outer fence (constructed by Military Police in 1945) to this racetrack-shaped inner fence, which today surrounds the obelisk and trinitite shelter. (Photo courtesy of White Sands Missile Range Public Affairs Office.)

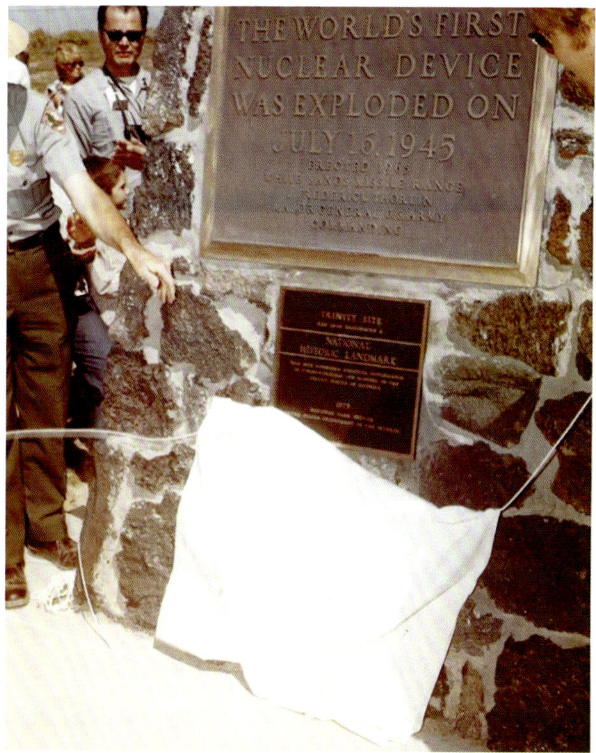

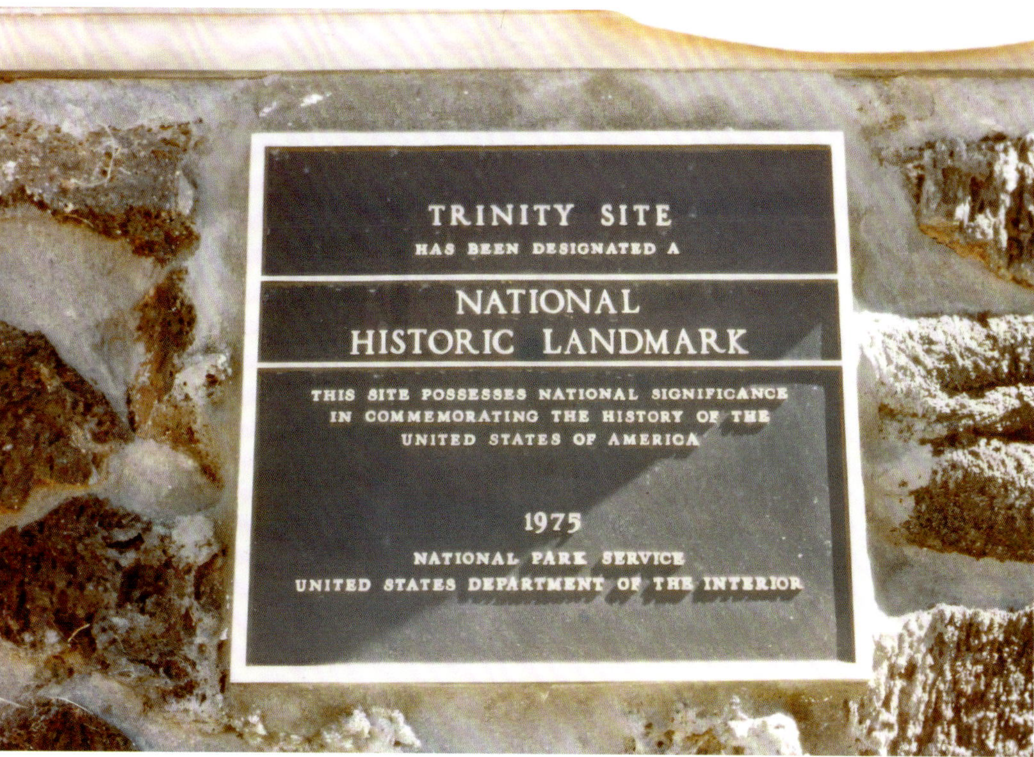

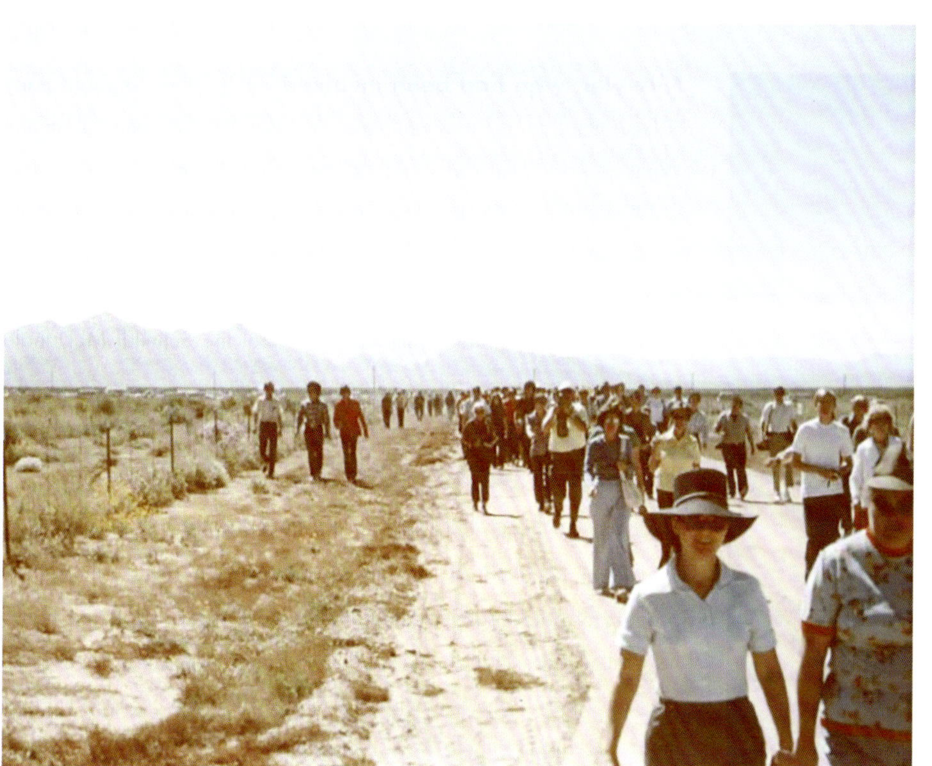

In 1971, Executive Order 11593, "Protection and enhancement of the cultural environment," created a legal obligation for White Sands and other federal entities to take stronger action in preserving historic sites under their care. A new era of cooperation slowly but surely began between the missile range and the National Park Service. At long last, in 1975, White Sands requested that Trinity Site be recognized as a National Historic Landmark.

At a dedication ceremony held on October 4, the day of the annual open house, a second plaque affixed to the obelisk was unveiled and presented to White Sands by a National Park Service representative. (Photos by Fred E. Mang Jr., National Park Service, US Department of the Interior.)

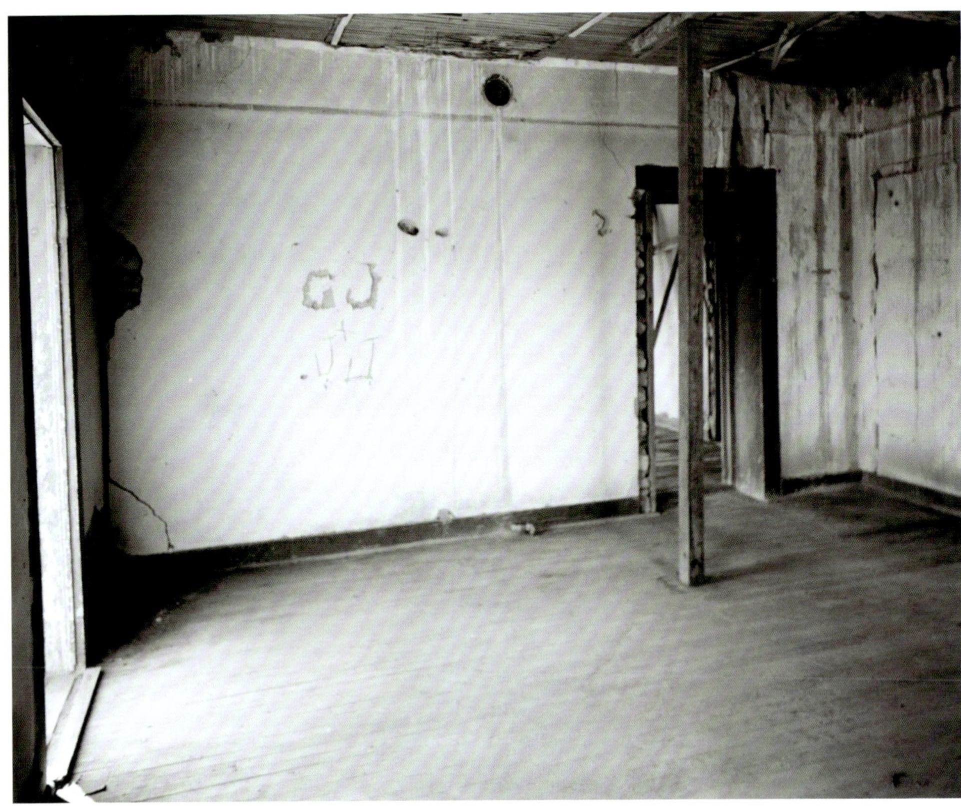

The site's new status did not immediately come with new attractions or restoration of the few remaining structures at Trinity. Besides ground zero, there was simply not much to see. Tours of the assembly room (*left*) in the Schmidt-McDonald Ranch House had been discontinued as the building became progressively more dilapidated. (Library of Congress, Prints & Photographs Division, HAER NM-1-A-42.)

In 1982, Maj. Gen. Niles Fulwyler became the sixteenth commander of White Sands Missile Range. Upon visiting the far northwest corner of the missile range, he was dismayed at the condition of the historic Trinity Site, especially the McDonald family's houses.

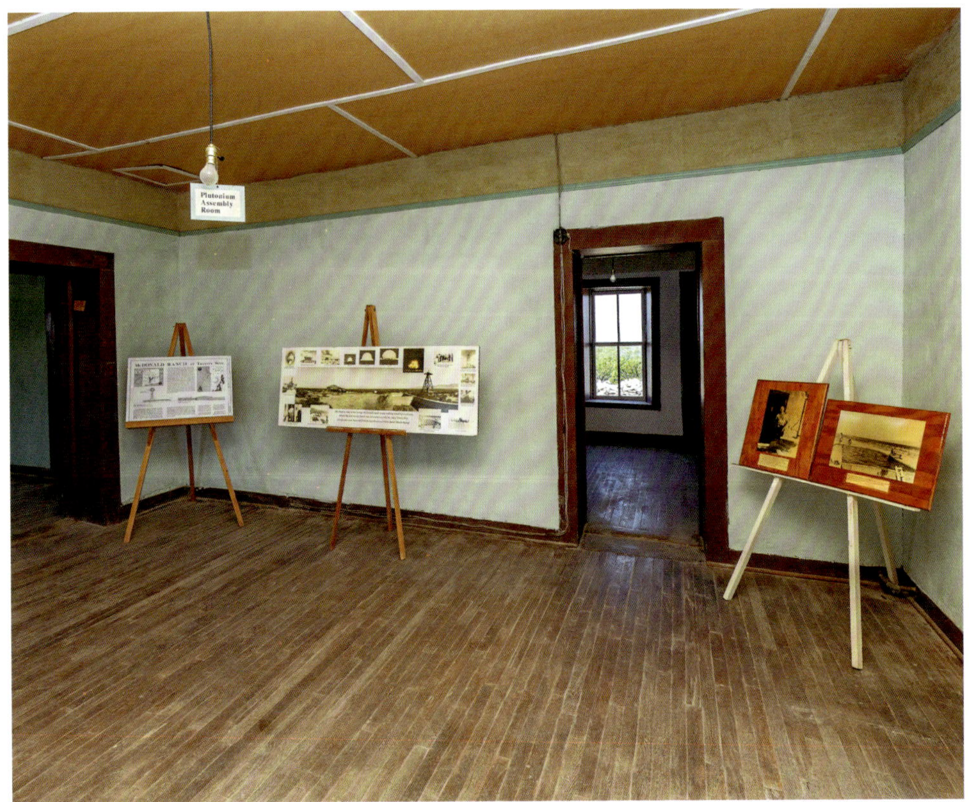

Fulwyler, who had been involved with nuclear weapons early in his military career, took an interest in Trinity that was unusual among White Sands commanders. He ordered both houses at base camp stabilized using missile range resources. Arranging a partnership between the Department of Energy (the successor to the AEC) and the Department of Army, with work to be overseen by the National Park Service, Fulwyler additionally saw to it that the Schmidt-McDonald Ranch House was returned—with historical accuracy—to its July 1945 condition. The restored landmark garnered record levels of interest in the Trinity Site open house in 1984.

However, by ten years later, on the fiftieth anniversary of the test, additional maintenance was already necessary to reverse the steady process of deterioration. After another several decades, the house had fallen again into poor condition. The missile range authorized major structural and cosmetic repairs in 2017, including preventative measures to curb future water intrusion. The photo at left shows the assembly room as of 2024.

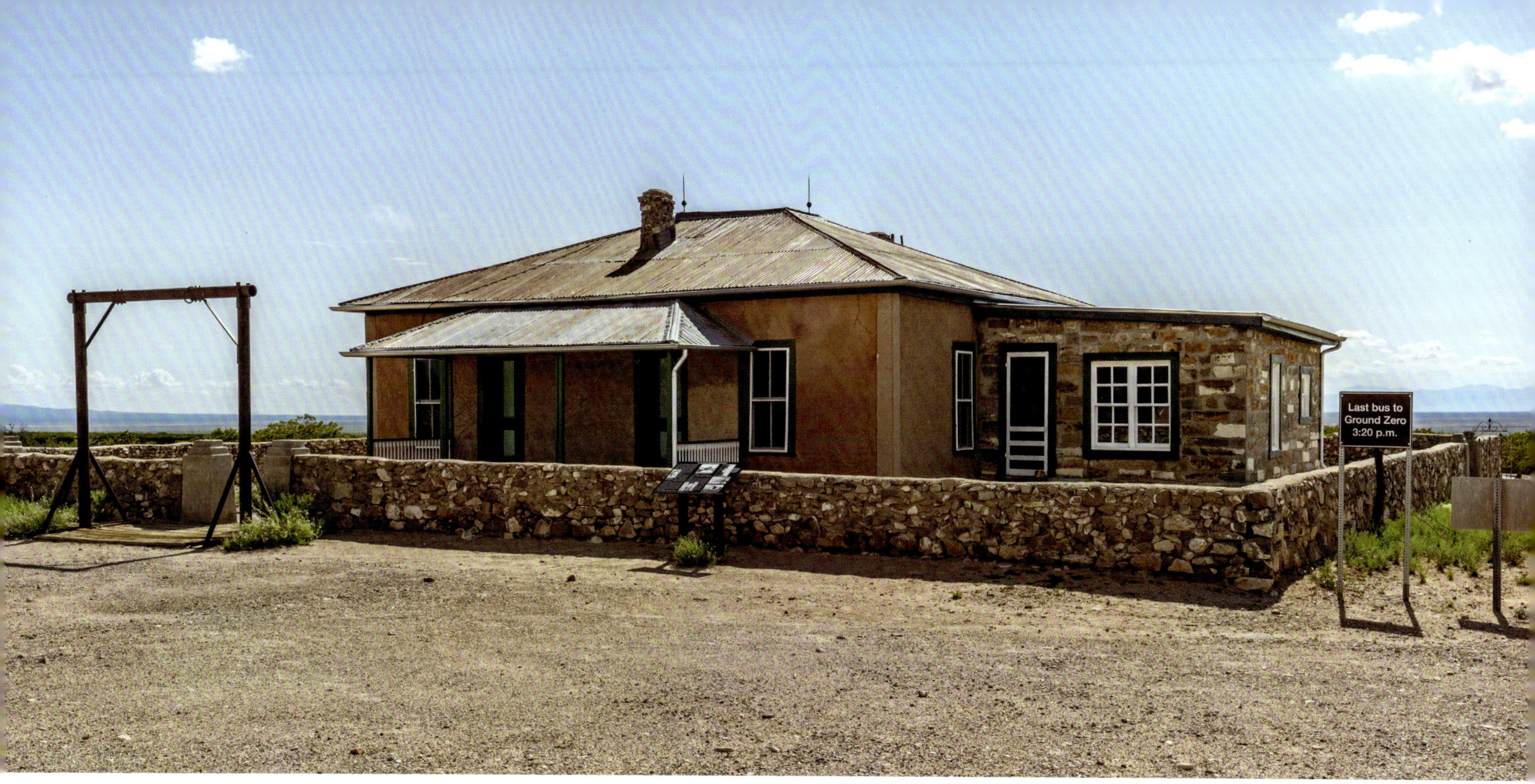

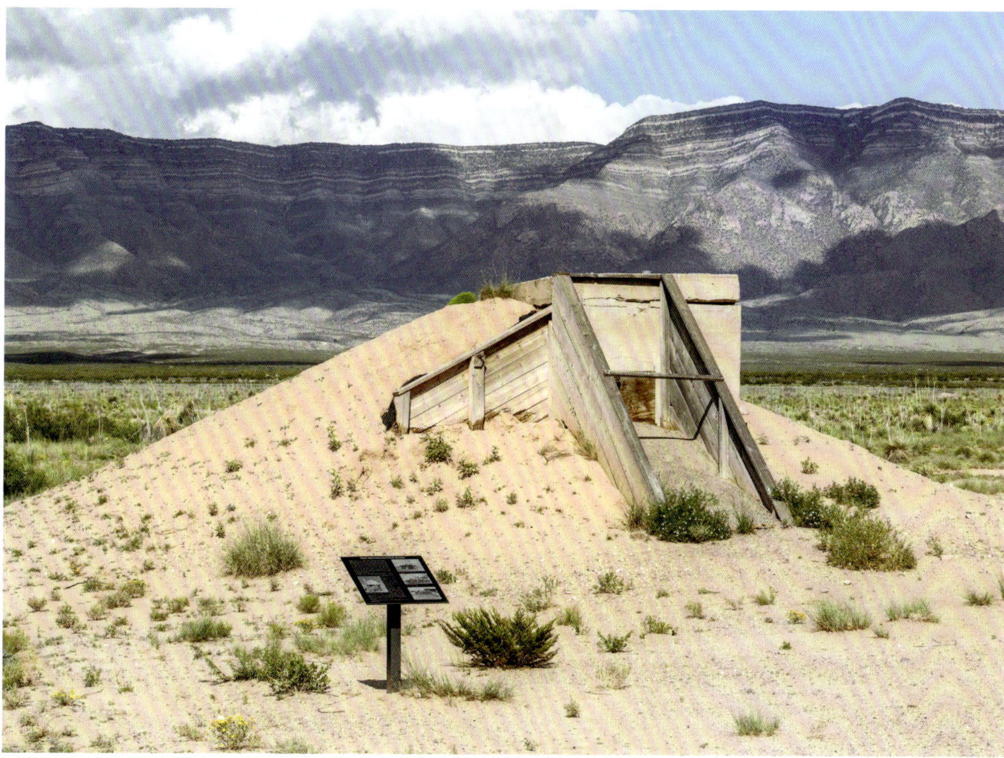

Informational exhibits featuring TR series photographs are now installed at the ranch house, the west 800 camera bunker (*left*), ground zero, and Jumbo.

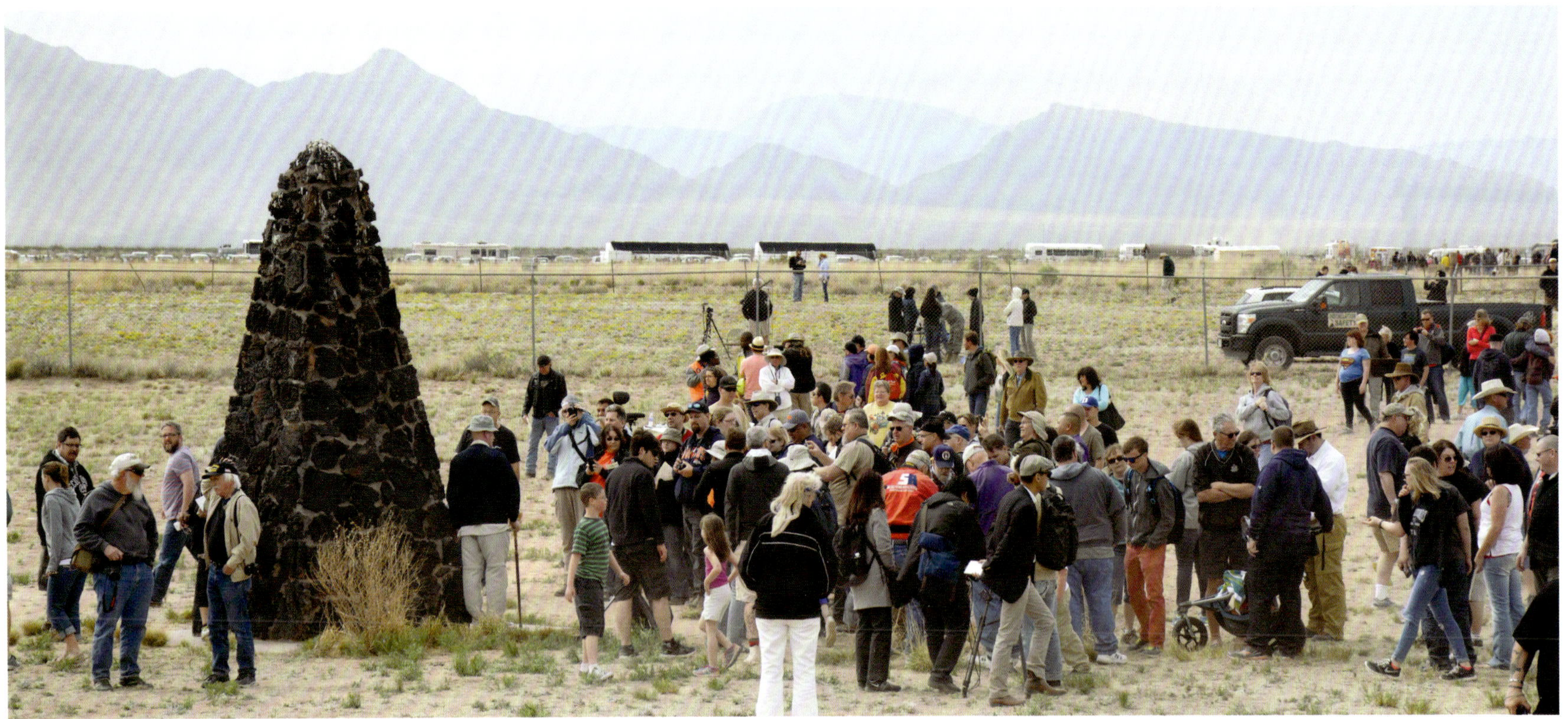

Since 1953, tens of thousands of visitors have come to Trinity from all corners of the world. Some arrive curious. Some are grieving. Some have come to protest nuclear weapons or to share their beliefs about the effects of fallout from the explosion. Many offer prayers for peace and healing. (Photo courtesy of White Sands Missile Range Public Affairs Office.)

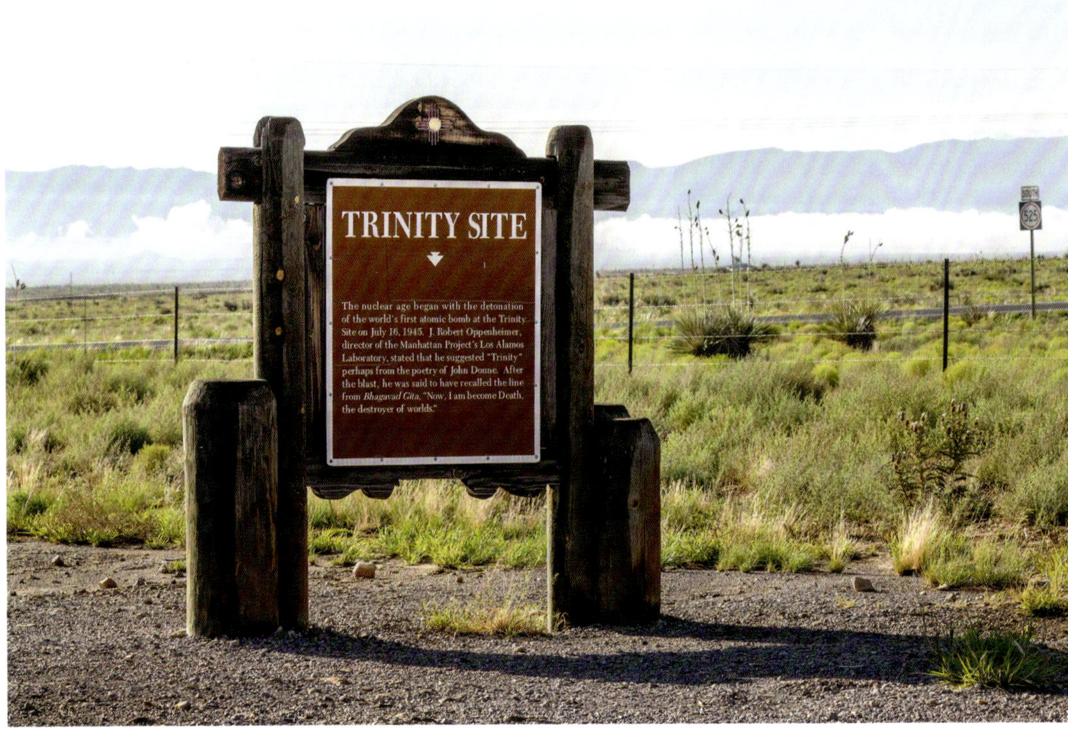

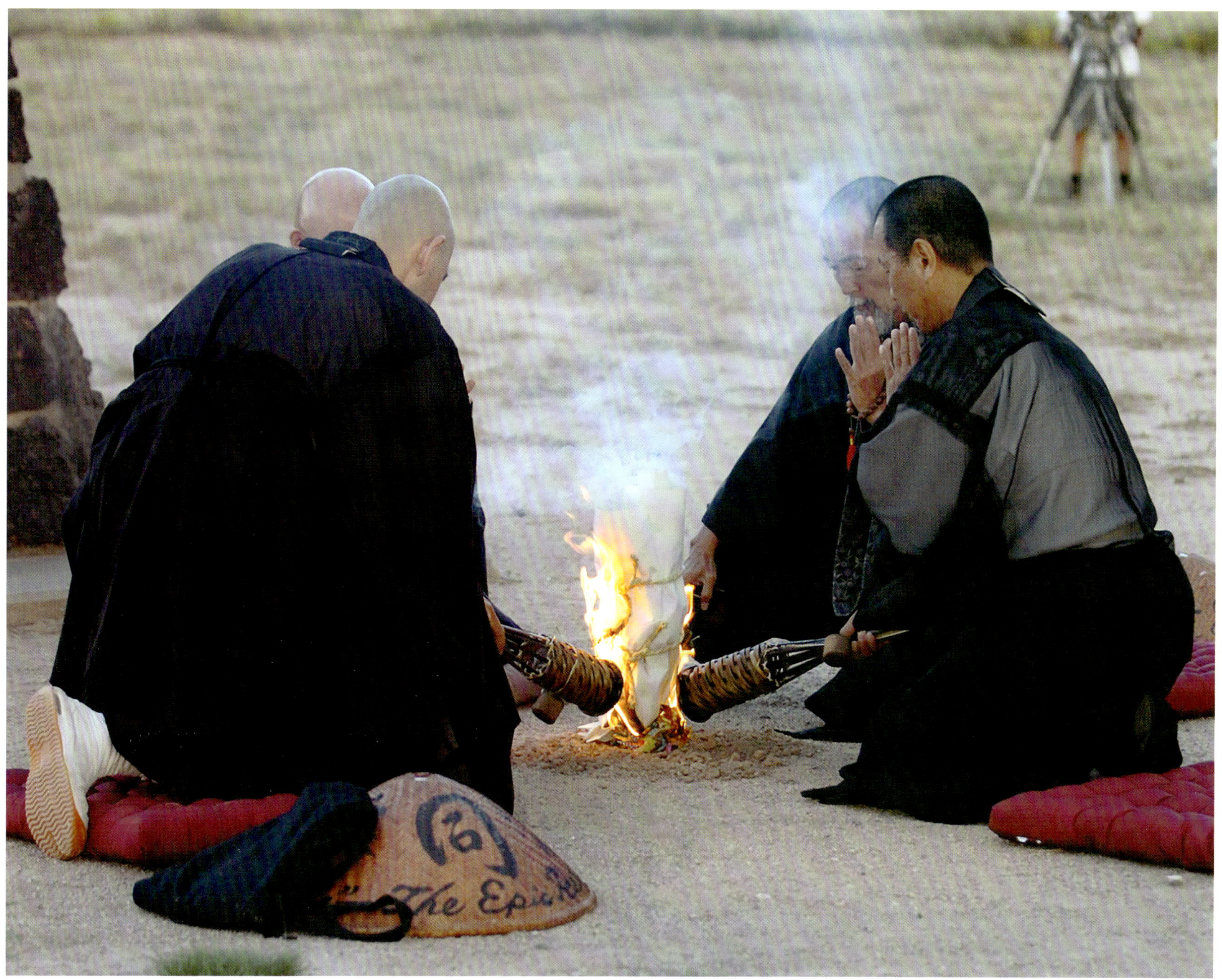

On the sixtieth anniversary of the test, a group of Japanese Buddhist monks began a 1,500-mile walk from San Francisco to Trinity Site. They brought with them a flame that had been lit from the fires at Hiroshima and carried around the world on silent peace marches since 1945. On August 9, 2005, at the site of the first atomic explosion, the flame that had been burning for sixty years—an important unit of time, according to their beliefs—was transferred to three torches and then extinguished. Using the torches, the monks together ignited a prayer cloth to signify the unification of Trinity and the two Japanese cities, extinguishing the torches as the cloth settled to embers. The ceremony was both an end and a beginning, meant to welcome a new cycle of continued peace. (AP Photo/*Alamogordo Daily News*, Ellis Neel.)

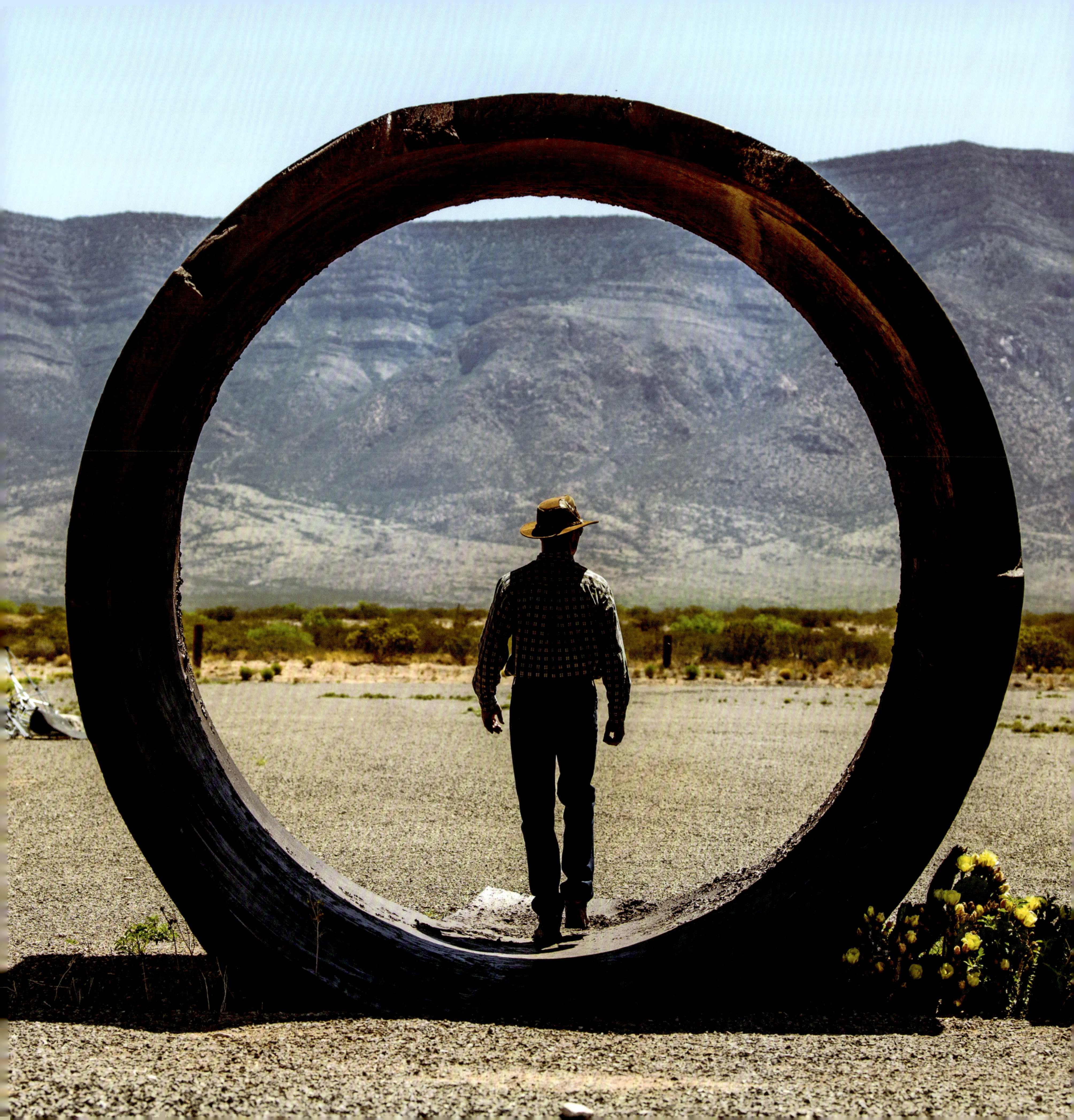

Today, the tradition of open houses continues. The black lava obelisk marks the exact location where the Gadget exploded, 100 feet above the ground. The restored Schmidt-McDonald Ranch House lies two miles away, down the road the plutonium core once traveled to the tower base. And Jumbo—moved to the Trinity Site parking lot in 1979—remains one of the most popular places for a photograph.

The allure is less material than one might imagine for a National Historic Landmark. In the emptiness, however, the ephemera are survived by a sense of magnitude. It finds form in the expansive stage of the quiet basin; the stark, stone horizon; and the same vast, New Mexico sky that for one breathtaking moment held the radiance of a thousand suns.

The world now shares in both the burden and promise of what was glimpsed on that July morning. The legacy of Trinity—the test and the place—will continue unfolding, as humankind chooses how to harness, and steward, the energy of the atom.

0060

17 August 1945.

The Secretary of War,
War Department,
Washington, D. C.

CLASSIFICATION CANCELLED
OR CHANGED TO
BY AUTHORITY OF Jack H. Kahn
BY B. Wise DATE 2-9-70

Dear Mr. Secretary:

The Interim Committee has asked us to report in some detail on the scope and program of future work in the field of atomic energy. One important phase of this work is the development of weapons; and since this is the problem which has dominated our war time activities, it is natural that in this field our ideas should be most definite and clear, and that we should be most confident of answering adequately the questions put to us by the committee. In examining these questions we have, however, come on certain quite general conclusions, whose implications for national policy would seem to be both more immediate and more profound than those of the detailed technical recommendations to be submitted. We, therefore, think it appropriate to present them to you at this time.

1. We are convinced that weapons quantitatively and qualitatively far more effective than now available will result from further work on these problems. This conviction is motivated not alone by analogy with past developments, but by specific projects to improve and multiply the existing weapons, and by the quite favorable technical prospects of the realization of the super bomb.

2. We have been unable to devise or propose effective military countermeasures for atomic weapons. Although we realize that future work may reveal possibilities at present obscure to us, it is our firm opinion that no military countermeasures will be found which will be adequately effective in preventing the delivery of atomic weapons.

The detailed technical report in preparation will document these conclusions, but hardly alter them.

3. We are not only unable to outline a program that would assure to this nation for the next decades hegemony in the field of atomic weapons; we are equally unable to insure that such hegemony, if

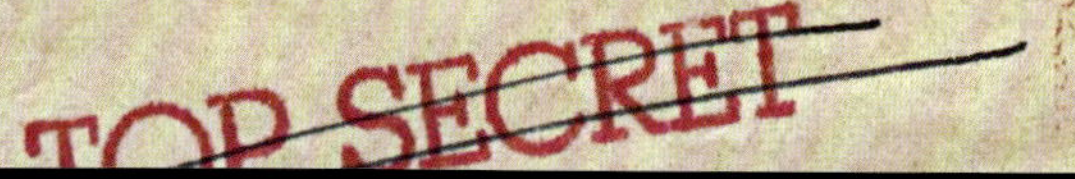

P3755

Afterword

August 17, 1945: "The development, in the years to come, of more effective atomic weapons, would appear to be a most natural element in any national policy of maintaining our military forces at great strength; nevertheless, we have grave doubts that this further development can contribute essentially or permanently to the prevention of war. We believe that the safety of this nation—as opposed to its ability to inflict damage on an enemy power—cannot lie wholly or even primarily in its scientific or technical prowess. It can be based only on making future wars impossible. It is our unanimous and urgent recommendation to you that, despite the present incomplete exploitation of technical possibilities in this field, all steps be taken, all necessary international arrangements be made, to this one end."

These are the words of my grandfather, J. Robert Oppenheimer, writing on behalf of the Scientific Panel of the Interim Committee on Nuclear Power to the secretary of war. The panel advised that the United States could not keep the science of fission secret, and there was no physical means of assuring safety from nuclear weapons in the hands of other nations. They urged acceptance of that reality and pursuit of cooperative measures to manage nuclear energy together, with a goal to end all wars.

We did not achieve the level of cooperation suggested by the Interim Committee scientists, and when the Soviet Union ended the American nuclear monopoly four years later, we plunged into an arms race the scientists warned against. Weapons became stronger and more versatile; at the peak of nuclear proliferation in the late 1980s, there were approximately 70,000 in existence.

More than once, we've been up to the brink of destruction. But we have not gone over it.

As the world continues to come to terms with the hope and peril of our new age, eighty years after it began, Trinity stands as a clear dividing line for humanity. We see in its receding light an age-old prophecy, that mastery of nature comes with a price. Like a bite taken from the apple of knowledge, the first nuclear detonation introduced a new reality: our science has become so powerful it could destroy us. Global war means total destruction now—a recognition that has paradoxically helped prevent it.

As nations have chosen how to build on the scientific advancements of the Manhattan Project, not only military uses of fission have flourished. Peaceful applications of nuclear technology have led to breakthroughs in cancer treatment, abundant carbon-free energy, and improvements in medical equipment and food safety and are yet to meet their full potential. The delicate balance that allows security and scientific progress to coexist in the nuclear age is supported by institutions like the United Nations' International Atomic Energy Agency, which promotes the safe and peaceful use of fission, and disarmament efforts like the Treaty on the Non-Proliferation of Nuclear Weapons. Such international agreements have fostered a degree of global stability that is close to the outcome that Oppenheimer and others hoped for in 1945.

Trinity spelled out our interconnectedness for all humans to understand. It revealed the necessity of global unity to our continued survival—a message that remains as important as ever in our increasingly complex society, where alongside the shared risks associated with nuclear technology, we also face threats tied to climate change, pandemics, artificial intelligence, energy poverty, and resource conflict. Each of these threats represents an opportunity to embrace wider cooperation—just as the demands of wartime, with its existential urgency, have inspired monumental feats of collaboration against a common enemy. If we can channel the same spirit of solidarity that once enabled us to unlock the energy of the atom, we will be better able to confront modern challenges together, using humanity's collective wisdom and ingenuity to envision a future we all want.

—Charles Oppenheimer,
Founder and Co-Executive Director
The Oppenheimer Project
2025

Acknowledgments

Since well before it was envisioned in 2006, many hands have contributed to the book presented here at last—in the eighty-first year since the Trinity atomic test. Deserving first and foremost recognition are the veterans of Trinity, and especially the photographers—Julian Mack, Berlyn Brixner, Ben Benjamin, Ernest Wallis, and their colleagues—who left behind so many records for successors at Los Alamos National Laboratory to preserve, interpret, and share. In the long interim between then and now, librarians like Dan Comstock, leaders like Henry Johnson, and those working before, after, and alongside them have stewarded archival materials like the TR series that make much of today's research possible, for this project and so many others.

Turning to my own Los Alamos National Laboratory colleagues and irreplaceable teammates in this endeavor, it is a privilege to begin by acknowledging historian Alan Carr, who had the excellent idea to pursue this project two decades ago and advocated to secure the necessary support to bring it into being. Thank you, Alan, for laying that foundation and for your expert counsel and collaboration during the creation process—from drafting introductions to The Countdown, Ground Zero, and Legacy chapters; to organizing a team visit to Trinity Site; to soliciting the forewords; to both arranging and conducting numerous peer reviews. To designer Paul Ziomek, who created the artwork and aesthetic that make this book truly special: I cannot thank you enough for your collaborative spirit in building out this volume, for your astounding patience through the tedium of revising pages, and for your thoughtful approach to every visual layer and moment in the story, which I was so grateful to follow along with and learn from.

The *Trinity* creative team would like to thank Brye Steeves, director of the Laboratory's National Security Research Center (NSRC), who championed and managed this project through every step of the process. We are also grateful to the NSRC's expert historians, digitization specialists, archivists, librarians, and consultants, who provided much-needed support along the way, with special appreciation for the contributions of Angie Piccolo, who located and scanned hundreds of images and documents from the Laboratory's vast collections and beyond; Rizwan Ali, retired founding director of the NSRC, who expertly retouched the more than 400 images reproduced in this book; John Moore, who was instrumental in the declassification of photographs publicly shared here for the first time; Cristy Pendergraft, who coordinated image licensing; and Octavio Ramos, Danny Alcazar, and Madeline Whitacre, who provided key editorial support in the earliest stages of project development.

Errors of any kind are the sole responsibility of the author, though their number has been greatly reduced through the efforts of a small army of peer reviewers, who offered valuable feedback on an early draft of the manuscript; their ranks include Jennifer Snead, Tom Kunkle, Jim Eckles, Earl Whitney, Jon Else, Roger Meade, Jim Kunetka, Jonathan Morgan, Steve Simon, Peter Kuran, Ellen McGehee, David Snead, Bill Godby, and Mark Paris. For their additional contributions during the writing process, I am grateful to Eckles, who generously shared knowledge and resources; Kuran, who identified long-overlooked footage of the 100-ton test and provided an excellent reference sheet detailing the many cameras used to record the Trinity detonation; McGehee, for historical consultation, especially on Project Alberta and the 509th Composite Group; and Kunkle, who offered a wealth of advice on technical details.

Special thanks are due as well to Scott Stearns of the US Army's White Sands Missile Range Public Affairs Office, who led the book team on a tour of Trinity Site that helped inform the final chapter, and to Ethan Frogget, head of the Laboratory's Multimedia Production Group, for creating a photographic record on this visit that brings the story into the present day. Thanks also to David Woodfin, who shot the modern photos of Trinity test artifacts that appear throughout the book—including the armored camera on the cover—in coordination with the Laboratory's Bradbury Science Museum, where these and other remnants of the Manhattan Project are preserved and exhibited.

Thank you to Gabriella Smith for assistance with layout and graphics and to Anne Jones for a thorough proofread of the full draft and assistance with the index.

As a final note, publication of *Trinity* would not have been possible without the Office of Classification and Controlled Information at Los Alamos National Laboratory, especially Diana Hollis and Maria Peña, along with derivative classifiers Karen LaRue, Albert Hsu, and John Moore. Their efforts continually support and enable the NSRC's mission of sharing the national nuclear history preserved in its collections.

Bibliography

"A-bombed Hiroshima photographed by Yoshito Matsushige, staff photographer for the *Chugoku Shimbun*." *Chugoku Shimbun*. August 6, 2024. https://www.chugoku-np.co.jp/stp/Edit/yoshito_matsushige/en/.

Anderson, Herbert L. Herbert Anderson to Kenneth Bainbridge, April 19, 1945. "The 100 Ton Shot Preparations – Addition No. 7 to Project TR Circular." US Department of Energy OpenNet, ALLA0006798. Office of Scientific and Technical Information, Oak Ridge, TN.

Arneson, R. Gordon. "Notes of the Interim Committee Meeting." June 1, 1945. Interim Committee on Atomic Energy--Notes of Meetings (May 9 to July 19, 1945). Subject Files. R. Gordon Arneson Papers (RG HST-RGA), NAID 333235044. Harry S. Truman Presidential Library and Museum, Independence, MO.

"Atomic Bomb Survivor Yoshito Matsushige Testimony | English-Dubbed | Hiroshima Peace Memorial Museum." Posted by Hiroshima Peace Memorial Museum, January 22, 2022. YouTube, 19 min., 19 sec. https://hpmm-db.jp/list/detail/?cate=testify_en&search_type=detail&data_id=13688.

Atomic Heritage Foundation. "Hiroshima and Nagasaki Missions – Planes and Crews." Published April 27, 2016. https://ahf.nuclearmuseum.org/ahf/history/hiroshima-and-nagasaki-missions-planes-crews/.

Bainbridge, Kenneth T. "1975: All in Our Time: A Foul and Awesome Display." *Bulletin of the Atomic Scientists* 76, no. 6 (2020): 374–380. https://doi.org/10.1080/00963402.2020.1847493.

Bainbridge, Kenneth T. *Trinity*. LA-6300-H. Los Alamos Scientific Laboratory, 1976. https://doi.org/10.2172/5306263.

Bainbridge, Kenneth T. "Prelude to Trinity." *Bulletin of the Atomic Scientists* 31, no. 4 (1975): 42–46. https://doi.org/10.1080/00963402.1975.11458229.

Bainbridge, Kenneth T. Kenneth Bainbridge to George Kistiakowsky, March 19, 1945. "Assistance Desired on 100 T Test." US Department of Energy OpenNet, ALLA0000922. Office of Scientific and Technical Information, Oak Ridge, TN.

Bainbridge, Kenneth T. Kenneth Bainbridge to All Concerned, March 14, 1945. "Photographic Service and Control of Photographic Records at Trinity." US Department of Energy OpenNet, ALLA0003335. Office of Scientific and Technical Information, Oak Ridge, TN.

Baty, Roy S., and Scott D. Ramsey. "On the Symmetry of Blast Waves." *Nuclear Technology* 207, suppl. 1 (2021): S335–S351. https://doi.org/10.1080/00295450.2021.1922263.

Bernstein, Barton J. "Roosevelt, Truman, and the Atomic Bomb, 1941–1945: A Reinterpretation." *Political Science Quarterly* 90, no. 1 (1975): 23–69. https://doi.org/10.2307/2148698.

Bethe, Hans A. Hans Bethe to J. R. Oppenheimer, March 27, 1944. "The Present Status of the Confining Sphere." US Department of Energy OpenNet, ALLA0001458. Office of Scientific and Technical Information, Oak Ridge, TN.

Bird, Kai, and Martin J. Sherwin. *American Prometheus: The Triumph and Tragedy of J. Robert Oppenheimer*. Knopf, 2005.

Blackburn, Richard A. Richard Blackburn to A. J. Frolich, April 24, 1946. "Demolition Operation at Trinity." US Department of Energy OpenNet, ALLA0003855. Office of Scientific and Technical Information, Oak Ridge, TN.

Bouville, André, Harold L. Beck, Kathleen M. Thiessen, et al. "The Methodology Used to Assess Doses from the First Nuclear Weapons Test (Trinity) to the Populations of New Mexico." *Health Physics* 119, no. 4 (2020): 400–427. https://doi.org/10.1097/HP.0000000000001331.

Bradbury, Norris E. Norris Bradbury to Personnel Concerned, July 9, 1945. "TR Hot Run." US Department of Energy OpenNet, NV0306039. Nuclear Testing Archive, Las Vegas, NV.

Brown, Eric N., and Dan L. Borovina. "The Trinity High-Explosive Implosion System: The Foundation for Precision Explosive Applications." *Nuclear Technology* 207, suppl. 1 (2021): S204–S221. https://doi.org/10.1080/00295450.2021.1913954.

Bulletin of the Atomic Scientists. "In Their Own Words: Trinity at 75." Published July 15, 2020. https://thebulletin.org/2020/07/in-their-own-words-trinity-at-75/.

Cahoon, Elizabeth K., Rui Zhang, Steven L. Simon, et al. "Projected Cancer Risks to Residents of New Mexico from Exposure to Trinity Radioactive Fallout." *Health Physics* 119, no. 4 (2020): 478–493. https://doi.org/10.1097/HP.0000000000001333.

Carlson, R. W. *Confinement of an Explosion by a Steel Vessel.* LA-390. September 14, 1945. https://sgp.fas.org/othergov/doe/lanl/lib-www/la-pubs/00330312.pdf.

Carr, Alan B. "Thirty Minutes before the Dawn." *Nuclear Technology* 207, suppl. 1 (2021): S1–S23. https://doi.org/10.1080/00295450.2021.1927625.

Eby, Nelson, Robert Hermes, Norman Charnley, and John A. Smoliga. "Trinitite—The Atomic Rock." *Geology Today* 26, no. 5 (2010): 180–185. https://doi.org/10.1111/j.1365-2451.2010.00767.x.

Eckles, Jim. "Jim Eckles' Interview." Interview by Cindy Kelly. *Voices of the Manhattan Project*, Atomic Heritage Foundation, December 7, 2017. YouTube, 2:28:18. https://ahf.nuclearmuseum.org/voices/oral-histories/jim-eckles-interview/.

Eckles, Jim. *Trinity: The History of an Atomic Bomb National Historic Landmark*. Published by the author, 2015.

Einstein, Albert. Albert Einstein to Franklin D. Roosevelt, August 2, 1939. Franklin D. Roosevelt Presidential Library and Museum. https://www.fdrlibrary.org/search?q=einstein+szilard+letter.

Else, Jon, dir. *The Day After Trinity.* Pyramid Films, 1981.

Farrell, Thomas F. Thomas Farrell to Leslie R. Groves, September 27, 1945. "Report on Overseas Operations - Atomic Bomb."

Exploratorium. "Nagasaki Journey." Accessed November 2024. https://annex.exploratorium.edu/nagasaki/related/journeyYamahata.html.

Fermi, Enrico. *My Observations during the Explosion at Trinity on July 16, 1945.* N.d. Atomic Archive. https://www.atomicarchive.com/resources/documents/trinity/fermi.html.

Frank, Richard B. *Downfall: The End of the Imperial Japanese Empire.* Random House, 1999.

Frisch, Otto R., and John A. Wheeler. "The Discovery of Fission." *Physics Today* 20, no. 11 (1967): 43–52. https://doi.org/10.1063/1.3034021.

Frisch, Otto R., and Rudolph Peierls. March 1940. "On the Construction of a 'Super-Bomb' Based on a Nuclear Chain Reaction in Uranium." Atomic Archive. https://www.atomicarchive.com/resources/documents/beginnings/frisch-peierls-2.html.

Frolich, A. J. A. Frolich to A. W. Betts, May 15, 1946. "Charge in Jumbo." US Department of Energy OpenNet, ALLA0003852. Office of Scientific and Technical Information, Oak Ridge, TN.

Glasstone, Samuel and Philip J. Dolan, eds. *The Effects of Nuclear Weapons.* 3rd ed. US Department of Defense and US Department of Energy, 1977.

Gosling, F. G. *The Manhattan Project: Making the Atomic Bomb.* DOE/MA-0001. US Department of Energy, 1999. https://doi.org/10.2172/303853.

Groves, Leslie R. "Some Recollections of July 16, 1945." *Bulletin of the Atomic Scientists* 26, no. 6 (1970): 21–27. https://doi.org/10.1080/00963402.1970.11457821.

Groves, Leslie R. *Now It Can Be Told: The Story of the Manhattan Project.* Harper, 1962.

Groves, Leslie R. Leslie Groves to George C. Marshall, July 30, 1945. National Security Archive. https://nsarchive.gwu.edu/document/28689-document-11-major-general-lr-groves-chief-staff-general-george-c-marshall-30-july.

Groves, Leslie R. Leslie Groves to George C. Marshall, August 7, 1944. "Atomic Fission Bombs – Present Status and Expected Progress." National Security Archive. https://nsarchive.gwu.edu/document/28502-document-4-memo-general-groves-chief-staff-marshall-atomic-fission-bombs-present.

Groves, Leslie R. "Military Policy Committee Minutes." May 5, 1943. National Security Archive. https://nsarchive.gwu.edu/document/28501-document-3-memorandum-leslie-r-grove-policy-meeting-5543-top-secret.

Hacker, Barton C. *The Dragon's Tail: Radiation Safety in the Manhattan Project, 1942–1946.* University of California Press, 1987.

Hanson, Susan K., and Warren J. Oldham. "Weapons Radiochemistry: Trinity and Beyond." *Nuclear Technology* 207, suppl. 1 (2021): S295–S308. https://doi.org/10.1080/00295450.2021.1951538.

Hempelmann, Louis. *Hazards of the 100 Ton Shot at Trinity.* May 18, 1946. Los Alamos National Laboratory, National Security Research Center, collection A-2020-019.

Hempelmann, Louis. Louis Hempelmann to Howard C. Bush, October 22, 1945. "Safety Instructions Concerning Military Personnel and Visitors in and around Crater Region." US Department of Energy OpenNet, NV0319657. Nuclear Testing Archive, Las Vegas, NV.

Hempelmann, Louis. *Visit of Reporters to Trinity.* September 29, 1945. Los Alamos National Laboratory, National Security Research Center, collection A-2020-019.

Hempelmann, Louis, and R. Watts. *Hazards of Trinity Experiment.* April 12, 1945. National Security Archive. https://nsarchive.gwu.edu/document/32277-document-4-louis-hempelmann-and-richard-watts-hazards-trinity-experiment-12-april.

Henderson, R. W. R. Henderson to A. J. Frolich, May 10, 1946. "Destruction of Jumbo — Trinity." US Department of Energy OpenNet, ALLA0003853. Office of Scientific and Technical Information, Oak Ridge, TN.

Henderson, R. W. R. Henderson to J. R. Oppenheimer, William Parsons, George Kistiakowsky, W. A. Stevens, and William Davalos, March 12, 1945.

Henderson, R. W. *Status Report of Jumbo.* January 1, 1945. US Department of Energy OpenNet, ALLA0004192. Office of Scientific and Technical Information, Oak Ridge, TN.

Henderson, R. W. *Status Report of Jumbo.* December 5, 1944. US Department of Energy OpenNet, ALLA0004140. Office of Scientific and Technical Information, Oak Ridge, TN.

Hersey, John. "A Reporter At Large: Hiroshima." *The New Yorker*, August 31, 1946.

Hirschfelder, Joseph O., and J. Magee. Joesph Hirschfelder and J. Magee to Kenneth Bainbridge, July 6, 1945. "Improbability of Danger from Active Material Falling from Cloud." US Department of Energy OpenNet, ALLA0000779. Office of Scientific and Technical Information, Oak Ridge, TN.

Historic American Engineering Record. "White Sands Missile Range, Trinity Site, Vicinity of Routes 13 & 20, White Sands, Dona Ana County, NM." Library of Congress Prints and Photographs Division, HAER NM,27-ALMOG.V,1A-. https://www.loc.gov/pictures/collection/hh/item/nm0139/.

Hoddeson, Lillian, Paul W. Henriksen, Roger A. Meade, et al. *Critical Assembly: A Technical History of Los Alamos during the Oppenheimer Years, 1943–1945.* Cambridge University Press, 1993. https://doi.org/10.1017/CBO9780511665400.

Hoffman, Joseph G. *Nuclear Explosion 16 July 1945: Health Physics Report on Radioactive Contamination throughout New Mexico following the Nuclear Explosion.* LA-626. Los Alamos Scientific Laboratory, February 20, 1947. https://sgp.fas.org/othergov/doe/lanl/lib-www/la-pubs/00348548.pdf.

Hubbard, Jack M. *July 16th Nuclear Explosion: Meteorological Report.* LA-357. September 10, 1945. US Department of Energy OpenNet, ALLAOSTI126089.

Hunt, Laura. "Atomic Flame Extinguished at Trinity Site." *Alamogordo News*, August 10, 2005. https://www.buddhistchannel.tv/index.php?id=64,1553,0,0,1,0.

Jenkins, Rupert, ed. *Nagasaki Journey: The Photographs of Yosuke Yamahata, August 10, 1945.* Pomegranate Artbooks, 1995.

Jones, Vincent C. *Manhattan: The Army and the Atomic Bomb.* US Army Center of Military History, 1985.

Katz, J. I. "Fermi at Trinity." *Nuclear Technology* 207, suppl. 1 (2021): S326–S334. https://doi.org/10.1080/00295450.2021.1927627.

Kistiakowsky, George B. "Trinity—A Reminiscence." *Bulletin of the Atomic Scientists* 36, no. 6 (1980): 19–22. https://doi.org/10.1080/00963402.1980.11458739.

Kistiakowsky, George B. George Kistiakowsky to J. R. Oppenheimer, William Parsons, Kenneth T. Bainbridge, Norris E. Bradbury, and W. A. Stevens, October 13, 1944. "Activities at Trinity." US Department of Energy OpenNet, ALLA0004059. Office of Scientific and Technical Information, Oak Ridge, TN.

Kistiakowsky, George B. George Kistiakowsky to J. R. Oppenheimer, Kenneth T. Bainbridge, Hans Bethe, Edward Teller, and William Parsons, May 24, 1944. "Conclusions Reached at the Meeting on Jumbo May 23, 1944." US Department of Energy, ALLA0001457. Office of Scientific and Technical Information, Oak Ridge, TN.

Kunetka, James. *The General and the Genius: Groves and Oppenheimer—The Unlikely Partnership That Built the Atom Bomb.* Regnery, 2015.

Kuran, Peter, dir. *Trinity and Beyond: The Atomic Bomb Movie.* Visual Concept Entertainment, 1995.

Laurence, William L. "Eyewitness Account of Bomb Test." *New York Times*, September 26, 1945.

Leet, L. Don. *July 16th Nuclear Explosion: Ground Vibrations.* LA-438. November 15, 1945. https://sgp.fas.org/othergov/doe/lanl/lib-www/la-pubs/00407592.pdf.

Loring, William S. *Birthplace of the Atomic Bomb: A Complete History of the Trinity Test Site.* McFarland, 2019.

Los Alamos National Laboratory. *Beginning of an Era: Los Alamos, 1943–1945.* 2nd ed. Los Alamos Historical Society, 2007.

Los Alamos Scientific Laboratory. "Sleeping Beauty Awakens." *The Atom*, May 1967.

Luis W. Alvarez letter to Ryokichi Sagane, August 9, 1945. Manuscripts, Archives, and Special Collections, Cage 1603. Washington State University Libraries, Pullman, WA.

Maag, Carl, and Steve Rohrer. *Project Trinity, 1945–1946.* DNA 6028F. Defense Nuclear Agency, 1982. https://apps.dtic.mil/sti/pdfs/ADA331688.pdf.

Mack, Julian E. *Semi-Popular Motion-Picture Record of the Trinity Explosion.* MDDC-221. Atomic Energy Commission, 1947. https://babel.hathitrust.org/cgi/pt?id=mdp.39015074121206&seq=1.

Mack, Julian E. *July 16th Nuclear Explosion: Space–Time Relationships.* LA-531. April 2, 1946. https://sgp.fas.org/othergov/doe/lanl/lib-www/la-pubs/00330363.pdf.

Mack, Julian E. Julian Mack to All Concerned, April 9, 1945. "Addition No. 4 to March 24, 1945, Project TR Circular." US Department of Energy OpenNet, NV0059745. Nuclear Testing Archive, Las Vegas, NV.

Manhattan Engineer District. *Auxiliary Activities.* Vol. 4 of *Book 1 – General.* In *Manhattan District History.* N.d. https://www.osti.gov/opennet/manhattan_district.

Manhattan Engineer District. *The Atomic Bombings of Hiroshima and Nagasaki.* 1946. https://digirepo.nlm.nih.gov/ext/dw/14110660R/PDF/14110660R.pdf.

Martz, Joseph C., Franz J. Freibert, and David L. Clark. "The Taming of Plutonium: Plutonium Metallurgy and the Manhattan Project." *Nuclear Technology* 207, suppl. 1 (2021): S266–S285. https://doi.org/10.1080/00295450.2021.1913035.

McGehee, Ellen D. "Commemorating Controversy: Place-Making at the Birthplace of the Bomb." PhD diss., University of New Mexico, 2015. https://digitalrepository.unm.edu/hist_etds/53/.

McKibben, Joseph L. *July 16 Nuclear Explosion: Relay Timing.* LA-435 (1947). http://www.gammaexplorer.com/lanlreports/lanl2_a/lib-www/la-pubs/00330360.html.

Meinzer, Oscar Edward, and Raleigh Frederick Hare. *Geology and Water Resources of Tularosa Basin, New Mexico.* Water Supply Paper 343. US Geological Survey, 1915. https://doi.org/10.3133/wsp343.

Merlan, Thomas. *Life at Trinity Base Camp.* Human Systems Research, 2001. https://wsmrmuseum.com/wp-content/uploads/2022/02/life_at_trinity_base_camp_HSR_2001.pdf.

Merlan, Thomas. *Trinity Experiments.* Human Systems Research, 1997. https://wsmrmuseum.com/wp-content/uploads/2022/02/trinity-experiments.reduced.pdf.

Morgan, Jonathan E. "The Origins of Blast-Loaded Vessels." *Nuclear Technology* 207, suppl. 1 (2021): S231–S265. https://doi.org/10.1080/00295450.2021.1909371.

New Mexico NHL Trinity Site. National Register of Historic Places and National Historic Landmarks Program Records: New Mexico. National Register of Historic Places and National Historic Landmarks Program Records. Records of the National Park Service (RG 79), NSAID 77847022. National Archives at College Park, MD.

Office of Scientific and Technical Information. "The Manhattan Project: An Interactive History." https://www.osti.gov/opennet/manhattan-project-history/Resources/about.htm.

Oppenheimer, J. R. J. R. Oppenheimer (for the Panel) to Henry L. Stimson, August 17, 1945. Box 291, folder 13, J. Robert Oppenheimer Papers, Manuscript Division, Library of Congress, Washington, DC.

Oppenheimer, J. R. "Recommendations on the Immediate Use of Nuclear Weapons." June 16, 1945. National Security Archive. https://nsarchive.gwu.edu/document/28526-document-25-memorandum-j-r-oppenheimer-recommendations-immediate-use-nuclear-weapons.

Oppenheimer, J. R. J. R. Oppenheimer to Kenneth Bainbridge, March 15, 1945. "Test Measurements Program for Trinity." US Department of Energy OpenNet, ALLA0000920. Office of Scientific and Technical Information, Oak Ridge, TN.

Oughterson, Ashley W., George V. Leroy, Averill A. Liebow, et al. *Medical Effects of Atomic Bombs: The Report of the Joint Commission for the Investigation of the Effects of the Atomic Bomb in Japan*, vol. 2. NP-3037. Army Institute of Pathology, April 19, 1951. https://doi.org/10.2172/4421016.

Palmer, T. O. *Evacuation Detachment at Trinity.* July 18, 1945. US Department of Energy OpenNet, NV0039285. Nuclear Testing Archive, Las Vegas, NV.

Penney, W. G. W. Penney to Kenneth Bainbridge, Hans Bethe, Lewis Fussel, Joseph Hirschfelder, J. R. Oppenheimer, and William Parsons, December 15, 1944. "The Heat of Combustion of Jumbo." US Department of Energy OpenNet, ALLA0003974. Office of Scientific and Technical Information, Oak Ridge, TN.

Ramsey, Norman, and Raymond L. Brin, eds. *Nuclear Weapons Engineering and Delivery.* Los Alamos Technical Series, volume 23. Los Alamos Scientific Laboratory, July 1946.

Rhodes, Richard. *The Making of the Atomic Bomb.* 25th anniversary ed. Simon & Schuster, 2012.

Selby, Hugh D., Susan K. Hanson, Daniel Meininger, et al. "A New Yield Assessment for the Trinity Nuclear Test, 75 Years Later." *Nuclear Technology* 207, suppl. 1 (2021): 321–325. https://doi.org/10.1080/00295450.2021.1932176.

Simon, Steven L., André Bouville, and Harold L. Beck. "Estimated Radiation Doses and Projected Cancer Risks for New Mexico Residents from Exposure to Radioactive Fallout from the Trinity Nuclear Test." *Nuclear Technology* 207, suppl. 1 (2021): S380–S396. https://doi.org/10.1080/00295450.2021.1918985.

Szasz, Ferenc Morton. *The Day the Sun Rose Twice: The Story of the Trinity Site Nuclear Explosion, July 16, 1945*. University of New Mexico Press, 1984.

Tibbets, Paul W. Jr. "I Think This Is the End of the War." N.d. https://www.thirteen.org/wnet/historyofus/web12/features/source/docs/C21.pdf.

"Trinity Films 1–4: Footage of the World's First Nuclear Weapons Test." Posted by Alan B. Carr, January 3, 2019. YouTube, 59 min., 19 sec. https://www.youtube.com/watch?v=pKLoF7Hwcas.

Warren, Stafford L. *Preliminary Report of Atomic Bomb Investigating Groups at Hiroshima and Nagasaki*. November 27, 1945. Stafford Warren Papers, University of California Los Angeles Libraries Special Collections, box 68, folder 7.

Weisskopf, Victor, J. Hoffman, Paul Aebersold, and Louis Hempelmann. Victor Weisskopf et al. to George Kistiakowsky, September 5, 1945. "Measurement of Blast, Radiation, Heat and Light and Radioactivity at Trinity." National Security Archive. https://nsarchive.gwu.edu/document/32288-document-15-inter-office-memorandum-v-weisskopf-j-hoffman-p-aebersold-and-l.

Wellerstein, Alex. "Counting the Dead at Hiroshima and Nagasaki." *Bulletin of the Atomic Scientists*. Published August 4, 2020. https://thebulletin.org/2020/08/counting-the-dead-at-hiroshima-and-nagasaki/.

White Sands Missile Range Museum. "Schmidt/McDonald Ranch House Historical Background Family Histories." Published February 2022. https://wsmrmuseum.com/wp-content/uploads/2022/02/Schmidt-McDonald-Ranch-at-Trinity.pdf.

Wilson, R. R. "A Recruit for Los Alamos." *Bulletin of the Atomic Scientists* 31, no. 3 (1975): 41–47. https://doi.org/10.1080/00963402.1975.11458215.

"Windows at Gallup, 235 Miles Away Rattle; No Loss of Lives." *Albuquerque Journal*, July 17, 1945.

Index

Page numbers in italics refer to illustrations.